Advances in Solar Energy

Volume 1 ● 1982

Advances in Solar Energy

An Annual Review of Research and Development

Volume 1 ● 1982

American Solar Energy Society, Inc.

New York   Boulder

NOTICE

ISSN 0731-8618
ISBN 0-89553-040-6

Library of Congress Cataloging applied for.

Director of Publications
American Solar Energy Society
110 West 34th Street
New York, NY 10001

Membership Services
American Solar Energy Society
1230 Grandview Avenue
Boulder, CO 80302

Publications Production Office
American Solar Energy Society
205B McDowell Hall
University of Delaware
Newark, DE 19711

Printed in the United States of America on neutral pH paper.

FOREWORD

The field of solar energy conversion has become an important discipline with a recognized potential to significantly contribute to the world supply of energy. It is diversified and encompasses a wide variety of disciplines — from mechanical engineering to physics, from biology to architecture, from ocean science to agriculture, from chemistry to atmospheric science, to name some of the major fields. It involves fields which have matured to the engineering aspects, such as the conversion of solar energy into heat or of wind into shaft work. It includes other fields in which more basic science research is necessary to unravel the micro-structures of nature, as, for example, for photovoltaic conversion or for certain bioengineering tasks.

Several of these fields have elements which have been common knowledge for centuries but sometimes forgotten at times of cheap energy supplies, while others have barely started with first studies. Most of the fields have seen during the last decade a substantial advance in sophistication, in theoretical understanding, in demonstrated feasibility, in developing hardware, in field testing, with some moving into a phase of initial commercialization.

As a field matures, need develops for a periodic, extensive survey on a larger sphere and level of completion than is usually covered in journals. These surveys are customarily the domain of "Advances in..." or other review volumes. Their aim is to assist the newcomer through a critical overview and a guide to who's who in the field. At the same time they assist the experienced researcher by providing a comprehensive and critical survey.

The state of development of solar energy conversion now demands such comprehensive reviews, and *Advances in Solar Energy* is designed to meet this need. The reviews will serve to bring together in a single document the results of years of work by many researchers in various locations, and also provide a comprehensive bibliography. There is sufficient knowledge in many facets of the field to warrant a detailed and comprehensive review, and to enlist the most qualified reviewer to present the material in a critical, yet not one-sided approach. We will proceed, therefore, according to the state of readiness of each of the topics, rather than along a sequential arrangement as is found in textbooks. Each of the topics shall be self-contained and may, in turn, reference not only original work but sub-topics reviewed in journals. As time progresses we will update reviews of subjects which, in the meantime, have advanced substantially.

Early volumes of this series will tend to emphasize more mature subjects. Later volumes may, in addition, start to treat areas that are not so mature, where the research is more current, and where implications of some results of the research may be more speculative.

The reader will find in the writings in this first volume of *Advances in Solar Energy* some variations in style and scope of the reviews. This is to be expected. Some of it will vanish as the field and this series mature; some of it will persist as we are serving a wide variety of readers.

To close this foreword, we express gratitude to the American Solar Energy Society for recognizing the timeliness and importance of this series and having offered to act as publisher, to Albert Henderson, as Director of the Publications Office, for having stimulated and encouraged the start-up, and to Barbara Bradley, as Publications Production Manager, for having put this volume together with patience and attention to the whole as well as to minute details.

Karl W. Böer
John A. Duffie
Co-Editors-in-Chief

TABLE OF CONTENTS

Radiometry—The Data   1
  Raymond J. Bahm

Solar Radiation Measurements: Calibration and Standardization
  Efforts   19
  G. A. Zerlaut

Biomass Pyrolysis: A Review of the Literature
  Part 1 — Carbohydrate Pyrolysis   61
  Michael Jerry Antal, Jr.

Recombinant Genetic Approaches for Efficient Ethanol Production   113
  H. W. Stokes
  S. Picataggio
  D. E. Eveleigh

Crystalline Silicon as a Material for Solar Cells   133
  M. Rodot

A Review of Large Wind Turbine Systems
  Large Wind Turbine Systems in the United States   175
  James J. Lerner

  Large Wind Turbine Systems Seen from the European Viewpoint   188
  H. Selzer

Controls in Solar Energy Systems   209
  C. Byron Winn

Passive and Hybrid Cooling Research   241
  John I. Yellott

Passive Solar Heating Research   265
  J. Douglas Balcomb

Subject Index   305

Author Index   311

Advances in Solar Energy. © 1983 American Solar Energy Society, Inc.

RADIOMETRY — THE DATA

RAYMOND J. BAHM

Raymond J. Bahm and Associates, Albuquerque, New Mexico 87120

Abstract

This chapter presents a summary of many of the important advances in the art and science of the use of solar radiation measurements. Emphasis is placed on those advances which will have the most impact in the utilization of solar energy. The work cited is mainly that done since 1977. The literature includes more comprehensive texts on solar radiation measurement,[1-3] a bibliography,[4] and a previous review of this field.[5]

USES OF SOLAR RADIATION DATA

Today most people are aware that solar energy is potentially a major energy source for the world. Already there are significant applications where solar energy is replacing other energy sources in a cost-effective manner. The most common of these are:

- domestic water heating;
- space heating of buildings;
- electric power for remote installations; and
- electric power for spacecraft

A variety of other applications are possible for the future.

One of the major benefits of the solar energy resource is that it is generally available at the location where the energy is needed. Thus, transmission lines or transportation of fuel is unnecessary. However, since the availability of solar radiation does vary from place to place and time to time, a knowledge of the expected future availability is necessary for the design of solar systems. Solar system designers are the largest group of users of solar radiation data.

Another important use of solar radiation resource data is economic planning. Global and national energy resource planning is beginning to include solar radiation availability as a significant factor.

Solar radiation provides both desirable and undesirable energy contributions to buildings.

Any designer of modern, high-rise buildings will analyze the projected solar energy gain of the building as part of the design process. Solar radiation data is a necessary input to this analysis.

Meteorologists and climatologists use solar radiation data in modeling the energy flows within the atmosphere. The energy input to this system is almost exclusively from the sun. Understanding the climate and weather requires a detailed knowledge of the solar radiation availability.

Agriculture is a major user of solar radiation data. Plants convert the available solar radiation by photosynthesis into chemical energy or biological energy. Models of agricultural productivity often include solar radiation measures. The irrigation requirements for crops are highly dependent upon evapotranspiration, which in turn is a function of solar radiation.[6] Brown et al., state that forecasts of solar radiation availabilities are used for scheduling irrigation.[7] The

Raymond J. Bahm is a principal in the firm of Raymond J. Bahm and Associates, which specializes in the collection, analysis and use of solar radiation data. He has worked on numerous projects for the U.S. Department of Energy and the State of New Mexico analyzing and evaluating measurements of the solar resource, and authored numerous papers in that area.

forecasting is important because the magnitude of the irrigation system requires many days of advance planning.

Solar radiation is a major factor causing the weathering of materials. The precise measurement of solar radiation and knowledge of its distribution is important to testing of the durability of materials.[8]

Recently, it has been recognized that high levels of solar radiation, especially in the ultraviolet, may be an important factor in causing human skin cancer.

Statistics on use of solar radiation data show that in 1979 alone the Environmental Data Service of the National Climatic Center responded to 1,714 separate requests for solar radiation data. These are in addition to the over 300 regular recipients of published solar radiation data.[9] Many of these are institutions which then process the data and pass on the results in a different form to groups of users.

TRENDS IN DATA COLLECTION AND USE

Today there is a strong trend toward improved data collection methods and instrumentation, resulting in more accurate, more meaningful, and more useful data. The major elements causing this trend are:

- much greater use of the data by a wider population of users;
- increasing availability of higher quality sensors;
- increased use of high quality sensors;
- improved data recording devices;
- more accurate calibration methods and more widespread knowledge about their use;
- increased use of digital computers and availability of solar radiation data in computerized form; and
- better data processing methods, including better methods for detecting and correcting errors.

Solar radiation has been measured regularly at many locations since the early 1900s. The earliest measures were attempts to determine the value of the ''solar constant''* and its possible variation by careful measures of the direct solar beam radiation at a variety of locations. Later measures were directed more toward the under-

standing of the energy available for the growth of crops and the energy inputs for weather phenomena. The most recent interest in the measurement of solar radiation has been for a better understanding of the energy impacts of climate on buildings and the development of solar energy systems.

Historically, solar radiation data have been summarized and published in the form of daily or monthly averages. However, the increased use of electronic digital computers has caused the use of detailed simulation models to be developed, and the demand for solar radiation data on an hourly basis has increased. Many researchers have demanded actual measures of hourly data in order to investigate the dynamic performance of systems in a realistic environment. Within the past year (1981), hourly solar radiation data have even become available for microcomputers on a floppy diskette.[10]

The basic instruments for the measurement of solar radiation are pyranometers, which measure the radiation incident on a horizontal surface, and pyrheliometers, which measure the beam radiation in a plane normal to that beam. Historically, a variety of other names have been applied to these instruments, but the type of measurement has been typically one of these two. Recent improvements, such as temperature compensation, have made these instruments more accurate and easier to use.

The best quality pyranometers are thermopile instruments. The stability of the calibration of instruments is very important, and has been a major problem for both manufacturers and users. The Precision Spectral Pyranometer (PSP) of the Eppley Laboratory is widely accepted as the best quality, commercially available instrument because of its demonstrated stability and other inherent characteristics. A new instrument was announced in 1981 by Kipp & Zonen, which is similar to the PSP, except that the manufacturer uses thick-film, mass-production techniques for the thermopile. If this instrument proves to be sufficiently stable, then it is likely it will be adopted widely if it maintains a lower cost.

Until the mid 1970s, few long-term, continuous measurements were made of the direct-beam solar radiation. The major problem had been the design of an instrument which would reliably track the diurnal and annual motions of the sun. The Eppley Tracker, which tracks the diurnal motion, has made beam measurements

---

* The solar constant is the flux of solar radiant energy across a unit area oriented normal to the solar beam at the mean earth-sun distance.

practical in some climates, but a number of problems still remain. These are:

- the difficulty in aligning the tracker axis parallel with the earth's polar axis to the necessary precision;
- the need for once or twice weekly manual adjustment of the tracker to account for the change of declination of the sun; and
- the inability to set or align the tracker when the sun is obscured by clouds.

New trackers have now been developed which alleviate the last two of these problems. These trackers are controlled by microprocessors and follow both the diurnal and annual motions of the sun. A problem with these trackers is that they are expensive.

Data collection, in the past, has been treated one parameter at a time. Temperature was recorded by one instrument, the data were processed in one way at one time, archived in one place, and published in one document. Wind was treated separately. Solar radiation data were still different. The data user had to search in different places for different parameters covering the same time period at the same location. It was up to the user to fill in missing data or to find complete data sets for matching time periods. Solar data were integrated over one time period related to solar time, and temperature data were integrated over another, related to local standard time. The result was that people relied heavily on statistically developed, representative-of-average data sets for design purposes.

Today, the data are beginning to be collected by one data recording system using standard times for all parameters and recording all the data on one computer-readable medium. Combined data sets prepared for specific purposes are beginning to emerge. The SOLMET data available for the U.S. is an example. These data provide a much improved dynamic picture of the environment for building and solar system designers.

Along with the coordinated data collection systems, better data processing methods are developing. Models are being used to check the reasonableness of the measured data. Two basic types of checks are used: the first is a check to see if the values are reasonable for the date of recording, and the second is the comparison of related measures by using a model which predicts one from the other. When these checks are applied during, or soon after, the measurement itself, problems with the instruments or recording system can often be detected before many data are lost.

Much of the need for measures of solar radiation is for the availability on surfaces which are nonhorizontal. The most convenient and generally adequate method for obtaining these data is by the use of models. The reason for this is that a significant portion of the solar energy available on nonhorizontal surfaces is reflected from the ground and other surroundings, and the reflectivity of these surroundings changes drastically even over very short distances. Thus, nonhorizontal measures of the energy incident on surfaces at one point are not necessarily representative of those at another point, even a short distance away. Models are used to transform the measures of the solar radiation incident on a horizontal surface to those on a nonhorizontal surface. The improved measurements mentioned in the foregoing paragraphs, and specialized measures of solar radiation on nonhorizontal surfaces where the reflectivity is known, are promoting the development of better models.

Another use of specialized measures of solar radiation is the use of "reference cells" for the testing of photovoltaic devices. A reference cell is a photovoltaic cell, usually from the same production process as those to be tested, which has been calibrated by a testing laboratory. The use of a reference cell is necessary because other instruments for measuring solar radiation do not have the same spectral response as the photovoltaic devices to be tested. The choice of a cell from the same production process is intended to minimize the effect of the natural spectral variations in solar radiation on the test results.

There has been an increased awareness of the importance of measurements of the spectral distribution of solar radiation. The need to estimate the solar radiation resource availability for photovoltaic devices has caused many persons to look for data which show the impact of the spectral characteristics of atmospheric absorption on the solar radiation resource availability for photovoltaic devices. There are a number of factors which must be considered. These are:

- most measures of solar radiation availability are made with instruments having equal sensitivity to the all parts of the solar spectrum;
- photovoltaic devices are sensitive to only a restricted part of the solar spectrum;

3

- different photovoltaic devices have different spectral sensitivities;
- the spectral transmission of the atmosphere varies with the varying climatic conditions, the major effects being absorption by water vapor, and absorption and scattering by atmospheric dust and gaseous molecules;

and

- some of the energy scattered from the solar beam reappears as diffuse solar radiation. This scattering is strongly dependent on the wavelength, and has its most rapid change in the region of the spectrum where the photovoltaic sensitivity is the greatest.

The resulting problems are:

- the accuracy of the resource estimates for flat-plate photovoltaics using the traditional measures will depend upon the atmospheric conditions at the time of use; and
- the accuracy of the resource estimates for concentrating photovoltaic systems will be even poorer than those for flat plate systems. Additional measures of the atmospheric characteristics are needed to improve the accuracy of the knowledge of the resource for these concentrating systems.

The increased development, testing, and application of mathematical models of the spectral characteristics of atmospheric transmission are due to the more widespread availability of computers and the decreasing cost of their use. These improved models are also showing us where improved instrumentation is needed and assisting in the definition of new instruments and measures.

Another trend is the decreasing use of the bimetallic strip type of solar radiation sensors. The general reason here is that these instruments tend to have poor accuracy.

EDUCATION, TRAINING, AND DISSEMINATION OF KNOWLEDGE

The recent interest in solar energy has created a much greater demand for solar radiation data than previously existed. More people are measuring solar radiation, more people are using the data, and there is a demand for more and better quality data.

The methods of measurement and the problems in using solar radiation data are not immediately obvious to the untrained user. A number of recent publications and workshops have helped eliminate the gap in that knowledge. Representative of these are:

(a) *The California Solar Data Manual.*[11] This publication provides general solar radiation data for the state of California, along with general formulas and other appropriate information for design of solar systems.

(b) *Radiation Measurement.*[12] This booklet describes instruments, their installation, calibration, and use. It is intended for the scientist or engineer who wants to set up a sophisticated solar radiation measuring station.

(c) *An Introduction to Meteorological Measurements and Data Handling for Solar Energy Applications.*[13] A compendium of chapters by many authors which serves as an introduction to solar radiation measurement.

(d) *Listing of Solar Radiation Measuring Equipment and Glossary.*[14] A listing of manufacturers as of 1976.

(e) *On the Nature and Distribution of Solar Radiation.*[15] An introduction to solar radiation and its atmospheric interaction.

(f) *Solar and Terrestrial Radiation.*[3] A text with emphasis on radiation instrumentation.

(g) Numerous solar radiation workshops have been held in the U.S. These have been sponsored by the U.S. Department of Energy, and most have been organized by one or more of the eight University Meteorological Research and Training sites.[16]

(h) There have been periodic review meetings of the U.S. Department of Energy contractors in the field of solar radiation resource assessment. These meetings have provided a center for information exchange and researcher interaction in the U.S. since 1975.

RESEARCH FACILITIES

A number of research facilities have recently been established in the U.S. which have provided, and will continue to provide, advances in the knowledge about solar radiation and its interaction with the atmosphere. Four of these are:

(1) The Solar Energy Research Institute (SERI), which has one entire division dedicated to solar radiation and solar energy resource evaluation. SERI is located in Golden, Colo., and was established in 1976.

(2) The eight University Meteorological Research and Training sites, which were established in 1977.[16] These are located at universities with atmospheric science units, as follows:

(a) The University of California at Davis

(b) The State University of New York at Albany

(c) The Georgia Institute of Technology at Atlanta

(d) The Solar Data Center, Trinity University at Antonio

(e) Oregon State University at Corvallis

(f) The University of Michigan at Ann Arbor

(g) The University of Alaska at Fairbanks

(h) The University of Hawaii at Honolulu

(3) The Solar Radiation Facility of NOAA Environmental Research Laboratory which was established at Boulder, Colo., in 1975. The primary purpose of this facility is to maintain standards and calibrate solar radiation measuring instruments for the U.S. government. Much of the current knowledge about the performance of solar radiation measuring instruments has come from research and calibration tests done there.[17]

(4) The DSET Laboratories at New River, Ariz., has recently established facilities for international solar radiation measuring instrument intercomparisons and calibration.[8]

PROCESSING, ARCHIVING, AND PUBLICATION

Solar radiation measurements for resource assessment are usually summed over hourly or daily time intervals, and then archived. The resource data customarily have been published as average monthly values. Some processing of the data is obviously needed between measurement and publication.

Undetected errors can contaminate large data bases, making them of poor quality or even unusable. Sometimes errors can be detected and partially corrected, even many years after they occur. The best time to detect measurement errors, however, is during or as soon after measurement as possible. An excellent example of a system to detect errors is given by Wendler et. al.[18]

The following sections illustrate some of the problems in processing, archiving, and publication. Corrective procedures are noted and procedures for the publication of the data in one specific format are identified. The methods developed, while solving problems, were applied to provide additional data sets generated by regression models. These sections provide good examples of potential problems and possible solutions.

REHABILITATION OF THE U.S. SOLAR RADIATION DATA BASE

This section describes a major effort to correct historical errors in the U.S. solar data base. The U.S. experience is not the first, nor will it be the last, of this type. This author has seen many other data bases with similar problems. Unfortunately, errors are often detected only when it is too late, and the usual solution is to discard the data. The rehabilitation methods are described here because important methods were developed for detecting and for correcting errors in solar radiation data.

The cause of the errors in the U.S. data base was two-fold. First, the lack of interest in the measurement program caused a situation where there was little support, and the training of those managing the program was inadequate. Second, a change in the manufacturing process for the instruments used created a situation where the calibration of some instruments changed with time. The most comprehensive description of this effort is given in Ref. 19. The following is excerpted from that reference. The long-term errors are described as:

(a) instrument sensitivity changes;

(b) improper calibration;

(c) wrong units used;

(d) changes in sensor environment or location;

(e) undocumented instrument changes;

(f) temperature sensitivity of calibration factor.

Besides these shortcomings, the solar radiation data were also referenced to two different international scales.

The rehabilitation process was divided into three phases:

(a) clean up and reformating;

(b) correction and quality control; and

(c) modeling and completion.

The clean-up and reformating phase consisted of:

(a) organizing the data and cataloging it;

(b) identifying and labeling any missing values;

(c) converting the entire solar data set to one time scale; and

(d) converting the data to metric units.

At the end of this phase, meteorological

data were merged with the solar radiation data and temperature corrections were applied to the solar data to compensate for the temperature coefficient of the older instruments.

The correction and quality control phase consisted of:

(a) reasonableness checks on the data. The following situations were flagged as potential errors and checked individually:

- solar radiation > 0 during night;
- temperature > 51.7°C
- dew point > dry bulb, and so on.

(b) Engineering corrections, including:

- calibration changes;
- solar radiation scale differences;
- midscale recorder chart setting;
- Parson's black paint degradation;
- so-called crossmatch problem;
- temperature response.

These engineering corrections were all based on the historical records of the instrument, station, or recorder, and were based on documented effects which could be defined and quantified. Figure 1 gives an idea of the data prior to the engineering corrections.

(c) Standard Year Irradiance (SYI) Corrections. A standard year irradiance model was developed by Hanson and Hoyt which calculated the irradiance on a clear day at solar noon.[19, 20] This model was used to finally adjust the solar radiation values to account for all errors which could not be indentified or corrected by other methods. The correction was applied to all of the data.

The modeling and completion phase consisted of the following steps:

(a) The filling of missing solar radiation values with values estimated from a regression model using sunshine and cloud cover data.*

(b) The addition of direct beam values generated from a model developed by Randall and Whitson.[22] Users should note that all the direct-beam values in the SOLMET data base are model-generated, not measured. This model is described elsewhere in this chapter.

The rehabilitated data base, termed SOLMET data, is now available from the

U.S. National Climatic Center on computer magnetic tape. It is described in Ref. 23. The rehabilitation process and models are described in Refs. 19 and 22.

SOLAR RADIATION DATA FROM THE REGRESSION MODEL

The regression model developed for filling the gaps in the rehabilitation process was very successful. There existed about 222 sites where sunshine data or cloud cover observations had been made, but where no solar radiation measurements had been made. It was decided to apply the regression model to these 222 sites using the regression constants from nearby solar radiation measurement locations to obtain synthetic or regression modeled solar radiation data for these locations. These data are now available in the SOLMET format. They are sometimes termed "ersatz," or Regression Modeled Data, to differentiate them from the measured data of the same format.

THE NEW 38-STATION U.S. SOLAR RADIATION MEASURING NETWORK

Many lessons were learned from the rehabilitation process and the examination of the network operation. A decision was made to revise the entire solar radiation measuring network, to install new instrumentation, to modernize the data recording and processing methods using computer technology, and to publish the data on a periodic basis.

The new solar radiation measuring network included:

(a) new instrumentation:

- pyranometers—measuring total global;
- pyrheliometers (at some locations)—measuring direct beam; and
- pyranometers with shadow bands (at some locations)—measuring diffuse.

(b) New data logging equipment with the following characteristics:

- data recorded directly on digital magnetic tape cassettes;
- measurements at one-minute intervals;
- clock contained in the datalogging system and;
- hourly integrated values printed by the datalogging system as the measurements were made.

---

* There is currently some indication that the version of the model used underestimated the solar radiation values for locations at the higher latitudes during cloudy days with snow on the ground. This is due to the lack of accounting for the increased albedo of the surface.[21]

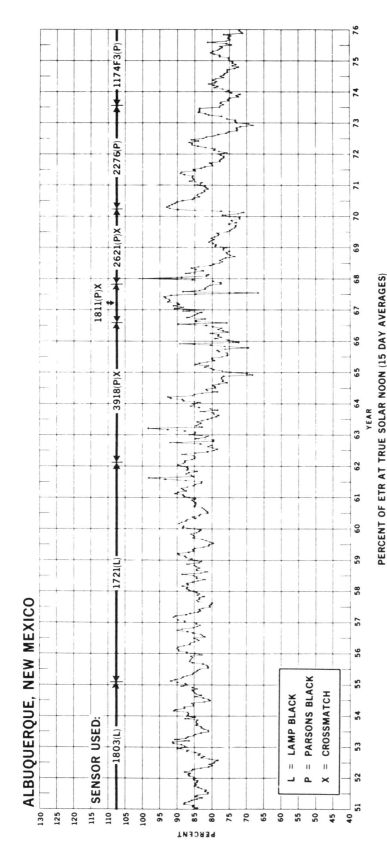

Fig. 1. Example of the solar radiation data from one station in the U.S. network prior to the Engineering corrections. Each point represents the average of approximately two weeks of solar noon measurements. The vertical axis is percent of extraterrestrial solar radiation.[19]

7

(c) A data processing system with the following characteristics:

- computerized data processing to minimize cost and time to publication; and
- automatic detection and flagging of potential errors.

The data processing is described in more detail in the following paragraphs.

The data processing system for the new U.S. Solar Radiation Monitoring Network evolved to suit the needs of the network equipment and the data recorded. The following is an overview of the processing. The complete processing system consists of 25 computer programs and over 16,000 lines of computer code. The following steps are included in the processing:[24]

(a) cassette editing. This consists of:

- reading the digital magnetic cassettes;
- checking for proper data format;
- range checks on the data values;
- combining the one-minute values into hourly sums; and
- writing the one-minute data into a standard microfiche report.

(b) Formatting and editing of hourly values:

- placing the data into a fixed format file with spaces for all possible data values whether they exist or not;
- adding values to replace missing data; and
- other edits as required.

(c) Comparison of measured global horizontal data with that from a clear sky model. The following conditions are identified and reviewed individually:

- modeled value < 120 Wh/m^2 and measured > 180 Wh/m^2;
- measured values exceeding the modeled by more than 5% during hours 11-14 local standard time; and
- measured values exceeding the modeled by more than 10% during other hours.

(d) Shadow band corrections are added to diffuse measurements using the method described in Ref. 12, pp. 30-31.

(e) Time shift of the data to correct for errors in time of measurement. In this process the data are fit to a curve using a spline to provide a smooth curve.

The data, through December of 1980, from the new network are archived on computer magnetic tape and are available in the SOLMET data format. The foregoing data processing from the 38 stations was discontinued with the January 1981 data.

A comparison of some of the data from the new network with the results of the models which were used in the rehabilitation process was made by Hall et al.[25] That analysis generally validates the models used. It identifies some weak areas as well. The reader should refer to Ref. 25 for more information.

A complete analysis of shadow band corrections in general is given by LeBaron et al. in Ref. 26. This includes a method for checking the corrections of calculated values by measurements, a description of several different types of corrections, and the results of experiments to validate the models given.

A SUBSET OF THE SOLMET DATA FOR SPECIAL PURPOSES

Solar radiation and meteorological data must be collected for a number of years in order to obtain a representative picture of the climate. The SOLMET data base of solar radiation and meteorological data has 23 years of data for each location. The cost of using the entire data set for one location, even with today's high-speed digital computers, is prohibitive for some users. Many users would like to have one "typical year" of data for the simulation of system performance. Since "typical" itself does not have a scientifically precise definition, it would also be desirable to have such a definition to use as a standard.

Such a data set would provide a means for estimating the typical or average performance of a nonlinear system at one location, or comparing the performance of different systems at one location or at a variety of locations. A project to define a typical meteorological year (TMY) and to prepare TMY data subsets of the SOLMET data was carried out at Sandia National Laboratories. The following are excerpts from Ref. 27, describing that project:

To be most useful to system designers, this data base should possess several characteristics. It should consist of hourly solar radiation and weather readings for a network of representative sites across the U.S. It should in some sense be "typical" of the long term data base and it should be of a year's duration. For a given site this data base could be reasonably called a "typical meteorological year" (TMY).

Defining the characteristics of a meteorological year which makes it "typical" is difficult; however, sensible properties of a TMY would seem to include the following:

(a) The meteorological measures of the TMY, that is, temperature, solar radiation, and wind, should have frequency distributions which are "close" to the long-term distributions.

(b) The sequences of the daily measures of the TMY should in some sense be "like" the sequences often registered at a given location.

(c) The relationships among the different measures for the TMY should be "like" the relationships observed in nature.

Briefly, the . . . approach adopted for selecting TMYs for a given station is as follows: a typical month for each of the twelve calendar months from the long-term data base (23 years for most stations) was chosen and then these twelve months (TMMS) were catenated to form TMYs. The data set generated to form the basis for the selection of a typical month consisted of thirteen daily indices calculated from the hourly values of dry bulb temperature, dew point temperature, wind velocity, and solar radiation. Monthly statistics were calculated for each index. Month/year combinations which had statistics that were "close" to the long term statistics were candidates for typical months. Final selection of a typical month included considerations of persistence of weather patterns.

The procedure for selecting a TMM consisted of two steps. The first step was to select five candidate years. The second step was to select the TMM from the candidate years.

In the succeeding discussion the term "year" refers to a month/year combination. That is, if May is the month under consideration, 1966 refers to May, 1966.

a) Selection of five candidate years

For each of the twelve calendar months the procedure involved selecting the five years that were "closest" to the composite of all 23 years. This was done by comparing the cumulative distribution function (CDF) for each year with the CDF for the long term composite of all 23 years for each of the 13 indices. (The CDF gives the proportion of values which are less than or equal to a specified value of an index.) . . . The statistic selected to measure the closeness of each year's CDF to the long term composite for a given index was the Finkelstein-Schafer (FS) statistic.

The FS values associated with important statistics would receive relatively larger weights than the less important statistics.

In the generation of these TMYs, it was determined that the three range statistics and the minimum of wind velocity were of little or no value in the selection process, so these statistics were omitted, . . . The actual weighting scheme used for the TMY's follows:

b) Final selection of TMM

The final selection of the TMM from the five candidate years involved examining statistics of persistence structure associated with mean daily dry bulb temperature and daily global solar radiation. The statistics examined were the FS statistic and the deviations for the monthly mean and median from the long term mean and median. Persistence was characterized by frequency and run length above and below fixed long term percentiles.

The TMY data set is now available on magnetic tape for the 26 locations described above. In addition to that, the same process has been applied to 208 of the regression modeled data sets. Those data are also available in the TMY format.

The TMY data set is now probably the most commonly hourly data set used for modeling of system performance.

FORECASTING OF SOLAR RADIATION

Most measurements of the solar radiation resource availability have as their goal the prediction of the solar radiation falling on some point of the earth at some time in the future. Long term averages of measurements give an estimate of the future amount to be expected. Forecasting is an attempt to improve that estimate by using other information, (for example, the recent history of solar radiation and other related parameters).

The forecasting of solar radiation has the potential to have an impact on several areas, including short-term planning of and load management for solar energy systems, irrigation scheduling, and agricultural management.

A workshop on the forecasting of solar radiation was held in February of 1981.[28] The participants discussed methods for forecasting solar radiation and some of the uses of these forecasts. Some of the recommendations of that workshop are:

a) NOAA should solicit outside advice, perhaps from international solar energy groups, to formulate its plans for disseminating one- and two-day predictions of global solar radiation. This review should include agreement on the usefulness of the forecasts, format, major product, public availability and association of solar forecasts with other associated weather elements like temperature.

b) The above ad-hoc committee should ensure the integrity of the product. It should especially try to

	Temperature						Wind		Solar
	Dry Bulb			Dew Point			Velocity		Radiation
	Max	Min	Mean	Max	Min	Mean	Max	Mean	
Wi:	1/24	1/24	2/24	1/24	1/24	2/14	2/24	2/24	12/24"

where the Wi are the weights for each parameter.

9

make the nature of the forecast as useful to users as is possible, (e.g. mean values, probabilities, etc.)

c) In particular, and if possible during the heating periods, the solar forecasts should be given together with a measure of heating needs, e.g. degree-days, and during cooling periods together with cooling needs.

Partly as a result of that workshop, a program of daily forecasts of solar radiation availability was initiated late in 1981 by the National Weather Service. These forecasts are available on the National Digital Facsimile Network as a map of the continental U.S. twice daily, giving forecasts twice daily for 24 and 48 hours ahead.[29] An example of one of these maps is shown in Fig. 2.

CIRCUMSOLAR RADIATION

The recent interest in highly concentrating solar energy systems, and the lack of detailed knowledge about the effects of the atmosphere on the energy available from within the solar aureole, were the major stimuli for a program of investigation of circumsolar radiation. The primary focus of the program, thus far, has been the development of design data for highly concentrating solar energy systems. However, there is considerable potential for using these data to develop a better understanding of the atmospheric optical phenomena.

The design of the instruments and data collection has been performed by the Lawrence Berkeley Laboratory of the University of California, Energy and Environment Division (LBL). Descriptions of this project are given in a number of reports from LBL,[30-34] from which the following are excerpts:

The purpose of this project is to provide measurements and analyses of the solar and circumsolar radiation for application to solar energy systems that employ lenses or mirrors to concentrate the incident sunlight. Circumsolar radiation results from the scattering of direct sunlight through small angles by atmospheric aerosols (e.g., dust, water droplets or ice crystals in thin clouds).

Concentrating solar energy systems will typically collect all of the direct solar radiation (that originating from the disk of the sun) plus some fraction of the circumsolar radiation. The exact fraction depends upon many factors, but primarily upon the angular size (field-of-view) of the receiver. A knowledge of the circumsolar radiation is then one factor in predicting or evaluating the performance of concentrating systems.

The project employs unique instrument systems (called Circumsolar Telescopes) that were designed and fabricated at LBL. The basic measurements are:

1) the "circumsolar scan," the brightness of the sun and circumsolar region as a function of angular distance from the center of the sun and

2) the usual "normal incidence" measurement of a pyrheliometer.

Both measurements are made for the entire solar

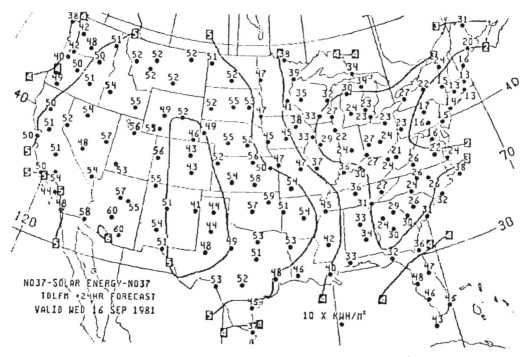

Fig. 2. An example of the NOAA solar radiation forecast maps.[29] The units are tenths of kWh/m², (divide by 10 to get kWh/m²).

spectrum, and (via colored filters) for eight essentially contiguous wavelength bands. Thus the measurements are applicable to systems in which the receiver is essentially wavelength-insensitive (e.g., central receiver) and to wavelength-sensitive systems (e.g., concentrating photovoltaics).

A secondary purpose of the project is to relate the data to the atmospheric processes that attenuate the solar radiation available to terrestrial solar energy systems.[32]

Data have been collected at different times at a number of different locations. These locations represent a variety of different climates. The locations, all within the U.S. include:

Boardman, Oreg.
Coalstrip, Mont.
Atlanta, Ga.
Albuquerque, N.M.
Ft. Hood, Tex.
Argonne, Ill.
Berkeley, Calif.
Edwards, Calif.
Barstow, Calif.

Some of the important conclusions from this project are:

1) A pyrheliometer of usual design (6 degree field-of-view) will yield an overestimate of the direct solar radiation (originating from the disk of the sun).

2) This overestimate is, on the average, a few percent of the total energy in the beam.

3) Data from nondesert areas have a generally higher fraction of the energy in the aureole than do the desert areas.

4) For sky conditions such as haze and cirrus clouds, a measure of the circumsolar radiation is necessary to understand the performance of an operating, highly concentrating system.[34]

An analysis of some of these circumsolar data and the development of a model to estimate the circumsolar radiation from solar and meteorological data was performed by Watt.[35] This report also includes data from the circumsolar telescope and a description of the atmospheric components causing the circumsolar radiation.

A method of expressing the distribution of circumsolar radiation intensity as a function of Gaussian distributions has been developed by C. N. Vittoe and F. Biggs, using the data from the circumsolar telescope.[36] This method provides a computationally simple analytic description of the intensity profile of the circumsolar radiation.

THE SOLAR CONSTANT MEASUREMENTS

The sun has a constant energy output, at least we treat the measurements of solar radiation at the surface of the earth as though it were constant. Historically, a major problem in solar radiation measurement has been establishing the value of the solar constant and defining a scale for the measurement.[8] Recently, however, measurements of the extraterrestrial solar radiation from earth-orbital spacecraft have refined this measurement and established a new value for the solar constant. The satellite measurements show some variability in the "solar constant" with time.

Two experiments are currently (1982) providing measurements of the solar constant. These are the Nimbus 7 spacecraft, Hickey[37] and the Solar Maximum Mission spacecraft, Willson.[38] Their respective values for the solar constant, 1372.6[37] and 1368.31[38] W/m², differ by only about 0.3%. Both experiments show that there are variations, over periods of the order of days, about the mean value, which exceed 0.2% of the mean value. Both also indicate that there currently appears to be a downward trend of the solar constant which is greater than 0.015% per year. However, since only two to three years of these measures are available, it is not yet possible to show how long such a trend might be expected to continue. One illustration of these data is shown in Fig. 3.[39]

These experiments are important because the research is now beginning to link solar variability with impacts on the climate at the earth's surface.[40, 41] Measurements of this type may ultimately be a major source of information for forecasters of solar radiation at the surface of the earth, and for other weather forecasters.

DIRECT, DIFFUSE, TOTAL SOLAR RADIATION AND THEIR INTERRELATIONSHIP

The most commonly measured solar radiation parameter is the total global radiation on a horizontal surface. The parameters most needed for energy design and analysis are the total global solar radiation on a tilted surface and the direct-beam solar radiation. Because of the lack of data in the latter forms, many people have worked on methods for estimating one from the other. These relationships are normally needed on an hourly, daily, and monthly average basis.

The most commonly used method was pub-

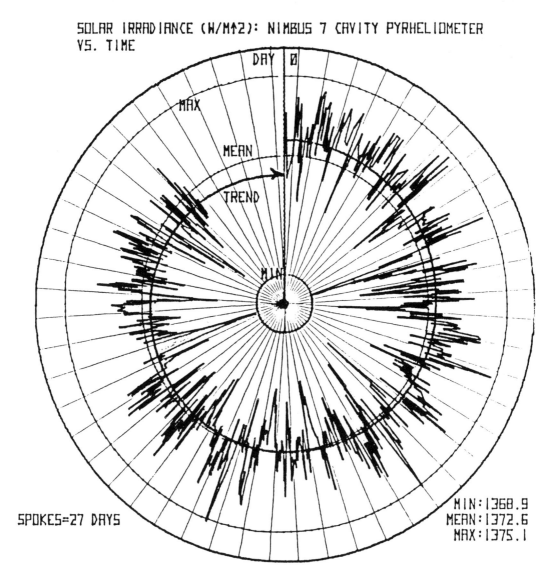

SOLAR IRRADIANCE (W/M↑2): NIMBUS 7 CAVITY PYRHELIOMETER VS. TIME

DAY 0

MAX

MEAN

TREND

MIN

SPOKES=27 DAYS

MIN:1368.9
MEAN:1372.6
MAX:1375.1

1404   DAY IRRADIANCE FROM NOV. 16, 1978 TO APRIL 15, 1982

Fig. 3. Illustration of Solar Irradiance Measurements from the Nimbus 7 Spacecraft. Time begins at day 0 (vertical) and increases clockwise. The data for 1247 days are plotted. Each spoke represents 27 days, (1404 days for 360 degrees). Intensity is expressed as increasing radius, with a suppressed zero. The center of the circle represents 1368 W/m², the outer circle is 1376 W/m². This is engineering level data and subject to revision. This plot compliments of John R. Hickey, Eppley Laboratory, NASA and NOAA/NESS: NASA/GSFC data, April 1982.[39]

lished originally by Liu and Jordan in 1960.[42] Since that time, the basic method has been:

a) applied to daily and hourly relationships,[22,43-45]

b) extended to surfaces not facing toward the equator,[46]

c) revised with new data,[45,47,48] and

d) compared to models of atmospheric transmission.[49]

The above references are not a complete list of all the work on this topic, but rather will give an idea of its extent.

The primary value in the Liu and Jordan

technique is that it allows the estimation of other parameters from just the total horizontal global radiation. While most of the work relating to this method has established its validity and usefulness for many purposes, a comprehensive analysis of the potential errors inherent in it has not been published. An approach to such an analysis was suggested in Ref. 49.

Engles et. al.,[50] suggest the use of a new parameter $Z = I_{th}/I_{th}^*$ instead of the more traditional K_t. The difference is that I_{th}^* is based on the transmission by a standard atmosphere, rather than on the extraterrestrial solar radiation. This has the advantage of emphasizing the variable atmospheric effects while minimizing the effects of transmission through the atmosphere.

Steinmuller[51] describes a technique for using data from two pyranometers oriented in different planes to estimate the energy available on a third plane.

THE ISOTROPIC DISTRIBUTION OF DIFFUSE SOLAR RADIATION

The simple methods for estimating the solar radiation on nonhorizontal surfaces require a knowledge of, or assumption about, the brightness of all points of the sky. The direct-beam intensity and direction are usually well known. The diffuse, however, is seldom well known. The total diffuse intensity is usually known, but the uniformity of the angular distribution of the diffuse sky brightness is not. It is common to assume that the diffuse is isotropic simply because the computation under other cases is very unwieldy. There has been little information on the uniformity of sky brightness until recently.

In 1977, Dave published model results showing that the isotropic-distribution assumption systematically underestimates the diffuse energy in many cases.[52] He also showed that this could be by as much as a factor of six.

Temps and Coulson published the results of measurements and a simple model which provided improved estimates of the diffuse distribution under a clear sky.[53]

Klutcher improved the Temps-Coulson model for partly cloudy conditions, and showed that the isotropic assumption fitted the overcast conditions for one location.[54]

Stewart, Spencer and Healey (1981) compared the Liu and Jordan, Temps and Coulson, and Klutcher models with measurements taken at the Solar Energy Meteorological Research and Training Site—Region 2. The models systematically overestimated insolation on the non-south surfaces.[55]

Valko gives some graphic examples of the instantaneous, nonistropic distributions of sky brightness and illustrates one method of displaying the data (Chapter 8, Ref. 13).

ATMOSPHERIC TRANSMISSION MODELS

Physical models of atmospheric transmission and detailed spectral measurements of the energy available at the earth's surface provide the most accurate understanding and prediction of the available solar radiation. The rapidly decreasing costs of digital computing have begun to make detailed physical models practical for many purposes. The following are some of the recent publications in this area.

Leckner provides a good but brief overview of the major atmospheric characteristics for a physical model in Chapter 3 of Ref. 13. Additional detail is given in Refs. 1 and 3. More comprehensive atmospheric data are available in Refs. 56 and 57 by Kondratyev, and in Ref. 58 by Kneizys et. al.

Leckner describes a model for the direct beam in Ref. 59. This model permits calculation of direct, diffuse, and global spectra and uses average values of ozone, turbidity, and water vapor as inputs.

King and Buckius describe models in Refs. 49 and 60. The first reference is a total spectrum model for diffuse. Results are compared to the measured data and the Liu-Jordan empirical model from Ref. 42. The second expresses the atmospheric transmission of the beam radiation with turbidity, humidity, and visibility as principal inputs.

Bird and Hulstrom describe models and their use in Refs. 61 and 62. The first describes Monte Carlo techniques. The latter compares several different models and their results and describes a model synthesized from these models.

The computer code LOWTRAN 5 is a sophisticated atmospheric transmission model.[58] LOWTRAN allows calculation of the spectral transmission of various slant paths through the atmosphere.

The LOWTRAN code was modified by Bird for direct-beam solar availability, and the latter version is called SOLTRAN 5.[63]

Hatfield et.al., describe a spectral model for both direct and diffuse solar radiation.[64] The principal input parameters to this model are geographical location, turbidity, water vapor, and ozone.

SPECTRAL AND INFRARED DATA

The spectral distribution of the extraterrestrial solar radiation is reasonably well known. However, absorption, scattering, and reflection within the earth's atmosphere can cause significant changes in the spectral content of the solar radiation by the time it reaches the surface of the earth. Theoretical models which explain these atmospheric effects are fairly well developed (see previous section of this chapter); however, there are still relatively few measured data from which to validate these models or to establish the average spectral characteristics of the solar radiation at the earth's surface.

The interest in making spectral measurements is increasing. A number of recent projects have contributed to the knowledge in this area. Among these are:

a) Measurements of the total horizontal radiation in six wavelength bands at three locations: Panama, Rockville, Md., and Barrow, Alaska. These data have been taken continuously since 1968 by the Smithsonian Radiation Biology Laboratory.[65]

b) Similar data currently being taken by the eight University Research and Training Sites.[16]

c) The direct-beam data being taken by the Circumsolar Telescope Program.[32]

d) Munroe and Shepherd[66] describe a unique instrument for measurement of the spectral content of solar radiation. It appears that this instrument is still under development.

e) DSET Laboratories[67] and SERI[68] have developed sophisticated, high-resolution instrument systems for measuring the solar spectrum.

f) Li-Cor has developed a portable spectroradiometer.[69]

One of the most important spectral effects on the visible portion of the solar spectrum is due to particulate scattering and absorption. This effect is often measured by a parameter termed turbidity. There exists a body of measurements of turbidity over the continental U.S. and elsewhere[70-74] which has been made by an instrument called the "sun photometer." Unfortunately, the stability and definition of this instrument are not well known and this has caused many questions on the validity of this data.

Most of the incoming solar energy received by the earth (0.2 to 4.0 micrometers wavelength) is returned from the earth back into space in the form reradiated energy at wavelengths of 5 to 40 micrometers. The interaction of the atmosphere with both incoming and outgoing radiation is a major factor in the weather. A new instrument has been developed which allows routine measurements of the spectral sky (atmospheric) emissivity within 6 spectral bands in the 8 to 20 micrometer region. Measurements have been made at a number of locations. The instrument and some of the measurements are described in Refs. 75-79.

THE AVAILABILITY OF SOLAR RADIATION DATA

The most common measure of solar radiation availability is the total horizontal global radiation, which is measured with a pyranometer in a horizontal position. These instruments read the instantaneous energy incident on a plane surface. The readings are normally integrated over some time period, hourly, daily, or monthly, and the resulting values are archived. These data are then published in the archived form or summarized further. The most common time interval for summary statistics is the monthly average value of the daily total.

Data have been collected at a wide variety of locations around the earth. The most complete global coverage in one reference is the study completed by Lof et. al. in 1966.[80]

The World Meteorological Organization collects and archives solar radiation data from members. These data are published by the A.I. Voeikov Main Geophysical Observatory, Leningrad, U.S.S.R., and are called the Solar Radiation and Radiation Balance Data from the World Network.

Most nations have some central meteorological organization which is assigned the responsibility for collecting and archiving solar radiation data. One place to obtain solar radiation data is to contact those organizations directly. Carter has published a list of 41 such countries, and names of stations, which are measuring solar radiation.[81]

Data for the U.S. are available from the National Climatic Center (NCC) at Asheville, N.C. There are many formats for these data and the requester should consider his needs carefully before contacting the NCC.

NEEDS FOR THE FUTURE

The more we learn about solar radiation, the more we become aware of the need for better knowledge and the gaps in the present knowledge. This section presents some areas where more work is needed.

There is a need for a comprehensive reference or text on the subject of solar radiation measurement and the use of those data. While there are some good references,[1-5] much has been learned since they were published. That information is scattered, as the reader of the current bibliography can see.

Better calibration methods are needed. The current capabilities for calibration accuracy are roughly:

a) Pyrheliometers—absolute instruments—0.5% under typical circumstances, 0.2% in very favorable situations,

b) Pyrheliometers—field instruments—1% to 2%, however, the real limitation here is the instruments' performance in the field and the relation of the measurement conditions to the conditions at the time of calibration. Another limitation is the training of the staff to take care of the instrument—the lack of care accelerates the problems.

c) Pyranometers—transfer of calibration from pyrheliometer to pyranometer is about 2%. However, the real problem here is the variability of sky conditions, the distribution of the diffuse brightness over the sky dome at both the time of calibration and the time of measurement, and the relative intensities of the beam and the diffuse energy. Since these parameters are quite variable, depending upon sky conditions, this means that the the measure itself, total global horizontal radiation, cannot be precisely quantified. Hence, there must remain a considerable amount of statistical variability in the measure, just because of its imprecise definition with respect to other standards. In order to obtain more accurate measurements, it will be necessary to define new measures, such as a combination of direct beam and an array of values of the diffuse as a function of position on the sky dome.

Better measurements of the spectral content of solar radiation are needed. Some of the uses of these data if they were available would be:

a) evaluation of the resource availability for flat plate and concentrating photovoltaics, as a function of the spectrally selective scattering of the beam due to atmospheric dust and other particulates, and

b) monitoring of the air pollution at various levels of the atmosphere by spectral measures of the solar radiation.

More information is needed on the errors to be expected from the many models discussed in this chapter. While many of these models and their resulting data are widely used, there is still very little information available to the average user on the accuracies to be expected.

Statistical methods which allow the users of the data to understand the quality of the data in terms of the performance of the solar energy systems which they are designing are needed. While these methods may exist, they are not commonly expressed in terms which the data users understand.

Finer geographic resolution of measured data is needed. Most system designers today must interpolate between points of measurement.

Most users of solar radiation data require associated weather data, especially temperature. Data bases need to be generated which include the temperature and other meteorological parameters along with the solar radiation data.

Better methods are needed for describing the nonuniformity of the brightness of the sky-dome, and for measuring it.

More comprehensive specifications of instruments for measuring solar radiation are needed. This includes the specification of idealized detectors for each type of measure currently made, as well as new measures such as the nonuniformity of the skydome.

Manufacturers need to supply better instructions for use with new instruments, including instructions for calibration, installation, maintenance, data recording, and data processing.

There is a need for a comprehensive philosophy, structure, and organization of the measurement process. Historically, solar radiation measurement has been related to the individual experimenters and their respective interests. Rather than a single goal, there appears to have been a variety of projects. The currently available data are still largely measurement-oriented rather than user-oriented. A user-oriented focus is needed. The following section contains suggestions by the author for such a program philosophy.

It is suggested that a primary goal of a resource assessment program shall be to provide a standardized method for predicting the availability of the solar resource at *any* given place over any given time period, and to improve the accuracy of that prediction. Such a procedure would necessarily include a wide variety of things such as forecasting, measurement, interpolation, modeling, a knowledge of errors and uncertainty, and their impact on the estimate.

Such a procedure should:

a) give values or sets of values where combined occurrence is important,

b) give an error band for those values, (and an uncertainty),

15

c) give the expected variation or distribution about those values,

d) give the same result no matter who uses the procedure,

e) give better values with better or more data,

f) be adopted by, authorized by, or otherwise approved by appropriate standards committees or organizations,

g) be acceptable in a court of law,

h) be traceable to accepted standard measures,

i) be testable and provable,

j) include long-term weather stability factors (e.g., pollution increases),

k) work for broad spectrum data (0.2 to 4.0 microns), for narrow spectrum (photovoltaic or biological) measures, or be confined to the visible part of the spectrum,

l) work for the variety of measures required, total global horizontal radiation, direct-beam radiation, diffuse radiation only, energy through a window, and so on,

m) work for new types of systems (e.g., multi-layer photovoltaic), and

n) incorporate the results of both measurements and modeling.

The existence of such a procedure will:

a) provide a precise definition of data quality,

b) provide a precise framework for specifying a data user's needs,

c) provide a framework for estimating the cost of attaining a specific data quality level,

d) provide a mechanism for evaluating solar system performance at any given location.

e) provide a clear mechanism for defining intermediate objectives in a resource assessment program.

This procedure does not have to be simple or easy to understand by all persons. It may be desirable to have a single agency, or group of agencies, provide a common and current data base for all users.

Most users of solar radiation data do not or cannot foresee a future need for those data sufficiently far enough in the future to make their own measurements. Only after the need comes is a request generated. The planning for and the measurement of solar radiation must be based on extrapolated demands, as well as foresight and scientific inquiry, since many years of measurements are normally required to satisfy a need.

RECOGNITION

Two persons, because of their positions in the U.S. Department of Energy, have played a major role in promoting many of the projects and programs described in the foregoing sections. They were architects of many of these projects, but were unable to participate in the research, and hence, seldom receive credit for their contribution. It is for this reason that we wish to recognize Michael R. Riches and Fred A. Koomanoff for their contributions to many of the advances in the measurement of solar radiation.

ACKNOWLEDGMENTS
The author wishes to thank P. Berdahl, J. Duffie, J. R. Hickey, and R. Cram for providing detailed and current information for inclusion in this chapter. Suggestions by the editors, Kinsell Coulson and J. Duffie were very helpful. The author is indebted to his wife Linda for assistance in preparing the manuscript.

BIBLIOGRAPHY
1. N. Robinson. 1966. *Solar radiation*. New York: Elsevier.
2. A. J. Drummond. 1970. *Precision radiometry: advances in geophysics*. Vol. 14. Edited by H. E. Landsberg and J. Van Mieghem. New York: Academic Press.
3. K. L. Coulson. 1975. *Solar and terrestrial radiation—methods and measurements*. New York: Academic Press.
4. S. Brazel. 1976. *A bibliography of solar energy instrumentation, measurement and network design*. Office of the State Climatologist for Arizona.
5. R. J. Bahm. 1980. Terrestrial solar radiation availability. In *Solar energy technology handbook*. Edited by W. C. Dickinson and P. N. Cheremisnoff. New York: Marcel Dekker.
6. G. C. Hargreaves. 1981. Climate and the third world agricultural development. *INTERCIENCIA* 6:4. Asociacion INTERCIENCIA Caracas, Venezuela.
7. M. E. Brown and J. R. Lambert. 1981. Daily solar radiation measurements and forecasts for application of agricultural crop simulation models. *Proceedings of the First Workshop on Terrestrial Solar Resource Forecasting and on Use of Satellites for Terrestrial Solar Resource Assessment*. Feb. 1981. Newark, Del.: American Section of the International Solar Energy Society, Inc. (AS/ISES), Univ. of Delaware, p. 114.
8. G. A. Zerlaut. 1982. Solar radiation measurements: Calibration and standardization efforts. *Advances in solar energy*. Vol. 1. March 1983. Edited by K. W. Böer and J. A. Duffie. Newark, Del: ASES, Univ. of Delaware, p. 19.
9. W. Kaszeta and F. Quinlan. 1981 (Mar.). Personal communication.

10. Microcomputer Design Tools, Inc. 1981. Product literature, Albuquerque, N.M.

11. P. Berdahl; D. Grether; M. Martin; and M. Wahlig. 1978. *California Solar Data Manual*. Report no. LBL-5971, NTIS.

12. J. Ronald Latimer. 1978. *Radiation measurement—international field year for the great lakes—technical manual series no. 2*. Edited by J. MacDowall. The Secretariat, Canadian National Committee for the Hydrological Decade. (Obtained from Supply and Services Canada, Hull, Quebec K1A 0S9.)

13. M. R. Riches, ed. 1980 (Oct.), *An introduction to meteorological measurements and data handling for solar energy applications*. Report no. DOE/ER-0084, NTIS.

14. E. A. Carter; S. A. Greenbaum; and A. M. Patel. 1976, *Listing of solar radiation measuring equipment and glossary*. Report no. ERDA/NASA/31293-76/3 or DOE/1024-1, NTIS.

15. D. Watt. 1978 (Mar.). *On the nature and distribution of solar radiation*. Report no. HCP/T2552-01, NTIS. Stock No. 016-000-00044-5.

16. R. L. Hulstrom. 1981. *Solar energy meteorological research and training site program—first annual report*. Report no. SERI/SP-642-947, NTIS.

17. E. C. Flowers. 1978. Solar Radiation Facility—report for 1978. NOAA/ERL-ARL Internal report.

18. G. Wendler and F. D. Eaton. 1980. Quality control for solar radiation data. *Solar Energy*. 25(2): 131–138.

19. F. T. Quinlan, ed. 1979. *SOLMET—Final Report*. Vol. 2. Environmental Data and Information Service. NCC., NOAA. U.S. Dept. of Commerce.

20. D. V. Hoyt. 1978. A model for the calculation of solar global insolation. *Solar Energy* 21(1):27–35.

21. J. Cotton. 1982 (Jan.). Air Resources Laboratory, NOAA, Silver Spring, Md. Personal communication.

22. C. M. Randall and M. E. Whitson. 1977. *Hourly insolation and meteorological data bases including improved direct insolation estimates*. Aerospace Report no. ATR-78(75592)-1, El Segundo, Calif.: Aerospace Corp.

23. *SOLMET—user's manual*. Vol. 1, 1977. Environmental Data and Information Service. NCC, NOAA. U.S. Dept. of Commerce.

24. Richard S. Cram. 1982 (Feb.). Chief, Primary Data Branch. Environmental Data and Information Service. NCC, NOAA. U.S. Dept. of Commerce. Personal communication.

25. I. J. Hall; R. R. Prarie; H. E. Anderson; and E. C. Boes. 1980. *Solar radiation model validation*. Report no. SAND 80-1755. Sandia National Laboratories.

26. B. A. LeBaron; W. A. Peterson; and Inge Dirmhirn. 1980. Corrections for diffuse irradiance measured with shadowbands. *Solar Energy* 25(1):1–14.

27. I. J. Hall; R. R. Prarie; H. E. Anderson; and E. C. Boes. 1981 (May). Generation of typical meteorological years for 26 SOLMET stations. Appendix to *Typical meteorological year—user's manual*. Asheville, N.C.: National Climatic Center.

28. R. J. Bahm, ed. 1981. *Satellites and forecasting of solar radiation, the proceedings of the first workshop on terrestrial solar resource forecasting and on use of satellites for terrestrial solar resource assessment*. Washington, D.C.; February, 1981. AS/ISES, Univ. of Delaware.

29. D. S. Cooley and J. S. Jensenius. 1981 (Sept.) Solar energy guidance. *National weather service technical procedures bulletin series no. 304*.

30. D. F. Grether; D. Evans; A. Hunt; and M. Wahlig. 1979 (June). *Applications of circumsolar measurements to concentrating collectors*. Lawrence Berkeley Laboratory report no. LBL-9412.

31. D. F. Grether; D. Evans; A. Hunt; and M. Wahlig. 1979 (Oct.). *Measurement and analysis of circumsolar radiation*. Lawrence Berkeley Laboratory report no. LBL-10243.

32. D. B. Evans; D. F. Grether; A. Hunt; and M. Wahlig. 1980 (Mar.), *The spectral character of circumsolar radiation*. Lawrence Berkeley Laboratory report no. LBL-10802.

33. D. F. Grether; D. Evans; A. Hunt; and M. Wahlig. 1980 (Oct.). *Measurement and analysis of circumsolar radiation*. Lawrence Berkeley Laboratory report no. LBL-11645.

34. D. F. Grether; D. Evans; A. Hunt; and M. Wahlig. 1981 (May). *The effect of circumsolar radiation on the accuracy of pyrheliometer measurements of the direct solar radiation*. Lawrence Berkeley Laboratory report no. LBL-12707.

35. A. D. Watt. 1980. (Apr.). *Circumsolar radiation*. Sandia National Laboratories report no. SAND 80-7009.

36. C. N. Vittoe and F. Biggs. 1981, Six-Gaussian representation of the angular-brightness distribution for solar radiation. *Solar Energy* 27(6):469–490.

37. J. R. Hickey et al. 1982. Extraterrestrial solar irradiance variability: two and one-half years of measurements from Nimbus 7. *Solar Energy* 28(5): 443–445.

38. R. C. Willson et al. 1981 (Feb. 13). Observations of solar irradiance variability. *Science* 211:700–702.

39. J. R. Hickey. 1982 (Apr.). Personal communication.

40. J. V. Evans. 1982 (Apr. 30). The sun's influence on the earth's atmosphere and interplanetary space. *Science* 216:467–474.

41. D. F. Salsbury. 1982 (Apr. 27). Don't blame the sun's 18-month chill for last winter's extra bite. *Christian Science Monitor*.

42. B.Y.H. Liu and R. C. Jordan. 1960. The interrelationship and characteristic distribution of direct, diffuse and total solar radiation. *Solar Energy* 4(3):1–19.

43. D. G. Erbs; S. A. Klein; and J. A. Duffie. 1982. Estimation of the diffuse radiation fraction for hourly, daily and monthly-average global radiation. *Solar Energy* 28(4):293–302.

44. M. Collares-Pereria and A. Rabl. 1979. The average distribution of solar radiation-correlations between diffuse and hemispherical and between daily and hourly insolation values. *Solar Energy* 22(2): 155–164.

45. M. Iqbal. 1980. Prediction of hourly diffuse solar radiation from measured hourly global radiation on a horizontal surface. *Solar Energy* 24(5):495–503.

46. S. A. Klein. 1977. Calculation of monthly average insolation on tilted surfaces. *Solar Energy* 19(4): 325–329.

47. S. A. Klein and J. C. Theilacker. 1980. An algorithm for calculating monthly-average radiation on inclined surfaces. *Journal of Solar Energy Engineering* 103:29–33.

48. J. K. Page. 1978. *Methods for estimation of solar energy on vertical and inclined surfaces*. Report no. BS-46. The Dept. of Building Science, University of Sheffield, Sheffield, England.

49. R. O. Buckius and R. King. 1978. Diffuse solar

radiation on a horizontal surface for a clear sky. *Solar Energy* 21(6):503–509.

50. J. D. Engels; S. M. Pollock; and J. A. Clark. 1981. Observations on the statistical nature of terrestrial irradiation. *Solar Energy* 26(1):91–92.

51. B. Steinmuller. 1980. The two-solarimeter method for insolation on inclined surfaces. *Solar Energy* 25(5):449–460.

52. J. V. Dave. 1977. Validity of the isotropic-distribution approximation in solar energy estimations. *Solar Energy* 19(4):331–333.

53. R. C. Temps and K. L. Coulson. 1977. Solar radiation upon slopes of different orientations. *Solar Energy* 19(2):179–184.

54. T. M. Klutcher. 1979. Evaluation of models to predict insolation on tilted surfaces. *Solar Energy* 23(2):111–114.

55. R. Stewart; D. Spencer; and J. Healey. 1981 (Aug.). An evaluation of models estimating insolation incident upon slopes of different orientations. *Proceedings of the International Solar Energy Society Congress*. Brighton England; Aug. 23–28, 1981. V3, 2391–2395.

56. K. Y. Kondratyev. 1969. *Radiation in the atmosphere*. New York: Academic Press.

57. K. Y. Kondratyev. 1973. *Radiation characteristics of the atmosphere and the earth's surface*. New Delhi: Amerind Publishing Co.

58. F. X. Kneizys; E. P. Shettle; W. O. Gallery; J. H. Chetwynd, Jr.; L. W. Abreu; J. E. A. Selby; R. W. Fenn; and R. A. McClatchey. 1980. (Feb.). *Atmospheric transmittance/radiance: computer code LOWTRAN 5*. Report no. AFGL-TR-80-0067. Hanscom AFB: Air Force Geophysics Laboratory.

59. B. Leckner. 1978. The spectral distribution of solar radiation at the earth's surface—elements of a model. *Solar Energy* 20(2):143–150.

60. R. King and R. O. Buckius. 1979. Direct transmittance for a clear sky. *Solar Energy* 22(3):297–301.

61. R. Bird and R. Hulstrom. 1979 (July). *Application of monte carlo techniques to insolation characterization and prediction*. Report no. SERI/RR-36-306. Denver, Colo.: Solar Energy Research Institute.

62. R. Bird and R. Hulstrom. 1981 (Feb.). *A simplified clear sky model for direct and diffuse insolation on horizontal surfaces*. Report no. SERI/TR-642-761. Denver, Colo.: Solar Energy Research Institute.

63. R. Bird. 1981. Solar Energy Research Institute. Denver Colo. Personal communication.

64. J. L. Hatfield; R. B. Giorgis, Jr.; and R. G. Flocchini. 1981. A simple solar radiation model for computing direct and diffuse spectral fluxes. *Solar Energy* 27(4):323–330.

65. W. H. Klein and B. Goldberg. *Solar radiation measurements/1968—*. Smithsonian Radiation Biology Laboratory, Rockville, M.D. Washington, D.C.: Smithsonian Institution.

66. M. Munroe and W. Shepherd. 1981. An assessment of the solar energy availability in different regions of the solar spectrum. *Solar Energy* 26(1):41–47.

67. G. A. Zerlaut and J. D. Maybee. 1981. Spectroradiometer measurements in support of photovoltaic device testing. Phoenix, Ariz.: DSET Laboratories.

68. R. E. Bird and R. L. Hulstrom. 1981. *Solar spectral measurements and modeling*. Report no. SERI/TR-642-1013, NTIS.

69. Li-Cor Inc. 1981. Lincoln, NE. Product literature.

70. J. T. Peterson and E. C. Flowers. 1978. Atmospheric turbidity across the Los Angeles basin. *Journal of Applied Meteorology* 17(4):428–436.

71. J. T. Peterson and E. C. Flowers. 1978. Urban-rural solar radiation and atmospheric turbidity measurement in the Los Angeles Basin. *Journal of Applied Meteorology* 17(11):1595–1609.

72. W. C. Malm and E. G. Walther. 1979. Reexamination of turbidity measurements near Page, Arizona and Navajo Generating Station. *Journal of Applied Meteorology* 18(7):953–955.

73. J. T. Peterson. 1979. Cooperative U.S.A.—U.S.S.R. atmospheric transparency measurements. *Bulletin of the American Meteorological Society* 60(9):1084–1085.

74. J.T. Peterson et al. 1981. Atmospheric turbidity over central North Carolina. *Journal of Applied Meteorology* 20(3):230–241.

75. M. Martin and P. Berdahl. 1978. Description of a spectral atmospheric radiation measuring instrument. *Proceedings of the Third Conference on Atmospheric Radiation*. Davis, Calif.; June, 1978, 148–149.

76. P. Berdahl and M. Martin. 1979. Spectral measurements of infrared sky radiance. *Proceedings of the 3rd National Passive Solar Conference*. San Jose, Calif.; Jan., 1979. AS/ISES, Univ. of Delaware.

77. P. Berdahl and M. Martin. 1979. Infrared radiative cooling. *Proceedings of the International Solar Energy Society Congress*. Atlanta, Ga.; May, 1979.

78. P. Berdahl and M. Martin. 1980. Spectral radiance of the clear midaltitude-summer atmosphere." *Proceedings of the 1980 International Radiation Symposium*. Fort Collins, Colo.; August, 1980.

79. P. Berdahl and R. Fromberg. 1981 (May). *An empirical method for estimating the thermal radiance of clear skies*. Preprint LBL-12720. Lawrence Berkeley Laboratory.

80. G. O. G. Lof; J. A. Duffie; and C. O. Smith. 1966. *World distribution of solar radiation, report no. 21*. Solar Energy Laboratory of the University of Wisconsin at Madison.

81. E. A. Carter and G. S. Banholzer, Jr. (1981) June). *A guide to world insolation data and monitoring networks*. Report no. SERI/TR-9119-1. Denver, Colo.: Solar Energy Research Institute.

18

Advances in Solar Energy. © 1983 American Solar Energy Society, Inc.

SOLAR RADIATION MEASUREMENTS: CALIBRATION AND STANDARDIZATION EFFORTS

G. A. ZERLAUT

DSET Laboratories, Inc., Phoenix, Arizona 85029

Abstract

This chapter reviews recent activities in the field of solar radiation measurements and includes discussions of historical problems, the genesis of various national and international activities dealing with studies to solve them, and current and planned activities in assessing various aspects of the measurement of solar irradiance. Emphasis is on the development of test methods and the promulgation of industry standards for the operation and calibration of solar radiometers used to determine irradiance, largely in support of solar device testing. The need for standardization at both the domestic and international levels is discussed, as are activities relating to the maintenance of the World Radiation Reference Scale in the U.S., and the relation of these activities to the internationally recognized pentannual pyrheliometric comparisons at Davos, Switzerland. Specific organizations mentioned are the World Meteorological Organization (WMO), the International Energy Agency (IEA), Subcommittee .02 on Environmental Parameters of ASTM Committee E-44 on Solar Energy Conversion, and Subcommittee 1 Climate of the International Standards Organization (ISO) Technical Committee 180 on Solar Energy. Specific attention is devoted to the characterization of pyranometers employed in the precision measurement of irradiance at angles other than horizontal.

INTRODUCTION

The state of the art of solar radiation measurements has been brought into sharp focus in the past three years. This has resulted largely from continued concern for the quality of data from the U.S. solar radiation network and, more recently, from the lack of confidence voiced in the calibration of pyranometers used for the precise determination of instantaneous solar irradiance on surfaces tilted from the horizontal. Solutions to these problems have been sought through a series of national and international seminars, workshops, round robin instrument comparisons, and most importantly, by the establishment of intercomparison protocols and procedures, and, largely in the U.S., by the development of consensus-developed instrument calibration and calibration-transfer standards.

It is quite appropriate that these recent investigations and the results of the various inquiries be summarized and reported in this, the first of an annual series that reviews the recent advances in solar energy technologies. It is not the author's intent to describe either the U.S.

solar radiation network or the various solar radiation instruments currently available. Reviews of both are to be found in the recently published *Solar Energy Technology Handbook*, sponsored by the American Section of the International Solar Energy Society.[1,2]

RECENTLY ELUCIDATED PROBLEMS IN SOLAR RADIATION MEASUREMENTS

Radiation Measurement Scales Prior to 1977
The historical problems that plagued the first U.S. network, which was operated by the National Weather Service, encompassed a range of

A graduate of the University of Michigan, Mr. Zerlaut is co-owner of DSET Laboratories, Inc., an internationally known materials weathering and solar device testing facility. He is chairman of ASTM Committee E-44 on Solar Energy Conversion, chairman of the U.S. TAG for ISO/TC180 on Solar Energy-Thermal Applications, and a member of numerous other professional organizations. Mr. Zerlaut is credited with more than 50 publications and holds the rights to six patents. He is listed in Who's Who in the West *and* Who's Who in Technology Today.

individual as well as generic instrument problems, poor calibration techniques, poor traceability, a lack of temperature compensation in the instruments deployed, and generally unsatisfactory maintenance.[1,3] Although calibration procedures and traceability were improved in the late 1950s when the U.S. adopted the International Pyrheliometric Scale in 1956 in support of the International Geophysical Year, the now well-known Parson's black deterioration in the Eppley Model 50 pyranometer caused further errors in the U.S. network data in the 1960s and into the early 1970s.[4] For those pyranometers that were neither replaced nor recalibrated, the quality of the data steadily worsened and confidence in the network data deteriorated further.

A significant improvement in the status of solar radiometry occurred when the U.S. adopted first the International Pyrheliometric Scale of 1956 (IPS-56) and then the Absolute Radiometric Scale (IPC-75), traceable to the World Radiation Reference (WRR) maintained for the World Meteorological Organization (WMO) by the Physikalisch-Meteorologisches Observatorium (PMO), Davos, Switzerland. The IPA-56 scale was about 2.5 to 3.5% below the Smithsonian International Scale of 1913 (SIS-13) and was the result of an agreement between world radiation experts to develop a compromise scale between SIS-13 and the earlier Angstrom scale of 1905 (AS-05), which differed from each other by about 5%. The IPS-56 scale was subsequently called into serious question when it was shown that the Angstrom pyrheliometers employed to maintain the scale differed among themselves by as much as 3%.[5,6]

Measurements Problems in Resource Assessment
The National Weather Service announced in early 1977 that its new NOAA solar radiation network would be calibrated and maintained to the WRR Absolute Scale defined by the absolute cavity radiometers compared in Davos at the 1975 International Pyrheliometric Conference, IPC IV. (This and subsequent regional and international intercomparisons are discussed in detail in a subsequent section.) The absolute WRR scale is 2.2% higher than the IPS-56 scale; thus, the old Smithsonian scale of 1913 (SIS-13) was only about 1% too high.

Although many of the instrumental problems of the former network have been solved by converting to Class I instruments, the Solar Radiation Laboratory of the National Weather

Service has recently begun to observe apparent deterioration in the sensitivity of the so-called World Meteorological Organization Class I pyranometers employed at the NOAA/NWS Solar Radiation Laboratory as a function of environmental exposure.[7] It was previously thought that Class I instruments suffered significant exposure-related degradation only under the high-ultraviolet/high-temperature exposure conditions of the southwestern desert, as reported by this author.[8]

A remaining problem for pyranometers employed in meteorology and resource assessment, one that can be solved by preselection of pyranometers (to obtain a truly Class I instrument in terms of the WMO classification), is the severity of deviations from the cosine law inherent in a large number of pyranometers. This deviation is variously described as cosine error or cosine correction. Simply stated, it is the ratio of the instrument response at a given angle of incidence (with respect to the receiver) to the product of the response at normal incidence and the cosine of the angle of incidence. Incident angle effects have often been denoted by the cosine response measured in two orthogonal planes, one of which passes through the electrical connector (north) of the pyranometer, and by a family of 360° azimuth response curves for a number of selected incident angles (that is, solar altitudes for a horizontally mounted pyranometer). These relationships are shown in Figs. 1 and 2. It is important to note that this

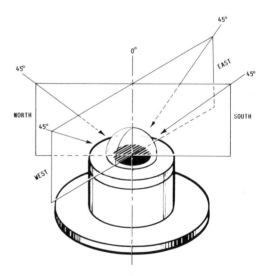

Fig. 1. A schematic of solar cosine response angles to plane of a pyranometer receiver in two orthogonal planes passing through the pyranometer axis.

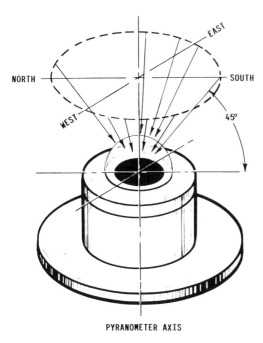

NORTH — SOUTH

EAST

WEST

45°

PYRANOMETER AXIS

Fig. 2. A schematic of solar azimuth response angles to the plane of a pyranometer's receiver at a given incident angle (for example, 45°).

definition has been challenged to the extent that it does not describe the actual range of incident angles of pyranometers in use at 0° horizontal and at south-facing tilts from the horizontal.

The essential consideration is that pyranometers exhibiting deviations from true cosine response are difficult to accurately calibrate in terms of a single instrument constant that is "good" for the range of solar altitudes encountered throughout the year. For the precision measurement of instantaneous irradiance required in performance evaluation of solar devices (see below), a number of instrument constants should be used, usually as file information in the computers employed. However, this procedure would be far too costly and cumbersome for daily continuous measurements of irradiance in support of meteorological and resource assessment applications.

Flowers[9] performs the calibration of pyranometers by obtaining the regression expression from several days of continuous ratioing to the NOAA/Solar Radiation Laboratory's reference pyranometer(s). At DSET Laboratories, Inc., we adjust the instrument constant of all meteorological instruments to a value that provides the same integrated response (total integrated irradiance) over a three-to-five-day period as obtained with our reference instrument, with the

test and reference instruments at the same tilt. This requires a reference instrument of small cosine, azimuth, and tilt sensitivity (in an absence of which precise characterization of the reference instrument is required).

The advantage of assigning an instrument constant resulting in a total integral equal to that of the reference instrument is one of seasonal accuracy. However, the method requires that the instrument be calibrated seasonally, or about four times a year. Furthermore, unless a variety of sky conditions is included, significant errors can result. In the NOAA regression technique, the reference is calibrated by the shade method and the instrument constant assigned is for a value of 50° incident angle. This virtually assures that the calibrations are all referenced, or normalized to 1,000 w/m².

Measurement Problems in Performance Testing of Solar Devices

Industry has historically expected, if not demanded, that thermal performance (efficiency) tests of solar thermal collectors tested to ASHRAE Standard 93–77 be accurate to within ±2%. There are often complaints when two test laboratories differ from each other by as little as 2% (or, only 50% of the permissible spread within the significance of ±2%). Yet we have shown that the probable error contributed to the efficiency expression

$$\eta = \frac{\dot{m}C_p\Delta T}{I_t} = F_R K_{\tau,\alpha}\alpha,\tau^*$$ (1)

by a presumably well calibrated pyranometer can easily be ±2% alone, and the possible error can exceed ±4%.[8] In Eq. (1), $\dot{m}$ is the mass flow rate, C_p is the specific heat, ΔT is the difference between inlet and outlet temperatures of the heat transfer fluid, I_t is the instantaneous solar irradiance, F_R is the plate efficiency factor, $K_{\tau,\alpha}$ is the incident angle modifier, α is the solar absorptance of the plate (receiver), and τ is the transmittance of the cover system.

---

* While it is generally agreed that the optical efficiency defined in Eq. (1) has no thermodynamic importance for solar collectors other than swimming pool collectors that may operate at near ambient temperatures, it is a most highly useful diagnostic tool for the test laboratory in assessing agreement between theory and observation.

These problems first came to public light when the results of the rather poorly conceived, NBS-sponsored round robin solar collector test program were scrutinized.[10] Two generically different collectors were tested by 21 laboratories (including 3 government laboratories) with the result that differences in the optical efficiencies exceeded an astounding 20%. Because the optical efficiency is the equivalent of the expression given in Eq. (1) for an inlet temperature approximating the ambient temperature, as defined by the Hottel-Whillier governing equation of the collector,[11] the ΔT expression in Eq. (1) is optimized and thermocouple errors in the measurement of ΔT are vanishingly small, even for those inexperienced laboratories that participated in the round robins. Thus, most of the errors resident in the discrepancies obtained for the optical efficiency of the two round robin collectors were due to the measurement of mass flow, m, and instantaneous solar irradiance, I_t. If one presumes that most of the laboratories determined mass flow by a simple gravimetric measurement of flow for the specified interval, one must conclude that a very significant portion of the errors in Eq. (1) are due to errors in the measurement of I_t, the solar irradiance. The authors cited[10] stated that a significant portion of the differences in optical efficiency could most likely be ascribed to problems in the measurement of solar irradiance.

Unfortunately, time has not solved the pyranometry problems inherent in the NBS round robin test sequence. A subsequent international round robin collector test program, managed under the auspices of the International Energy Agency,[12] resulted in serious, but somewhat smaller, differences between laboratories. More recently, results of comparative collector performance tests performed in solar simulators have shown similar disparities in the measurement of optical efficiency.[13]

Subsequent investigations have indicated that these solar irradiance measurement errors resulted from (1) use of WMO Class II pyranometers that have subsequently been shown to exhibit tilt errors of as much as 10 to 12%, (2) use of Class I and II pyranometers that had not been calibrated in some time, and (3) use of Class I and II pyranometers that exhibited significant deviations from the cosine law (which becomes increasingly important when testing collectors at a fixed tilt angle as opposed to altazimuth tracking of the sun).

THE NEED FOR CALIBRATION AND MEASUREMENT STANDARDS

Webster's *New Collegiate Dictionary* defines *standard* n. as "something set up and established by authority as a rule for the measure of quantity . . . value, or quality." Within this context, it has become obvious that *standard methods* of calibrating solar radiation instruments are urgently required. Experience has shown that *standard practices,* or *standard operating procedures,* are also required if accurate and meaningful solar irradiance measurements are to be obtained.

Confidence in the calibration and maintenance of calibration of the U.S. network has largely been restored by the thoroughness of the NOAA/Solar Radiation Laboratory of the National Weather Service (NWS/SRL) during the past several years. Standardized procedures are nevertheless still required for meteorological applications in order that pyranometer calibration techniques employed by the NWS/SRL in support of the U.S. network, and those employed by organizations in support of certain regional and state networks, yield equivalent and comparable results.

Perhaps even more critical is the need for standardized procedures for the calibration of pyranometers employed in the instantaneous measurement of solar irradiance. In the absence of either precise calibrations or intercomparison procedures, agreement between laboratories on the thermal efficiency characteristics of solar collectors will be fortuitous at best. Yet, as will be shown in this chapter, the calibration procedures required for meteorological applications and for laboratory testing of solar devices are necessarily different. Solar test laboratories around the world have relied too readily on meteorological calibrations; they have incorrectly and improperly taken refuge in inappropriate, but otherwise quite correct, meteorological-type calibrations for use in the thermal efficiency testing of solar collectors. Calibration authorities, being largely supported by meteorological applications, have been slow to accept the distinction between the calibration requirements for resource assessment and those for the precise determination of instantaneous solar irradiance at various angles of incidence to the plane of a solar device/receiver—the latter requiring much greater accuracy in the sense of being a "physical optics" measurement.

Standardization efforts are also needed in

the field use of pyrheliometers and pyranometers. Here the distinction between the requirements for resource assessment and precision performance testing diminishes by comparison. Daily maintenance, such as cleaning and alignment checks (for pyrheliometers), is required for both applications. Interferences with the field of view must be minimal for both resource assessment and performance testing. Alignment of the pyranometer's receiver with the spirit level is important to meteorological applications and is critical to the accurate assessment of instantaneous irradiance in performance testing.

MAINTAINING THE WORLD RADIATION REFERENCE SCALE (WRR)

Absolute Cavity Radiometry

History
Absolute cavity pyrheliometers were derived from early satellite cavity radiometers, known as active cavity radiometers (ACRs), having fields of view of approximately 80°.[14] Subsequent spacecraft cavity radiometers were employed to measure the solar constant. They were manufactured by the Jet Propulsion Laboratory of the California Institute of Technology and the Eppley Laboratory, Inc. Early terrestrial versions of these spacecraft radiometers were the Kendall PACRAD (Primary Absolute Cavity Radiometers), manufactured both by the Eppley Laboratory and the Jet Propulsion Laboratory,[15] and the Willson ACR (Active Cavity Radiometer), manufactured by JPL and California Measurements, Inc.[16]

The state-of-the-art absolute cavity radiometers currently manufactured in the U.S. are the Eppley Model H-F Absolute Cavity Pyrheliometer,[17] manufactured by the Eppley Laboratory, Inc.* and the TMI Mark VI Kendall Radiometer,[18] manufactured by Technical Measurements, Inc.† Additionally, the PMO-6 Absolute Radiometer,[19] manufactured by Compagnie Industrielle Radioelectrique,‡ is available. Two other absolute cavity radiometers were registered participants at the International Pyrheliometer Comparisons (IPC IV and V) held in Davos, Switzerland in October 1975 and 1980; they are the PVS absolute cavity radiometer designed by

Prof. Yu. A. Sklarov of Saratov University (the U.S.S.R.), and the CROM-series radiometers designed by Prof. D. Crommelynck of the Royal Meteorological Institute (Belgium).

Most absolute cavity radiometers are, for the most part, optically and thermodynamically similar and their "terrestrial" versions are utilized throughout the world as primary reference instruments. They are characterized by generally identical pyrheliometer-type viewing optics, are self-calibrating and "absolute" in the sense of employing electrical substitution for the solar radiation induced detector signal, and generally require continuous attention while in operation.

Physical Principles and Construction
A schematic description of a generalized absolute cavity pyrheliometer is presented in Fig. 3. The construction is centered upon two opposing inverted cones that, as the critical element of the cavities in which they reside, are the absorber, or receiver, surfaces of the radiometer. The cones, in turn, reradiate to the sensitive ele-

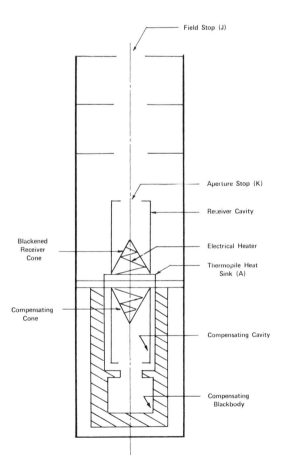

Fig. 3. A schematic of a typical absolute cavity pyrheliometer.

* 12 Sheffield Avenue, Newport, RI 02840.
† Box 838, LaCanada, CA 91011.
‡ Bundesigasse 16, CH-3001, Berne, Switzerland.

ment of the detector (which may be either a wire wound thermopile or a platinum resistance thermometer).

As shown in Fig. 3, the cold junctions of both the receiver and the compensating thermopiles are heat sunk (A), and the compensating cavity (rear cavity) is viewed by a heat-sunk black body. The receiver, or forward, cone and the cylinder in which it resides are wound with an electrical heater. In the Eppley model H-F cavity radiometer, the rear, or compensating cavity and cone, is also wound with an electrical heater. Both the PMO-series cavities employed at the World Radiation Center-Davos and the Willson ACR use an electrically calibrated, differential heat flux transducer with platinum resistance thermometers as the detector.

The entire receiver/compensation cavity pair is mounted in a collimator tube which limits the field of view as a function of the field and aperture stops (J and K in Fig. 3). The geometrical considerations that apply to pyrheliometer design are described by Fig. 4, where

Z_0 (the opening angle) $= \tan^{-1} R/L$ (2)
Z_p (the slope angle) $= \tan^{-1} [(R - r)/L]$ (3)
Z (the limit angle) $= \tan^{-1} [(R + r)/L]$ (4)

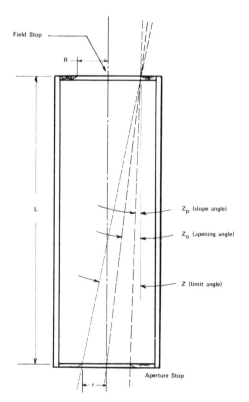

Fig. 4. The occluding geometry of pyrheliometers.

and the field of view is $2Z_0$, or twice the opening angle. The half angle values recommended by the World Meteorological Organization (WMO) are $Z_0 \leq 4°$ and $1° \leq Z_p \leq 2°$. The instruments that have been intercompared, both in the Davos and New River intercomparisons, have the collimator angles shown in Table 1.

Operation

Absolute cavity radiometers, often called self-calibrating radiometers because of their electrical substitution mode of operation, may be operated in either the passive or active mode. In the passive mode, the front heater is adjusted to produce the same signal from the detector as when illuminated with the field stop open and the heater completely off. Irradiance is computed simply as the power per unit area (of receiver). In the active mode, servo-type controls are used to adjust the heater power to a level necesary to maintain a specific temperature of heat flux level during exposure. The power is continuously monitored and the power required to reach the higher preset level provides the power equivalent of the solar irradiance. The Willson ACR and the WRC PMO-series radiometers are operated in the active mode.[16,19]

Most absolute cavity radiometers are operated in the so-called PACRAD mode. That is, they use the front heater to electrically substitute the heat flux incident on the detector. However, the Eppley H-F cavity pyrheliometer is capable of using the rear heater to duplicate the power equivalent of the solar irradiance on the front cavity. This is done with the field stop continuously open, that is, unshuttered, and the power to the rear heater is adjusted so as to exactly null the signal developed as a result of the solar heating of the front cavity. This is the so-called Angstrom mode. This technique is amenable to active operation insofar as the rear heater is easily servo-controlled so as to exactly null the signal resulting from solar irradiation of the front cavity.[17]

Instrument	Model	Slope Angle	Opening Angle
Eppley	H-F	0.804°	2.50°
TMI	Mark VI	0.761°	2.46°
ACR	V; VI	0.939°	2.49°
Eppley	EPAC	0.816°	2.50°
PMO	PMO-2	0.741°	2.43°

Table 1 Optical Geometry of Absolute Cavities Intercompared Both in the New River and Davos Comparisons

It should be noted that all absolute cavity radiometers require characterization due to (1) the small but still significant nonequivalence between electrical and radiant heating, (2) the absorptivity of the cavity, (3) a stray light factor for the pyrheliometer tube, and (4) the voltage drop in the cavity lead wires. While these are all small, they may all be measured based on first principles, wherein the equation for the passive PACRAD mode is

$$H_S = kI(V - IR_c) \qquad (5)$$

where H_s is the solar irradiance, I is the measured heater current, V is the measured heater voltage, R_c is the measured lead correction resistance (and IR_c is the voltage loss), and k is for the various instrument constants (including the area of the receiver). For active PACRAD operation, the relationship is

$$H_s = k(I_o V_o - I_i V_i) \qquad (6)$$

where $I_o V_o$ is the power applied to the heater with the cavity shaded and $I_i V_i$ is the power with the cavity irradiated.

There are two mechanisms to determine the constant of an absolute cavity radiometer: (1) characterization by the laborious measurement of the various instrument constants embodied in k, and (2) by comparison to a fully characterized instrument or group of instruments. There is a considerable body of opinion claiming that only full characterization is appropriate, and that comparisons to determine k do not help to maintain the absolute radiation scale (WRR). The intercomparisons at IPC IV and V (Davos) and NRIP I-V (New River, AZ, DSET) made no attempts to adjust the k of participating instruments. It goes without saying, however, that instruments exhibiting significant deviations from, for example, the mean of a large group of instruments with which they have been historically compared, may very well be overdue for a complete recharacterization, that is, to redefine k.

World Meteorological Organization and PMOD (Davos)

The creation of the United Nations provided a new framework for international collaboration in various areas, including technical and scientific fields. Consequently, in Washington, D.C. in 1947, the World Meteorological Convention was adopted, establishing a new organization

founded on a formal agreement among governments. This convention was ratified by a large number of countries, and in 1951 the new World Meteorological Organization (WMO) replaced the former organization. In December 1951, the General Assembly of the United Nations approved an agreement between the United Nations and the WMO by which the latter was recognized as a specialized agency of the United Nations.[20] Its purpose is to establish uniform procedures for the measurement of meteorological parameters, to provide for the calibration of instrumentation within the worldwide network of solar radiation measurements, to provide for and host international intercomparisons, to develop special instrumentation (such as absolute cavity radiometers and sun photometers), and to measure the spectral aspects of the atmosphere and its relation to worldwide weather patterns.

The World Meteorological Organization is organized into two world radiation centers located at Davos, Switzerland and Leningrad, U.S.S.R., and into a number of regional centers. The U.S. is in Region IV, for which regional centers are located in Toronto, Canada and Boulder, Colorado (the regional network is given in Appendix II.)

World radiation centers are equipped with at least three of the most accurate primary standard pyrheliometers available, as well as auxiliary radiometers and measuring equipment for the maintenance of the WRR Scale of 1975. In addition, they serve as centers of interregional and international comparisons of radiation, follow closely all developments in the measurement of solar radiation, employ qualified scientists with wide experience in solar radiation, and undertake the training of specialists in solar radiation.[21]

The World Radiation Center (WRC) in Davos is operated for WMO by the Physikalisch-Meteorologisches Observatorium (PMOD). As noted previously, one of the responsibilities of PMOD is the maintenance of the internationally accepted radiation scale, currently the Absolute Radiometric Scale defined by a group of absolute cavities at PMOD. It is also the responsibility of PMOD to make these radiometers and the scientific facilities of the WRC available for comparison with the primary standard radiometers of the regional centers, of the official meteorological institutes of various member countries, and of various other worldwide scientific laboratories working in solar radiometry. This is ac-

Type	Serial No.	Designer	Owner
PACRAD (Original)	III	J. Kendall, Sr.	JPL, Pasadena, Calif.
ACR (Active cavity)	311	R. C. Willson	R.C. Willson, Altadena, Calif.
ACR (Active cavity)	701	R. C. Willson	CSIRO, Australia
Eppley (PACRAD type)	11402	Kendall/Eppley Lab.	Eppley Lab., Newport, R.I.
Eppley (PACRAD type)	12843	Kendall/Eppley Lab.	NOAA/NWS, Boulder, Colo.
Eppley (PACRAD type)	13617	Kendall/Eppley Lab.	U. Bergen, Norway
PMO (Active type)	2	R. W. Brusa	PMO/WRC, Davos, Switzerland
PMO (Active type)	3	R. W. Brusa	PMO/WRC, Davos, Switzerland
PMO (Active type)	5	R. W. Brusa	PMO/WRC, Davos, Switzerland
PVS	5	Yu. A. Skliarov	U. Saratov, U.S.S.R.
TMI (PACRAD type)	67502	J. Kendall/C.M. Berdahl, Sr.	NOAA/NWS, Boulder, Colo.

Table 2   Absolute Cavity Radiometers compared at IPC IV (Davos, November 1975)

complished primarily through the convening of International Pyrheliometer Comparisons at Davos on a frequency of about once every five years.

International Comparison of Regional Pyrheliometers and Absolute Cavity Radiometers

The Fourth IPC (1975)

IPC IV was held in October 1975 at the World Radiation Center of PMOD in Davos, Switzerland. Thirty-four Angstrom and one Abbott (silver disk) pyrheliometers, one Linke-Fuessner and one Eppley Model NIP pyrheliometer, and eleven absolute cavity radiometers were compared over a five-day period. The eleven absolute cavity radiometers first compared at IPC IV[22] are of principal importance to the maintenance of the absolute WRR scale; they are listed in Table 2. (It will be observed that the terrestrial version of the Eppley Model H-F dual cavity radiometer was not available for IPC IV.)

Because PACRAD III, manufactured by the Jet Propulsion Laboratory, was observed to be very stable and was the only absolute cavity radiometer that participated in IPC III (1970), it was chosen as the reference instrument for IPC

IV. The results of the nearly 2,000 instantaneous observations are presented in Table 3. The results are given as the mean ratio of the irradiance measured by the stated instrument to that measured by PACRAD III, for the number of observations shown.

The weighted average of all instruments for all observations was 1.0017, with a standard error of only 0.18% between instruments. It is thus seen that the excellent agreement between all cavity radiometers justified the establishment of the absolute World Radiation Reference (WRR) scale as a result of IPC IV.

The absolute WRR scale is in reality based on the fact that the cavity radiometers employed in IPC IV (and subsequent comparisons) are fundamentally related to Standard International (SI) units by the physical standards of length, temperature, and electrical measurements of the various official national standards institutes, associations, and organizations of the country of manufacture. These intercomparisons establish the relationship between all such instruments and they, when in excellent agreement, thereby establish the scale defined by the intercomparisons. Traceability to the absolute WRR scale is achieved and maintained either by regular, periodic participation in such intercomparisons,

Instrument	Owner	Mean Ratio	SDS	Number
PACRAD III	JPL, USA	1.0000	(Reference Instrument)	1993
ACR 311	Willson, USA	1.0027	0.0042	1993
ACR 701	CSIRO, Austr.	1.0010	0.0079	1992
EPAC 11402	Eppley, USA	1.0020	0.0022	1642
EPAC 12843	NOAA, USA	1.0039	0.0016	980
EPAC 13617	U. Bergen, Norway	0.9987	0.0019	1610
PMO-2	PMOD, Switz.	1.0029	0.0028	1218
PMO-3	PMOD, Switz.	1.0052	0.0032	1813
PMO-5	PMOD, Switz.	1.0017	0.0039	1974
PVS-5	U. Saratov, USSR	0.9971	0.0045	27
TMI 67502	NOAA, USA	1.0004	0.0021	1678

*Adapted from Ref. 22.

Table 3   Summary of Results of Comparison of Absolute Radiometers in IPC IV*

or by substantial comparison to an absolute cavity radiometer that is itself periodically intercompared with one or more international or national absolute radiometers having similar histories.

The Fifth IPC (1980)

Twenty-four absolute cavity radiometers and thirty-six pyrheliometers (Angstrom, Model NIP, and so on) were intercompared on nine different days at IPC V held in early October 1980 at WRC/PMO in Davos, Switzerland. Because it was desirable to establish as the reference instrument for IPC V a cavity maintained at the WRC/Davos, the PMO-2 absolute radiometer that had particpated in IPC IV was chosen. The instruments, their owners, and a summary of results are presented in Table 4.[23] Of the twenty-four absolute radiometers, the six absolute cavity radiometers starred (*) participated in IPC IV. Of these, PMO-2 and EPAC 13617 exhibited essentially no change over the five intervening years when normalized to PACRAD III.

Again, excellent agreement between absolute instruments was achieved, with an average ratio (to PMO-2) of 0.9995 and a standard deviation between instruments of only ±0.26%.

MAINTAINING THE ABSOLUTE WRR SCALE IN THE U.S. (THE NEW RIVER INTERCOMPARISONS)

Genesis of the New River Intercomparisons

Largely as a result of our early assessment of the disparate results of the first National Bureau of Standards-sponsored round robin thermal performance tests of solar collectors,[10] and our own independent observations of pyranometer-to-pyranometer differences (even for so-called WMO Class I instruments), DSET Laboratories purchased an Eppley Model H-F absolute cavity pyrheliometer in January 1977. (DSET obtained the first terrestrial version of the Hickey-Frieden absolute cavity radiometer that was developed for the Nimbus satellite series. In early 1979, we obtained a second Model H-F absolute cavity radiometer.)

By summer of 1977, we began periodic shading disk calibrations of our family of Eppley Model PSP pyranometers at normal incidence, that is, in the exact mode at which they were mounted to test solar collectors for thermal performance. Transfers were made directly from the H-F cavity radiometer.

Because of the need for a U.S. reference base both independent of, and with greater fre-

Instrument	Serial	Owner	Mean Ratio	SDS	Number
PMO	*PMO-2	PMOD, WRC	1.00000 (Reference Instrument)		162
PMO	*PMO-5	PMOD, WRC	0.99740	0.09	78
PMO	PMO-6D	Obs. Hamburg	0.99315	0.07	81
PMO	PMO-6G	NBS, USA	1.00400	0.06	81
PMO	PMO-6C	PMOD/WRC	1.00071	0.07	81
Eplab-HF	14915	Eppley USA	1.00111	0.05	162
Eplab-HF	15744	NTI, Boras, Sweden	0.99922	0.07	96
Eplab-HF	17142	DSET, USA	1.00106	0.06	150
Eplab-HF	18747	A.E.S., Canada	0.99790	0.20	88
Eplab-HF	19744	CNR-IEA, Italy	0.99874	0.04	96
Eplab-HF	19746	Met. Inst., Hungary	1.00034	0.06	150
EPAC-Kendall	13219	Met. Inst., India	0.99532	0.12	18
EPAC-Kendall	*13617	U. Bergen, Norway	0.99617	0.93	77
PACRAD	II	JPL, USA	0.99952	0.13	108
PACRAD	* III	JPL/PMOD	0.99779	0.10	150
TMI-MK VI	67401	TMI, USA	1.00101	0.07	96
TMI-MK VI	*67502	NOAA/SRF, USA	1.00057	0.06	96
TMI-MK VI	67604	Met. Off., UK	0.99749	0.04	96
TMI-MK VI	67702	JPL, USA	1.00105	0.07	96
TMI-MK VI	67814	SERI, USA	0.99800	0.04	96
TMI-MK VI	68016	Met. Serv., France	0.99837	0.07	96
ACR IV	401	JPL, USA (Willson)	1.00426	0.07	75
ACR IV	403	JPL, USA (Willson)	1.00553	0.09	75
ACR	* 701	C.S.I.R.O., Australia	0.99232	0.14	45
CROM	2L	Crommelynch, Belgium	0.99599	0.08	45
CROM	3R	Crommelynch, Belgium	0.99876	0.44	41

†Adapted from Ref. 23.
*Participated in IPC IV (1975).

Table 4   Summary of Results of Comparison of Absolute Radiometers in IPC V†

| Intercomparison | Date | Registered Instruments | | Solar Observations |
		New	Total	
NRIP I	Nov. 1-5, 1978	14	14	315
NRIP II	May 2-5, 1979	4	12	441
NRIP III	Nov. 5-9, 1979	4	17	550
NRIP IV	Nov. 17-19, 1980	4	15	465
NRIP V	May 3-5, 1982	1	9	—

Table 5   NRIP Instrument Breakdown

quency than, the WRC/PMOD-hosted comparisons at Davos every five years, as well as the superior environmental conditions that prevail at DSET's New River, Arizona site,* we proposed hosting the first of five national, regional, and international intercomparisons of absolute cavity radiometers, commencing in November 1978. These three-to-five-day intercomparisons, known as the New River Intercomparisons of Absolute Cavity Pyrheliometers (NRIPs), have, except for NRIP V, been cosponsored by DSET Laboratories, the Solar Energy Research Institute, the National Weather Service's Solar Radiation Laboratory (NOAA), and the Department of Energy. In May 1982, in conjunction with NRIP V, the New River intercomparisons became self-supporting on a participating fee basis. The success of the NRIPs has been largely due to the helpful support and enthusiasm of Mr. Edwin Flowers of NWS/SRL (NOAA), Mr. John Hickey of the Eppley Laboratory, Mr. James Kendall of the Jet Propulsion Laboratory, and Mr. Chester Wells of the Solar Energy Research Institute.

Operation of the NRIPs
More than 27,000 individual instantaneous irradiance readings were accumulated by 27 different absolute cavity radiometers in NRIP I through IV, as shown in Table 5. (The data had not been compiled from NRIP V at press time.)

Dates for the intercomparisons are selected to ensure the greatest probability of cloud- and haze-free conditions. The DSET Laboratories site at New River, Arizona has been generally agreed upon by the experimenters because of the combination of moderate temperatures, low average relative humidity, clear sky conditions, and high average daily sunshine for the times of year chosen.

Outdoor laboratory facilities were constructed specially for the NRIP experiments. They consist of three permanent double-tiered

instrument benches, with the upper table serving as the cavity mounting deck, which in turn shades a lower electronic shelf that also doubles as working desk space. Each table accommodates eight experimenters, making it possible to intercompare 24 cavity radiometers at a time. A photograph of NRIP III absolute cavity radiometers and participants is shown in Fig. 5.

Instantaneous readings were organized into 10-minute sequences of 21 readings at 30-second intervals (ACR-type instruments usually obtained 11 readings at 1-minute intervals). Readings for all instruments are sensed within less than 1 second of each other to minimize scatter between instruments due to rapidly changing sky conditions. Calibrations were usually performed both before and after each experimental sequence with the apertures shuttered. A complete description of the NRIP comparisons has been published by Estey and Seaman.[24]

Summary of Results of NRIP I through NRIP V
Results in terms of the total number of instantaneous readings N for those experiments in which the instrument participated have been compiled by the author,[25] and are given in Table 6. Complete results have been compiled by Estey and Seaman.[24] The standard deviation (S.D.) shown is the statistical average of the S.D. computed for each intercomparison in which the instrument was involved. The standard deviation for all readings was ±0.0022. If the 3 active cavity radiometers are removed from the population, the S.D. for the remaining 23 instruments is a very low value of ±0.0016. Since the Eppley Model H-F and the TMI Mark VI absolute cavity instruments are the principal absolute cavity radiometers being manufactured and marketed today, it is instructive to examine their statistics separately. Their results, as well as those of all instrument types, are shown in Table 7.

The excellent agreement obtained between all absolute cavity instruments (with the agreement being to within 0.25% for more than 27,000 instantaneous measurements) is attributed prin-

---

* Near Phoenix

28

Fig. 5. A photograph of some of the participants and instruments for NRIP III.

Model	SN	Experimenter	NRIPs	N	Mean	S.D.
EPAC	11399	AES, Canada	1,3	40	0.9997	± 0.0021
*EPAC	11402	The Eppley Lab.	1	25	1.0008	0.0018
*EPAC	12843	NOAA (Boulder)	3,4	50	0.9968	0.0042
‡†H-F	14915	The Eppley Lab.	1,2,3,4,5	77	1.0003	0.0006
H-F	14917	NASA-Lewis	2	22	0.9995	0.0004
*H-F	15745	NOAA (SRL)	1,2,3,4	97	0.9970	0.0007
‡†H-F	17142	DSET Labs.	2,3,4,5	81	1.0013	0.0010
†H-F	18747	AES, Canada	3	25	0.9990	0.0007
H-F	18748	The Eppley Lab.	3,4	44	1.0026	0.0009
‡ACR	501	Boeing Aerospace	1,5	12	1.0042	0.0035
ACR	601	NOAA	3	24	1.0067	0.0028
ACR	1104	Lawrence Berkeley	4	31	1.0079	0.0028
†*PM02	—	PMOD/Davos, Switz.	3	29	1.0008	0.0007
MK VI	001	JPL	2,3,4	74	1.0027	0.0017
†MK VI	67401	Technical Meas.	1,2,3,4	92	0.9992	0.0010
‡†*MK VI	67502	NOAA (SRL)	1,2,3,4,5	97	1.0004	0.0010
MK VI	67603	Sandia Labs	1,2,3	59	0.9996	0.0013
†MK VI	67702	JPL	1,2,3,4	91	1.0006	0.0007
MK VI	67706	So. Cal. Edison	1,2,3,4	93	1.0000	0.0011
MK VI	67707	Utah State Univ.	1	15	1.0000	0.0007
MK VI	67811	Sandia Labs.	2	11	0.9997	0.0006
MK VI	67812	Ga. Inst. Tech.	1,3	40	0.9980	0.0007
‡†MK VI	67814	SERI	1,2,3,4,5	97	0.9986	0.0006
‡MK VI	68017	SERI	4,5	27	0.9989	0.0006
MK VI	68018	SERI	4	31	1.0010	0.0008
MK VI	68020	Martin Marietta	4	31	0.9990	0.0006

*Also compared in IPC IV, Davos, November 1975.
†Also compared in IPC V, Davos, October 1980.
‡Compared in NRIP V, but results not included in the Mean.

Table 6   Results of NRIP I Through NRIP V With Ratio of Each Instrument to Group Mean

cipally to (1) the excellent quality of currently available absolute cavity instruments, (2) the quality of the environmental conditions obtained at New River for the times of the year chosen for the intercomparisons, and (3) the improved techniques with which the experimenters themselves operate their respective instruments as a result of the experience gained. Six instruments that were used in all four NRIP comparisons provide data to investigate secular drift over the two-year period. Estey[24] computed regression lines from rank values versus elapsed time and from the slope and found that the mean drift was only −0.04% per year. Thus, there was no measurable long-term drift.

The advantages of the NRIP intercomparisons relate to the de facto maintenance of the SI Radiometric Scale (WRR) in the U.S. by this peer group of instruments, to the reference quality of the intercomparisons for citation and traceability purposes, and to the availability of representatives of the peer grouping for the quality control purposes of any given organization.

Attention is directed to those instruments listed in Table 5 that were intercompared at the 1975 International Pyrheliometric Conference IV held at the World Radiation Center in Davos, which are starred with an asterisk. Further, eight absolute cavity radiometers that participated in the NRIP series also participated in IPC V held in Davos in October 1980 and are also identified.

Comparison of the NRIPs with IPC IV and V
Unlike the WMO-sanctioned intercomparisons held every five years at Davos where only the cavity itself is compared, the New River inter-

Model	Manufacturer	Type	No. of Instruments	N	Mean X	S.D.
H-F	The Eppley Lab.	Dual Cavity	6	346	0.9998	0.0008
Mark VI	Technical Meas.	Cavity	13	758	0.9999	0.0010
ACR	JPL & Radiomet.*	Active Cavity	3	67	1.0068	0.0029
EPAC	The Eppley Lab.	Cavity	3	105	0.9985	0.0031
PMO2	PMOD (Davos)	Cavity	1	29	1.0008	0.0007

*No longer in business

Table 7   Results of NRIP I Through IV as a Function of Instrument Type

comparisons compare the cavity and its associated electronic data logging and readout equipment. The advantages of the NRIP-type comparisons are that the actual equipment compared is the same as employed by the participants when they use the absolute cavity radiometers as primary reference instruments in their own laboratories. Hence, the performance of individual instruments compared in NRIPs has greater reference significance, if not greater credibility, than the same instruments compared at Davos.

Wells[26] has discussed these differences in detail in terms of the advantages and disadvantages of each approach. He concluded that the principal advantages of the New River Intercomparisons were the following:

- Because instrumentation, techniques, and operators are essentially unchanged between those in end-use applications and those of the intercomparisons, the significance and traceability of the intercomparison may very well have more significance when the instrument is employed outside the framework of the peer grouping.
- There are no difficulties in compatibility between the electronics of a given radiometer and the data accumulation techniques of the NRIPs.
- Operational problems are either weather related or they affect only a specific instrument at a time.
- Protocol changes are readily made to accommodate new or different instruments.

Wells cites some disadvantages of the New River intercomparisons, the most significant of which follow:

- A large investment in time of each operator is required to initially record the solar irradiance of each instrument.
- The format of this data for subsequent computer analysis is also very labor intensive.
- There may be slight timing errors caused by the operators not reading their instruments simultaneously.

Advantages of the WRC (IPC) method and protocols follow:

- The data are collected with significantly less recording errors, and with little human effort.
- First (nonquality controlled) results can be distributed within hours of the comparisons.
- Operators are freer to watch for subtle instru-

ment problems without the tedium of continuous data readings.

The principal disadvantages of the WRC/IPC methods follow:

- The data acquisition and sequence of operational steps employed in the IPCs are not necessarily the methods employed in the normal use of the absolute radiometers.
- As a result, the question of whether or not the data from the IPCs (that is, the irradiance values) be identical to those which would obtain in normal use.
- New or different instruments are more difficult to accommodate.
- Changes in procedure/protocol are more difficult and changes in zero or system gain cannot be determined at the end of each run, so linear corrections with time cannot be determined and applied.

CURRENT NATIONAL AND INTERNATIONAL ACTIVITIES

ASTM Committee E-44 on Solar Energy Conversion

The American Society for Testing and Materials, the largest voluntary consensus standards organization in the world, organized Committee E-44 on Solar Energy Conversion in June 1978.[27] It is organized into 12 subcommittees dealing with Nomenclature, Environmental Parameters, Safety, Materials Performance, Heating and Cooling Subsystems and Systems, Process Heating and Thermal Conversion Power Systems, Photovoltaic Electric Power Systems, Wind Driven Power Systems, Ocean Thermal Power Systems (inactive), Biomass Conversion Systems, Advanced Energy Systems (inactive), Passive Heating and Cooling Systems, and Environmental and Society Impact of Solar Energy Conversion Systems. The Subcommittee on Environmental Parameters, E-44.02, is charged with the identification of environmental parameters and the establishment of standard measuring and reporting procedures for data pertinent to solar energy conversion. As such, it is currently concerned with the promulgation of seven standards dealing with (1) the calibration of pyranometers and pyrheliometers (five standards), and (2) the development of standard global and direct-beam solar irradiance spectral distributions for Air Mass 1.5 conditions (two standards).

Two of the five standards dealing with the calibration of solar radiation measuring devices became ASTM Standards in May of 1981. They are E 816, Standard Method for Calibration of Secondary Reference Pyrheliometers and Pyranometers for Field Use, and E 824, Standard Method for Transfer of Calibration from Reference to Field Pyranometers. (These methods are discussed later.) Two draft standards dealing with the calibration of pyranometers by the shading disk technique with axes horizontal and axes tilted, respectively, have only one technical area of contention remaining to be resolved by the subcommittee before being advanced to full society level balloting.

Current activities of the task group dealing with solar radiometry are involved primarily with the development of standard methods to characterize pyranometers. This draft document, entitled "Standard Method for Calibration of Reference Pyranometers with Respect to Cosine, Azimuth and Tilt Errors," represents the indoor laboratory approach to the problem. (Various aspects of this problem are covered later.) The task group working on this document and its advisors from throughout the solar radiation community increasingly believe that a companion method is required for the establishment of incident angle and tilt effects by methods that use natural sunshine outdoors.

It is worth noting that the two standard spectral energy distributions of sunlight for global and direct-beam irradiance at Air Mass 1.5 have been approved by society ballot as ASTM Standards. (Their numerical designations have not yet been assigned.)

Interested parties may obtain copies of ASTM Standards, and for the purposes of review only, copies of ASTM draft documents that are in consensus development, by writing ASTM, 1916 Race Street, Philadelphia, PA 19103.

International Energy Agency (IEA) Activities

The IEA

In recognition of the need for a coordinated effort to resolve certain energy problems facing the petroleum importing countries, member countries of OECD (Organization for Economic Cooperation and Development) established in 1974 the International Energy Agency (IEA). The IEA is an international treaty organization whose funding is obtained from the member countries.

A Solar Heating and Cooling Program was developed by IEA in 1975 as one of 16 energy technology areas of concern to OECD. The five tasks and the country responsible for each are listed in Table 8.

Initial Activities in Pyranometry

As a consequence of results of an international round robin program on testing thermal solar collectors sponsored by Task III,[12] an intercomparison of 21 pyranometers was held at WRC/PMOD in Davos in March 1980.[28] Nine Eppley Model PSPs, ten Kipp and Zonen CM5s, one Kipp and Zonen CM10, and one Schenk instruments were referenced to WRC's 6703A (PMOD) pyranometer at 0° horizontal.

The results of this intercomparison were quite disturbing to the participants.[8,28] The average deviation between the Kipp and Zonen and Eppley PSP instruments was 3% and 2%, respectively, using instrument constants supplied with the instruments. The maximum deviation for these two groups was an alarming 10.2% and 6.3%, respectively. These results prompted a series of round robin investigations to determine the cause of the disparities.

Round Robin I Pyranometer Comparisons

It was generally agreed that these disparities could not be explained by differences in the characteristics of the primary absolute cavity reference instruments to which they were traceable, insofar as the absolute cavity radiometers all agreed to within a standard deviation of about

Task	Subject	Operating Agent
I	Investigation of the Performance of Solar Heating and Cooling Systems	Denmark
II	Coordination of R & D on Solar Heating and Cooling Components	Japan
III	Performance Testing of Solar Collectors	Germany
IV	Development of an Insolation Handbook and Instrument Package	U.S.A.
V	Use of Existing Meteorological Information for Solar Energy Application	Sweden

Table 8   IEA/Solar Heating and Cooling Program Tasks

Manufacturer	SN	Ratio	Owner
Eppley	14806F	0.9378	National Bureau of Standards
Eppley	19129F	0.9468	DSET Laboratories, Inc.
Kipp & Zonen	774120	0.9159	Kernforschungsanlage, Jülich

Table 9   Round Robin Pyranometers

±0.002 solar constants as can be shown in Tables 4 and 6. It was subsequently agreed by the owners of three of the instruments involved that a separate round robin intercomparison would be performed in an attempt to explain the reasons for these discrepancies. The instruments, their owners, and the ratios to PMOD are presented in Table 9. The calibrations were performed in order by DSET Laboratories, NOAA's Solar Radiation Facility (Boulder), and the Eppley Laboratory.

Calibrations were referenced to instruments whose calibrations were traceable to previously compared absolute cavity radiometers, or pyranometers were directly calibrated with such absolute cavity radiometers by the shading disk method. In DSET's experiments, field instruments are calibrated by the shading disk method referenced directly to the Eppley Model H-F cavity SN 17142 at a tilt defined by normal incidence for the particular season. This was done to conform to the need to calibrate under the end-use conditions of solar device testing on altazimuth, sun tracking mounts.

Horizontal Shading Disk: The three laboratories performed calibrations both by the shading disk method and by transfer from a reference pyranometer that had itself been calibrated by the shading disk technique. Selected results are shown in Table 10.

Excellent agreement between labs was obtained for the PSPs when calibrated against absolute cavity pyrheliometers by the shading disk method, even though DSET uses a 30 sec/30 sec and NOAA a 5 min/6 min sequence for the shaded/unshaded measurements. The DSET shading calibrations were performed at an average solar elevation of 64° (as opposed to 60° for the NOAA measurements). The Eppley calibrations were obtained during the winter months when the average solar elevation was only 25°.

Horizontal Transfer of Calibration: Good agreement was obtained for the horizontal measurements referenced against Model PSP pyranometers. The agreement between DSET and NOAA was exceptionally good, with the values differing by only 0.17, 0.21, and 0.41% for SN 19129, SN 14806, and SN 774120, respectively. The Eppley integrating hemisphere calibrations averaged about 1.3% higher than the outdoor calibrations. It is noted that the reference pyranometers at Eppley and NOAA were their respective primary working standard pyranometers. However, at DSET, the NBS instrument (SN 14806) was referenced against the "*horizontal* shading disk" calibration of the DSET instrument (SN 19129), the DSET instrument (SN 19129) was referenced against the "*horizontal* shading disk" calibration of the NBS instrument (SN 14806), and the value for the K & Z instrument (SN 774120) was the average obtained when referenced against SN 19129 and SN 14806.

Calibrations at Tilt: Excellent agreement was obtained between Eppley and DSET in normal

			Instrument Constants, (in μV/W $\cdot$ m^{-2})		
			Eppley PSPs		K & Z
Test Mode	Lab	Reference	SN 19129	SN 14806	SN 774120
Horizontal	DSET	ACP 64° sun	10.427	9.843	—
Shade disk	NOAA	ACP 60° sun	10.500	9.840	—
	EPLAB	ACP 25° sun	10.290	9.290	—
Horizontal	DSET	PSP outdoor	10.570	9.910	12.820
	NOAA	PSP outdoor	10.588	9.889	12.873
	EPLAB	PSP hemisph.	10.640	10.070	13.090
Tilt, normal	DSET	ACP 30°	10.410	9.832	—
	DSET	ACP 60°	10.330	—	—
	EPLAB	ACP 60°	10.340	—	—
Tilt, off	DSET	ACP 30°	10.470	9.837	—
normal	NOAA	PSP 40°	10.496	9.884	—

Table 10   Summary of DSET/NOAA/EPPLEY Results

incidence calibrations of SN 19129 by the shading disk method. The DSET data were obtained at a tilt of 30° (summer months) and the Eppley data were obtained at a tilt of 60° (early winter). On return to DSET, PSP SN 19129 was recalibrated by the shade method at normal incidence, and a value of 10.33 μV/Wm^{-2} was obtained. It is important to note that the tilt required to achieve normal incidence in the second DSET shading disk calibration (then wintertime) of SN 19129 was 60° from the horizontal—identical to that obtained at 60° tilt during the wintertime normal incidence calibration at the Eppley Laboratory.

Pyranometer SN 19129 was subsequently recalibrated at a tilt of about 30° by the shading disk technique referenced to the Model H-F Absolute Cavity Pyrheliometer. Its new value was 10.37 μV/Wm^{-2}, indicating an approximately 0.5% tilt effect.

Rehabilitation of Davos Results

The ratio between the radiation measured by each of the three round robin instruments to that measured by the Davos reference pyranometer PMOD/SN 6703A[28] is given in Table 11, along with the average instrument constant determined by DSET and NOAA, and cosine and temperature corrected values. As noted from Table 11, the average deviation from the reference instrument was 6.6% (column 3) and after recalibration in this study, was still 3.1% (column 5). It is additionally instructive to employ the cosine and temperature compensation corrections for the Davos data as defined by the declination of −6.37° and the solar noon sun elevation of 37.1° for Davos on March 5, 1980, and a temperature of 0°C. These corrections are taken from DSET data and the report by E. Flowers of NOAA presented by Riches et al.;[29] they are presented in columns 6 and 7 of Table 11. The temperature

correction for PSP SN 14806 is unity based on the difference between 26°C (the nameplate temperature) and 0°C as determined by its compensation curve. No correction was made for the Kipp and Zonen instrument since we have no knowledge of the temperature at which the "original" instrument constant was determined.

We have shown that agreement between the three instruments compared to PMOD 6703A can be significantly improved using carefully determined instrument constants, and can be further improved by employing cosine and temperature corrections. As seen from column 7 in Table 11, the three corrected instruments agree to within 0.2% with each other, although they differ from PMOD/SN 6703A by about 1.6%. PMOD's reference has since been determined to have been in error by approximately 2.5%, and had a more conservative estimate of the mean solar altitude been employed for the Davos comparisons, that is, 34° to 35° rather than 37°, the mean value for all three corrected instruments (last column of Table 11) would have approached even more closely the error now known to be in the PMOD reference.

The Second IEA Round Robin

A second round robin pyranometer study was initiated by Task V on Meteorology in early 1981. Twenty-two instruments were first compared by the National Atmospheric Radiation Center (NARC) of Canada, by the U.S. National Oceanic and Atmospheric Administration Solar Radiation Facility (NOAA/SRF), and the U.S. Department of Energy's Solar Energy Research Institute (SERI).

Results of the horizontal calibrations performed by NARC and NOAA/SRF are presented for several selected instruments in Table 12.[29] The NOAA/SRF results are referenced to the NARC values. Also, the original PMOD

| Instrument | Davos | | RR Recal. | | Cosine Correct. | Cos/Temp Correct. |
	Nameplate I.C.*	Ratio	New I.C.*	Ratio		
KZ/SN 774120	13.70	0.9159	12.84	0.9772	0.9834	0.9834†
EP/SN 14806	10.02	0.9378	9.84	0.9550	0.9871	0.9871
EP/SN 19129	10.76	0.9468	10.46	0.9740	0.9768	0.9827
Average		0.9335		0.9687	0.9824	0.9844
S.D.		0.0130		0.0098	0.0043	0.0019

* μV/W · m^{-2}.
† Temperature correction not applied.

Table 11 Cosine and Temperature Corrections to the Davos Comparison Ratios

SN		Calibration	Ratio of Stated Calibrations to that of NARC	
			SRF	PMOD Rev.†
PSP	Organization	NARC*	NARC	NARC
14806	NBS, USA	9.66	1.026	0.998
15834	SNTI, Sweden	8.74	1.026	0.991
16692	TIL, Denmark	9.55	1.017	0.992
17750	NARC, Canada	9.24	1.034	1.007
17823	KFA, Jülich, FRG	8.67	1.037	1.002
K&Z CM6				
75-2438	ITE, Stuttgart, FRG	10.45	1.025	1.049
77-4120	KFA, Jülich, FRG	12.56	1.025	1.061
78-4750	Facultie Blytech., Belgium	10.81	1.001	1.039

* In $\mu V / W \cdot m^{-2}$.
† Original PMOD values (Ref. 28) multiplied by 1.026.

Table 12   Selected Results of IEA Round Robin II Pyranometer Calibrations and Comparisons

values* were multiplied by the factor 1.026 and ratioed to the NARC calibration values. This scalar corresponds to the 2.6% increase in the calibration factor assigned by PMOD to the reference pyranometer that was employed in the original Davos comparisons.[28]

While the difference between the revised PMOD results and the Canadian NARC values is less than between the NOAA/SRF and NARC values, it is thought that the agreement between PMOD and NARC in these results is largely fortuitous.[29] This is ascribed to the wide difference between the techniques employed in the two cases: the NARC calibrations were performed by direct comparison to a reference pyranometer in an integrating sphere, and the PMOD values were obtained outdoors at low sun angles with an uneven envelope of snow in the north field of view of the pyranometers (including the PMOD reference).

The SERI studies were largely confined to comparing the irradiance measured by test instruments to PSP SN 17860 as a function of the angle of incidence of the direct-beam component at tilts from the horizontal of 0°, 20°, 40°, 60°, and 90°. Selected results are shown in Figs. 6 through 8.

The disparate results of RR II represented by the differences between calibration results and between pyranometers themselves simply amplify the need to fully characterize pyranometers that will be employed in testing solar collectors and in determining instantaneous solar irradiance.

Conclusions

Results of the two IEA-sponsored round robin comparisons indicate the sensitivity of transferring calibrations from one pyranometer to another under conditions where small errors due to deviations from cosine response, failures to account for the temperature dependence of instrument constants, small tilt effects, and normal aging processes, all can conspire to cause significant errors when employing the best pyranometers available for precision, instantaneous measurements of solar irradiance.

International Standards Organization (ISO) Activities

Founded in 1947 and headquartered in Geneva, the International Standards Organization issues voluntary International Standards (IS) on nearly every technical field associated with world commerce. New ISO standards are derived, whenever possible, from existing national standards used by its member countries. However, methods are developed within the international technical community when necessary, and usually involve interlaboratory testing prior to consensus. Documents developed by the various ISO committees are voted upon by the member countries of any given technical committee, and finally by all ISO member countries. Negative votes are resolved to give as high a degree of consensus as is possible.

International standards are important to international trade as one of the principal mechanisms for reducing certain trade barriers. First, they result in and promote a common language for the establishment of trade, and secondly,

*From Ref. 28.

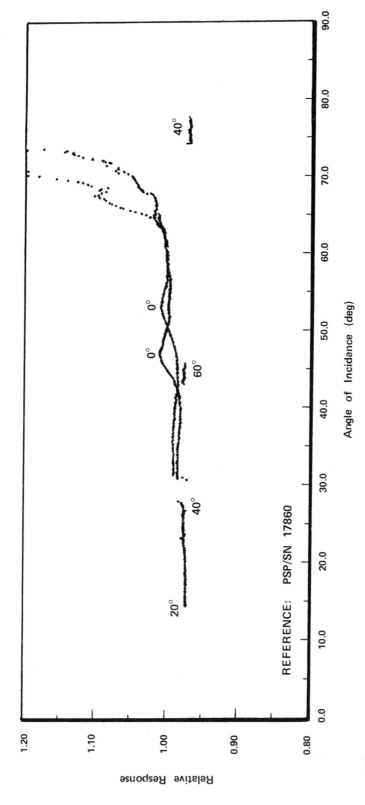

Fig. 6. The ratio of irradiance of Eppley PSP SN 17750 to reference PSP as a function of angle of incidence at several tilt angles (adapted from Ref. 29).

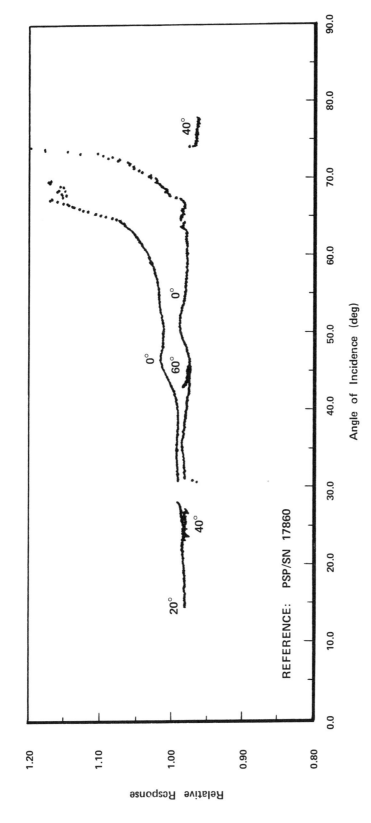

Fig. 7. The ratio of irradiance of Eppley PSP SN 17823 to reference PSP as a function of angle of incidence at several tilt angles (adapted from Ref. 29).

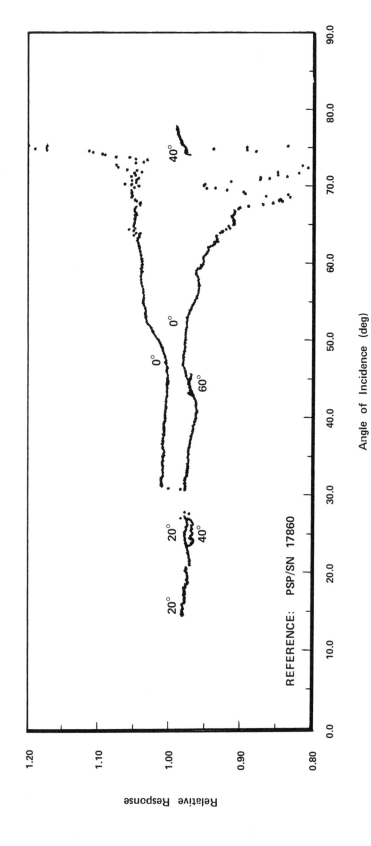

Fig. 8. The ratio of irradiance of Kipp and Zonen CM5 SN 78-4750 to reference PSP as a function of angle of incidence at several tilt angles (adapted from Ref. 29).

they reduce the extent to which products must meet a plethora of often disparate requirements.

The Technical Committee on Solar Energy (Thermal Applications), ISO/TC180, was organized at an international meeting held in Sydney, Australia in May 1981, and is one of approximately 160 currently active technical committees of the International Organization for Standardization (ISO). Its secretariat is held by the Standards Association of Australia (SAA), which is the ISO member body in Australia. The organization meeting, attended by official representatives from Australia, Canada, Sweden, Germany, France, Israel, New Zealand, Japan, Kenya, the United Kingdom, and the United States, undertook the development of a program of work for one working group and four subcommittees. (It is noted that in addition to the countries attending the organization meeting, other participating (P) members were Austria, Belgium, Brazil, Denmark, Greece, Italy, Jamaica, Norway, Rumania, South Africa, Switzerland, and the U.S.S.R.)

The committee breakdown and the secretariats responsible for the individual programs of work are:

Working Group (WG) 1	Nomenclature: Canada (SCC)
Subcommittee (SC) 1	Climate: Germany (DIN)
Subcommittee (SC) 2	Materials: France (AFNOR)
Subcommittee (SC) 3	Component-Thermal Performance/Reliability and Durability (Colors): Israel (SII)
Subcommittee (SC) 4	Systems-Thermal Performance/Reliability and Durability: U.S.A. (ANSI)

The program of work for the Subcommittee on Climate (SC 1) includes four high priority items:

(1) specification and classification of instruments for measuring global and direct-beam solar radiation;
(2) calibration of a primary reference pyranometer using a primary reference pyrheliometer;
(3) characterization of the primary reference pyranometer; and
(4) transfer of calibration from primary reference pyranometers to field pyranometers.

Secondary priority has been given to four other standards development areas:

(1) transfer of calibration from primary reference pyrheliometers to field pyrheliometers;
(2) standard practice for use of field pyranometers;
(3) standard reference solar spectrum for direct-beam and global radiation; and
(4) standard practice for measurement of meteorological data in solar applications.

Under the auspices of the American National Standards Institute (ANSI), the ISO member body for the United States, a U.S. Technical Advisory Group (TAG) was organized in the spring of 1981 to represent the technical positions of U.S. industry and the U.S. solar community of interest in ISO/TC180 activities. The U.S. TAG is made up of technical experts selected from and appointed by the various standards organizations with pertinent interests in the fields represented by the TC180 program of work. The U.S. TAG currently has 15 members which represent the American Society of Heating, Refrigeration and Air Conditioning Engineers (ASHRAE), the American Society for Testing and Materials (ASTM Committee E-44), the Solar Energy Industries Association (SEIA), and the American Section of the International Solar Energy Society (AS/ISES). This representation ensures that the U.S. delegations to technical committee meetings and meetings of its subcommittees and working groups are properly briefed on U.S. positions as represented by their constituent organizations.

The U.S. TAG is organized into Subcommittee Advisory Groups (SCAGs) which possess the pertinent technical expertise of the TAG for each of the ISO/TC180 subcommittees and working groups, respectively. It is the responsibility of each SCAG to ascertain and develop a U.S. consensus position for each item of work on each of the subcommittees and working groups within TC180; the necessary requirement is that these SCAGs develop positions essentially independent of the position of the member body of any other country. It is the responsibility of the delegation to International Standards meetings, selected by the TAG, to make whatever compromises are necessary and expedient as long as they are to the best possible interest of that country's industry position.

Since the Subcommittee Advisory Group of the U.S. TAG for SC1—Climate consists largely of representation from ASTM's Committee E-44, the liaison between the U.S.—developed documents within the relevant domestic consensus standards developing group, namely ASTM's Subcommittee E-44.02, and the U.S. TAG is a straightforward and highly relevant process.

CALIBRATION STANDARDS FOR PYRHELIOMETRY

Future Comparisons of Absolute Cavity Radiometers

While the WMO-sanctioned International Pyrheliometer Comparisons can be expected to continue at Davos on a frequency of once every five years, it is the consensus of the participants of the New River intercomparisons (NRIPs) that they should be continued on a frequency of every two years. (NRIP VI is currently scheduled for October 31 through November 4, 1983.)

The importance of the U.S.-based NRIPs relates primarily to their being increasingly referenced and cited for traceability purposes in the U.S. It has been agreed by the participants of NRIPs I through V that official recognition by WMO of the New River Intercomparisons of Absolute Cavity Pyrheliometers is desirable, and that broader participation by both South and Central American countries, as well as the U.S. and Canada, should be sought. However, in an absence of U.S. federal or WMO financial support, it would be necessary to charge a nominal registration fee to all participants, which might very well be a barrier to general recognition of the NRIPs by other countries in the hemisphere.

Nonetheless, the protocols developed for the NRIP activities have in themselves become de facto standard procedures—procedures that most likely will be offered eventually to a consensus standards organization for formal standardization.

Transfer of Calibration from Reference to Field Pyrheliometers

The calibration of reference and field pyrheliometers is now covered by ASTM Standard E-816 entitled "Calibration of Secondary Reference Pyrheliometers and Pyrheliometers for Field Use."[30] The standard is limited to pyrheliometers with field angles of 5° to 6°; for calibration of secondary reference pyrheliometers, the primary reference instrument must be a self-calibrating, absolute cavity radiometer that is directly traceable to the World Radiation Reference (Absolute Radiation Scale) through partici-

pation in IPC IV or V, Davos, held in 1975 and 1980, or in one of the NRIP series intercomparisons held in New River, Arizona. A secondary reference pyrheliometer so calibrated may be employed to transfer calibrations to pyrheliometers for field use. Neither primary nor secondary reference instruments may be field pyrheliometers and their exposure to sunlight must be limited to calibration or intercomparison activities.

The Standard discusses such interferences and cautions as sky conditions, comparisons between instruments of different opening angles and spectral responses, and wind-induced errors.

The procedure calls for performing simultaneous, instantaneous measurements of solar irradiance with identically tracked reference and field pyrheliometers. Measurements are taken every 30 seconds for 10 minutes to produce a 21-value measurement set. Five test sets are required on each of three days with data required both two hours before and after solar noon. The instrument constant of the test instrument is obtained simply by dividing the voltage output of the test instrument by the irradiance determined by the reference pyrheliometer.

CALIBRATION STANDARDS FOR PYRANOMETRY

Problems in Pyranometry

Bahm and Nakos[31] have recently completed a comprehensive review of the calibration of solar radiation instruments and have covered a wide range of pertinent subjects. These include a discussion of principal instrument characteristics such as linearity and stability of response, incident angle and spectral response characteristics, temperature effects, and the possible interactions between these characteristics.

A number of the problems that are associated with uncertainties in the determination of irradiance that are peculiar to solar collector testing are discussed in the following paragraphs.

Incident Angle Effects

Cosine and azimuth errors, or deviations, are simply two descriptions of the one effect of deviation from Lambert's cosine law at incident angles greater than 0°. Some researchers prefer to describe cosine response as the deviations from the cosine law in two orthogonal planes, each passing through the axis of the pyranometer, with azimuth response being the deviations from Lambert's cosine law at a given incident angle, or solar altitude, when the horizontally mounted pyranometer is turned through 360° (see Figs. 1 and 2). Other workers simply refer to cosine response in terms of the deviation from the cosine law at the incident angle defined at solar noon for a given time of year, or solar altitude. Azimuth response then is a set of response factors that track the pyranometer circumferentially for each incident angle of interest. Such azimuthal plots may then be used to characterize the pyranometer for all of the solar altitudes it will encounter throughout the year, whether mounted at horizontal or at various tilts from horizontal.

The laboratory assessment of cosine and azimuthal correlation factors may be performed in accordance with draft procedures under consideration by ASTM Committee E-44's subcommittee E-44.02 on Environmental Parameters, draft document No. 142.* In the draft method, both the cosine and azimuthal corrections are determined on the apparatus depicted in Fig. 9. For cosine response measurements, the pyranometer is mounted with axis vertical on the turntable and is positioned such that the source can be swung in a vertical plane through an arc centered on the pyranometer's receiver (or, entrance aperture). With the source intensity held constant, the response is measured. The cosine error is computed as the ratio between the response at any angle divided by the product of the response on axis (at normal incidence) and the cosine of the angle of concern. Employing the same apparatus, the pyranometer is positioned on the turntable, which is a precision rotating stage, such that its vertical axis is precisely colinear with the axis of rotation of the stage. The source is swung in a vertical plane to a predetermined incident angle (representing the altitude of the sun for a horizontally mounted pyranometer). With the source intensity held constant, the pyranometer response is measured as a function of the azimuth position through the azimuthal angles $-180°$ to $0°$ to $+180°$, where $0°$ represents true south with the electrical connector positioned at $\pm 180°$. The azimuth response is normalized to $0°$.

Azimuthal response plots may be developed

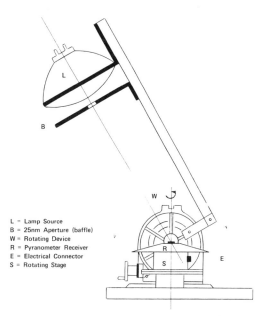

L = Lamp Source
B = 25nm Aperture (baffle)
W = Rotating Device
R = Pyranometer Receiver
E = Electrical Connector
S = Rotating Stage

Fig. 9. Apparatus for measurement of cosine and azimuth response correction.

in both normal-normal and polar coordinates. Polar coordinates, in turn, may be done in several ways. The plot shown in Fig. 10 represents isopleths of azimuth response for a Kipp and Zonen Model CM5 pyranometer, courtesy of U.K. Meteorological Office.[32] A different polar representation is given in Fig. 11, where all data are normalized to the normal incident pyranometer response.

Actually, there is considerable disagreement within the solar radiation community regarding the appropriateness of the proposed indoor characterization, as well as the actual utilizability of such data. Flowers[33] has stated that similar, carefully done laboratory experiments by himself, as well as others, have produced results that do not agree with results of outdoor experiments for certain instruments. He has pointed out that the source beam in Fig. 9 must be well collimated and must cover the glass domes (since they contribute to cosine/azimuth response errors). The precise alignment of the pyranometer axis and the source beam for all angles of incidence of interest is not a trivial problem, and it is important that the color temperature of the source be as close to 5,700 k as possible.

Andersson et al.[34] have performed cosine and azimuth corrections to pyranometers and have reported reasonable agreement between laboratory and field behavior. Cosine deviations for an Eppley PSP and a Kipp and Zonen Model

---

*This is a working document number and is not the designation the eventual standard will be given.

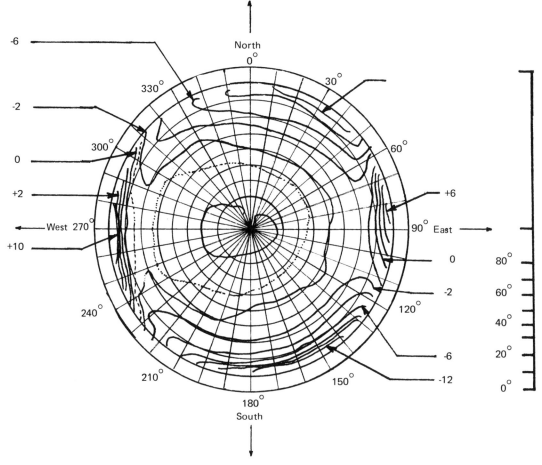

Fig. 10. The combined azimuth and elevation response for Kipp & Zonen SN 763154 (CM5). (percentage response relative to zenith measurement, Ref. 32.)

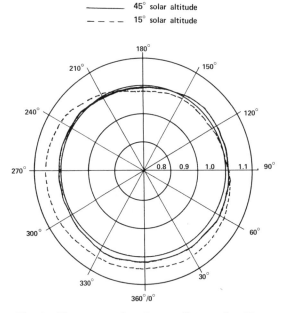

——— 45° solar altitude
– – – 15° solar altitude

Fig. 11. The response in polar coordinates of an Eppley PSP in terms of normal incidence response.

CM5 are shown in Figs. 12 and 13, respectively; azimuth response at incident angles of 45° and 75°, corresponding to solar altitudes of 45° and 15°, are shown in Figs. 14 and 15, respectively, for these two pyranometers.

One of the problems elucidated at the author's facility is the possible nonappropriateness of the symmetrical characterization of pyranometers by indoor techniques—no matter how accurate the family of azimuthal plots might be. Unless the complete set of characterization data are used as file information in the computer employed to reduce solar collector efficiency test data, simple incident angle, or even azimuthal, plots of the response error as a function season (at the range of solar altitudes throughout a test day for a given time of year) may be much more useful. For normal incident testing of solar collectors using an altazimuth follow-the-sun mount (as performed at DSET and several other test laboratories), the problem reduces to the ac-

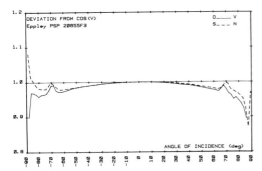

Fig. 12. The deviation from cosine response (relative) for PSP SN 20655. (courtesy Ref. 34.)

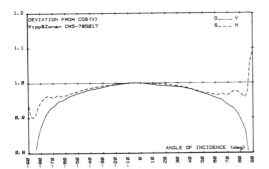

Fig. 13. The deviation from cosine response (relative) for Kipp & Zonen CM5 SN 785017. (courtesy Ref. 34.)

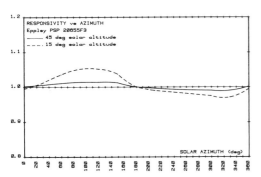

Fig. 14. Responsivity as a function of azimuth relative to south 180° for PSP/SN 20655. (courtesy Ref. 34.)

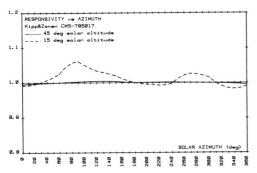

Fig. 15. Responsivity as a function of azimuth relative to south 180° for Kipp and Zonen CM5/SN 785017. (courtesy Ref. 34.)

curate calibration of the pyranometer at normal incidence and the characterization of the pyranometer as a function of angle of incidence in two orthogonal planes of the pyranometer, at the tilts defined by the general time of year. This simplified cosine calibration is required for bidirectional incident angle modifiers [$K_{\tau\alpha}$ in Eq. (1)] required by ASHRAE Standard 93-77.[11] A more complete characterization of the pyranometer is required for determination of the incident angle modifiers of flat-plate collectors outside the two orthogonal planes of the collector.

Such incident angle plots have been performed at the Solar Energy Research Institute in support of the IEA activities discussed above[29] (Figs. 6 through 8).

Data on incident angle response errors are now provided in DSET's calibration reports in the format presented in Figs. 16 through 18. These graphs plot the ratio of the test pyranometer's irradiance to that of the mean of three reference instruments (shown in Fig. 19) versus incident angle at 0° horizontal and a tilt from horizontal of 45°. (Negative values represent morning and positive represent afternoon data.) While the near-normal, uncorrected irradiance of PSP "A" (Fig. 16) is within less than a percent of the mean of the three reference instruments, it will be observed that the instrument has significant cosine error at incident angles of ±20°. Thus, even when corrected to unity, this pyranometer would result in the propagation of significant errors if used to test solar collectors at a fixed angle (or to determine incident angle modifiers), errors that would be on the order of 4% for the 60° incident angle point required by the test standard.[11] By contrast, PSP "B" (Fig. 17) has a very flat response and correction of its instrument constant by the observed 2% will result in a pyranometer having a high degree of reliability for the measurement of instantaneous irradiance at tilts to 45° from the horizontal. On the other hand, the plot for PSP "C" (Fig. 18) shows a linear decreasing relationship from −50° to +50°, which indicates that the plane of this PSP's receiver and its "spirit level" may not be exactly parallel. Other peculiarities may also be observed in plotting data in this manner. For example, if the two sides of the plot are asymmetrical (that is, if one is flat and the other curved, or if they intercept the normal incident ordinate at different points), the domes may not be centered exactly over the center of the circular receiver, or vice versa.[33]

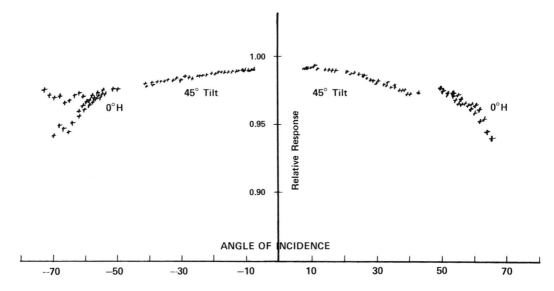

Fig. 16. Deviations from cosine law for PSP ''A'' compared to mean of three reference PSPs.

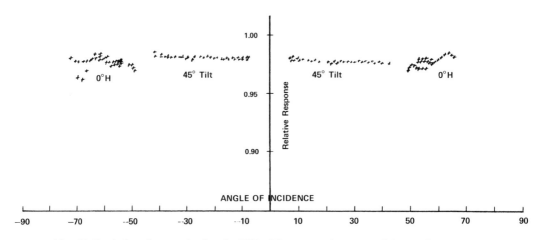

Fig. 17. Deviations from cosine law for PSP ''B'' compared to mean of three reference PSPs.

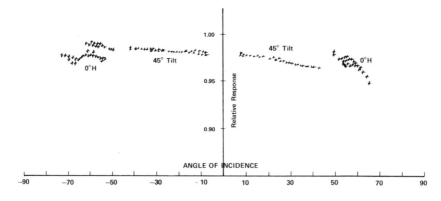

Fig. 18. Deviations from cosine law for PSP ''C'' compared to mean of three reference PSPs.

44

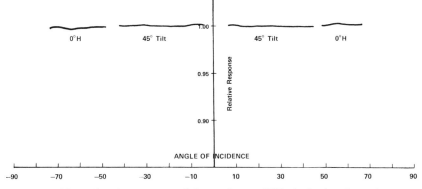

Fig. 19. Mean of cosine response of three reference PSPs (ratioed to themselves).

Tilt Effects

The effect associated with changes in pyranometer sensitivity as a function of tilt has become well known over the past several years. In the field, tilt can be separated from cosine response errors by accurate shading disk calibrations at solar noon when the solar elevation is greatest (during summer months in the Northern Hemisphere), compared to solar noon calibrations at other times of the years that define the tilt at normal incidence. Although one could perform normal incident shading disk calibrations at lower morning and afternoon solar elevations, considerable errors will be introduced if the irradiance levels are less than about 70% of the near-horizontal values.

Tilt errors of as high as 13 to 15% have been observed for certain so-called WMO Class II pyranometers—largely believed to be due to thermodynamic cooling of the thermopile hot junction as a result of an increase in convective heat transfer at tilts from the horizontal. Although Norris[35] has reported tilt errors of 11% for Eppley PSPs at vertical, or 90° orientations, most researchers have observed errors of only about 1 to 2% at high tilts. In our own studies, we observed response errors of only about 0.5% for PSPs tilted at 45° from the horizontal. Loxsom and Hogan[36] have also identified another type of tilt effect as a result of solar heating of the pyranometer body due to the noneffectiveness of the sunshade at moderate solar elevations. They claim to have solved this problem by placing a white, insulating foam jacket around the pyranometer body.

ASTM Committee E-44 draft document No. 142, entitled "Calibration of Reference Pyranometers with Respect to Cosine, Azimuth and Tilt Errors," contains a prescriptive procedure for measuring the effect of tilt from the horizontal on the response characteristics of pyranometers. This method provides for mounting the pyranometer under test and a strong source of solar radiation at opposite ends of a rigid tilt table, or bar, as shown in Fig. 20. With the tilt fixture assembled in a darkened room, the pyranometer and source are swung in a vertical plane. Pyranometer response readings are taken at appropriate angles of tilt and the tilt correction is computed as the ratio of the response at tilt to the response at exactly zero degrees horizontal, that is, with axis vertical. The cognizant ASTM subcommittee agrees on the need for performing the tilt test at three irradiance levels ($\sim$500, 750, and 1000 W/m²), and also agrees on the need for matching the color temperature of the sun as closely as possible. However, there is general disagreement as to the requirement for a diaphragm (noted in Fig. 20) to just illuminate the pyranometer dome. The most recent version of this draft standard cautions that the selection of source is critical insofar as it is essential that the radiant characteristics of the source's burner, or filament, do not change as a function of gravitational effects at tilt. The method now requires that the source be monitored during the response measurements with a silicon, solar-type cell detector not subject to tilt effects.

Andersson et al.[34] has measured the tilt response of several pyranometers employing a similar tilt table. Selected data are presented in Figs. 21 through 23. They showed that PSPs have little error associated with tilt effects. At tilts greater than 45°, they observed nearly 2% tilt error in the two Kipp and Zonen CM5 instruments, and nearly 4% error in the two Schenk Star pyranometers measurement. Employing an enclosed tilt box otherwise similar to that depicted in Fig. 20, Flowers[33] measured the tilt response of a number of pyranometers at an ir-

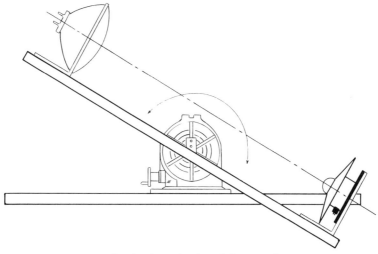

Fig. 20. The apparatus for the determination of tilt corrections to pyranometers.

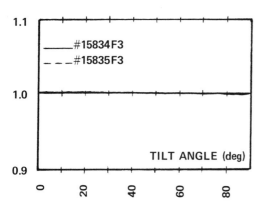

Fig. 21. Variations in response of Eppley PSP pyranometers with tilt from horizontal (adapted from Ref. 34).

radiance level of about 450 W/m². Typical data are presented in Fig. 24 for Eppley PSPs, Kipp and Zonen CM5s, and Spectrolab SR73s. These data do not agree in sign (that is, direction) with those of Andersson[34] for Eppley PSPs, but it is noted that again the magnitudes are small and the difference might be accounted for by the significant difference in irradiance levels employed by the two authors.

Temperature Response Problems
in Pyranometry

For a number of years, the Eppley Laboratory has furnished temperature response curves with every Model NIP pyrheliometer and PSP pyranometer sold. We at DSET Laboratories, Inc. regularly employ this data in our solar collector

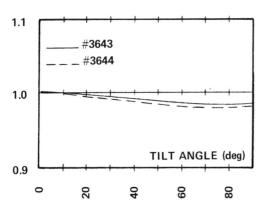

Fig. 22. Variations in response of Kipp & Zonen CM5 pyranometers with tilt from horizontal (adapted from Ref. 34).

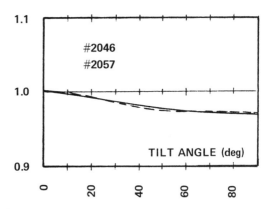

Fig. 23. Variations in response of Schenk Star 8101 pyranometers with tilt from horizontal (adapted from Ref. 34).

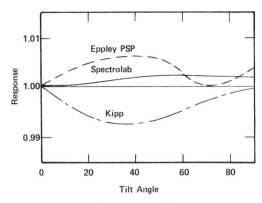

Fig. 24. Variations in response of three pyranometers at several tilts from horizontal (adapted from Ref. 33).

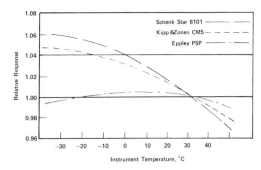

Fig. 26. The temperature dependence of the instrument constant of three pyranometers (adapted from Refs. 33 and 34).

testing procedures by adjusting pyranometer calibration values throughout each day to account for changing ambient temperature. A typical sensitivity plot is presented in Fig. 25 for DSET's working standard Eppley PSP, SN 19129. Although we have learned that few laboratories take the trouble to adjust the sensitivity of their field instruments during collector testing, we have called attention to the need for using this fundamental characteristic of nearly all pyranometers in precision irradiance measurements, whether temperature compensated or not. Andersson[34] and Flowers[33] have both measured the response changes in pyranometers as a function of temperature by heating the constantly irradiated instrument in a temperature-controlled chamber. Their data are in excellent agreement for the Eppley PSP, Kipp and Zonen CM5, and Schenk Star pyranometers studied, and are typified by the plots presented in Fig. 26.

Linearity of Response
Most researchers using pyranometers assume that they are dealing with radiometers whose instrument constants are linear with irradiance level in the range of interest. This is not necessarily true, as shown by Andersson et al.[34] While

the maximum variation in the sensitivity observed in PSPs by the authors cited was on the order of only a couple of tenths of a percent between 200 and 1,000 W/m², the two Schenk Star and two Kipp and Zonen CM5 pyranometers exhibited a decrease in sensitivity of about 5% in the same range.

Linearity effects, combined with potential cosine deviation at the higher incident angles required by ASHRAE Standard 93-77,[11] and the combination of disproportionate convective cooling at low irradiance levels and tilt errors at concomitant high tilts from horizontal, can conspire to create very significant errors in the determination of the incident angle modifier, $\kappa_{\tau\alpha}$ Eq. (1), required in thermal performance testing of solar collectors.

Calibration of Primary Reference Pyranometers

General Considerations
Until such time as an *absolute cavity pyranometer* is available having a 180° solid angle field of view and a cosine response approaching theoretical at high angles of incidence (that is, low solar elevations), the only technique available for calibration of primary reference pyranometers is comparison to absolute cavity pyrheliometers or normal incident pyrheliometers. Transfer of calibration from pyrheliometers to pyranometers can only be accomplished by occulting the field of view of the pyranometer—either by a view-limiting, pyrheliometer comparison tube that occludes the sky, or by a shade disk that occludes the sun. In the first case, transfer is direct so long as the slope and opening angles* of the collimator closely agree with the primary reference pyranometer. In the shade method, the compari-

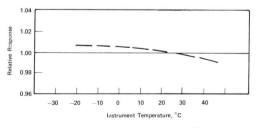

Fig. 25. The temperature dependence of instrument constant of PSP/SN 19129. (courtesy The Eppley Laboratory.)

---

*See Eqs. (2–4) and Fig. 4.

son is indirect and is made by dividing the difference between the pyranometer's voltage response for the unshaded and shaded conditions by the irradiance measured with the primary instrument.

While there are pros and cons to each method, the pyranometer comparison tube method has fallen into some disfavor and the shading disk approach is the more or less official technique prescribed by the World Meteorological organization and its constituent participants.

Occulting Tube Approach

A typical pyrheliometer comparison tube consists of a blackened enclosure for mounting and optically aligning the pyranometer exactly beneath a baffled occulting tube. The apparatus used at DSET for calibrating photovoltaic standard reference cells is shown in Fig. 27.†

While this method has the advantage of permitting direct transfer of instrument constants from the primary reference pyrheliometers, it has a number of disadvantages, chief among which are solar-heating-induced drift of pyranometer response, optical misalignment between the occulting tube and pyranometer receiver, and optical interaction (for example, stray light) between the domes and pyranometer case and the view-limiting aperture stop (K) and enclosure. (It is noted that the DSET pyrheliometer comparison tube employed in photovoltaic reference cell calibrations is equipped with a rotating axle that permits precise determination of the cell's cosine response to beam radiation. Such a device is impractical for domed pyranometers due to the above mentioned alignment and interaction problems.)

Although a ventilated occulting tube calibration facility could most likely be constructed to accurately transfer calibration to a specific pyranometer design, such as, for example, the Eppley PSP, residual optical alignment and vignetting effects could very well produce errors equal to or greater than those experienced in the proper application of the shading disk techniques.

Shade Disk Approach

While the shading disk technique for transferring calibration to pyranometers from pyrheliometers is specified by WMO as the acceptable method,[21]

---

†Because the diffuse portion of the sky is blue and spectrally quite different from the yellow beam radiation, PV reference cells designed for PV concentrator testing must be calibrated in a sky occulting tube only.

and has been employed throughout the world for many years, no procedural, method-type standards have been promulgated by user nations to the best of the author's knowledge. It was for this reason that two draft standard methods were prepared in ASTM Subcommittee E44.02 to deal with shading disk calibrations of pyranometers both with axis vertical (horizontal receiver) and axis tilted. They are:

- ASTM Committee E-44 Draft Document 104, Standard Method for Calibration of Reference Pyranometers with Axis Vertical by the Shading Method, and
- ASTM Committee E-44 Draft Document 141, Calibration of Reference Pyranometers with Axis Tilted by the Shading Method

These methods require calibration to the WRR Scale, also known as the Absolute Radiometric Scale defined at IPC-75. (Traceability to the International Pyrheliometric Scale of 1956 is not permitted.)

In both methods, two types of calibrations are specified: Type I, referenced to a self-calibrating absolute cavity pyrheliometer and Type II, referenced to a secondary reference pyrheliometer which must itself be calibrated by transfer from an absolute cavity in accordance with ASTM Standard E-816.[30] In both draft methods, direct traceability is required to either IPC IV or V, to one of the New River Intercomparisons of Absolute Cavity Pyrheliometer (NRIPs I-V), to any future intercomparisons of comparable reference quality, or to any absolute cavity having participated in those intercomparisons.

The draft methods were substantially taken from an informal consensus developed during discussions with a number of experts in the U.S. and other countries. They require mounting the pyrheliometer on either an equitorial or altazimuth sun-following tracker, and mounting the pyranometer on the appropriate test stand equipped with a shade disk. The geometrical relation is shown in Figs. 28 and 29 for axis vertical and tilted, respectively. (Although the shading disk technique can be performed at any incident angle on the tilted pyranometer receiver, Fig. 28 shows the specific case of normal incidence (0°) at the tilt angle.) An essential requirement in both methods is that the shadow cast by the disk completely covers the dome of the pyranometer. A fundamental requirement is that the field of view occulted by the shade disk must be

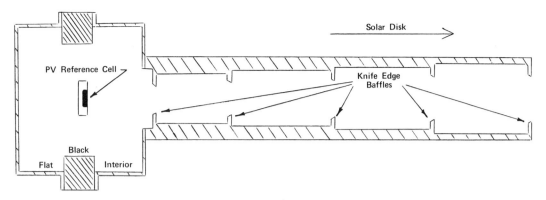

Fig. 27. A pyrheliometer comparison tube for calibration of photovoltaic reference cells.

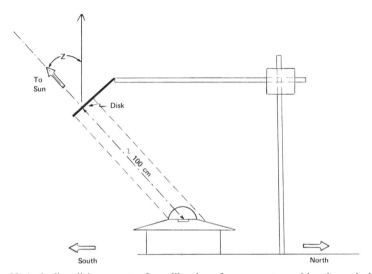

Fig. 28 A shading disk apparatus for calibration of pyranometers with axis vertical.

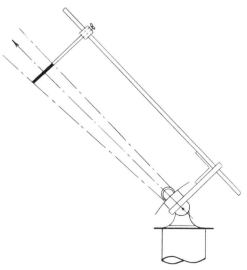

Fig. 29. A shading disk apparatus for calibration of pyranometers with axis tilted (normal incident mode shown).

essentially equal to that of the pyrheliometer employed as the reference radiometer. Since Eppley Model NIPs have traditionally been employed as reference pyrheliometers (especially prior to the wide availability of absolute cavity radiometers), the field of view used, being twice the opening angle (see Fig. 4) is 5.7°. Practical considerations dictated that a 10-cm diameter disk, located 100 cm from the center of the pyranometer's receiver, both adequately shadowed the domes and provided a field of view of 5.7° referenced to the center of the receiver.

It is important to understand that such a shadow only poorly simulates the point-source nature of the sun and while the opening angles (fields of view) can be matched to the reference pyrheliometer, the slope angle cannot [see Eqs. (2) and (3) and Fig. 4]. This implies that large amounts of circumsolar radiation can result in significant, but small, errors in transfer of cali-

49

bration by this method,* and accounts for the need to ensure that no clouds are within a 30° solid angle of the sun and that calibrations are performed only under minimum acceptable sky conditions represented by a direct-beam component of 0.80 or greater.†

The methods require that the pyranometer is alternately shaded and unshaded, and that the difference between the two response signals is divided by the product of the beam irradiance (determined by the reference pyrheliometer) and the cosine of the incident angle of the direct beam component:

$$k = \frac{V_u - V_s}{I_d \cos\theta} \tag{7}$$

where k is the instrument constant in $\mu V \cdot W^{-1} cm^2$; V_u and V_s are the voltage signals of the pyranometer when unshaded and shaded, respectively; I_d is the direct-beam irradiance; and θ is the angle of incidence of the direct beam to the plane of the pyranometer's receiver.

For Draft Method 104 with axis vertical, the sun's zenith angle z and the cosine angle are essentially identical. In this case,

$$\cos \theta = \cos z = \sin \phi \cdot \sin \delta + \cos \phi \cdot \cos \delta \cos \omega \tag{8}$$

where z is the sun's zenith angle; ϕ is the station latitude; δ is the solar declination, $\delta = 23.45 \sin [9863 (n + 283.4)]$; and ω is the hour angle from solar noon, determined as $\pm 15°$ per hour on each side of solar noon.

For Draft Method 141 with axis tilted, the angle of incidence is computed as:

$$\begin{aligned} \cos \theta = &[\sin \delta \sin \phi \cos \beta] - [\sin \delta \cos \phi \\ &\sin \beta \cos \gamma] \\ &+[\cos \delta \cos \phi \cos \beta \cos \omega] \\ &+[\cos \delta \sin \phi \sin \beta \cos \gamma \cos \omega] \\ &+[\cos \delta \sin \beta \sin \gamma \sin \omega] \end{aligned} \tag{9}$$

where, ϕ is the latitude; δ is the declination (defined above); β is the tilt angle from the horizontal; γ is the azimuth angle with zero being due south, (east negative and west positive); and ω is the hour angle, defined above. For the special case of shading disk calibrations at normal incidence, Eq. (9) reduces to Eq. (10).

$$k = \frac{V_u - V_s}{I_d}. \tag{10}$$

Historically, the shade/unshade timing sequence has varied from laboratory to laboratory. In the U.S., the Eppley Laboratory, Inc., the NOAA Solar Radiation Facility at Boulder, and DSET Laboratories, Inc., have all used different sequences. These include 6 min unshaded/5 min shaded (Curve A in Fig. 30), 1 min unshaded/1 min shaded (Curve B), and 30 sec unshaded/30 sec shaded (Curve C).

The two ASTM draft standards call for 30 sec unshaded/30 sec shaded for pyranometers having a time constant of about 1 second (such as for Eppley PSPs). However, as a result of some uncertainties identified by DSET Laboratories during the November 1980 DSET pyranometer comparisons (which are given below), the entire question of shading disk calibration was discussed by several of the participants* of the recently completed NRIP V (May 1982). The discussion centered around a suggestion by C. Wells of SERI that the shading disk timing sequence should represent a 90 to 95% illuminated (unshaded) duty cycle. Because the Second DSET Comparisons of Pyranometers was scheduled to commence immediately following NRIP V, we at DSET Laboratories initiated an investigation of the entire question of the timing requirements for the shade versus unshade condition.

If one performs a simple energy balance on an idealized pyranometer, the timing sequence required to provide a calibration that is optimized in terms of accuracy can easily be determined. Consider the energy incident on the thermopile's receiver as shown in Fig. 31. For the steady state, fully illuminated condition:

$$V_u = k (I_D + I_S + Q_G + Q_B) \tag{11}$$

where V_u is the response signal generated by the pyranometer, k is its unknown instrument constant, I_D is the beam irradiance measured by the pyranometer, I_S is the unknown contribution to V_u from sky irradiance, Q_G is the unknown infrared irradiance of the receiver due to heating of the glass dome(s), and Q_M is the thermal contribution to the thermopile from solar heating of the pyranometer body. Likewise, for the steady state shaded condition,

---

*Most experts agree, however, that these errors are smaller than those resident in the occulting tube method.

†Several experts, including the author, believe that the value should be at least 0.85.

*E. Flowers (NOAA/SRF), J. Hickey (Eppley Labs), C. Wells and D. Myers (SERI) and the author.

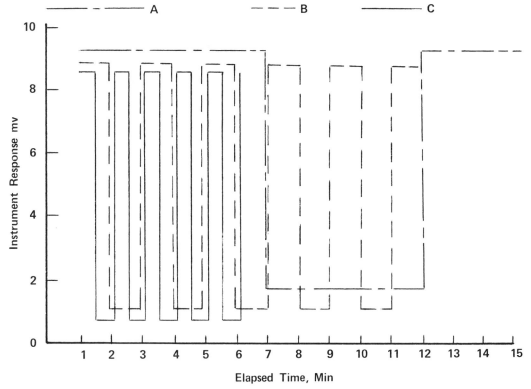

Fig. 30. Typical shade/unshade timing sequences employed in the U.S.

$$V_S = k(I_S + Q_M). \qquad (12)$$

Substituting these equations in Eq. (10), we obtain

$$kI_B = k(I_D + I_S + Q_G + Q_M) - k(I_S + Q_M) \quad (13)$$

where I_B is the beam irradiance measured by the reference pyrheliometer. Simplifying Eq. (13), we obtain

$$I_B = I_D + Q_G \qquad (14)$$

and we observe that an error is propagated the extent of which is dependent on the thermal contribution from the glass domes in the unshaded portion of the sequence. Realizing that a pyranometer, as is true of all thermal devices, possesses multiple time constants,† it becomes obvious that to eliminate the contribution of Q_G in Eq. (14), we needed to determine V_S in the nonsteady state condition immediately after the thermopile reaches a steady state defined by 20 time constants, but prior to the onset of linear decay of signal due to the cooling glass dome (and certainly prior to reaching the overall instrument steady state).

---

†With the smallest and most important being the thermopile/receiver time constant.

The Eppley Laboratory specifies that the time constant of PSPs is approximately 1 sec; we have verified that claim for all three reference pyranometers used in both the November 1981 and May 1982 DSET pyranometer comparisons. These were determined as the time required to reach 1/e of the final steady state illuminated value after being shaded for about 60 sec. We then plotted the signal from all four pyranometers as a function of illumination condition (shaded versus unshaded), every 10 sec. A typical plot is shown in Cycle 1 and 2 of Fig. 32. It will be observed that the illuminated pyranometer reaches a steady state much more quickly than does the shaded instrument, and even after 3 min the shaded thermopile response is decreasing in a linear fashion. Since 20 time constants will give 99.5% of the final thermoelectric response value, the shade/unshade sequence presented in the solid curve of Fig. 32 was chosen for the shade calibrations performed in support of the May 1982 DSET pyranometer calibrations.

For the purpose of comparison, DSET's primary reference pyranometer SN 19129 was calibrated by three of the four techniques specified in Figs. 30 and 32. (The initial soak period of 30 min followed by one "dry" cycle in which no

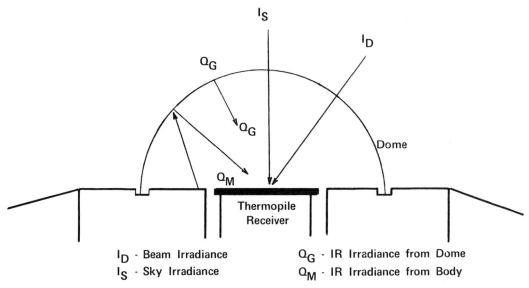

Fig. 31. A schematic of energy on a pyranometer's receiver.

data were taken is required to precondition the thermal state of the pyranometer.) Results are presented in Table 13. The calibration value obtained using the 6 min/5 min sequence (Curve A, Fig. 30) is believed to be in error simply because of the contribution by the hot glass dome. That is, it is allowed to cool off in the shade condition and subtraction of V_s from V_u creates the error defined by Eq. (14). As seen from Table 13, a 1% difference occurred between Methods A and C, neither of which is thought to be correct on the basis of these analyses.

It is instructive to note that the November 1981 normal incident shade calibration of SN 19129 was 10.33 $V \cdot W^{-1} m^2$, essentially within 0.1% of the value presented for Method C (in Table 13). Method C was the technique employed in the November 1981 DSET comparisons.

Transfer of Calibration from Reference to Field Pyranometers

We have discussed the many problems associated with the accurate calibration and characterization of primary reference pyranometers. The transfer of calibration from reference to field instruments is a less complicated problem and is now covered by ASTM Standard E-824 entitled "Standard Method for Transfer of Calibration from Reference to Field Pyranometers."[37] The procedure requires traceability to the WRR and the Absolute Radiometric Scale, as does the companion transfer method for pyrehliometers discussed above.[30]

The method is applicable to field pyranometers without regard to the photoreceptor employed, but the primary reference pyranometer must meet the requirements for the so-called Class I pyranometer specified by WMO, as shown in Appendix I.[21] Two types of calibrations are covered; Type I using natural sunshine, and Type II employing an integrating sphere and an artificial light source. Although specifications for the integrating sphere and light source are not encompassed by the standard method, guidance is given for their use.[38-41] Transfer of calibrations can be performed at tilt as well as horizontal. The principal requirement is that the reference pyranometer shall have been calibrated at tilt or be thoroughly characterized in this respect.

The procedure for transferring calibration is essentially identical to that for pyrheliometers, ASTM Standard E-816.[30] It consists of performing a minimum of fifteen 21-point data sets that are taken over a period of 10 to 20 min (every 30 sec to 1 min). The sets must span an entire day from nearly sunrise to sunset to provide the instrument constant of the field pyranometer which is computed by dividing the ratio of average (or integrated) voltage signal for the solar altitude (period) of interest by the irradiance for the same period determined by the primary reference pyranometer.

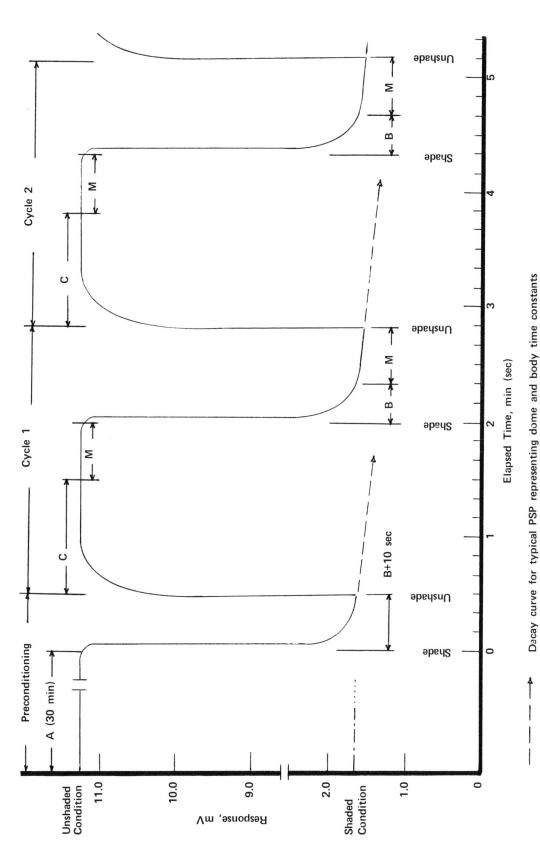

Fig. 32. Shade/unshade timing sequence for shading disk calibration of pyranometers (actual numbers shown are for a typical Eppley PSP with a first time constant of 1 sec).

Decay curve for typical PSP representing dome and body time constants

53

Method	Timing (sec) B[1]	C[1]	M[1]	June 1982 Calibration Value, k (μV · W^{-1} · cm^2)
A[2]	300	360	30	10.25
C[2]	30	30	2	10.34
[2,3]	20	60	30	10.30
—	10	60	30	10.36
—	30	90	30	10.31

[1] These values refer to the timing sequences shown in Fig. 32.
[2] These methods refer to diagrams presented in Fig. 30.
[3] This is the preferred timing sequence.

Table 13 Shade Calibration Values for SN 19129 as a Function of Timing Sequence

RESULTS OF FIRST AND SECOND NEW RIVER COMPARISONS OF PYRHELIOMETERS AND PYRANOMETERS

Rationale

An intercomparison of pyrheliometers and pyranometers was suggested by the participants of NRIP IV held November 17–19, 1980 at DSET Laboratories, Inc.'s facilities in New River, Arizona. The purpose of the proposed intercomparison was to elucidate the extent to which a set of instruments, presumably manufactured and calibrated in the U.S., showed the same deviation from each other as observed in the March 1980 Davos comparison of pyranometers from various manufacturers and different calibration histories.[8,28] It was not suggested by the call for these comparisons that Model NIP pyrheliometers would exhibit the disparities between instruments in a group that was observed for the pyranometers in the Davos comparisons. Rather, it was believed that a definitive experiment would effectively separate the question of pyrheliometer calibrations from the questions being raised concerning pyranometer calibrations.

Experimental

Pyrheliometers

Eight Eppley Model NIP pyrheliometers were mounted to an altazimuth sun-tracking platform employing two sets of matched, shaded solar cells. The two paired detector trackers were mounted in the plane of receivers at right angles to each other, with one set for azimuth and the other for the elevation mode. The tracking accuracy required for the pyrheliometers was achieved.

The pyrheliometers were calibrated directly against DSET's Eppley Model H-F absolute cavity radiometer, SN 17142, in accordance with ASTM Standard E-816.[30] The overall test set up is pictured in Fig. 33 and shows the pyrheliometer, the pyranometers mounted at a tilt of 45°S, the Model H-F absolute cavity, and the DSET Solar Scanning Spectroradiometer.*

Solar observations were made simultaneously with all eight pyrheliometers and the H-F absolute cavity radiometer. Data were acquired, digitized, and recorded on an Esterline Angus PD2064 data logger interfaced to a Three Phoenix magnetic tape transceiver. Analysis was accomplished using a Data General NOVA 3D computer. Solar observations were determined every 30 sec and were organized into 10 min sets of 21 measurements each. Data were taken in accordance with the schedule presented in Table 14.

Pyranometers

Fourteen pyranometers were mounted on 15-in. centers on a common platform. They were leveled with the platform mounted in an exactly horizontal plane by affixing the platform to a multiposition, multitilt exposure rack, as shown in the left figure of the photograph (Fig. 33). Leveling was accomplished with the spirit level that is an integral part of each pyranometer. The 0° horizontal and 45° tilt south-facing tests were performed on this test facility. The platform containing the exactly coaligned pyranometer was removed from the multitilt exposure rack and placed on a large, precision, altazimuth, sun-tracking mount for the normal incident portion of the comparisons.

The comparisons were organized in 10 min data sets in an identical manner to the pyrheliometer comparisons described in the previous section. They were performed in exact compliance with ASTM Standard E-824[37] and were organized as shown in Table 15.

Whereas the primary reference for the pyrheliometer comparisons was DSET's Model H-F absolute cavity pyrheliometer, the references for the pyranometer comparisons were three Eppley PSPs selected on the basis of their histories and employment as reference instruments elsewhere: DSET Laboratories, Inc. SN 19129; NOAA/SRF's SN 19917; and the Eppley

---

*The SSSR was used to take direct beam spectral measurements throughout the day on the third day of the comparison.

Fig. 33. Photograph of test arrangement for pyrheliometer and pyranometer comparisons showing the Eppley Model H-F absolute cavity and the DSET solar scanning spectroradiometer.

Table 14 Organization of Pyrheliometer Comparisons

Date	Number of Sets (Readings)	Times (Solar)	°C Average Temperature
11-02-81	5 (105)	1327-1444	29.3°
11-03-81	12 (252)	1009-1407	28.5°
11-12-81	16 (336)	0900-1416	26.2°

Laboratory's SN 21017. The reference was the mean irradiance determined on the basis first of the instrument constants, k, determined by the respective owners, and secondly by k values determined by DSET Laboratories, Inc. at the three tilt modes employed. The DSET calibrations were performed by the shade method in accordance with the procedures described above, Method C of Fig. 30. The primary reference pyrheliometer for the shade calibrations was DSET's Eppley Model H-F absolute cavity radiometer. The calibration values determined by DSET for the three reference instruments, compared to original values determined by the owner, are presented in Table 16. These values have been corrected for temperature difference between the ambient temperature during shade calibration and the temperature of the comparison at the respective tilt modes. This was done by ratioing using techniques described above.

Results

Pyrheliometer Comparison
The mean ratio of each Model NIP pyrheliometer to the mean of all NIPs and to the irradiance determined by the Model H-F absolute cavity* is presented in Table 17.

The average agreement with "true" irradiance as measured by the Model H-F absolute cavity was 0.993, or ±0.7% (SD 0.012). If instru-

---

*From which new instrument constants were computed.

Table 15   Organization of Pyranometer Comparisons

Mode	Number of Sets & Observ.*	Dates	Apparent Solar Times	Average Temperature (°C)
0° Horizontal	95 (1995)	11-02-81	1135-1438 hr	28.6
		11-03-81	1006-1408 hr	28.8
		11-17-81	0945-1529 hr	28.2
		11-18-81	0918-1510 hr	24.5
45° South	84 (1764)	11-12-81	0900-1437 hr	26.5
		11-13-81	0915-1455 hr	27.9
		11-16-81	0900-1455 hr	26.7
Normal incidence	24 (504)	11-23-81	0940-1432 hr	23.1

* Instantaneous observations.

	Part I	Part II Ics		
Instrument	IC	0° H	45° S	NI
DSET 19129	10.386[1]	10.300	10.309	10.369
NOAA 19917	10.110[2]	10.204	9.962	9.935
Eppley 21017	9,900[3]	9.594	9.775	9.833

[1] At normal incidence October 1981.
[2] At horizontal fall 1981.
[3] At horizontal in integrating hemisphere.

Table 16   Instrument Constants Employed for the Three Reference Instruments in the Part II (Shade Method) Comparisons

ments NIPs SN 11755 and 17608 are removed from the analysis, this widely divergent group of instruments in terms of date of manufacture (SN number) was a remarkable 0.999, or ±0.1% (SD 0.005) in a total of 693 observations for each instrument (4158 readings).

Pyranometers
All pyranometers but one were Eppley Model PSPs; SN 4041 was a photovoltaic-type pyranometer.† The mean ratio of each pyranometer to the mean of the group, and the mean ratio of each to the mean irradiance established by the three reference instruments, are given in Tables 18 through 20 for normal incidence, 0° horizontal and 45° S, respectively. It should be noted that the three reference PSPs (SN 19129, 19917, and 21017) are each referenced to the mean of the three and a new instrument constant is computed for each (in the last column of all three tables) as for all other pyranometers.

The computer file data were interrogated to produce ratios of each instrument's irradiance to

---

†Because of its disparate behavior and the fact that only one unit participated, the type is not given in order to not prejudice others of its type and generic class.

---

the mean of the three reference PSPs for all data taken at 0° horizontal and 45° south. The resulting plots showed certain PSPs to be comparatively insensitive to incident angle and others to have differences of as much as 5% between ±60° and normal incidence. Typical plots were presented for instruments SN 10155, 14317, 17178, and the mean of the three reference instruments (Figs. 16 through 19).

The shading disk calibrations of the three reference PSPs agree to within greater than 99% of the values furnished by the participant at the most nearly corresponding tilt mode (compare the values in column 1 with the corresponding values in columns 3, SRF, and 5, DSET and Eppley). The Eppley integrating hemisphere result more closely matches the DSET normal incident result in both geometry (uniformity of dome illumination) and value.

Examination of the mean ratios of each pyranometer to the lot mean shows that those pyranometers with recent instrument constants listed as traceable to NRIP II-IV or IPC V exhibited an average irradiance that was 1.008 times the mean of all instruments, compared to an average irradiance of all others that was 0.979 times the mean (SN 4041 was not included in the second set, but was included in the lot mean). The difference of 2.9% is attributed largely to differences in calibration techniques (although some deterioration in instrument sensitivity most likely occurred with several of the instruments in the second group). Even more important is the observation that the six pyranometers that were deemed recently calibrated traceable to one of the absolute cavity pyrheliometers in NRIP I-IV or IPC V Davos gave mean irradiance values that were within 0.8% of the mean irradiance of the three reference PSPs, whereas the

Test NIP		Mean Ratio to Mean of Lot		Mean Ratio to Mean of 3 Ref. PSPs		
SN	Old k*	Ratio	SD	Ratio	SD	New k*
11755	7.89	0.977	0.002	0.970	0.004	7.66
12016	8.00	1.014	0.002	1.007	0.005	8.05
13169	6.74†	1.004	0.002	0.997	0.003	6.73
13384	6.56	1.004	0.002	0.997	0.003	6.54
13707	7.20	1.008	0.002	1.001	0.004	7.21
14306	8.09	0.999	0.001	0.992	0.003	8.03
17608	7.70	0.985	0.008	0.979	0.010	7.54
17664†	8.43†	1.008	0.001	1.002	0.004	8.44

* $\mu V \cdot W^{-1} \cdot m^2$.
† Owner calibrations, all others original Eppley labels.

Table 17   Mean Ratio of Each NIP Pyrheliometer to the Average and to H-F SN 17142 Absolute Cavity

Test Pyranometers		Mean Ratio to Lot Mean		Mean Ratio to Mean of 3 Ref. PSPs		
SN	Old k	Ratio	SD	Ratio	SD	New k
10155	7.11	0.998	0.005	0.986	0.005	7.01
14317	7.20	1.001	0.013	0.989	0.012	7.12
14439*	7.82	0.995	0.006	0.984	0.005	7.69
14898	9.33	0.983	0.006	0.971	0.007	9.06
15446	10.58	0.956	0.003	0.945	0.001	10.00
17173	10.30	0.994	0.004	0.982	0.002	10.12
17860*	7.91	1.009	0.004	0.997	0.002	7.89
17897*	8.25	0.995	0.004	0.984	0.004	8.12
18864	10.89	0.946	0.003	0.935	0.002	10.18
19129* (DSET)	10.37	1.008	0.004	0.996	0.002	10.33
19917* (SRF)	9.94	1.016	0.003	1.005	0.002	9.98
21017* (Eppley)	9.83	1.011	0.003	0.999	0.001	9.83
21070	9.88	0.971	0.007	0.960	0.006	9.49
4041	11.50	1.118	0.039	1.105	0.043	12.71

*Instruments that have presumed traceability to NRIP II-IV and IPC V because of previous known histories.

Table 18   Comparison of Pyranometers at Normal Incidence

average irradiance of the remaining seven pyranometers was lower by 3.7% (not including the photovoltaic instrument in the second set).

One of the more interesting observations is the spread in sensitivity of certain pyranometers as a function of tilt mode. Because we know that most PSPs are relatively insensitive to tilt (to 45° or greater), most of the disparities are due to incident angle effects resulting from changing illumination of the domes, receiver, and so on.

The efficacy of transferring calibration from reference to field pyranometers in accordance with ASTM Standard E-824 has been established. However, transfer of calibration to field

pyranometers should be from the mean of three reference instruments that are themselves in close agreement at all angles of incidence. Three reference pyranometers are required to minimize deviations from cosine law (inherent in all pyranometers to some extent) that would most likely result in significantly greater uncertainties if only one of the instruments were employed.

Pyranometers ought not to be calibrated at 0° horizontal (either by the shade method or by transfer) during the wintertime at the higher latitudes of the Northern Hemisphere (for example, above 30° north latitude), unless the requirement exists for precision measurements at 0° hori-

Test Pyranometers		Mean Ratio to Lot Mean		Mean Ratio to Mean of 3 Ref. PSPs		
SN	Old k	Ratio	SD	Ratio	SD	New k
10155	7.11	0.995	0.008	0.976	0.008	6.94
14317	7.20	0.984	0.008	0.966	0.008	6.96
14439*	7.82	1.007	0.006	0.988	0.004	7.73
14898	9.33	0.966	0.005	0.948	0.007	8.85
15446	10.59	0.956	0.007	0.939	0.008	9.93
17173	10.30	0.995	0.005	0.976	0.005	10.06
17860*	7.91	0.988	0.006	0.970	0.007	7.67
17897*	8.25	0.998	0.006	0.980	0.003	8.08
18864	10.89	0.966	0.005	0.949	0.004	10.33
19129* (DSET)	10.30	1.024	0.006	1.005	0.002	10.36
19917* (SRF)	10.20	1.003	0.005	0.984	0.004	10.04
21017* (Eppley)	9.59	1.029	0.005	1.010	0.002	9.69
21070	9.88	0.984	0.006	0.966	0.005	9.54
4041	11.50	1.104	0.044	1.084	0.047	12.47

*Instruments that have presumed traceability to NRIP II-IV and IPC V because of previous known histories.

Table 19   Comparison of Pyranometers at 0° Horizontal

Test Pyranometers		Mean Ratio to Lot Mean		Mean Ration to Mean of 3 Ref. PSPs		
SN	Old k	Ratio	SD	Ratio	SD	New k
10155	7.11	0.995	0.004	0.977	0.006	6.94
14317	7.20	1.003	0.004	0.984	0.006	7.09
14439*	7.82	1.002	0.004	0.983	0.003	7.69
14898	7.33	0.977	0.003	0.959	0.006	8.95
15446	10.58	0.957	0.003	0.940	0.005	9.94
17173	10.30	0.996	0.009	0.978	0.010	10.07
17860*	7.91	1.003	0.002	0.984	0.004	7.79
17897*	8.25	0.996	0.004	0.978	0.003	8.07
18864	10.89	0.953	0.003	0.936	0.002	10.19
19129* (DSET)	10.31	1.018	0.005	0.999	0.004	10.30
19917* (SRF)	9.96	1.019	0.004	1.000	0.003	9.96
21017* (Eppley)	9.78	1.018	0.002	0.999	0.003	9.77
21070	9.88	0.977	0.004	0.959	0.003	9.48
4041	11.50	1.086	0.019	1.066	0.022	12.26

* Instruments that have presumed traceability to NRIP II-IV and IPC V because of previous known histories.

Table 20   Comparison of Pyranometers at a Tilt of 45° South

zontal in the wintertime. This is due to the fact that at incident angles of 65°, the uncertainties in both the shade method and in transfer are significantly increased. Furthermore, summertime calibrations at high tilt angles in the more southern latitudes (for example, below 50° north latitude) will likely result in significant uncertainties for identical reasons.

REFERENCES

1. D.H. Lufkin. 1980. New U.S. network for solar radiation measurements, *Solar energy technology handbook, part A: engineering fundamentals.* W. C. Dickinson and P. M. Cheremisinoff, eds., New York/Basel: Marcel Dekker, Inc., pp 167–172.
2. D. Anson. 1980. Instrumentation for solar radiation measurements. *Solar energy technology handbook, part A: engineering fundamentals.* W. C. Dickinson and P. M. Cheremisinoff, eds., New York/Basel: Marcel Dekker, Inc.; pp. 173–199.
3. K.J. Hanson. 1973. Comments on the quality of the NWS pyranometer network from 1954 to the present. *Proceedings of the 1973 Solar Energy Data Workshop.* Nov. 1973.
4. E.C. Flowers. 1973. The "so-called" Parson's black problem with old-style Eppley pyranometers. *Proceedings of the 1973 Solar Energy Data Workshop.* Nov. 1973.
5. M.P. Thekaehara. 1976. Solar radiation measurement: techniques and instrumentation. *Solar Energy* 18 (No. 4):309–325.
6. A.J. Drummond. 1966. Pyrheliometric calibrations at the Eppley Laboratory. Eppley Memorandum. Nov. 10, 1966.
7. E.C. Flowers, 1982 (Mar.). Personal communication.
8. G.A. Zerlaut. 1981. Why standard pyranometer calibrations are inappropriate for solar collector testing. *Proceedings, of the Annual Meeting of the American Section of the International Solar Energy Society.* Philadelphia, Pa.; May 29, 1981.
9. E.C. Flowers. 1981. Calibration and comparison of pyranometers and pyrheliometers. *Proceedings, Workshop on Accurate Measurement of Solar Radiation.* AS/ISES Annual Meeting. Philadelphia, Pa.; May 29, 1981.
10. E.R. Streed et al. 1978. Results and analysis of a round robin test program for liquid-heating flatplate solar collectors. NBS-975. Washington, D.C.
11. ASHRAE Standard 93-77. 1977. American Society of Heating, Refrigeration and Air Conditioning Engineers.
12. H.D. Talarek. 1979 (Dec.). Testing of liquid-heating flat-plate collectors based on standard procedures. KFA-IKP-113-4/79. Kernforschungs-anlage-Jülich GmbH. Presented by E. Streed to ASTM Committee E-44 as an IEA progress report.
13. W. Ley. 1979 (Dec.). Survey of solar simulator test facilities and initial results of IEA round robin tests using solar simulators. IEA Task III Report on Performance Testing of Solar Collectors. Cologne, W. Germany: Deutsche Forschungs- und Versuchsanstalt fur Luft- und Raumfahrt ev. (DFVLR).
14. J.A. Plamondon. 1969. *JPL Space Programs Summary for 1969* III:37–59.
15. J.M. Kendall, Sr. 1969 (July). Primary absolute cavity radiometer. Jet Propulsion Laboratory Technical Report 32-1396.
16. R.C. Willson. 1973 (Apr.). Active cavity radiometer. *Applied Optics* 12:810–17.
17. The Eppley Laboratory. 1977. The self-calibrating sensor of the electric satellite pyrheliometer (ESP) program. *Proceedings of the Annual Meeting of AS/ISES.* I. Orlando, Fla.; June 6–10, 1977.
18. J.M. Kendall, Sr. and C.W. Berdahl. 1970 (May). Two black bodies of high accuracy. *Applied Optics.* 9 (No. 5):1082–1091.
19. Doc. T586. PMO-6 Absolute Radiometer (brochure). Compagnie Industrielle Radioelectrique (CIRI). Sept. 1980.
20. Dr. Nels Johnson. Personal communication. International Division, NOAA, Rockville, Md.
21. *Measurement of Radiation and Sunshine, Guide to Meteorological Instrument and Observing Practices.* Chap. 9. 4th ed. WMO-No. 8, TP. 3. Geneva.
22. Anon. 1977. Pyrheliometer comparisons 1975, results and symposium, IPC IV. Working Report WR No. 58. Davos: World Radiation Center.

23. Anon. 1981 (Feb.). Pyrheliometer comparisons 1980, results and symposium, IPC V. Working report, WR No. 94. Davos: World Radiation Center.

24. R.S. Estey and C.H. Seaman. 1971 (July). Four absolute cavity radiometer (pyrheliometer) intercomparisons at New River, Arizona. JPL Publication 81-60. Jet Propulsion Laboratory.

25. G.A. Zerlaut. The calibration of pyrheliometers and pyranometers for testing photovoltaic devices. *Solar Cell* (in publication). Also *Proceedings*. Commercial Photovoltaic Measurements Workshop. Vail, Colo.; July 27–29, 1981.

26. C.V. Wells. An experiment to compare data acquisition methods employed in the IPC V and New River comparisons. In Ref. 23 (WRC-D WR 58).

27. G.A. Zerlaut. 1981 (Apr.). The challenge of solar energy—ASTM Committee E-44. *ASTM Standardization News*.

28. H.D. Talarek and C. Fröhlich. 1980 (June). Results of a pyranometer comparison. IEA/PMOD (WRC) Report. March 5,6, 1980.

29. M.R. Riches; T.L. Stoffel; and C.V. Wells. 1981 (Apr.). International Energy Agency Conference on Pyranometer Measurements. Draft Report SERI/TR-642-1156. Solar Energy Research Institute.

30. ASTM Standard E-816. Calibration of secondary reference pyrheliometers and pyrheliometers for field use. *Annual book of ASTM standards*. Part 41. American Society for Testing and Materials. Philadelphia, Pa.

31. R.J. Bahm and J.C. Nakos. 1979 (Nov.). The calibration of solar radiation measuring instruments. Report No. BER-1(79) DOE-684-1. DOE Contract No. EM-78-S-04-5336.

32. J. McGreggor. Personal communication. Cardiff University, Cardiff, Wales, UK.

33. E. Flowers. Personal communication. NOAA/SRF, Boulder, Colo.

34. H.E.B. Andersson; L. Liedquist; J. Lindblad; and L.A. Norsten. 1981. Calibration of pyranometers. Report SP-RAPP 1981:7. Boras, Sweden: Statens Provningsanstalt (National Testing Institute). March 6, 1981.

35. D.J. Norris. 1974. Calibration of pyranometers in inclined and inverted positions. *Solar Energy* 16:53–55.

36. F. Loxsom and W.D. Hogan. 1981. Measurement of irradiation on tilted surfaces. In *Accurate measurement of solar radiation*. Proceedings of *Workshop, 1981 Annual Meeting American Section/ISES*. Philadelphia, Pa.; May 29, 1981.

37. ASTM Standard E-824. Standard method for transfer of calibration from reference to field pyranometers. *Annual Book of ASTM Standards*. Part 41. American Society for Testing and Materials. Philadelphia, Pa. 1966.

38. Augustus N. Hill. Calibration of solar radiation equipment at the U.S. Weather Bureau. *Solar Energy* 10(No. 4):1–4.

39. J.R. Latimer. 1966. Calibration program of the canadian meteorological service. *Solar Energy* 10(No. 4):4–7.

40. J.R. Latimer. 1964. An integrating sphere for pyranometer calibration. *Journal of Applied Meteorology*. 3:323–6.

41. A.J. Drummond and H.W. Greer. 1966. An integrating hemisphere (artificial sky) for the calibration of meteorological pyranometers. *Solar Energy* 10(No. 4):7–11.

Appendix I
The Classification of Accuracy of Radiometers *

	Sensitivity (mW cm^{-2})	% Stability	% Temperature	% Selectivity	% Linearity	Aperture	Time Constant (max)	Cosine % Response	Azimuth % Response	Errors in Auxiliary Equipment		Chrono-meter
											%	
Reference standard pyrheliometer	± 0.2	± 0.2	± 0.2	± 1	± 0.5	(1)	25 s	—	—	0.1 unit	0.1	0.1 s
Secondary Instruments												
1st class pyrheliometer	± 0.4	± 1	± 1	± 1	± 1	(1)	25 s	—	—	0.1 unit	0.2	0.3 s
2nd class pyrheliometer	± 0.5	± 2	± 2	± 2	± 2	(1)	1 min	—	—	0.1 unit	± 1	—
										Errors in recording apparatus		
1st class pyranometer	± 0.1	± 1	± 1	± 1	± 1	—	25 s	± 3	± 3	0.3		
2nd class pyranometer (2)	± 0.5	± 2	± 2	± 2	± 2	—	1 min	± 5–7	± 5–7	± 1		
3rd class pyranometer	± 1.0	± 5	± 5	± 5	± 3	—	4 min	± 10	± 10	± 3		

*Adapted from Ref. 21.

Appendix II
Radiation Centers*

1. World Radiation Centers

   Davos, Switzerland
   Voejkov, U.S.S.R.

2. Regional Radiation Centers

   Region I

   Khartoum, Sudan
   Kinshasa, Zaire
   Pretoria, South Africa
   Tunis, Tunisia
   Tamanrasset, Algeria

   Region II

   Poona, India
   Tokyo, Japan

   Region III

   Buenos Aires, Argentina

   Region IV

   Toronto, Canada
   Boulder, Colo., United States of America

   Region V

   Aspendale, Australia

   Region VI

   Davos, Switzerland
   Kew, United Kingdom of Great Britain and
      Northern Ireland
   Leningrad, Union of Soviet Socialist Republics
   Stockholm, Sweden
   Uccle, Belgium

*Adapted from Ref. 21.

Advances in Solar Energy. © 1983 American Solar Energy Society, Inc.

BIOMASS PYROLYSIS: A REVIEW OF THE LITERATURE
PART 1—CARBOHYDRATE PYROLYSIS

MICHAEL JERRY ANTAL, JR.

Coral Industries Professor of Renewable Energy Resources, University of Hawaii, Honolulu, Hawaii 96822

This paper is dedicated to Martin Summerfield, Emeritus Professor of Aeronautical Engineering, Princeton University.

Abstract

A normative review of the literature describing the products, mechanisms and rates of carbohydrate pyrolysis is presented. The role of a complex sequence of competing solid and vapor phase pyrolysis pathways is elucidated.

INTRODUCTION

The decade of the 1970s bore witness to an accelerating crescendo of public concern over the energy crisis. This concern was manifest in daily newspaper headlines, dramatic increases in federal expenditures for research and development focused on energy production, and renewed efforts by industry to transform coal, oil shale, and other low quality fossil fuels into premium synfuels. As in a great symphony, the long crescendo was abruptly interrupted early in the 1980s and the tranquil interlude which has followed may have caused many to wonder if the problem has somehow gone away.

Unfortunately, the fluid fuels crisis confronting the industrialized world has not gone away. The problem is best viewed in the context of M. King Hubbert's research and publications,[1-4] which clearly foresaw the energy crisis seventeen years before its debut, and presently foretell the steady diminution of proved reserves of oil and natural gas throughout the industrialized world. As viewed by Hubbert, the fluid fuels crisis is not so much manifest in a brief interruption of the flow of oil and gas from the Persian Gulf, as it is seen in the steady march to exhaustion of a precious natural resource. As an alternative to increased reliance on imports or

NOTATION

A = Pre-exponential constant (apparent frequency factor).
Cr I = Crystalline index; see Ref. 138.
E = Apparent activation energy (kcal/gmol).
f_o = Orientation factor or degree of stretch; see Ref. 60.
k = Rate constant: $k = A \exp(-E/RT)$.
t = Time (sec).
T = Temperature
w (t) = Time dependent sample weight (g).
w_i = Initial sample weight (g).
w_f = Final sample weight (g).
α = Degree of conversion: $\alpha = (w_i - w(t))/(w_i - w_t)$.
β = Heating rate (°C/min).
Δ = Increase: ΔCO is increase in CO mass yield due to gas phase pyrolysis.

Michael Jerry Antal, Jr. graduated Summa Cum Laude with Highest Distinction in Physics and High Distinction in Mathematics from Dartmouth College in 1969. He earned an MS in Applied Physics in 1970, and a PhD in Applied Mathematics in 1973, both from Harvard University. After completing his graduate work, Dr. Antal spent two years as a theorist with the Thermonuclear Weapons Physics Group of the Los Alamos Scientific Laboratory, and six years with Princeton University. At Princeton he was a member of the faculty of the Mechanical and Aerospace Engineering Department and Director of the Renewable Resources Research Laboratory (RRRL). In 1981 Dr. Antal was invited to assume the newly endowed Coral Industries Distinguished Professor of Renewable Energy Resources Chair with the University of Hawaii. In 1982 the RRRL moved from Princeton to Honolulu, and is now actively engaged in research on the pyrolysis of biomass materials, and high temperature solar thermal energy utilization.

other forms of nonrenewable fossil fuels, the conversion of the renewable biomass resource into fluid fuels represents one potentially attractive solution to the energy problem of the industrialized world.

Developing countries face a totally different kind of energy crisis[5]. Unable to afford costly oil and gas, these countries have long relied on the combustion of locally available wood and other biomass to meet their need for fuels. Inefficient stoves and other combustion equipment, combined with a gross neglect of reforestation and good soil management practice, have lead to the deforestation of vast areas of Africa and Asia. For these countries, the energy crisis requires the development of more efficient means for utilizing biomass as a fuel, as well as a commitment to renew the resource using modern forestry techniques.

Thus, both the developing world and the industrialized nations require new technologies which efficiently utilize the biomass resource. Thermochemical processes are thought to have great promise as a means for efficiently and economically converting biomass into higher value fuels. Pyrolysis lies at the heart of all thermochemical fuel conversion processes. Consequently, the theory and practice of biomass pyrolysis has become the focus of intensive research and development activities throughout the world.

Magnitude and Availability of the Resource

Simple calculations evince the impressive potential of biomass as a fuel resource. Each year worldwide photosynthetic activity stores 17 times as much energy in plant life as is consumed by all the nations of the world.[6] In the U.S., terrestrial biomass production roughly equals energy consumption.[7] Unfortunately, these order of magnitude estimates have little significance because they do not account for the difficulty of collecting the biomass and converting it to more useful fuel forms. Greater care is required to obtain a meaningful projection of the contribution biomass could make to the U.S.A.'s energy supply.

Estimates of the magnitude of the available biomass resource have been published by Inman,[8] Alich and Witwer,[9] Anderson,[10] Burwell,[11] Poole,[12] Poole and Williams,[7] and more recently by the Office of Technology Assess-

ment (OTA) of the U.S. Congress.[13] These estimates range from 1.6 to over 17 quads per year. Table 1 summarizes the results of the OTA study, which are now widely accepted. Values given in Table 1 indicate that biomass can become a major contributor to our supply of fluid fuels if efficient conversion processes are developed.

The need for the development of efficient conversion processes is well illustrated by the work of Pimentel.[14] Behind Pimentel's conclusion that biomass can play only a minor role in meeting the U.S.A.'s demand for energy lies the consistent assumption of grossly inefficient conversion processes. Pyrolysis processes can achieve thermal (first law) conversion efficiencies approaching 100% which permits biomass to make a far larger contribution to our energy supply than indicated by the work of Pimentel. The future ability of the biomass resource to play a major role in the energy economy of the industrialized world is thus largely contingent upon the development of highly efficient, thermochemical conversion processes. Needless to say, the availability of such processes will be of great benefit to the developing world as well.

Physical and Chemical Characteristics of the Resource

The chemistry of biomass materials has been the subject of intense scientific research for many centuries. The chemical composition, gross physical, molecular, and atomic structure of most biomass materials (with the exception of municipal refuse, manures, and sewage sludge) have been relatively well characterized. However, much of this research was accomplished by scientists interested in the life sciences. Consequently, the large body of accumulated knowledge is not always pertinent to the needs of fuel scientists. For example, no standard classification system, paralleling the ASTM classification system for coal, exists for biomass materials.

Source	Quads/yr
Wood	10
Grasses	5
Crop residues	1
Biogas from manure	0.3
Ethanol from grains	0.3
Agricultural Processing Wastes	0.2
Total	17

Table 1 OTA "High Development" Year 2000 Bioenergy Supply[13]

Thus, it is not generally known whether the gasification characteristics of corncob material more closely resemble hardwood, sotfwood, cellulose, or cow manure. Moreover, no standard methods for biomass analysis or assays exist, although it is customary to use coal analyses (ultimate and proximate analysis) for biomass materials. Consequently, it is not surprising that biomass fuels are so poorly understood and play such a small role in the energy economies of industrialized nations.

The intent of this section is to summarize the relevant physical and chemical characteristics of biomass fuels. These characteristics offer important guidance for the design of thermal conversion systems. Since no classification scheme based on the thermal properties of biomass exists, the following discussion relies on popular characterizations of the resource.

Refuse Material

The composition and combustion characteristics of municipal solid wastes (MSW) have been reviewed by many authors.[15-18] In general, cellulosic materials (paper) constitute the bulk of the MSW resource. It is worthwhile to note that the characteristics of MSW vary significantly according to the time of year, affluence, local demography, and so on of the MSW source.[18]

Proximate analyses of MSW are rarely reported. A typical ultimate analysis is given in Table 2. The table illustrates several of the important characteristics of MSW—raw MSW is high in moisture, and low in sulfur and nitrogen as compared to fossil fuels. Although its chlorine content is rather high, tests in St. Louis showed

that two-thirds of the chlorine is in the form of "inert" inorganic chlorides which do not form HC1 during combustion.[18] The calorific value of raw MSW (10.2 kJ/g or 4,400 Btu/lb) is low; however, it does not compare unfavorably to peat or lignite. Dry, *organic* MSW has a calorific value of 19.3 kJ/g (8,300 Btu/lb), exceeding the heating value of some lignites.

Although proximate analyses of MSW are generally not available, research in my laboratory has measured the volatile matter content of several types of newsprint (a major constituent of refuse). Surprisingly, over 80% of the newsprint is volatile matter (according to the ASTM D271-58 classification test for coal) in contrast to about 10% for oil shale, 40% for many coals, and 60% for peat.

Trace metals present in the ash of MSW may be a cause for some concern. The concentrations of cadmium, copper, lead, and zinc are higher in MSW than typical coals.[18-21] In some cases, the ash of MSW and other organic materials may actually be an economical source of metals.[22] The environmental impact of trace metals in MSW must be weighed against its low sulfur and nitrogen content before any firm conclusions can be reached.

Woody Material

The most important physical characteristics of wood are its density, moisture content, and calorific value. Wood density (specific gravity) can vary by as much as 0.38 to 0.75 from species to species. Moisture content considerably affects wood density. There are two measures of a wood's moisture content (m.c.):

m.c. (dry basis) =
$$\frac{\text{weight of moisture lost during drying}}{\text{weight of dry wood}}$$

m.c. (wet basis) =
$$\frac{\text{weight of moisture lost during drying}}{\text{original weight of wood}}.$$

Clearly it is possible for the m.c. (dry basis) to exceed 100%. Green wood is typically 40% moisture (wet basis) with a calorific value of 11.6 kJ/g (5,000 Btu/lb). Dry wood has a typical calorific value of 20 kJ/g (8,600 Btu/lb); however, the presence of extraneous substances in a particular wood (resins, terpenes, fats, phenols, tannins, and so on) can have a considerable effect on its calorific value. The reader is referred to Ref. 23 for a more complete discussion of the density, moisture content, and calorific value of

Component	Analysis (as Received) % by Weight	Analysis (Dry Basis) % by Weight
Moisture	25.1	0.0
Carbon	25.2	33.5
Hydrogen	3.2	4.3
Oxygen	18.8	25.2
Nitrogen	0.4	0.5
Chlorine (organic 0.16), (inorganic 0.14)	0.3	0.4
Sulfur	0.1	0.1
Metal	8.7	11.6
Glass, ceramics	12.2	16.3
Ash	6.0	8.1
Total	100.0	100.0
Higher heating value, HHV	4,400 Btu/lb	5,600 Btu/lb

Table 2 Ultimate Analysis and Heating Value for Typical Mixed Municipal Solid Waste[18]

common woods. Other physical properties of wood (specific heat, hygroscopicity, and so on) are discussed by Wenzl.[24]

The chemistry of wood can only be briefly touched on here. The reader is referred to the classic texts of Wenzl,[24] Goldstein,[25] Wise and Jahn,[26] and Browning[27] for a more thorough treatment. It is traditional to discuss the chemistry of wood in terms of isolated components (cellulose, hemicellulose, lignin, and extraneous components) derived from "summative wood analysis"[24] using well-known analytic methods. These components are not chemically uniform; nevertheless, they offer a surprising insight into the chemical structure of various wood species. The following paragraphs summarize the chemistry of each of these components in turn.

Cellulose is the main component of the cell wall, with the elementary chemical formula $(C_6H_{10}O_5)_n$. Over a century of research has shown cellulose to be a linear polysaccharide composed of anhydroglucose units connected to each other by $1{\rightarrow}4{-}\beta$-glucosidic linkages (see Fig. 1). Cellulose is characterized by an orderly, largely crystalline structure. The functional groups present in the cellulose molecule have a significant effect on its chemical and physical properties. The principal functional groups in pure cellulose are the hydroxyl groups; thus, cellulose behaves as a polyalcohol when undergoing oxidation.[24] The term "alpha" cellulose refers to cellulose extracted from wood using a carefully defined analytic technique. Alpha cellulose constitutes about 50% of the weight of most woods. The text by Dorée[28] gives an excellent discussion of cellulose chemistry.

All noncellulosic polysaccharides and related substances comprise the "hemicelluloses." These hemicelluloses are further characterized as polyuronides, mannan, galactan, xylan, or araban, which are by no means homopolymer complexes. The hemicelluloses are sometimes characterized as pentosans, hexosans, or wood polyoses. They are readily hydrolyzable. Figure 2 portrays the structural formulas of the monomeric sugars that play a role in

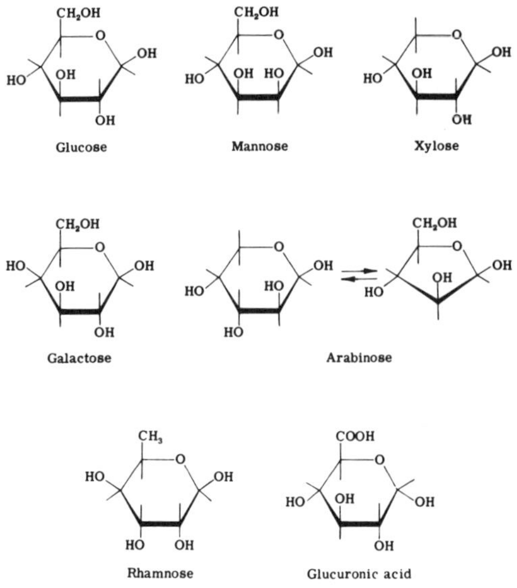

Fig. 2. Monomeric Sugars.[24]

the structures of cellulose and hemicellulose. Glucose is the prevalent monomer in the structure of cellulose, mannose in the hemicellulose component of coniferous ("soft") woods, and xylose in deciduous ("hard") woods. Hemicelluloses constitute 20% to 35% of the weight of most woods.

The chemistry and physics of lignin have been the subject of intense interest in wood research for several decades. References 29–32 provide an insight into recent research on lignin. Plant lignin is almost insoluble, and its chemical isolation is accompanied by marked changes in its molecular structure. Lignin is known to have an aromatic structure composed of phenylpropane units, as illustrated in Fig. 3. Coniferous wood lignins contain exclusively guaicylpropyl units; whereas, deciduous wood lignins contain both guaiacyl- and syringyl-propyl units (see Fig. 3). Figure 4 displays a proposed representative structure for lignin.[33] Lignin contents of most woods range from about 15% to 30% by weight.

The extraneous components of wood include aliphatic, aromatic and alicyclic compounds, hydrocarbons, alcohols, ketones and various acids, esters, phenolic compounds, sterols, tannins, essential oils, resins, terpenes, and so on. These components are easily extracted from the wood using organic solvents, or (sometimes) water. They form the basis of the "naval stores" industry and historically have been an important source of chemicals. The

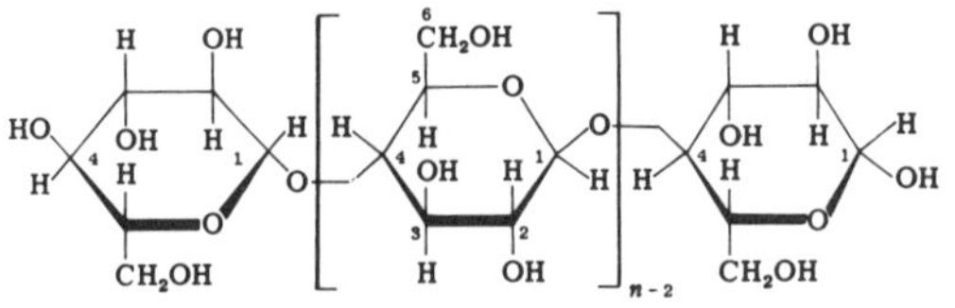

Fig. 1. Cellulose Structure.[24]

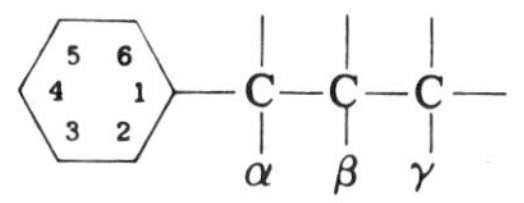

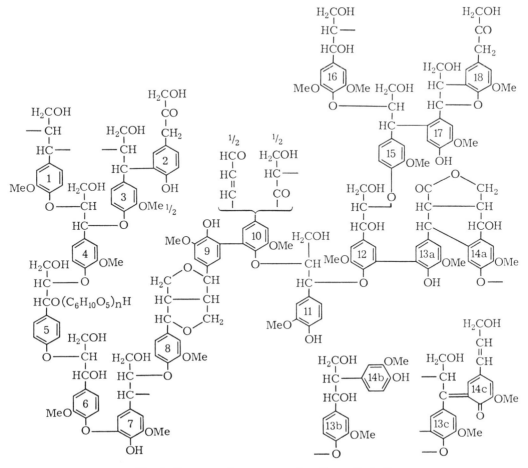

Fig. 3a. Phenylpropane Unit.[24]

Guaiacyl   Syringyl

Fig. 3b. Guaiacyl and Syringyl Units.[24]

woods. In comparison with fossil fuels, wood is very low in sulfur, nitrogen, and ash. The calorific value of wood compares favorably with subbituminous C or B coals. Based on ultimate analysis data, wood is a premium solid fuel.

Although proximate analysis data are not widely available, research in my laboratory and elsewhere has shown that wood loses between 70% to 80% of its weight by low temperature (<500°C) pyrolysis reactions. Thus, the volatile matter content of wood exceeds that of coal, lignite, and peat.

reader is referred to Refs. 34–36 for a more complete discussion of the extraneous components of wood.

Like MSW, proximate analyses of wood are not widely available. Table 3 contains a listing of representative ultimate analyses of various

Agricultural Material

With few exceptions, agricultural wastes have not been considered to be a fuel resource due to other competitive uses for the material; consequently, even less information is available on their thermal properties than for wood or MSW. As indicated in Table 4, agricultural materials are generally higher in cellulose and hemicellulose, and lower in lignin than most woods. Since lignin is much more refractory than cellulose or

Fig. 4. Representative structure of coniferous lignin.[33]

65

	Percent by Weight						Heating Value Btu/lb	
	C	H	S	O	N	ash	Higher	Lower
White Cedar	48.8	6.37	—	44.46	—	0.37	8400	7780
Douglas Fir	52.3	6.3	—	40.5	0.1	0.8	9050	8438
Pitch Pine	59.0	7.19	—	32.68	—	1.13	11320	10620
White Ash	49.73	6.93	—	43.04	—	0.30	8920	8246
Elm	50.35	6.57	—	42.34	—	0.74	8810	8171
Maple	50.64	6.02	—	41.74	0.25	1.35	8580	7995
Black Oak	48.78	6.09	—	44.98	—	0.15	8180	7587

Table 3 Typical Analyses of Dry Wood[37]

hemicellulose, agricultural material is probably better suited for thermal conversion processes than wood.

To the best of this author's knowledge, very little information is available on the proximate and ultimate analyses of agricultural materials. Table 5 gives an ultimate analysis of a red corn-cob sample.[38] The material is low in ash, moisture, and sulfur. Its calorific value compares favorably with lignite. Research in my laboratory indicates that volatile matter content is about 80%, suggesting that it is a most desirable feedstock for thermal conversion processes.

Manure and Sewage Material
Ultimate analyses of feedlot cattle manures and sewage sludge are given in Tables 6 and 7. The ash content of these materials is much higher than air classified MSW, wood, or agricultural wastes. Consequently, the calorific value (typically 14 kJ/g or 6,000 Btu/lb on a dry basis) and volatile matter content (typically 60%) is less than other organic materials. Also the moisture, sulfur, and nitrogen contents are comparatively high, suggesting that manures and sewage may not be as well suited for pyrolytic processes as the other biomass materials.

Comparison of Biomass with Solid Fossil Fuels
The calorific value, moisture content, sulfur content, and volatile matter content are widely accepted figures of merit of the quality of a solid fossil fuel. The following paragraphs offer a comparison of biomass to common solid fossil fuels based on the data presented in earlier sections of this review.

The calorific value of lignite is less than 19.3 kJ/g, subbituminous C coal ranges from 19.3 to 22.1 kJ/g (8,300 to 9,500 Btu/lb), and subbituminous B coal ranges from 22.1 to 25.6 kJ/g (9,500 to 11,000 Btu/lb). On this basis, mixed municipal refuse with about 10.2 kJ/g (4,400 Btu/lb) is comparable to lignite, and dry woods (18.6 to 26.3 kJ/g or 8,000 to 11,300 Btu/lb) are comparable to subbituminous C or B coals. The calorific value of agricultural residues is comparable to a good grade of lignite. Manure is also comparable to lignite. Based on calorific value, peat is less valuable than wood but compares favorably with refuse and manure. Of the fuels considered here, raw oil shale is the least desirable, having a calorific value of only 2.8 kJ/g (1,200 Btu/lb).

Moisture comprises typically 45% of the weight of lignite, 30% of subbituminous C, 24% of subbituminous B, and 17% subbituminous A coals. On this basis, refuse compares favorably to a subbituminous A coal. Fresh wood contains between 25 and 50% moisture, whereas air dried wood contains between 10% and 15%. Consequently, air dried wood also compares favorably to a subbituminous A coal. Manure varies between 8% and 37% moisture, comparing favorably with subbituminous coals. Peat is typically

Lignocellulose	α-Cellulose %	Lignin %	Hemicellulose %
Jute	c. 65	19–20	15
Manila hemp	65	14	20
Esparto grass	65	20.0	15
Rye straw (Norman)	55	19.5	32
Oat straw (Norman)	53	18.5	32

Table 4 Composition of Various Agricultural Materials[28]

Parameter	Results
% Carbon	44.96%
% Hydrogen	5.78%
% Nitrogen	2.42%
% Oxygen	42.53%
% Sulfur	Not detectable
% Chlorine	0.29%
% Ash	4.02%
% Moisture	0.55%
Btu/lb.	7215 Btu/lb.

Table 5 Ultimate Analysis of Red Corn Cob[38]

	Experimental Feedlot Manures		Commercial Feedlot Manures
	A	B	
Moisture content (wt %)[a]	9.6	7.6	15.3–36.7
Ultimate analysis (wt %)[b]			
Carbon	42.6	39.7	35.1–39.6
Hydrogen	5.5	4.8	5.3–5.9
Nitrogen	2.8	2.9	2.5–3.1
Sulfur	. . .	0.5	0.4–0.6
Ash	24.9	26.6	23.5–29.2
Heating value (Btu/lb)[b,c]	6350	5860	5750–6730

[a] As received basis.
[b] Dry basis.
[c] Gross.

Table 6 Characterizations of Feedlot Cattle Manures [39]

35% or more moisture and ranks with lignite as a relatively wet fuel. Oil shale has little or no moisture associated with it.

The sulfur content of U.S. coals varies between 0.2% and 7.0%. Low-sulfur coal (less than 1% sulfur) is found primarily in the western U.S. The sulfur content of refuse and wood is very low and varies between 0.0% and 0.1%, agricultural residues contain no sulfur, and manure contains 0.3% sulfur on a *dry* basis. These fuels clearly outclass coal as a low-sulfur fuel. With a sulfur content of 2.3% on a *dry* basis, only sewage sludge remotely resembles coal in pollutant value. Peat and oil shale are also low in sulfur; however, both have significant amounts of nitrogen and are potential sources of NO_x pollution. Manure and agricultural residues also suffer from this problem.

The volatile matter content of a solid fuel is a good indicator of the ease with which the fuel can be converted to a liquid or gaseous fuel by thermochemical processes. On a dry basis, the volatile matter content of U.S. coals is less than 50%. Most organic materials are composed of 70% to 80% volatile matter, and some contain more than 90% volatiles. Consequently, the

Constituent	% by Weight
Moisture —	96% as raw thickened sludge
	80% as sludge cake
Carbon	44.0 (dry basis)
Hydrogen	5.6
Oxygen	14.3
Nitrogen	2.8
Sulfur	2.3
Chlorine	1.0
Ash	30.0

Table 7 Typical Analysis of Sewage Sludge [40]

volatile matter content of biomass suggests that it should be a much better feedstock for pyrolytic conversion processes than coal. Peat has a volatile matter content of about 70% (moisture, ash-free basis), falling between coal and biomass in ease of volatilization. Oil shale contains only about 10% volatile matter and is poorly suited for synfuel production.

Finally, among common solid fuels, biomass has a very low ash content. This low ash content contributes significantly to the ease of utilization of biomass materials.

In summary, the following is a ranking of the *intrinsic* suitability of common solid fuels for conversion to synthetic liquid and gaseous fuels by thermal processes:

1. biomass (most suitable for thermal conversion),
2. peat,
3. lignite,
4. coal,
5. oil shale (least suitable for thermal conversion).

Recognizing biomass to be the progenitor of peat, peat the progenitor of lignite, lignite the progenitor of subbituminous, bituminous, and semibituminous coals which are ultimately the progenitors of anthracite, it is not surprising to find biomass the most favored feedstock for synthetic fuel production, followed by peat, lignite, subbituminous, bituminous and semibituminous coals, and finally anthracite. The only rationale for considering oil shale as a conversion feedstock is the fact that conversion results primarily in the formation of a liquid hydrocarbon fuel at low temperatures. Apart from the liquid form of the primary conversion products, oil shale is the least desirable solid fuel available in the U.S. today.

Goals

In recent years, many reviews,[37,41,42] books,[43-49] and journals[50] emphasizing biomass conversion have appeared. In spite of this wealth of material, certain specific questions regarding biomass pyrolysis chemistry remain unanswered. These questions, and their answers, are the focus of this review.

Results of the comparison of biomass with conventional fossil fuels given above lead one to inquire why pyrolytic processes have not been widely adopted as a means for converting biomass into more useful fluid fuels. The answer involves two important observations: (1) the pyrolysis of biomass typically results in a distribution of solid, liquid, and gaseous products and (2) as fuels, these products presently command a low value in the marketplace. To the extent that the specificity of pyrolytic processes can be improved, emphasizing the formation of high value products, their role in the marketplace will grow. If pyrolytic processes continue to emphasize the production of low value char, "oil," and gas, their role in the marketplace will remain small.

The primary goal of this review is to detail the complex pyrolysis chemistry of biomass materials in order to identify chemical pathways which selectively lead to the formation of a very limited number of high value products. A second goal of this review is to translate the sometimes obscure research of the cellulose, carbohydrate, wood, and food chemists into a language which can be better appreciated by the engineering community. A detailed knowledge of the intricate pathways of biomass pyrolysis is prerequisite to the development of improved pyrolytic processes by the engineering community.

Although these goals have not been emphasized by earlier reviewers, their value is well known to the community of scientists concerned with biomass pyrolysis. For example, to summarize the SERI-sponsored Specialist's Workshop on Biomass Pyrolysis,[51] Diebold[52] presents in Fig. 5 a global reaction scheme for biomass pyrolysis. Pathways leading to olefin formation, identified in Fig. 5, are of great interest to SERI scientists[52] and other researchers.[53] The goal of this review is to present a series of pictures, similar to Fig. 5, which identify the detailed mechanisms of biomass pyrolysis in order to uncover new combinations of process conditions which lead to the selective production of a few high value products.

Scope

Typically, two genre of reviews exist in the literature, the encyclopedic review and the normative review. This review is normative in that it attempts to identify major concepts or principles, and the research from which those principles evolved. Because the review is not encyclopedic, no attempt has been made to cite all the papers that have appeared on biomass pyrolysis during recent years. In addition, because the context of this review is pyrolytic fuel production from renewable resources, little attention has been given to oxidative phenomenon.

Part 1 of this review presents a brief discussion of the experimental methods used to study biomass pyrolysis phenomena, followed by a review of the pyrolysis of carbohydrate materials. The pyrolysis of lignin and whole biomass materials, together with commercial methods of thermochemical conversion, will be covered in Part 2 (to appear in Volume 3 of *Advances in Solar Energy*).

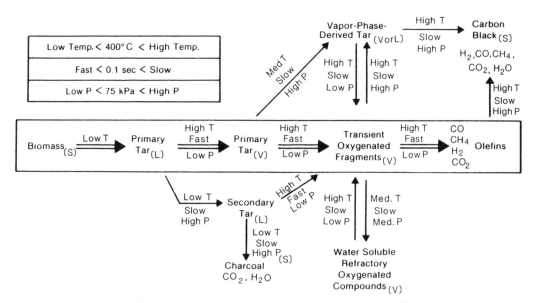

Fig. 5. Overview of the Pathways of Biomass Pyrolysis.[52]

EXPERIMENTAL METHODS

A surprisingly large variety of techniques have been used to study the pyrolysis of biomass materials. These techniques can be conveniently categorized in terms of their data-handling capability. Some provide data on an essentially continuous basis, whereas others generate data intermittently. Furthermore, some methods are aimed at the study of primary pyrolysis phenomena, whereas others focus on secondary pyrolysis phenomena. Each approach manifests some strength in the acquisition of certain data but often gains this strength from compromises made in other areas. The following subsections briefly review the various experimental methods used to study biomass pyrolysis and they attempt to identify both the strengths and the weaknesses of those methods.

Discontinuous Methods

Discontinuous methods usually attempt to heat biomass materials rapidly to a chosen temperature. After a chosen period of time at that temperature, the pyrolysis reactions are frozen by rapidly cooling the sample and apparatus. Following the experiment, the products are collected and subject to extensive analysis.

The chief advantage of this approach is the opportunity it provides to perform extensive quantitative chemical analysis of the products following the experiment. A second advantage is its ability to accommodate conditions (high pressures, temperatures, and/or heating rates) which would be difficult to effect using some other method.

The chief disadvantage of the approach is that the data often evidence significant scatter, perhaps reflecting some difficulty in controlling the extraordinary conditions often studied using these methods. The data scatter necessarily introduces ambiguities into the conclusions drawn from the experiments. A second weakness is the impossibility of instantaneously heating the sample to a preselected temperature and instantaneously cooling it after a chosen period of time. The finite rise and quench times can introduce nonnegligible errors into rate measurements and product distributions for materials as reactive as cellulose and hemicellulose. A third weakness is the method's inability to decouple the effects of the primary (solid-phase) from the secondary (gas-phase) pyrolysis phenomenon. Thus, the products obtained from discontinuous studies of primary pyrolysis phenomenon can manifest the effects of gas-phase pyrolysis as well.

It should be understood that the foregoing caveats do not necessarily apply to every apparatus discussed in ths subsection. Nevertheless, these caveats offer some insight into the compromises which are often made in the design of this class of experiments.

Primary Pyrolysis Studies

The simplest apparatus for the study of biomass pyrolysis involves a Pyrex or quartz reactor vessel, containing a weighed sample of material. The material is introduced into a furnace at a preselected temperature. After a chosen period of time, the reactor vessel is removed, allowed to cool, and weighed. Alternatively, the biomass may be introduced and removed from a preheated vessel after a chosen period of time. The vessel may operate at vacuum or provide for a flow of gas over the sample. If a gas flow is incorporated into the reactor design, it may be used to carry volatile matter derived from the pyrolyzing solid into a higher temperature flow reactor designed to study secondary pyrolysis phenomena. Occasionally, isothermal weight loss is continuously monitored using thermogravimetry.

Many researchers [54-63] have incorporated this generic apparatus in its simplest form into the design of incisive experiments which have offered detailed insight into the phenomenon of cellulose pyrolysis. Lipska and Parker[64] introduced an elegant variation into this class of apparatus by using a fluidized bed of sand to enhance heat transfer to the sample. Further improvements in heat transfer and high temperature capabilities were introduced by Howard and his colleagues [65,66] and Gavalas.[67] They wrapped the sample in a folded stainless steel screen and rapidly heated it by electricity to a chosen temperature for a selected period of time. This apparatus, used extensively in the study of coal pyrolysis,[68] has the unusual capability of acquiring pyrolysis data at very high pressures and temperatures.

To study the effects of extraordinary heating rates on solid phase pyrolysis phenomenon, a number of investigators [69-74] have employed flash radiative heating of the sample. This method emphasized the energy flux density to the sample, but sacrificed control of the sample temperature. The work of Hopkins et al.[74] illustrates this approach. They employed a Xenon flash tube, which produced a peak flux density of about 8,000 W/cm^2, to vaporize small samples of various biomass materials in vacuum. The use of radiative heating in vacuum minimized the effects of secondary reactions on the pyrolysis product distribution. The pyrolysis products were quantitatively collected and analyzed. Results from this experiment, and others in its class, have offered important insight into one high temperature pathway for cellulose pyrolysis.

Secondary Pyrolysis Studies

As noted earlier, volatile matter derived from primary pyrolysis of the biomass sample can be entrained by a flow of gas and carried into a higher temperature tubular reactor. The controlled variation of gas-phase temperature, residence time, pressure, and volatile dilution can be used to study the secondary (gas-phase) pyrolysis phenomenon of biomass-derived volatile matter.[75-82] Data from this class of reactors offer insight into the effects of reactor conditions on gaseous and condensable product distributions. Because the solid sample is introduced into the reactor in the batch mode and because gas flow in the reactor is almost always laminar, this class of reactor is known as a semibatch, tubular, laminar flow reactor (SBTLFR).

Apart from the weaknesses mentioned earlier, the SBTLFR has been criticized because the gas-phase residence time reflects the effects of laminar (as opposed to plug) flow. Also, the wide variety of free radical diffusivities could lead to a situation where the flow field was fully laminar for one class of species (such as $C_2H_5\cdot$) and plug flow for another (such as $H\cdot$). In addition, the well-known potential effects of wall reactions and departures of the reactor from true laminar flow have been conjectured to undermine the SBTLFR's ability to study gas-phase pyrolysis phenomenon. Fortunately, detailed studies, using a tubular laminar flow reactor, of the products and rates of pyrolysis of t-butyl alcohol and acetaldehyde (which are well known in the literature) have allayed many of these fears.[83] In all cases, good agreement has been found between the accepted rate and product data, and comparable rate and product data derived using a tubular laminar flow reactor.

A true weakness of the SBTLFR is the variation in concentration of the volatile matter in the carrier gas stream during the course of the experiment. This variation is due to the changing rate of pyrolysis of the solid sample, which acts as the source of volatile matter during the experiment, and is generic in this class of reactors. The changing concentration of volatile matter in the carrier gas stream considerably complicates potential studies of gas-phase pyrolysis mechanisms and reaction orders. Nevertheless, the SBTLFR has shed considerable light on the gas-phase pyrolysis phenomenon of biomass-derived volatile matter.

Continuous Methods

Continuous methods avoid many of the problems of discontinuous methods, but only by sacrificing some of their chief advantages. Many continuous methods of analysis heat the sample at a constant rate from a low temperature to a temperature sufficiently high to volatilize the sample. During the heating period, a few physical parameters (such as weight loss or heat uptake) or chemical properties (such as free radical concentrations in the sample or the rate of evolution of a species with a particular atomic weight) are continuously recorded. At the conclusion of the experiment, some of the products may be collected and subjected to further analysis.

The chief advantage of this approach is that the temperature/time history of the sample can be both well known and mathematically well characterized. No ambiguities are introduced into the experiment as a result of a poorly defined temperature/time history during heatup and quench, as is the case with the discontinuous methods discussed earlier. A second advantage is the high quality of the data obtained from an experiment. Three-significant-figure accuracy can be realized using these techniques. A third advantage is the continuous record of data obtained from an experiment. Thus, no ambiguities exist in the functional dependence of the data on time, and time derivatives of the data can be employed in the analysis of results.

The chief disadvantage of this method is its inability to provide measurements of the more complex physical and chemical changes which occur during an experiment. These measurements cannot usually be made in real time and often involve sequential analysis by complex chemical instrumentation. A second disadvantage is the inability of the method to easily accommodate high heating rates, pressures, and other extraordinary conditions. This weakness is sometimes hidden by the beguiling ability of data acquired under extraordinary conditions to appear accurate, when in fact it is flawed by imprecise temperature measurements at high heating rates or pressures, by slow response time of the instrument, or by some hidden compromise in the experimental design. A third weakness of the technique is the inability of existing theory to interpret the results of nonisothermal experiments involving complex pyrolysis phenomenon.

Again, this reviewer emphasizes that the foregoing discussion is general in scope and does not necessarily apply in its entirety to each of the experimental techniques described in the following paragraphs.

Primary Pyrolysis Studies

In his ground-breaking research, Broido[84-86] relied on the evolving methods and instrumentation of thermal analysis to gain an insight into the mechanism and rates of cellulose pyrolysis. Thermogravimetric analysis (TGA), differential thermal analysis (DTA), and differential scanning calorimetry (DSC) have been the subjects of many reviews[87-93] and will be discussed here only briefly.

TGA involves continuous weighing of a sample that is enclosed in a furnace capable of being programmed to rise in temperature at a chosen, constant rate to a preselected final value. Instrumentation for DTA records the difference in temperature between the sample and an inert reference while the two are heated at a constant rate within a furnace. DSC instrumentation is similar to DTA; however, DSC provides a signal which can be calibrated to be a direct measure of the rate of heat evolution or consumption by the sample as it is heated. Instrumentation for TGA, DTA, and DSC is often very expensive, but can appear to be beguilingly simple in its operation. However, only a few instruments (perhaps no more than two) are suited for detailed kinetic studies, and data obtained from these instruments requires considerable insight and scrutiny if it is to be used to develop a kinetic model.

In addition to the research of Broido,[84-86] the research of Shafizadeh,[54] Madorsky,[55-57] Basch and Lewin,[60,94,95] Halpern and Patai,[96] Tang,[97,98] and Arseneau[99,100] has employed conventional methods of thermal analysis in their studies of biomass pyrolysis. Antal[101] modified a standard Dupont 951 Thermal Gravimetric Analyzer to accept steam as a working environment and later

embedded a pair of microreactors in a DSC to study biomass pyrolysis at elevated pressures.[102] At the present time, the utility of thermal analysis instrumentation is limited by difficulties encountered in interpreting nonisothermal kinetics; however, some progress in this area has been reported.[103,104]

The novel use of an ESR spectrometer to follow the formation of free radicals during the low temperature pyrolysis of cotton cellulose was described by Arthur and his colleagues.[105-108] Similar studies of the solid state transitions of lignin using NMR and IR instruments have been reported by Hatakeyama and his colleagues.[109,110] In these studies, measurements of the spectra were made in a discontinuous fashion; however, the temperature resolution was sufficiently good to warrant their inclusion in this section.

Secondary Pyrolysis Studies

With a few exceptions, continuous methods for the study of secondary pyrolysis phenomena resemble the discontinuous methods discussed earlier. The chief difference between the two methods is the selection of a detector capable of continuously analyzing the effluent of the gas-phase reactor for a selected product or physical property.

McCarter[80,81] reported the use of a combustible gas detector together with a mass-flow meter and an infrared, water vapor analyzer to continuously monitor the effluent of a high temperature SBTLFR. Similar studies were described by Baker[82] using infrared CO and CO_2 detectors. Shafizadeh has studied the effective heat content of various forest fuels[111] using an instrument which detects the evolution of combustible carbon carried by the volatile matter evolved from pyrolyzing biomass.[112] His laboratory has also reported the development of a new combustible gas detector which measures the amount of oxygen required to achieve complete combustion of a stream of pyrolysis gases.[113]

A different approach to the study of gas-phase pyrolysis phenomena involves a departure from the use of batch reactors. Milne[114-116] has described the early results of exciting research on the pyrolysis of various biomass materials in steam produced by a stoichiometric hydrogen/oxygen flame. A sophisticated MS system was used to continuously detect and study the pyrolysis products, and provisions are being made to continuously introduce biomass into the flame.

Similarly, Deglise, Lede, and their colleagues[73] have reported the results of a continuously fed, radiant flash pyrolysis reactor, which subjects the evolved volatile matter to higher temperature, gas-phase cracking conditions prior to quench and product analysis. Because this work does not include provisions for real time analysis of the pyrolysis products, it might also be classed as a discontinuous technique. Finally, Hopkins et al.[117] have described the development of a continuously fed, spouted bed, radiant flash pyrolysis reactor which is able to decouple the gas-phase temperature from the pyrolyzing solid's temperature by as much as 200°C. Volatile matter evolved from the hot, rapidly pyrolyzing solids is immediately quenched and carried out of the reactor. The ability of this reactor to preserve the integrity of very reactive anhydrosugars evolved as the primary products of cellulose pyrolysis has been proven; consequently, the reactor has the potential to serve as a continuous source of primary volatile matter. Reactors of this kind may enable the development of continuous, tubular, laminar flow reactors, which have the potential of circumventing the problems of the SBTLFR discussed earlier and providing a good insight into the mechanisms of the gas-phase pyrolysis reactions.

Hybrid Methods

Apart from expense, there is nothing to prohibit the hybridization of discontinuous and continuous methods of analysis. For example, a combination TGA/DTA/FTIR/MS system could shed considerable light on the products, mechanisms, and rates of the primary, solid phase, pyrolysis reactions. Similarly, a SBTLFR/FTIR could offer better insight into the gas-phase pyrolysis chemistry of biomass-derived volatile matter. To this reviewer the development of such methods constitutes one of the most promising fields of future research.

PYROLYSIS OF CELLULOSIC CARBOHYDRATES

Because cellulose comprises about 50% by weight of most biomass materials, its pyrolytic properties command special attention. A wealth of knowledge on cellulose pyrolysis exists in the literature of the combustion community and has been summarized in several earlier reviews.[118-125] In this review, the prevalent pyrolysis temperature is emphasized as a means for understanding the hierarchy of cellulose-pyrolysis phenom-

ena. Below 250°C, volatilization of the cellulose polymer is slow, and its pyrolysis behavior is affected markedly by the substrate's fine structure. Above 250°C, the cellulose polymer begins to decompose rapidly, forming condensable sirups,[126] in addition to the carbon dioxide, carbon monoxide, and water obtained at lower temperatures. Above 500°C, the volatile sirups begin to undergo gas-phase pyrolysis, producing a variety of light, permanent gases. No clear-cut demarcation of phenomena occurs at these particular temperatures; nevertheless, they provide a convenient means for grouping and discussing the dominant phenomena.

Low Temperature Phenomena

As will be discussed later, the presence of trace quantities of extraneous materials usually has a pronounced effect on the observed pyrolysis phenomena of cellulosic materials. With few exceptions, the research reviewed in this section emphasized studies of very pure cellulose, attempting to exclude the catalytic influence of trace quantities of ash or other extraneous material.

Aside from the gross influence of extraneous materials on pyrolysis, other effects are manifest at low temperatures which can be masked by rapid weight loss at higher temperatures. The rapid drop in the degree of polymerization (DP) of cellulose during pyrolysis, and the effects of fine structure and cross linking on pyrolysis rates and products,[127] offer a special insight into the pyrolysis mechanism. This would be difficult to gain from the results of higher temperature experiments. The following sections discuss these effects in detail.

Products

Below 250°C, the major products of cellulose pyrolysis are H_2O, CO, CO_2, and char.[54-57,127,128] Even at temperatures as low as 70°C, cellulose decomposes,[129] generating a gas composed of 90 to 95% CO_2 and 5 to 10% H_2 (by volume). Major[128] and Shafizadeh[54] have measured the rates of formation of CO and CO_2 in N_2 and air at 170°C. Shafizadeh[54] showed that in air the rate of glycosidic bond scission (DP reduction) approximately equaled the rate of formation of CO plus CO_2 in moles per glucose unit. But in N_2, the rate of bond scission greatly exceeded the rates of CO and CO_2 evolution.

Measurements of the formation of carbonyl and carboxyl groups in the cellulose substrate by

Major[128] showed no evidence of carboxyl groups and a constant carbonyl group content after 20 hours for pyrolysis in N_2. Similarly, Shafizadeh[122] studied the pyrolytic decarboxylation of carboxylcellulose (carboxyl groups substituted at C1, C2, C3, and C6). the sample oxidized at C2 and C3 evidenced decarboxylation at 190°C, but otherwise decarboxylation was not observed. These findings would seem to imply that the formation of CO_2 may not be the result of a simple decarboxylation reaction.

At somewhat higher temperatures in vacuum, Madorsky and his co-workers[55-57] found that the rates of formation of H_2O, CO, and CO_2 gradually decreased. Moreover, the rate of sirup formation gradually increased until 20% weight loss was achieved; whereupon, the relative proportions stabilized (see Table 8). These results are in accord with the findings of Golova,[130] who observed little levoglucosan formation (5 to 20%) during the first 5 to 10% weight loss when the cellulose DP was reduced to about 200. After this, a higher yield of levoglucosan was obtained.

Careful studies by Patai and Halpern[131] of the pyrolysis of four different cellulose samples showed that the residue remaining after 20 to 30% weight loss was composed primarily of oligosaccharides. From this they concluded that the early stages of weight loss do not involve a destruction of the carbohydrate structure of cellulose.

Experiment	Temperature	Weight of sample	Duration For step	Duration Cumulative	Volatilization For step	Volatilization Cumulative	V_{pyr} (Tar)	V_{25} (H_2O)	V_{-80} (CO_2)	V_{-190} (CO)
	°C	mg	min	min	%	%	%	%	%	%
					Cotton					
1	250	49.1	300	300	2.7	2.7	26.3	53.9	12.0	7.8
2a	280	83.3	38	38	3.3	3.3	33.9	50.7	10.5	4.9
2b	280	83.3	58	96	7.0	10.3	55.9	36.1	6.0	2.0
2c	280	83.3	53	149	9.1	19.4	63.8	29.6	5.4	1.2
2d	280	83.3	100	249	15.9	35.3	. . .	. . .	. . .	. . .
2e	280	83.3	77	326	9.8	45.1	70.2	23.6	4.7	1.5
2f	280	83.3	180	506	16.7	61.8	. . .	. . .	. . .	. . .
2g	280	83.3	155	661	7.8	69.6	70.8	21.9	5.5	1.8
3	321	39.5	9	9	22.0	22.0	65.2	28.2	4.3	2.3
4	397	13.8	159	159	91.4	91.4	69.6	22.5	6.2	1.7
					Hydrocellulose					
5a	280	80.0	35	35	4.3	4.3	40.6	49.7	6.7	3.0
5b	280	80.0	86	121	8.1	12.4	59.0	32.3	6.9	1.8
5c	280	80.0	72	193	9.3	21.7	62.4	31.5	4.7	1.4
5d	280	80.0	130	323	17.3	39.0	. . .	. . .	. . .	. . .
5e	280	80.0	120	443	13.1	52.1	73.7	20.9	4.0	1.4
5f	280	80.0	185	628	13.5	65.6	76.7	18.0	3.7	1.6
					Viscose Rayon					
6a	280	85.6	7	7	3.3	3.3	19.6	61.7	11.7	7.0
6b	280	85.6	19	26	3.7	7.0	24.4	59.1	10.5	6.0
6c	280	85.6	22	48	5.3	12.3	39.5	48.4	8.7	3.4
6d	280	85.6	24	72	6.2	18.5	51.8	37.9	7.8	2.5
6e	280	85.6	65	137	16.3	34.8	. . .	. . .	. . .	. . .
6f	280	85.6	32	169	6.2	41.0	52.6	37.1	8.1	2.2
7	295	12.5	57	57	45.3	45.3	48.2	40.8	7.8	3.2
				Cotton Impregnated with Na_2CO_3 (7%)						
8a	251	10.0	180	180	36.8	36.8	6.1	61.0	19.6	13.3
8b	291	10.0	30	210	14.6	51.4	15.3	54.6	22.7	7.4
				Cotton Impregnated with NaCl (8%)						
9	280	7.1	120	120	49.3	49.3	12.2	59.3	14.9	13.6

Cotton impregnated with Na_2CO_3 degraded initially so fast that it was not convenient to start experiment at 291° C. Instead, the sample was first heated for 180 min at 251°, with a resulting loss of 36.8%; the residue was then heated for another 30 min at 291°, with a further loss of 14.6%. Total loss in weight was 51.4%.

Table 8 Pyrolysis of Cellulose in a Vacuum

• Reduction in DP

In 1957, Golova and her co-workers[130,132] reported that the DP of cellulose undergoes a rapid drop from above 1,000 to a steady state value of about 200 during the early stages of pyrolysis. Paucault and Sauret[133] confirmed these findings and showed that the reduction in DP follows a zero order rate law with an activation energy of 25.8 kcal/g mol. Paucault and Sauret speculated that the bond scissions occur at the boundaries between the crystalline and amorphous regions of the cellulose. Eleven years later, Fung[134] reported similar findings but concluded that the reduction in DP followed a first order rate law with an activation energy of 35.4 kcal/g mol. However, as noted by Broido,[135] the data of Fung, Paucault and Sauret, and Golova may have been flawed by the presence of ash in their samples of cellulose.

Working with relatively pure cellulose, Halpern and Patai[58] found that at temperatures between 250° and 300°C the DP dropped from 1,800 ± 180 to 200 very quickly, and continued to fall to 160 with a 10 to 15% weight loss. Thereafter, the DP remained constant to 60% weight loss. Broido et al.[136] extended these studies, showing that during heating, Whatman No. 541 filter paper exhibited a drop in DP from 2,650 to 600 at 175°C with no weight loss. The DP further declined to 375 with a 1.02% weight loss, and to 205 with a 15.5% weight loss. Similar studies with an amorphous (decrystallized) cellulose revealed only a small decrease in DP at 175°C. However, as the decrystallized cellulose regained crystallinity at 175°C, its DP dropped more rapidly and ultimately stabilized at a lower value than was obtained with the neat Whatman No. 541 filter paper. Because the decrease in DP of the decrystallized cellulose followed the development of crystallinity, Broido et al.[136] concluded that bond rupture occurs at points of maximum strain in the polymer along the boundary where the crystalline and amorphous regions meet (see also Ref. 137).

Apparently in disagreement with Broido,[136,137] Shafizadeh and Bradbury[54] attribute the rapid decrease in cellulose DP during pyrolysis to a random bond scission process. Their measurements indicate that in N_2 the bond scission reaction initially follows a zero order rate law with an activation energy of 27 kcal/g mol.

This is in good agreement with the work of Paucault and Sauret.[133]

• Effects of Crystallinity

Based on theoretical considerations, Kilzer and Broido[84] argued in 1965 that dehydration of the cellulose polymer would occur preferentially in the crystalline regions. Five years later, Broido and Weinstein[59,138] showed that decrystallized cellulose formed less char than ordinary cellulose[138] and that its initial rate of weight loss exceeded more crystalline cellulose. This finding was in agreement with Patai and Halpern,[131] who observed more rapid weight loss with completely noncrystalline, ball-milled cellulose than more crystalline samples. Because the measured crystallinity index (Cr I) of cellulose showed little drop even when weight loss reached 59%, Weinstein and Broido[59] argued that unzipping of the polymer must proceed along chains through both the ordered and disordered regions of the material.

Halpern and Patai[58] also found that cellulose crystallinity remains constant to 60% weight loss. From this they concluded that the crystalline regions of the material were the major source of levoglucosan. In agreement with Weinstein and Broido,[59] Patai and Halpern[131] noted that unzipping can occur in chains which are located in amorphous regions of the cellulose. Based on pyrolysis experiments at 150°C, Shimazu and Sterling[139] also concluded that the low temperature breakdown of the cellulose chains occurs primarily in the amorphous regions.

Basch and Lewin[60] confirm this point of view, showing that the initial rapid weight loss due to dehydration of the cellulose polymer by cross linking reactions correlates with the percentage of less ordered regions (% LOR = 100 − Cr I). When the rate of pyrolysis at 251°C in vacuum was plotted against $(\% \text{ LOR})(f_o)$ $(DP)^{-1/2}$, where f_o is the cellulose orientation factor, a straight line was obtained with a correlation coefficient R of 0.94 for most of the samples studied. Moreover, Basch and Lewin,[60] in agreement with Golova[130,132] and Halpern and Patai,[58] showed that no levoglucosan was found in the condensate collected after 150 min with 4% weight loss of the starting cellulose. The activation energy also showed some correlation with crystallinity, reaching a projected value of 61.2 kcal/g mol for crystalline cellulose and 29.6 kcal/g mol for the noncrystalline material.

Basch and Lewin associate the high activation energy with levoglucosan formation in the crystalline regions.

Basch and Lewin[60] used their data to support the pyrolysis mechanism of Chatterjee and Conrad,[140,141] who posited an initiation reaction (random chain cleavage) followed by a propagation reaction (stepwise depolymerization) as the pyrolysis mechanism. A plot of (weight loss)$^{1/2}$ versus time resulted in a straight line as suggested by the mechanism of Chatterjee and Conrad.[140,141] The initiation step exhibited an activation energy of about 30 kcal/g mol and appeared to be rate limiting at the low temperatures used in these studies. Extrapolation of the (weight loss)$^{1/2}$ versus time plot to zero time resulted in nonzero values for the initial rate of weight loss, which were proportional to the % LOR of the sample. However, Basch and Lewin[60] appear to have assumed the mathematical identity $A^2 - B^2 \overset{?}{=} (A - B)^2$ in their treatment of the data, which would lead to errors in their subsequent analysis of the results. If correct, the supposed proportionality between the initial rate of weight loss and the % LOR of the sample would seem to offer further support for the hypothesis that the initial weight loss is due to cross linking in the amorphous regions of the cellulose.

Kato and Kamorita[142] offer an interesting insight into the effects of crystallinity on the formation of volatile compounds. Comparing the pyrolysis products of crystalline and amorphous cellulose between 200 and 300°C, the yields of furfural and 5-hydroxy-methyl furfural were observed to increase more rapidly with time during the pyrolysis of the amorphous cellulose sample, whereas at 500°C higher yields of acetaldehyde were obtained from the crystalline cellulose. These results provide further evidence for the action of different reaction mechanisms during the pyrolysis of crystalline and noncrystalline cellulose.

• *Influence of Orientation*
Continuing their extensive studies of the effects of fine structure on cellulose pyrolysis, Basch and Lewin[61,62] investigated the effects of orientation f_0 on the rate of pyrolysis.[62] The cellulose samples studied differed only in their degree of stretch (orientation). Basch and Lewin found that the rate of the initial reaction, described as thermal cross linking in the LOR, increased with increasing orientation according to a first order

rate law. However, the extent of reaction decreased. To explain these results, Basch and Lewin[62] note that with low orientation the distance between chains is comparatively large; hence, the rate of cross linking is low. However, with low orientation at elevated temperatures, the chain is more free to align itself to favor cross linking, thus increasing the overall extent of the cross linking reaction.

Basch and Lewin[62] show that the char residue formed from the fast initial weight loss reaction is given by 1.6 times the initial weight loss. A second source of char formation in the LOR is the slower bulk decomposition due to chain scission (discussed earlier). The rate of this reaction increases with increased orientation.

In a parallel study, Basch and Lewin[143] established that the degree of increase in the measured crystallinity of cellulose during thermal annealing at 200°C in vacuum is inversely related to orientation. Samples with the lowest orientation evidenced the greatest rise in crystallinity during the annealing process. Other researchers[131,139] have also reported a rise in the crystallinity of cellulose samples during heating.

• *Effects of Cross Linking*
Back and his co-workers[144,145] established the importance of rapid thermal auto cross linking between 230° and 320°C as a thermal hardening mechanism. This counteracts the thermal softening of cellulosic fibers due to second order transitions at 25° and 210°C. Back[145] associates an activation energy of 20 to 30 kcal/g mol with the cross linking reaction, in rough agreement with the findings of Basch and Lewin.[60]

Rodrig, Basch, and Lewin[94] artificially cross linked cellulose fibers using formaldehyde and found that formaldehyde cross linking enlarged the distances between chains in the LOR. This reduces the opportunity for thermal auto cross linking and results in a lower initial weight loss and a smaller char residue.

The mechanism of the cross linking reactions remains an open question. In 1966 Arthur and Hinojosa[105] reported the formation of free radicals in cellulose heated to 250°C. Later studies[106-108] of the irradiation of cellulose with γ radiation at 23°C in vacuum and air evidenced chain cleavage, the formation trapped free radical sites, and the evolution of H_2, CO, and CO_2. Arthur[108] speculates that cross linking could occur following the cleavage of a hydrogen atom from a carbon atom. The free hydrogen atom could strip a hydrogen atom from an adjacent chain, leaving two chains with highly reactive sites that could interact to form an intermolecular bond or cross link.

The findings of Arthur have received support from the research of Hon,[146,147] who detected the formation of free radicals in cellulose irradiated with ultraviolet light ($\lambda < 2800$ Å). Scission of the polymer chain was also observed, resulting in the formation of H_2, CO, CO_2, and H_2O.

In contrast with these studies, Reeves,[148] Mehta,[149] Parikh,[150] and Barker and Vail[151] all attribute the formation of cross links in cellulose to a carbonium ion mechanism. Although there is some dispute regarding the details of the mechanism,[149,150] these scientists do not consider any other explanation for their results.

• *Char Formation*
Because the rates of formation of carbon-rich char are very slow below 250°C, a discussion of char formation will be discussed later in this review.

Summary and Critique
The many contributions of Basch and Lewin to the field of cellulose pyrolysis have been reviewed in Refs. 95 and 127. Unfortunately, no similar review of the elegant research of Broido is available, and there seems to have been no resolution of the contrary findings of the two laboratories. One feels some concern that the decrystallized ("swelled") cellulose used by Broido in his studies of the effects of crystallinity

on cellulose pyrolysis [59,84,136-138] may have introduced extraneous, hidden variables into the problem whose influence may not have been recognized. For example, did the swelling procedure increase the distance between chains, and thereby artificially reduce the rate of cross linking reactions in the decrystallized cellulose, as was observed to occur with the formaldehyde-cross-linked cellulose?[94] Studies of the weight loss behavior of Avicel PH 102 microcrystalline cellulose and a variety of grades of Whatman filter paper in my laboratory[101,102] reveal significantly more weight loss by the microcrystalline cellulose than the less crystalline filter paper. This result supports the findings of Basch and Lewin.

A resolution of this issue would be an important contribution to the field, since a knowledge of the influence of crystallinity on levoglucosan formation could justify a search for pretreatment processes which increase (or decrease, if Broido's conjecture is correct) the crystallinity of cellulose in order to realize higher yields of levoglucosan.

In 1963, Broido and Kilzer[135] prepared an incisive review of prior research on cellulose pyrolysis. They note that the dramatic effects of ash concentrations exceeding 0.1% call into question the results and conclusions of Golova and all other researchers who did not work with very pure cellulose. It is interesting to speculate on whether the presence of ash may influence the mechanism of pyrolysis by enhancing the role of free radical reactions over carbonium ion chemistry (or vice versa). A resolution of this issue would clearly be of benefit to the field.

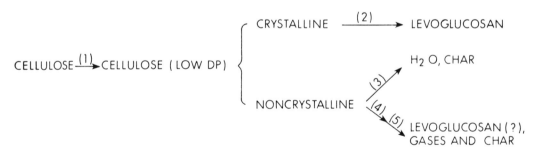

(1) Sharp drop in DP to 200 preceding weight loss and occurring at crystalline boundaries, $E \cong 27$ kcal/gmol, zero order reaction
(2) Levoglucosan peels off crystal face, $E \cong 60$ kcal/gmol
(3) Fast thermal auto cross linking, char = 1.6 times the initial weight loss
(4) Rate-limiting initiation step of slower bulk reaction, $E \cong 30$ kcal/gmol
(5) Propagation step of slower bulk reaction, char = 20% of noncrystalline mass

Fig. 6. Low temperature pathways of cellulose pyrolysis.

Based on the foregoing discussion, Fig. 6 summarizes a consensus view of the low temperature mechanism of cellulose pyrolysis. Because a complete understanding of the degradation pathways has not yet been established, the mechanism portrayed in Fig. 6 should be regarded as tentative and subject to future modification.

Moderate Temperature Phenomenon

Above 250°C, cellulose rapidly undergoes complete degradation by pyrolysis, forming a variety of permanent gases, condensable liquids, and char. The complexity of the pyrolysis phenomena at moderate temperatures can mask some of the effects observed at lower temperatures, and makes great demands on an experimentalist's skills in trying to gain insight into the mechanisms and rate laws governing pyrolysis. The following sections emphasize major issues and points of agreement among various investigators, especially concerning the mechanisms and kinetics of levoglucosan formation.

Products

The products of cellulose pyrolysis are often characterized as primary or secondary, according to whether their immediate precursor was the cellulose substrate or another product of cellulose pyrolysis. Because the secondary products are primarily noncellulosic carbohydrates, a discussion of their pyrolysis chemistry will be deferred until later. The primary products of cellulose pyrolysis can be classified as volatile or nonvolatile, and each class has been the object

of considerable interest and scrutiny over the years.

• Volatile Products

As early as 1918, Pictet and Sarasen [152] identified levoglucosan as a major product of cellulose pyrolysis. More recently, many investigators have used GC, [153-160] GC-MS, [161,162] and other techniques [122] to identify the products of cellulose pyrolysis at moderate temperatures. Figure 7 summarizes the findings of Wodley [160] who isolated 59 compounds with molecular weights less than 150. However, Wodley was only able to identify 37 of the compounds. Because some of these compounds were aromatic, Wodley concluded that not all were primary products of cellulose pyrolysis.

Several of the investigators [154-160] note that the products of cellulose pyrolysis are essentially the same as those of levoglucosan pyrolysis. From this fact they conclude that the decomposition of cellulose probably results in the formation of levoglucosan as an intermediate, which further decomposes to yield the observed pyrolysis products. However, Wodley [160] notes that furfural alcohol, butyrolactone, phenol, o-cresol, m-cresol, p-cresol, 2,5-dimethylphenol, and one unidentified compound which were present in the products of cellulose pyrolysis, were absent from the product slate of levoglucosan pyrolysis. Since Wodley does not appear to have worked with very pure cellulose, the significance of these differences is not clear.

Five of the research teams [153,154,159-161,175] observed that the formation of identified pyrolysis

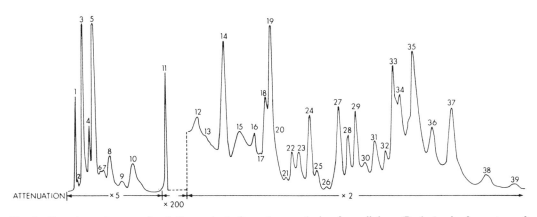

Fig. 7. Gas chromatogram of volatile products from the pyrolysis of ∝-cellulose (Peak 1: air; 2: pentene; 3: acetaldehyde; 4: furan; 5: acetone; 6: 2 Methylfuran; 7: butyraldehyde; 8: methyl ethyl ketone; 9: 2,5 dimethylfuran + methyl vinyl ketone; 10: 2,3-butanedione; 11: water; 12: cyclopentanone; 13: cyclooctatetrene; 14: acetolpyruvaldehyde; 18: acetic acid; 19: furfural; 23: 5 methyl-2 furfural; 24: furfural alcohol; 25: butyrolactone; 28: 2-hydroxy-3-methyl-3-cyclopenten-1-one; 30: phenol + O-cresol; 31: P-cresol + 2,5-dimethylphenol + M-cresol; 32: 3,4, dimethylphenol; 38: 5-hydroxymethylfurfural).

77

products is unaffected by temperature within the ranges studied. This finding offers further support to the hypothesis that levoglucosan may be the common intermediate in the formation of all volatiles obtained from cellulose pyrolysis. Further details of the product distribution will be given and discussed later.

• Nonvolatile Products

The solid carbonaceous residue of cellulose pyrolysis, known as char, has the representative elemental analysis[163] 86% C, 2.4% H, and 1.6% O. As early as 1937, Smith and Howard[164] noted that the char contains aromatic structures. More recently, Shafizadeh[165] has described the influence of temperature on the surface area and the concentration of free radicals in the char. Both values peak at temperatures between 500° and 600°C, and Shafizadeh notes that the char formed under these conditions is pyrophoric and highly reactive. Shafizadeh's observations are in harmony with the earlier, extensive studies of Rensfelt[166] who noted the extraordinary gasification rates of biomass chars relative to coal and peat derived chars. Reference 47 provides an extensive review of other properties of biomass derived chars.

The dependence of char yield on the temperature of pyrolysis was strikingly illustrated by the work of Broido and Nelson[163] who pretreated cellulose samples at temperatures ranging between 230 and 275°C prior to pyrolysis at 350°C. Depending upon the temperature and duration of the pretreatment, the char yield varied from 11.0 to 27.6% by weight. They showed that this behavior could easily be explained by a competitive degradation mechanism involving depolymerization and dehydration reactions. The depolymerization reaction (2) formed no char; whereas, the dehydration reaction (1) formed 36% by weight char. Furthermore, Broido and Nelson[163] showed the ratio of rates $k_1/k_2 = (2.05 \times 10^6) \exp(-8,900/T)$.

Mechanisms

At moderate temperatures the effects of fine structure on the mechanisms of cellulose pyrolysis can be masked by the very rapid transformation of the polymer into volatile matter and char.[55] Thus, the effects of crystallinity and orientation are not usually considered; however, the depolymerization step cannot be ignored because it can be rate limiting under some circumstances.

Shafizadeh[120] has postulated an excellent global model for the chemistry of cellulose pyrolysis, summarized in Fig. 8. As a global model, Fig. 8 merely outlines the pathways available for the thermal degradation of cellulose. It does not purport to fully detail the complex series of concurrent and consecutive reactions which occur during cellulose pyrolysis. Nor does it indicate the sensitivity of the various reaction pathways to the pyrolysis temperature and rate of heating, the ambient atmosphere and pressure, or the presence of additives or other impurities. Nevertheless, Fig. 8 provides an excellent context for the following discussion of reaction mechanisms.

• Reaction 1: Dehydration and Charring

Reaction 1 progresses more rapidly than reactions 2 and 3 at low temperatures, forming char, CO, CO_2, and H_2O from the cellulose substrate. The relationships between the rates of CO, CO_2, and H_2O formation and the rate of glycosidic bond scission were discussed above. Similarly, the role of dehydration reactions in the thermal auto cross linking of cellulose was also described above. The mechanism of these reactions may involve either free radical or carbonium ion intermediates, as reviewed earlier. A global rate law for the charring reaction 1 will be given in the following section.

• Reaction 2: Transglycosylation and Levoglucosan Formation

There are two viewpoints expressed in the literature concerning the mechanism of formation of levoglucosan. The first asserts that the glycosidic bonds are ruptured homolytically and that depolymerization proceeds by a free radical mechanism. The second regards the transglycosylation mechanism to be heterolytic with depolymerization proceeding by a carbonium ion intermediate. Both points of view are strongly supported by experimental evidence, as discussed in the following paragraphs.

■ Homolytic Bond Cleavage. In 1957, Golova[130] postulated that the decomposition of

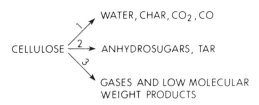

Fig. 8. Competing pathways for cellulose pyrolysis.[120]

cellulose proceeds in two stages: reduction in DP to about 200, followed by a chain process involving rapid decomposition of the polymeric macroradical by scission of the monomeric end unit. However, an attempt to detect the posited existence of a biradical precursor of levoglucosan failed.[167] Other experiments with the vacuum pyrolysis of tri-O-methylcellulose showed that blockage of the C-6 hydroxyl group stabilized the posited free radicals and lead to the formation of a 1, 4 anhydro ring.[168] Golova also conjectured that levoglucosan formation is favored in regions of high packing density.[169]

Madorsky[58] discussed the pyrolysis mechanism in terms of a random dehydration reaction along the chain, together with simultaneous thermal scissions of the C-O bonds in the chain to form levoglucosan, H_2O, CO, and CO_2. Based on his extensive experience with the thermal degradation of polymers, Madorsky[57] asserts that a high yield of monomer from a polymer chain indicates an unzipping sequence which is primarily the result of free radical reactions. Madorsky associates an absence of free radicals with the formation of various fragments and varying amounts of monomer.[170]

Further support for the role of free radicals in levoglucosan formation is given by Arthur and Hinojosa[105] who observed the rate of formation of free radicals at 250°C by studying the ESR spectra of cotton cellulose. Arthur and Hinojosa state that the ESR spectra are characteristic of spectra obtained from cellulose undergoing chain cleavage at the 1-4 glycosidic link and obtain a value of 33 kcal/g mol for the activation energy of the bond scission reaction.

Perhaps the most compelling evidence supporting the free radical mechanism was given by Kislitsyn et al.,[171] who introduced a free radical inhibitor (di-β-naphthylphenylenediamine) into the cellulose macromolecule and observed its effects on the product distribution. Kislitsyn found that the introduction of 2.25 mole % inhibitor reduced levoglucosan formation by more than 50%, and 4.5 mole % inhibitor suppressed formation almost entirely (less than 15% levoglucosan). From this the authors concluded that the rupture of the glycosidic bond is homolytic and that thermal depolymerization proceeds by a free radical mechanism. Their proposed free radical mechanism[172] is given in Fig. 9 which concurs with that proposed by Pakhomov[173] some years earlier. In a similar vein, Golova[174] showed that the addition of D-glucose to cellu-

lose reduces the yield of levoglucosan, presumably by blocking free radical centers.

The arguments for the role of a free radical mechanism in cellulose pyrolysis have been summarized in a superb, critical review of the literature by Golova,[175] which is available in English translation. Golova's critique of the arguments favoring heterolytic reaction mechanisms is beyond the scope of this review, but is strongly recommended to the interested reader. In this review, Golova hints of a possible differentiation of roles, with free radical reactions responsible for the formation of levoglucosan from the crystalline regions of cellulose and heterolytic reactions active in the formation volatiles from the amorphous regions. In view of the findings of Kato and Komorita,[142] this observation takes on special meaning.

■ **Heterolytic Bond Cleavage.** In a ground-breaking paper, Kilzer and Broido[84] proposed that the mechanism of levoglucosan formation involves a concerted displacement. This results in the formation of 1,4 anhydro-α-D-glucopyranose which quickly rearranges to form either 1,6-anhydro-β-D-glucopyranose (levoglucosan) or 1,6-anhydro-β-D-glucofuranose (see Fig. 10).

Shortly after the publication of Kilzer and Broido's work,[84] Gardiner[157] and Bryne et al.[158] described an extensive series of experiments on

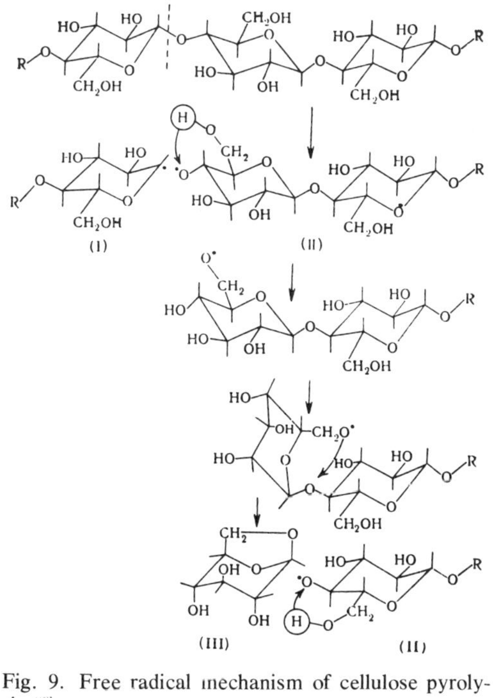

Fig. 9. Free radical mechanism of cellulose pyrolysis.[172]

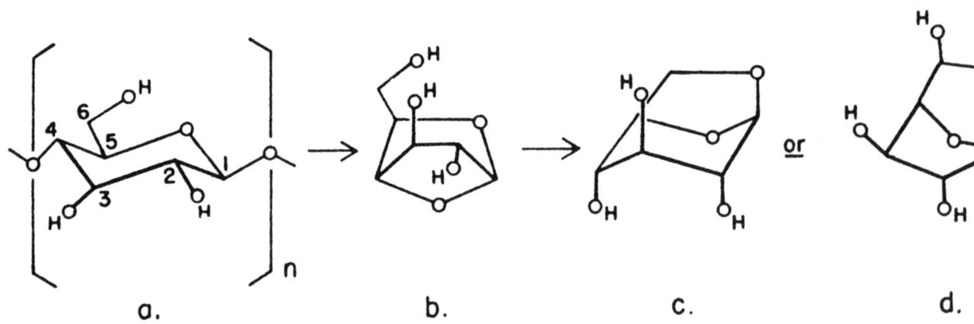

Fig. 10. Proposed mechanism for the depolymerization of cellulose.[84]
  A. — The Glucopyranose Monomer;
  B. — 1,4-Anhydro-α-D-Glucopyranose;
  C. — 1,6-Anhydro-β-D-Glucopyranose (Levoglucosan);
  D. — 1,6-Anhydro-β-D-Glucofuranose.

the pyrolysis of selected hexoses and derived di-, tri-, and polysaccharides including cellulose. Noting that the pyrolysis products resemble those derived from acid-catalyzed reactions in aqueous solutions, Byrne et al.[158] proposed two general pyrolysis modes (see Fig. 11). The first involves concerted displacements and the formation of an intermediate 1,2-anhydroglucose unit. This unit decomposes to form either 5-(hydroxymethyl) furfural, 1,6-anhydro-β-D-glucofura-

nose, or levoglucosan following an initial chair to boat (C1 to B1) conformation change of the original glucose unit. The second mode results in the formation of volatile carbonyl compounds or condensation products (char) by an irreversible carbonium ion mechanism.

Fig. 12 summarizes the exhaustive studies of Shafizadeh on the pyrolysis of cellulose [176-178] and a great variety of substituted glycosides, as well as the corresponding unsubstituted compounds. Shafizadeh asserts that the transglycosylation mechanism must be heterolytic because it was shown to be influenced by variations in the electron density produced through the introduction of various substituents on the aglycone.[122] In agreement with Gardiner,[157] Shafizadeh notes that the transglycosylation reaction requires a change in conformation of the molecule and increased flexibility. This could be achieved at higher temperatures by the breaking of hydrogen bonds and a glass transition.[122]

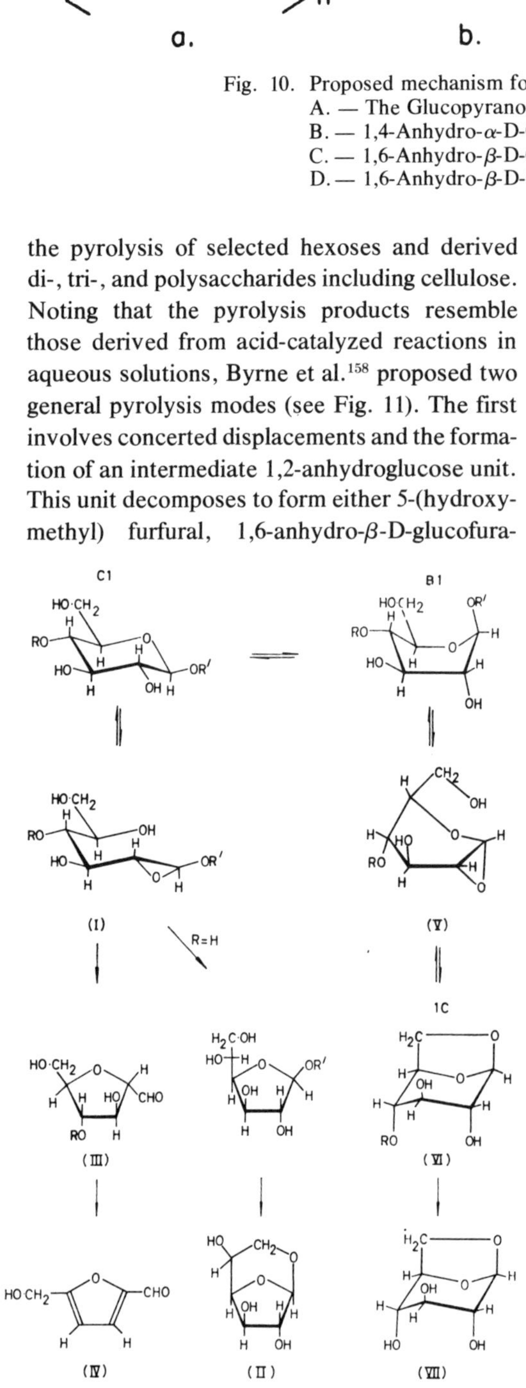

Fig. 11a. Mechanism of formation of anhydroglucoses and furfurals.[158]

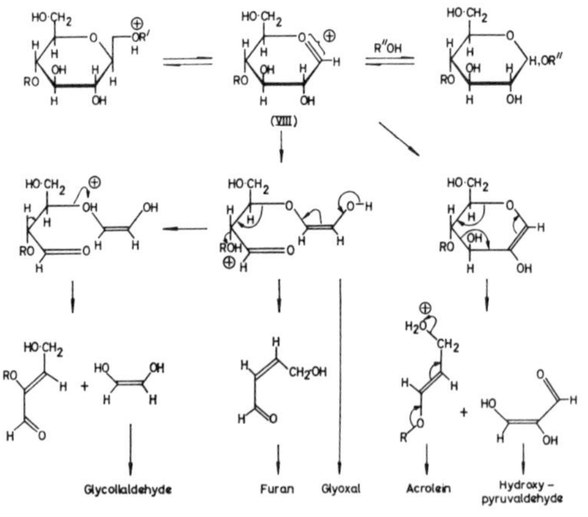

Fig. 11b. Mechanism of formation of carbonyl compounds.[158]

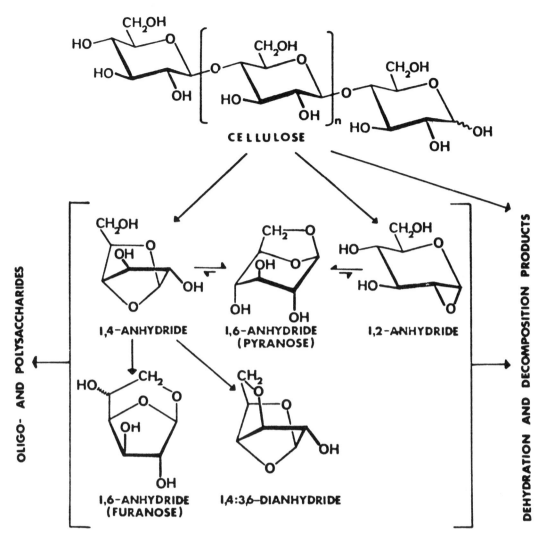

Fig. 12. Pyrolysis of cellulose to anhydro sugars and other compounds by transglycosylation reactions.[122]

■ Yields of Levoglucosan. Extensive studies by Shafizadeh et al.[177] indicate that the cellulose substrate has a very significant influence on the formation of sirups and levoglucosan. The pyrolysis of two lots of Whatman CF 11 cellulose powder in vacuum inexplicably resulted in yields of 29% levoglucosan and 58% sirup, and 39% levoglucosan and 69% sirup. Similarly, the pyrolysis of cotton hydrocellulose formed 58% levoglucosan and 85% sirup, whereas only 14% levoglucosan and 46% sirup was derived from a cotton fabric.[177]

Shafizadeh et al.[177] attempted to improve the yields by washing the cellulose materials in boiling distilled water, but met with no success. However, a dilute acid wash followed by distilled water washes to neutrality resulted in a dramatic improvement in sirup and levoglucosan yields. Shafizadeh et al. were unable to deter-

mine if the improvement was due to (1) the removal of inorganic impurities from the cellulose, (2) the addition of trace amounts of acid to the cellulose, or (3) some change in the crystallinity or fine structure of the cellulose, or all three.[177] In a subsequent paper, Shafizadeh et al.[178] show that pyrolysis of cellulose in 0.1 MPa of flowing nitrogen at 500°C reduces the sirup yield to 60% from the 82% obtained under vacuum.

■ DTA Data. DTA studies of cellulose reveal one major endotherm beginning at about 300°C (β = 10°C/min) due primarily to levoglucosan formation and volatilization, and a minor endotherm at 100°C due to desorption of water.[96] Many workers have noted an exotherm at higher temperatures, but this exotherm is probably the result of secondary reactions and not representative of primary pyrolysis phenomena.[99] Halpern and Patai[96] noted an inverse relationship between endothermicity and char formation, which has been confirmed by more recent research.[102] The DTA signal is also known to be influenced strongly by the presence of additives.[96]

• Reaction 3: Fission and Disproportionation
Evidence for this reaction (see Fig. 8) is less convincing than the wealth of data supporting the dehydration and transglycosylation reactions. In Ref. 120, Shafizadeh notes that even under the most ideal conditions, the evolution of levoglucosan is accompanied by the formation of light volatile products (mainly H_2O, CO, and CO_2). The persistence of these gases at higher temperatures and the maintenance of their molar ratios over a temperature range of 100° to 280°C is used by Shafizadeh[120] to posit the existence of an additional competitive pathway. However, these arguments would seem to provide further support for pathway 1 (dehydration and charring) discussed earlier.

Shafizadeh[120] also provides considerable insight into the fission and disproportionation reactions of levoglucosan, but this only offers support for reaction 4 (see Fig. 13) without shedding any light on the high temperature pathway. The different influences of basic and acidic catalysts on the product slate of cellulose pyrolysis can also be explained by their influence on the pyrolysis of levoglucosan.[122] Hence, these catalysts seem to affect reactions 4 and 5 (Fig. 13), without indicating the existence of a high temperature pathway.

The pyrolysis mechanism given in Fig. 8 has been included in this review because the author believes evidence supporting the existence of reaction 3 is available from the results of very high temperature pyrolysis studies, as discussed below.

Kinetics
As early as 1969, Halpern and Patai[96] discussed the futility of attempts to employ a simple rate law in modeling the complex mechanism of cellulose pyrolysis. Their results led them to con-

clude that if a simple rate law were presumed, the evaluation of activation energies and other kinetic constants would be influenced more by the mathematical method of analysis than the actual chemistry. These caveats have been underscored recently by the research of Cardwell and Luner[179-181] who observed a continuous variation in the values of the apparent activation energy and order of cellulose pyrolysis by using sophisticated multiple heating rate methods of kinetic analysis.

Other hidden complexities of cellulose pyrolysis can easily flaw a rate measurement. For example, Broido and Kilzer[135] pointed to the often unrecognized potentially dramatic effects of ash and other impurities on the results of many early pyrolysis studies. Moreover, most moderate temperature studies make no attempt to account for the potential effects of fine structure on their results.

The foregoing situation is aggravated by the fact that above 300°C the rates of cellulose pyrolysis are extremely rapid, forcing most investigators to explore a relatively narrow temperature range in order to avoid the experimental difficulties associated with quantitative measurements of very rapidly changing physical and chemical parameters. The determination of kinetic parameters over a narrow temperature range necessarily introduces a large uncertainty into the values of the calculated activation energy and the other rate constants.[83] In addition, studies made over a narrow temperature range rarely offer insight into a complex pyrolysis mechanism because the relative roles of the various pathways often cannot be altered or made evident without significant changes in temperature.

A clear example of these difficulties can be found in the analysis of data obtained from a set of elegant experiments performed by Lipska and Parker.[64] They studied the temperature range 250 to 300°C and concluded that cellulose undergoes pyrolysis in three stages: (1) a rapid initial decomposition followed by (2) a zero-order volatilization reaction with activation energy E = 42 kcal/g mol and (3) a first-order char forming reaction with E = 42 kcal/g mol. Reexamining the data of Lipska and Parker in the light of a chain reaction mechanism, Chatterjee[141] found activation energies of 49 kcal/g mol for the initiation reaction and 42 kcal/g mol for the propagation reaction. Roberts[182] also examined the data of Lipska and Parker. He concluded that their data indicated an activation energy of 48 kcal/g

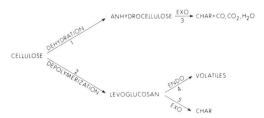

Fig. 13. Cellulose pyrolysis mechanism.

mol for the second stage volatilization reaction and 42 kcal/g mol for the third stage char formation reaction. Thus, a single set of well-characterized experimental data led to three different sets of kinetic data. It is ironic that Lipska and Parker[64] observed more char formation from preheated cellulose in their experiments, yet none of the three kinetic models used to examine the data could account for this fact.

In an attempt to avoid some of the problems associated with kinetic studies over narrow temperature ranges, Antal et al.[101] studied cellulose pyrolysis in the temperature range of 320° to 460°C using TGA. Because the model was intended for engineering studies, a simple rate law was employed which did not account for the influence of temperature on char formation. Kinetic parameters were obtained which resulted in a good fit of calculated weight loss curves with experimental data. Unfortunately, it is now known that measurements of temperature associated with the higher heating rate weight loss curves were flawed by the poor heat transfer characteristics of the TGA. Consequently, the reported activation energy is too low. This result illustrates one difficulty which can be encountered in attempting to extend the temperature range over which pyrolysis rate data is obtained.

Table 9 summarizes the values of E for cellulose pyrolysis obtained by a variety of workers using small (< 1 g) samples.[101] When a complex rate law was employed, the value cited for the sirup-forming depolymerization step was listed in Table 9. The wide range in values for E exhib-

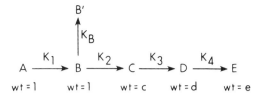

Fig. 14. Cellulose pyrolysis reaction scheme. For definition of nomenclature see Ref. 85.

ited in the table may reflect (1) the complex pyrolysis mechanism which cannot be adequately simulated by a simple rate law, (2) the possible influence of fine structure, ash, or other impurities on the pyrolysis rate, and/or (3) the narrow temperature range examined by many of the experiments.

A minimal requirement for any kinetic model of cellulose pyrolysis would seem to be the ability to account for variations in char yield with temperature and heating rate. The following section reviews the development of such a model.

• *Rate Laws Based on a Competitive Mechanism*
Although a great many researchers had pointed to the importance of competitive reactions in the thermal degradation of cellulose prior to 1971, Arseneau[99,100] and Broido and Weinstein[85] independently announced the development of competitive rate laws describing cellulose pyrolysis at the Third International Conference on Thermal Analysis. The complex mechanism of Broido and Weinstein[85] (see Fig. 14) contained

Sample	Reference	Experiment	E(kcal/gmol)
Cellulose	Akita and Kase [183]	TGA, TC in vacuum	53.5
Cotton	Madorsky, Hart and Straus [55]	TGA, TC in vacuum	50.0
Cellulose	Ramiah[184]	TGA, TC in vacuum	36.0-60.0
Cellulose	Tang [97]	TGA, TC in vacuum	56.0
Cellulose	Tang and Neil [98]	TGA, TC in vacuum	53.0-56.0
Cellulose	Arseneau [99]	TGA, Flowing N_2	45.4
Wood	Browne and Tang[185]	TGA, Flowing N_2	35.8
Cotton	Chatterjee and Conrad[140]	TGA, Flowing He	33.0
Cotton	Mack and Donaldson [186]	TGA, Flowing N_2	48.8
Cellulose	Lipska and Parker [64]	Fluidized bed	42.0
Cellulose	Chatterjee (data of Lipska and Parker) [141]	Fluidized bed	42.0
Cellulose	Lipska and Wodley [159]	Fluidized bed	42.0
Cellulose	McCarter [81]	Evolved gas	40.5
Cellulose	Murphy [187]	Evolved gas	39.4
Cellulose	Martin [69]	Radiation	30.0
Cellulose	Shivadev and Emmons [188]	Radiation	26.0
Cellulose	Lewellen, Peters and Howard [65]	Electrically heated screen	33.4

Table 9 Pyrolysis Kinetics Derived from Experiments Utilizing Small (< 1 g) Samples.[101]

five steps with detailed and elegantly justified stoichiometric parameters, but only indicated the temperature dependence of k_B: $k_B = (3.15 \times 10^{19}) \exp(-54900/RT)$ min^{-1}. Figure 15 portrays the mechanism of Arseneau and Stanwick [100] who cited values of 36.2 kcal/g mol for the activation energy associated with anhydrocellulose formation and 45.4 kcal/g mol for levoglucosan formation.

Four years later Broido and Nelson [163] analyzed data on char formation and concluded that the ratio of rate constants is $k_B/k_2 = (2.05 \times 10^6) \exp(-17,800/RT)$. Broido's final attempt [86] to analyze data taken from a 1,000 hour vacuum pyrolysis experiment [85] at 226°C resulted in values $k_B = (1.70 \times 10^{15}) \exp(-52,940/RT)$ and $k_2 = (1.60 \times 10^{10}) \exp(-40,500/RT)$ sec.$^{-1}$. However, Broido notes that the data also permits values $k_2 = (1.94 \times 10^{14}) \exp(-41,300/RT)$ or $k_2 = (3.0 \times 10^{17}) \exp(-49,000/RT)$. The stoichiometry of Broido's equations is consistent with the elimination of one mole of CO, one mole of CO_2, three moles of H_2O, and two moles of CH_3OH per mole of cellobiose. This is in good agreement with an ultimate analysis of the char residue from his experiment.

In 1979 Shafizadeh's laboratory [63] reported kinetic data for the mechanism given in Fig. 15. The values of E associated with k_v and k_c given in Fig. 15 enjoy relatively good agreement with the comparable values of Arseneau. [99] However, they are much lower than the values given by Broido. [86] Bradbury [63] presumes first-order rate laws throughout, which would seem to contradict the findings of earlier workers. [64,101] Unlike many other studies, Bradbury [63] explicitly exhibits the agreement of the solutions of his rate equations with his experimental data. The observed good agreement gives substantial credence to the model. Unpublished research in my laboratory on the pyrolysis of Avicel PH 102

microcrystalline cellulose appears to substantiate the findings of Bradbury and furnishes additional support for the model.

Models of this kind are often subject to the criticism that they contain a sufficient number of free variables to adequately fit any experimental curve. Thus, it is asserted that a good agreement of the model with experimental data lacks fundamental significance. In answer to this criticism, Broido [86] notes that the shape of the weight loss curve places severe limitations on the values of the rate constants. The experience of my laboratory [189] indicates that when the pyrolysis mechanism is relatively well known, good agreement of the kinetic model with weight loss measurements obtained over a variety of heating rates occurs only after a unique set of values for the rate constants has been found. Moreover, our experience indicates that, in general, if the "wrong" mechanism is chosen, good agreement cannot be obtained between the kinetic model and the experimental data in spite of the potentially large number of free variables available for curve fitting.

• *Other Rate Laws*

In spite of its inability to account for the effects of temperature on char yield, the chain reaction mechanism of Chatterjee and Conrad [140] has gained considerable acceptance in the literature. [60] This mechanism posits a two-step process with glycosidic bond scission as the initiation step and levoglucosan formation as the propagation step. Unfortunately, the experimental work of Chatterjee and Conrad [140] was flawed by their choice of a slow heat up rate (3°C/min) to isothermal conditions. This must have enhanced the formation of anhydrocellulose (and char) at the expense of the posited chain reaction sequence which produced levoglucosan. Moreover, their data covered a very narrow temperature range (270° to 310°C), and no agreement of the kinetic model with experimental weight loss curves was presented. Because of its focus on levoglucosan formation, the model remains of interest; however, it must be modified to account for char formation and tested against weight loss data obtained over a wide range of temperatures before it can be regarded as useful for engineering studies.

Dollimore and Holt [190] reported an interesting study of the pyrolysis of four types of cellulose in nitrogen at a slow heating rate (4°C/min) to 900°C, and isothermal pyrolyses between 260°

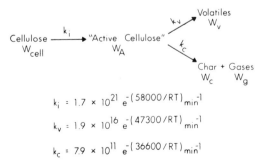

$$k_i = 1.7 \times 10^{21} \, e^{-(58000/RT)} \, \text{min}^{-1}$$

$$k_v = 1.9 \times 10^{16} \, e^{-(47300/RT)} \, \text{min}^{-1}$$

$$k_c = 7.9 \times 10^{11} \, e^{-(36600/RT)} \, \text{min}^{-1}$$

Fig. 15. Pyrolysis model for cellulose. [63]

and 285°C. Analyses of the isothermal data lead the authors to conclude that the Avrami-Erofeev equation

$$[-\ln (1 - \alpha)]^{1/n} = kt$$

with n = 2 best fit the isothermal data. This form of the Avrami-Erofeev equation reflects a model reaction based on random nucleation and nucleus growth in cylindrical particles. Dollimore and Holt[190] obtained the value 34.4 kcal/g mol for the activation energy, in close agreement with many other studies. More recent research by Fairbridge et al.[191] and Kala et al.[192] has also pointed to the utility of the Avrami-Erofeev equation in modeling cellulose and cellobiose pyrolyses. Unfortunately, these models are also flawed by (1) an inability to account for the effects of temperature on char formation and (2) rate measurements limited to a very narrow temperature range. The agreement of the Avrami-Erofeev equation with isothermal data over a narrow range of temperatures is not convincing, and no comparison was given between the model and weight loss curves over a variety of temperatures. It would have been especially interesting to see the agreement of the model with the weight loss curves of Dollimore and Holt[190] obtained to 900°C.

Recently, several authors have pointed to the role of the compensation effect in the pyrolysis of cellulosic materials.[191,193-195] Research evidencing the compensation effect has typically employed a single, simple rate law to interpret the experimental data. Consequently, the observed compensation effect may have been simply an artifact of overly simplistic theoretical methods of kinetic analysis. It would be interesting to see if more sophisticated kinetic treatments also evidence a compensation effect.

Summary and Critique
Evidence presented in this section strongly suggests that it should be possible to quantitatively convert cellulose into levoglucosan and related anhydrosugars. Recognizing the inherent value of anhydrosugars, this pyrolysis pathway necessarily merits considerable attention. Unfortunately, almost every question concerning the formation of levoglucosan appears to be a subject of debate in the literature.

Compelling evidence has been offered which supports the role of both free radical chemistry and carbonium ion chemistry in the formation of levoglucosan. If instead free radicals are responsible for parasitic char formation (as contended by some researchers), then the addition of a free radical trapping agent might be used to enhance levoglucosan formation in a functioning reactor. Conversely, if free radicals are requisite for levoglucosan formation, the addition of an appropriate free radical source could be used to augment the rate of levoglucosan formation and reduce the role of parasitic side reactions. Clearly, the field would benefit from a conclusive resolution of this controversy.

Similarly, it is known that the fine structure of the cellulose substrate dramatically influences the yield of levoglucosan. However, the details of this influence are not well understood. For example, why does acid washing improve the yields of levoglucosan obtained from most cellulosic materials? Similarly, why does the presence of trace quantities of ash or other additives dramatically reduce the yield of levoglucosan? Are these two phenomena related? Is either phenomenon related to the role of free radicals or carbonium ion chemistry in the formation of levoglucosan? Are free radical reactions responsible for the formation of levoglucosan from crystalline cellulose, and are concerted displacement or carbonium ion mechanisms active in the formation of furfural and related compounds from amorphous cellulose? A resolution of these questions could be a key element in the design of technically sophisticated biomass conversion processes.

Finally, a rate law describing levoglucosan formation has not been established. In view of the foregoing comments, the lack of a good rate law is not surprising. Because an accurate rate law is needed to facilitate reactor design, definitive kinetic studies which encompass and describe the range of external influences on levoglucosan formation would be of great benefit to the field.

High Temperature Phenomenon
To secure larger reactor throughputs and minimize char formation, scientists studying the conversion of biomass to more valuable fuels and chemicals have emphasized the high temperature pyrolysis phenomenon in their research. Unfortunately, pyrolysis chemistry of biomass materials at temperatures above 500°C has not received much attention in the literature. Apart from research related to nuclear weapons effects, most information on the high temperature

pyrolysis of cellulose has been published only recently. In this section, the high temperature, solid phase pyrolysis of cellulose is reviewed, reflecting the fact that the gas-phase (secondary) pyrolysis reactions involve noncellulosic carbohydrates—typically anhydrosugars. A discussion of the pyrolysis of noncellulosic carbohydrates is given below.

Products

Berkowitz-Mattuck and Noguchi,[70] Martin,[69] and Lincoln[71] all note the decrease in char formation from cellulose with increasing heating rate. Because Martin[69] observed negligible char formation at very high heating rates, he concluded that at least one pathway exists for cellulose pyrolysis which does not include char formation.

Berkowitz-Mattuck and Noguchi[70] note the presence of CO, CO_2, CH_4, levoglucosan, and at least 12 polar organic compounds in the volatile products with radiant flux densities of 20 to 100 W/cm^2. Martin[69] found CO, CO_2, and H_2O to be the earliest volatile products. Increasing radiant flux densities effected increasing yields of saturated and unsaturated aldehydes, ethylene, hydrogen, and sirup while decreasing yields of H_2O and CO_2. Martin[69] asserts levoglucosan to be the only product of the higher temperature pyrolysis pathway, concluding that the presence of H_2, CH_4, C_2H_4, C_2H_6, and other hydrocarbons in the gaseous products is a result of secondary, heterogeneous pyrolysis reactions between the levoglucosan vapors and the hot char.

In his studies of the effects of the intensity of radiant energy on cellulose pyrolysis, Lincoln[71] notes an increase in the mass ratio of CO to CO_2 formation from 0.33 at 6 W/cm^2 to 9.2 at 12,000 W/cm^2 (peak). Lincoln attributes this increase to the fact that flash heating indiscriminately breaks the cellulose polymer into smaller pieces. Baker[82] confirms this finding and speculates that the favorable influence of heating rate on CO formation is due to increasing fracture of the glycosidic ring. The shock tube studies of Eventova et al.[196] provide further evidence for the enhanced formation of CO and H_2 at higher temperatures, but the data includes much scatter.

Milne and Soltys[114-116] have reported the results of some very exciting research involving the flash pyrolysis of a great variety of biomass materials in hot steam (to 900°C or more). This work parallels that of Schulten[197,198] but will in-volve studies of a wide range of experimental conditions intended to elucidate the high temperature pyrolysis pathways. A summary of some of Milne's data,[115] which includes the effects of secondary reactions, is given in Fig. 16. Interestingly, Milne[115] notes that all the higher mass species seem to evolve nearly simultaneously as though they arose from a common intermediate.

Mechanisms

The chief question to be answered is whether the higher temperature products are formed as a result of reactions 1 and 2 portrayed in Fig. 8, or if a new degradation pathway becomes active at very high temperatures. Because levoglucosan is very unstable at these elevated temperatures, this is a difficult question to answer conclusively. In Ref. 122, Shafizadeh refers to reaction 3 (see Fig. 8) as the high temperature pathway for cellulose pyrolysis, but the evidence he presents to justify this designation is inconclusive.

Table 10 summarizes the mass ratios of the gaseous products CO, CH_4, and C_2H_4 derived from the high temperature pyrolysis of various biomass materials. With the exception of a few studies,[67,69,74,117] most of the data given in Table 10 does not distinguish between solid and gas-phase pyrolysis phenomenon. The logarithmic ratio has been selected as a variable of interest due to the results of earlier research in my laboratory.[77]

With a few key exceptions, values of ln (CO/CH_4) for cellulose pyrolysis given in Table 10 range between 1.9 and 2.7. Similarly, values of ln (CO/C_2H_4) range between 2.2 and 2.7. However, the experiments of Martin[69] and Hopkins et al.[74] involving flash heating of the solid substrate by a high radiative flux depart from these ranges, exhibit values as high as 4.6 and 4.2 for ln (CO/CH_4) and ln (CO/C_2H_4), respectively. The experiments of Hopkins et al.[74] did not involve pure cellulose because the cellulose did not absorb sufficient light to undergo pyrolysis during the brief flash of the xenon bulb. Nevertheless, their results for cellulose and biomass (composed largely of cellulose) corroborate the results of Martin[69] and indicate unusually high yields of CO from the flash pyrolysis of cellulose.

It might be supposed that these abnormally high relative yields of CO were derived from the high temperature gas-phase pyrolysis of cellulose-derived volatile matter. However, the results of Hopkins et al.[74] were obtained in hard

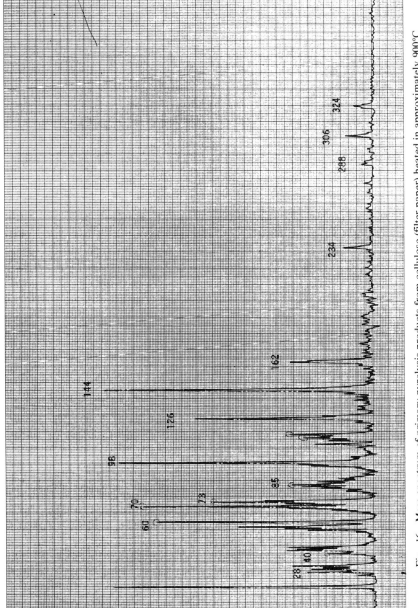

Fig. 16. Mass spectrum of primary pyrolysis products from cellulose (filter paper) heated in approximately 900°C steam/argon.[116] Numbers in the Figure represent major mass to charge ratios in the 16.5 eV spectrum.

Author	Material	Temperature (°C)	gas (mass %)	ln of CO/CH$_4$	ln of CO/C$_2$H$_4$
	Electric Furnace Heating				
Antal, et al [202]	Cellulose	750	84	1.9	2.2
	Kraft Lignin	750	36	0.6	2.1
	D-Mannose	750	65	1.9	2.4
Hajaligol, et al [66]	Cellulose	750	36	2.7	2.7
	Cellulose	1000	47	2.2	2.3
Mudge, et al [263]	Wood (with K$_2$CO$_3$)	650	—	1.5	3.9
	Wood (no catalyst)	650	—	2.2	3.7
Deglise, et al [204]	Dry Beech	700	—	1.8	2.1
	Dry Beech	1000	—	1.9	2.6
Lede, et al [73]	Douglas Pine Sawdust	700	52	1.8	2.1
	Douglas Pine Sawdust	1000	62	1.9	2.7
Diebold & Scanlon (cyclone) [52]	Softwood Sawdust		34	2.1	2.3
	Birch Flour		—	2.1	2.2
Halligan, et al (fluid bed) [205]	Bovine Waste	728	—	1.4	1.8
	Bovine Waste	796	—	1.1	1.1
Iatridis & Gavalas (screen) [67]	Kraft Lignin	400	—	0.8	4.1
	Kraft Lignin	650	—	0.6	3.3
	Radiant Heating	Flux (W/cm^2)			
Antal, et al [202]	Cellulose	70	—	2.1	2.3
	Corn Cob	70	—	2.3	2.3
	Hardwood	70	46	1.9	2.4
Hopkins, et al [117]	Cellulose	200	26	2.6	2.2
	Kraft Paper	200	55	2.3	2.3
Lede, et al [73]	Douglas Pine Sawdust	60		2.9	3.2
	Douglas Pine Sawdust		—	3.1	3.1
	Douglas Pine Sawdust	75		2.6	3.7
Martin [69]	Cellulose (8 sec)	17	—	3.1	3.1
	Cellulose (4 sec)	46	—	2.6	2.6
	Cellulose (.965 sec)	46	—	4.4	3.8
	Cellulose (.49 sec)	46	—	4.6	4.2
Hopkins, et al [74]	Lignin	8000	42	5.1	5.1
	Lignin	8000	36	4.5	4.9
	Redwood	8000	57	4.3	4.4
	Redwood	8000	36	3.6	4.0
	Dextrose	8000	35	3.6	—
	Dextrose	8000	21	2.8	3.6
	D-cellobiose	8000	36	3.2	2.7
	D-cellobiose	8000	34	3.0	2.6
	Kraft paper	8000	30	3.5	3.8
	Kraft paper	8000	20	3.2	3.6
	Leucaena	8000	63	4.3	4.6
	Corn Cob	8000	10	3.1	—
	Corn Cob	8000	8	3.1	2.7
	Calatropis Procera	8000	28	3.7	4.6
	Newsprint	8000	28	3.4	3.7
	Cow Manure	8000	47	4.3	—

Note: "—" indicates that the gas yield was not listed or if found in the third column, that no C$_2$H$_4$ was observed in the pyrolysis products.

Table 10　A Summary of Reported Carbon Monoxide, Methane, and Ethylene Mass Ratios[74]

vacuum, thereby minimizing secondary reactions. Moreover, research on the gas-phase pyrolysis of cellulose-derived volatile matter evidenced a gradual *decline* in the values of ln (ΔCO/ΔCH$_4$) from 2.5 at 500°C to 1.8 at 750°C. Similarly, the values of ln (ΔCO/ΔC$_2$H$_4$) *declined* from 2.6 to 2.0 from 600° to 750°C. Recalling that C$_2$H$_4$ is not a primary product of the low and moderate temperature pyrolysis pathways (reactions 1 and 2 of Fig. 8), and noting that

higher temperatures would cause a decline (rather than a rise) in the values of the two logarithmic ratios if the gaseous products were the result of gas-phase secondary pyrolysis reactions, it seems necessary to conclude that the high values of the logarithmic ratios cannot be the result of gas-phase pyrolysis of levoglucosan derived from reaction pathway 2 in Fig. 8. In other words, pathway 2 cannot account for the product distribution given in Table 10.

Can pathway 1 of Fig. 8 account for the product distribution? As summarized by Hopkins et al.,[74] the low temperature, solid-phase pyrolysis of cellulose and related carbohydrates results in values of $\ln (CO/CO_2)$ ranging from -2.1 to $+0.16$. Thus, the low temperature pathway forms very little CO relative to CO_2 and cannot account for the high yields reported by Martin[69] and Hopkins et al.[74] The inescapable conclusion is that a third, high temperature pyrolysis pathway must exist in order to explain the results of Martin[69] and Hopkins et al.[74] This pathway must result in the direct formation of CO from the high temperature fracture of the glycosidic ring. Presumably, this pathway is reaction 3 in Fig. 8 referred to by Shafizadeh[122] as the fission reaction.

In summary, the trends evidenced in Table 10 are very difficult to explain without postulating the existence of a high temperature pyrolysis pathway involving a catastrophic fragmentation of the glycosidic ring into CO and H_2. Clearly, it would be desirable to confirm this hypothesis by the acquisition of more compelling experimental data.

Kinetics

High temperature pyrolysis kinetics must necessarily reflect the rates of the lower temperature pathways discussed above. Consequently, rate laws describing high temperature pyrolysis phenomena must necessarily incorporate the competitive reaction scheme discussed earlier. In addition, the elucidation of a comprehensive high temperature rate law also requires a determination of the rate law associated with the postulated high temperature fission pathway. Unfortunately, no studies of the effects of temperature on this reaction are available. Moreover, the mathematical complexity of three competitive pathways, combined with the recognition that the primary products of the high temperature pathway (CO and H_2) are also products of the low temperature pathway (CO) and the moderate temperature pathway (if levoglucosan undergoes further pyrolysis in the gas phase), suggests some of the difficulties which must be overcome before a rate law for the high temperature pathway can be determined.

Summary and Critique

Research described in this section indicates that the dominant pathway for cellulose pyrolysis at higher temperatures is the same as that at moderate temperatures, and leads to levoglucosan formation. Some evidence for a third, high temperature pathway was presented, but its products do not appear to be of great interest. Consequently, the questions raised earlier also pertain to high temperature pyrolysis. In addition, it would be of academic interest to establish or disprove the existence of a high temperature pathway.

It is important to recall that the focus of this section has been the solid-phase pyrolysis of cellulose. Gas-phase pyrolysis phenomena will be treated later.

Effects of Various Parameters

Earlier in this review, emphasis was given to the effects of temperature on the products and mechanisms of cellulose pyrolysis. A typical experiment involved a very small, pure cellulose sample (often < 10 mg) heated at a modest rate in vacuum or an inert environment to a relatively high temperature. Modern engineering practice seeks to maximize reactor throughput and reaction specificity by manipulating a wide variety of conditions within the reactor. For this reason, it is of interest to identify the effects of various reaction parameters, such as heating rate, pressure, particle size, ambient gas environment, and additives (catalysts) on the products, pathways, and rates of cellulose pyrolysis.

Heating Rate

Because cellulose is so reactive, it is difficult to heat it rapidly enough to cause isothermal pyrolysis at temperatures in excess of about 600°C. Consequently, above about 500°C the effect of heating rate can be of greater interest to an engineer than the effects of temperature. In general, a higher heating rate favors the higher temperature transglycosylation reactions over the lower temperature charring reactions. A quantitative measure of the effects of heating rate on the relative roles of the two reactions was given above, and especially in Fig. 15. If the temperature/time history of the particle within the reactor is known, the rate equations given in Fig. 15 can be integrated to specify the yields of char and anhydrosugars resulting from the solid-phase pyrolysis reactions.

Pressure

The effects of pressure on biomass pyrolysis have been described in a series of papers by Mok and Antal.[79,102] In general, increasing the pressure dramatically reduces the yield of volatile materials and enhances the formation of char. An increase in ambient pressure from 0.1 MPa to 2.5 MPa increases the char yield from 12% to 22% by weight.[102] Among gaseous products, the formation of CO_2 and H_2 is favored by increasing pressure, while the yields of CO, CH_4, C_2H_4, C_2H_6, and C_3H_6 are reduced. Under a high pressure (0.5 MPa), an increased residence time reduced the yield of CO and C_3 hydrocarbons, and enhanced the formation of CO_2 and C_2H_6[79] in the product stream.

As shown in Fig. 17, a strong correlation exists between the heat of reaction (ΔH_{pyr}) and the char yield. An increase in char yield is associated with an increase in reaction exothermicity. Consequently, increasing pressure is also associated with increasing exothermicity of the pyrolysis reactions. However, the relationship is complicated by the heats of reaction of the gas-phase pyrolysis reactions, which are also influenced by pressure. Because increasing pressure favors exothermicity in the gas-phase reactions, a DSC may detect equal exothermicity under high pressure and high flow rate conditions, or moderate pressure and low flow rate conditions.

Mok and Antal[102] have proposed the pyrolysis mechanism given in Fig. 18 to explain the effects of pressure on the products and heats of cellulose pyrolysis reactions. Increasing pressure favors the char forming pathways 4 and 6, whereas decreasing pressure favors the volatile forming pathways 3 and 5. Reference 102 pro-

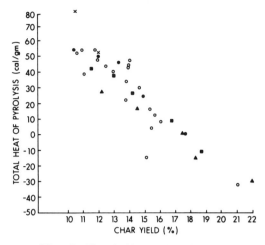

Fig. 17. Plot of ΔH_{pyr} vs char yields.[102]

vides a more detailed discussion and justification of the mechanism given in Fig. 18.

Particle Size

Using both the techniques of thermogravimetry and DSC, Arseneau[99] was able to observe the appearance of a high temperature exotherm as cellulose samples of increasing thickness were studied. This exotherm was attributed to the formation of secondary char resulting from the decomposition of volatile matter (levoglucosan) which was unable to rapidly escape from the thicker cellulose samples. Thus, the pyrolysis of larger particles of cellulose should evolve more char and heat at the expense of volatiles formation. However, as noted by Martin,[69] volatile products evolved at the center of a hot, pyrolyz-

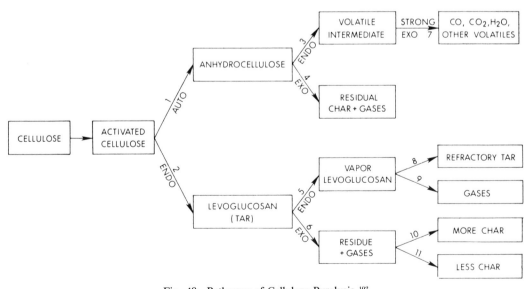

Fig. 18. Pathways of Cellulose Pyrolysis.[102]

ing biomass particle must pass through the hot char layer as they exit the particle. Heterogeneous reactions are likely to oxidize the char layer by reducing the oxygen-rich volatile matter, thus gasifying the char. These secondary reactions may mitigate the influence of particle size on the formation of secondary char.

Gaseous Environment

Biomass pyrolysis can be accomplished practically in a variety of gaseous environments. Steam is thought to be the most practical alternative; however, hydrogen, methane, and recycled gas have also been studied.

The effects of steam on cellulose pyrolysis have been studied, but not yet well understood. The early work of Wiegerink[199] and Waller et al.[200] indicated that the thermal breakdown of cellulose is accelerated by the presence of water. These findings were partially supported by Hon[147] who found that moisture contents ranging from 0 to 5% and > 7% increase the formation of free radicals in photo-irradiated cellulose. However, Hon[147] also found a dramatic decrease in free radical formation for moisture content in the range 5% to 7%. Shimazu and Sterling[139] reported that between 100° and 150°C, moist cellulose breaks down less rapidly than dry cellulose, and they speculate that the presence of water quenches free radicals by an unspecified mechanism. Arthur[105,107] confirmed this speculation, asserting that water reacts with free radicals and decreases their concentration.

Two kinetic studies of the effects of steam on cellulose pyrolysis have appeared in the literature.[101,201] The low temperature results of Stamm[201] evidenced much more rapid degradation of cellulose under steaming conditions, reportedly lowering the apparent activation energy by 50%. The moderate temperature results of Antal et al.[101] evidenced a negligible effect of steam on the pyrolysis rate.

In summary, steam is reported to increase, decrease, and have no effect on the rate of cellulose pyrolysis. Recognizing the controversy surrounding the mechanism of cellulose pyrolysis, this lack of agreement may not be surprising. Nevertheless, it suggests how much remains to be learned about the details of cellulose pyrolysis.

Ash and Additives

The dramatic effects of trace amounts of ash on cellulose pyrolysis were noted earlier. Additives, known as fire retardants, have received an extraordinary amount of attention in the research literature of the combustion community. Unfortunately, much of this literature is not germane to the thrust of this review since the role of fire retardants is to catalytically enhance the formation of nonflammable char at the expense of flammable volatiles production. In many countries of the world, char (carbon) is thought to be a low value commodity because ample supplies of coal are available. Consequently, in those countries the transformation of biomass into another, low value, solid fuel is of little interest. However, for those countries which lack a substantial coal resource, catalysis of the char forming pathway may be commercially interesting and worthy of further study.

The efficacy of various additives as flame retardants has been well summarized in the reviews of Molton and Demmitt,[123] and Shafizadeh.[120] As reviewed by Shafizadeh,[120] acid catalysts strongly influence the dehydration reactions, enhance the formation of levoglucosenone and a variety of furan derivatives, and promote the production of char. Alkaline catalysts favor fission and disproportionation reactions, enhancing the yields of glyoxal, acetaldehyde, and low molecular weight carbonyl compounds, as well as char.[120]

In a recent report, Milne[115] describes the dramatic formation of 50% char from rapidly heated cellulose impregnated with 10% (by weight) K^+ ions derived from K_2CO_3 or KOH. The char formation was accompanied by the prompt evolution of low molecular weight gases. Thus, the alkaline catalyzed, rapid heating of cellulosic materials may be an attractive source of char.

Summary and Critique

Low heating rates, increased pressure, and the presence of ash or additives all enhance the formation of char at the expense of volatile production. The relative ease of transforming biomass into charcoal justifies an exploration of the possibilities for using charcoal to replace fluid fuels.

Although inconclusive, some evidence suggests that the chemical composition of the surrounding gaseous environment can be used to influence the pyrolysis mechanism and enhance the formation of higher value products. Basic research on the influence of steam, hydrogen, and methane on the pyrolysis pathways merits increased attention.

PYROLYSIS OF NONCELLULOSIC CARBOHYDRATES

Although the literature describing the pyrolysis of noncellulosic carbohydrates is less extensive than that available on cellulose pyrolysis, a substantial body of knowledge exists. Several fine reviews of the pyrolysis of various carbohydrates,[206-211] as well as reviews of properties of the anhydrosugar pyrolysis products,[212-215] are available. The important role of pyrolysis reactions in the cooking of foods should not be overlooked.[211] However, the emphasis of this section is on the high temperature thermochemical conversion of noncellulosic carbohydrates, which has received less attention in the literature.

The response of noncellulosic carbohydrates to the action of heat is not dissimilar to that of cellulose. Noncellulosic carbohydrates undergo thermal degradation at lower temperatures than cellulose, but the pyrolysis phenomenology is very similar. For this reason, in the following sections, the pyrolysis chemistry is presented in terms of low temperature ($<200°C$), moderate temperature ($200°$ to $500°C$), and high temperature ($>500°C$) phenomena.

Low Temperature Phenomena

Shafizadeh and his colleagues have reported extensive studies of the thermodynamic properties of anhydrosugars[216-218] and more complex carbohydrates.[219] These studies have established the existence of at least one crystal transition for many anhydrosugars.[216,217] They have also noted increased values of the specific heat c_p together with a soft and waxy appearance of the anhydrosugar crystal following the transition.[217] Closely related sugars do not manifest these crystal transitions but simply melt upon heating.[217] Shafizadeh speculates that this behavior of anhydrosugars results from their globular shape, which facilitates the formation of a plastic phase at higher temperatures.[218] Much information on the crystalline structure of the anhydrosugars[220] and other properties[221-225] is also available.

Products

At temperatures below 200°C, levoglucosan readily undergoes polymerization to form a dextrin.[226-232] The presence of α-D-(1→6), β-D-(1→6), α-D-(1→4), and β-D-(1→2) linkages have been noted in the polymer.[230] Similarly, 1,6 anhydro-β-D-galactopyranose (see Fig. 19 which displays the structures of all the anhydrosugar

discussed in this review) undergoes polymerization to form a high molecular weight, branched polysaccharide with a mixture of α and β linkages.[233] A primary volatile product of cellulose pyrolysis, 3-deoxy-D-erythro-hexosulose, rapidly degrades at temperatures as low as 100°C. It forms a complex polymer[234] as well as H_2O, CO_2, CO, furan derivatives, and low molecular weight carbonyl compounds.

The formation of glucan from α-D-glucose, both in the presence[235-237] and absence[238-241] of acid catalysts, has been extensively studied. Again the polymer is highly branched, containing both (1→4) and (1→6) linkages.[236,237] Other pyrolysis products of D-glucose[238,242] and D-xylose[243] have been reported by Sugisawa, and Kato and Komorita. In the pyrolysis products of D-glucose, D-xylose, and α-cellulose, Kato and Komorita[243] have detected the presence of 3-deoxyglycosones which were proposed as intermediates in the pyrolytic formation of furfurals from sugars. At 194°C sucrose decomposes to a mixture of glucose, disaccharides, oligosaccharides, and polysaccharides.[244] Glucose is the major product of hydrolysis of both the oligosaccharides and the polysaccharides.[244]

The low temperature pyrolysis of starch is the basis of the pyrodextrin industry;[206-210] consequently, much study has been devoted to the dextrinization of starch. The early X-ray diffraction studies of Katz[245,246] revealed that above 180°C the diffraction patterns of various granular starches lost their sharpness and indicated the formation of an amorphous material above 210°C. Results of various DTA studies of starch pyrolysis,[247-250] as reviewed by Greenwood,[210] were largely inconclusive. The chemical structure of dextrins produced from amylose[251-254] and starch[209,240,252,255-258] have been elucidated by several laboratories. Kerr and Cleveland[252] showed that the extent of dextrinization of amylose was dependent upon its crystallinity and other physical properties. The presence of α-D-(1→6), β-D-(1→6), and β-D-(1→2) linkages, as well as 1,6-anhydro-β-D-glucopyranose end groups, were detected by Thompson and Wolfrom[253] in a dextrin formed from amylose. Similarly, transglycosylation reactions during the pyrolysis of starch give rise to a highly branched dextrin.[240,252,257,258]

Mechanisms and Kinetics

In their exceptional review of the 1,6-anhydro derivatives of aldohexoses, Cerny and Stanek[215]

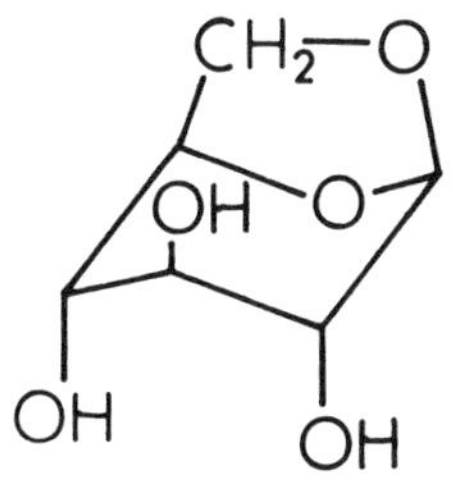

I

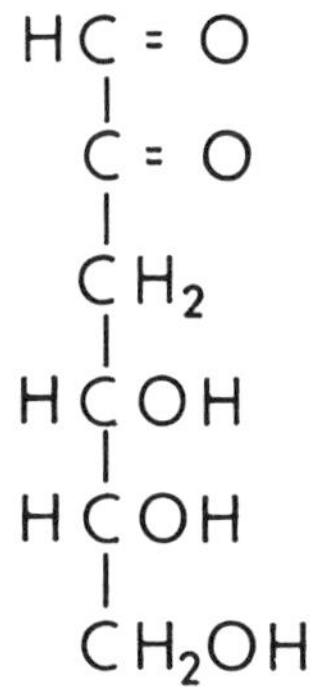

V

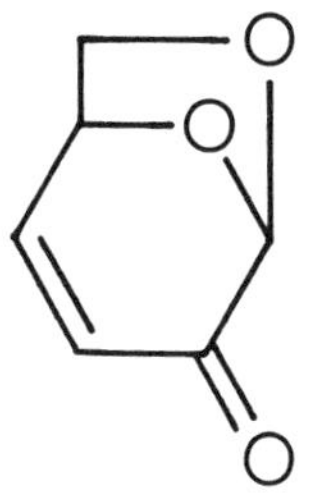

II

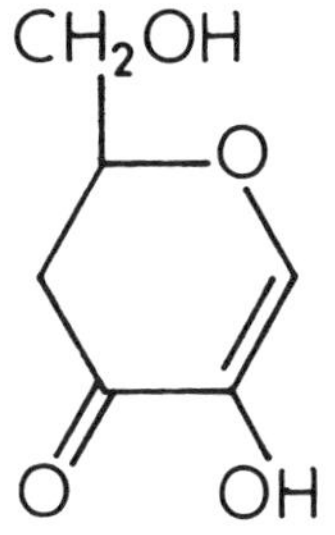

VI

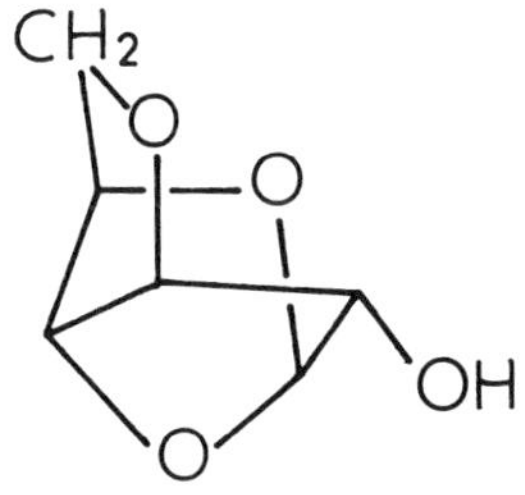

III

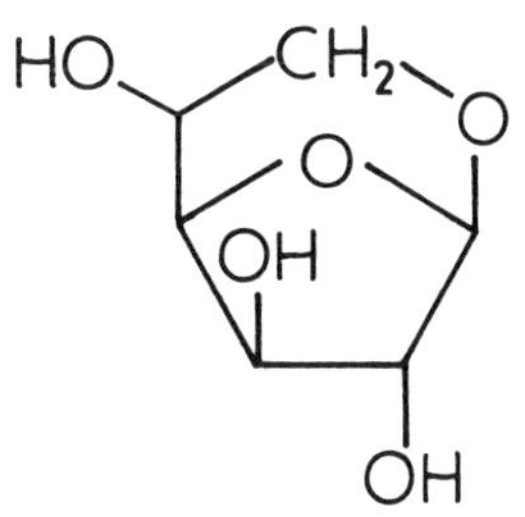

VII

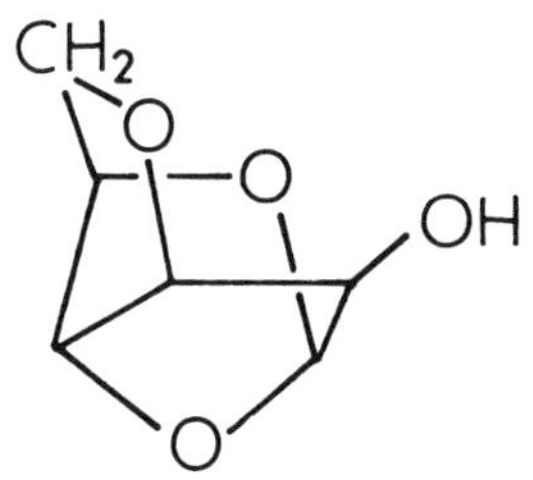

IV

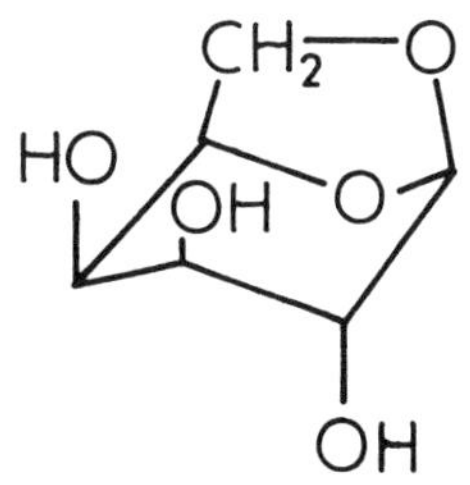

VIII

Fig. 19. Structures of various anhydrosugars.

I Levoglucosan
II Levoglucosenone
III 1,4:3,6 dianhydro-D-glucopyranose
IV 1,4:3,6 dianhydro-D-mannopyranose
V 3-deoxy-D-erythro-hexosulose
VI 1,5-anhydro-4-deoxy-D-glycero-hex-1-en-3-ulose
VII 1,6 anhydro-β-D-glucofuranose
VIII 1,6 anhydro-β-D-galactopyranose

discuss in detail the mechanisms of anhydro-sugar formation and polymerization. The authors note that no radical polymerization of levoglucosan has been observed (except in the presence of maleic anhydride).

Orsi[259] has reported studies of the thermal decomposition of glucose and fructose heated at 6°C/min. Up to 240°C, competing reactions which form oligo- and polysaccharides by polymerization, and a brownish coloring matter were observed. Above 240°C, both the polysaccharide and the coloring matter decomposed into an insoluble brown material. Fructose exhibited similar behavior at lower temperatures. For both sugars the apparent activation energy decreased from an initial value of 26 kcal/g mol to 18 kcal/g mol, and then returned to the initial value of 26 kcal/g mol.

The mechanism of sucrose polymerization, elucidated by Richards and Shafizadeh,[244] is given in Fig. 20. At 190°C, the rate of formation of the polysaccharide was shown to satisfy a first-order rate law involving the fructose anhydride F*:

$$G - F \rightarrow G + F* \qquad \text{(slow, rate determining)}$$
$$G - F + F* \rightarrow F - G - F \text{(fast)}$$

where G and F are the glucose and fructose constituents of sucrose. Richards and Shafizadeh[244] note that no gas was evolved at 194°C and that the polysaccharide decomposed upon further heating to 303°C. Thermogravimetric studies of the pyrolysis of maltose, lactose, cellobiose, and melibiose have been described by Weill et al.[260] who reported that both acid hydrolysis and thermogravimetry yielded identical values of the apparent activation energy for both lactose and cellobiose. They attributed this similarity to the rate-limiting formation of a carbonium ion by cleavage of the glycosidic bond as the common step in both cases.

The mechanisms of dextrin formation from amylose have been reviewed by Greenwood.[210] Figure 21 illustrates the formation of a 1,6-anhydro end group, followed by the reaction of this end group with an adjacent hydroxyl group to form an ether linkage.[254] This mechanism is posited to play a major role in dextrin formation from dry, nonacidified starch, whereas a hydrolysis-reversion mechanism followed by recombination is thought to be responsible for dextrinization under acidic conditions.[254,210]

Moderate Temperature Phenomena

Above 200°C, all noncellulosic carbohydrates undergo rapid decomposition, producing a variety of anhydrosugars, light volatile organic compounds, H_2O, CO_2, CO, and char. Although the seemingly endless variety of mono-, di-, tri-, oligo-, and polysaccharides available for pyrolysis studies might provide an infinitude of products and mechanisms, the actual picture is relatively simple. The reader will be interested to note the many similarities to cellulose pyrolysis at moderate temperatures discussed above.

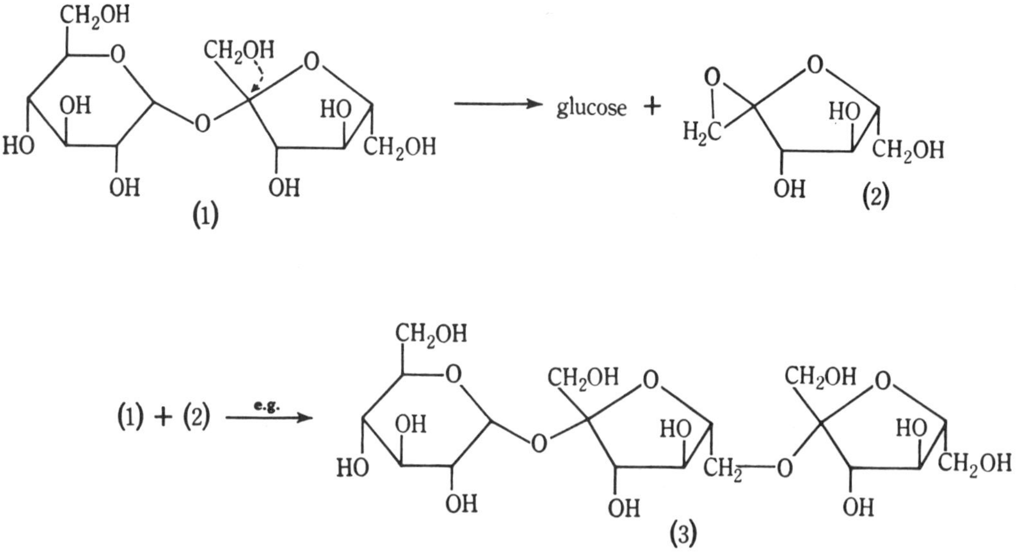

Fig. 20. Mechanism of Sucrose Polymerization.[244]

94

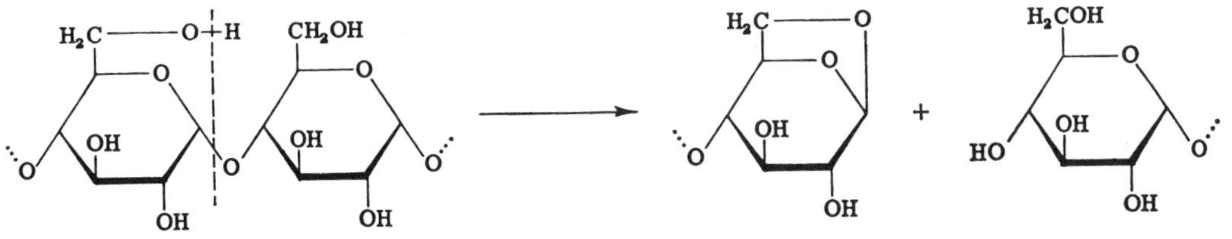

Fig. 21a. Mechanism of 1,6 anhydro end group formation from amylose.[254]

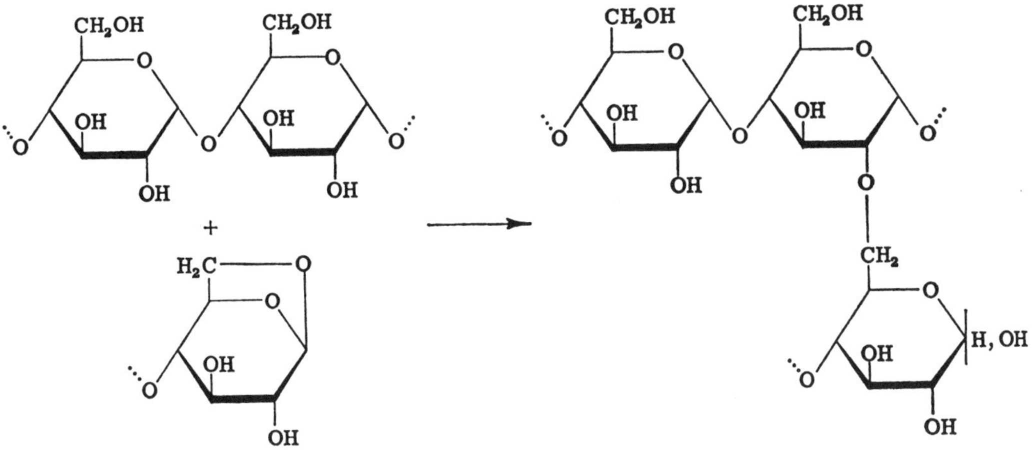

Fig. 21b. Mechanism of ether linkage formation.[254]

Products

• Anhydrosugars and Related Monomers

At low pressures, the anhydrosugars distill when heated to temperatures above 200°C and can be quantitatively recovered.[261] However, increased pressure or the presence of catalysts can cause the anhydrosugars to rapidly decompose in either the solid or the vapor phase.

A thought provoking series of experiments by Broido et al.[261,262] demonstrated the catalytic vapor-phase dehydration of levoglucosan to levoglucosenone over $NaHSO_4$ at 295° to 335°C. Shafizadeh et al.[263] have shown that $ZnCl$ catalyzes the exothermic decomposion of levoglucosan to char, H_2O, CO_2, and 2-furaldehyde between 120° to 180°C, whereas NaOH catalyzes the exothermic formation of various carbonyl compounds, CO_2, and char. Later, work by Shafizadeh et al.[264] confirmed the earlier work of Broido et al.[261] and indicated that the formation of levoglucosenone from levoglucosan is favored by Arrhenius acids (but not Lewis acids, which catalyze the decomposition and charring of levoglucosenone) and higher pressures. Shafizadeh et al.[264] were able to obtain an 11% yield of levoglucosenone from 5 mg of H_3PO_4-treated cellulose, but the yield fell to 2% when 10 g

of treated cellulose were pyrolyzed. Although Shafizadeh et al.[264] ascribe this decrease to heat transfer limitations, the known fragility of levoglucosenone (see below) would suggest mass-transfer limitations leading to pronounced heterogeneous interactions of levoglucosenone with the hot char layer (which it must pass through as it exits the solid matrix) as the source of the problem. After treatment with 1.7 to 5.6% H_3PO_4, Shafizadeh et al.[264] were able to obtain a 5 to 6% yield of levoglucosenone from kraft paper. Using this result, the authors describe a method for producing 90 to 95% pure levoglucosenone from 20 g batches of newsprint.

In 1965, Bedford and Gardiner[265] reported the formation of two 1,4:3,6-dianhydro-D-hexopyranoses (see Fig. 19) from amylose, 3,6-anhydro-D-glucose and D-mannose. The yields of III (Fig. 19) from amylose and 3,6-anhydro-D-glucose were 1.26% and 34.8%, respectively, whereas the yield of IV from D-mannose was 2.3%. Similar results were reported by Shafizadeh et al.[266] who also found that III sublimes at 100°C, melts at 127°C, and vaporizes at 190°C, leaving little or no char. When mixed with diphenyl phosphate, an Arrhenius acid, the pyrolysis of III results in the formation of 3,6 anhydro-D-glucose and a mix-

95

ture of oligo- and polysaccharides. These ultimately decompose to form char. In the vapor phase, III undergoes dehydration in the presence of phosphoric acid on glass wool to form levoglucosenone, 2-furaldehyde, and 5 methyl-2 furaldehyde.[266]

The early work of Halpern et al.[262] established the propensity of levoglucosenone to polymerize at room temperature. In contrast with these results, Shafizadeh and Chin[267] found that levoglucosenone evaporates completely at 100 to 150°C and begins to decompose rapidly above 400°C. At 500°C, in flowing N_2 at 0.1 MPa, 72% of the levoglucosenone distilled and formed 94% of the tar recovered. The apparent disagreement of these results with those of Halpern et al.[262] may be explained by the stabilizing effect of dilution obtained when levoglucosenone enters the gas phase.

Shafizadeh et al.[268] have also described the formation and thermal properties of 1,5-anhydro-4-deoxy-D-glycero-hex-1-en-3-ulose (VI in Fig. 19), which was obtained with a 1.4% yield from cellulose. The authors found that VI can be crystallized, melts at 98°C, can be distilled in vacuum and quantitatively recovered, is stable in acids, but decomposes in the presence of bases. They also report the yields of VI from a wide variety of carbohydrate materials.

• *Mono and Disaccharides*

In 1967, Hounimer and Patai first reported the results of an extensive series of elegant experiments involving the pyrolysis of D-glucose labeled with ^{14}C at various positions.[269,270] The authors noted gross differences in the amounts of products formed in vacuum between 175° and 275°C, depending upon the presence of unidentified contaminants. The dominant contributor to CO and CO_2 formation was the C1 atom along with a lesser presence of the C2 atom at lower temperatures. The aqueous product contained increasing amounts of organic, volatile products with increasing temperature, reaching 14% by weight at 275°C. Thus, water was found to be the primary condensable product. Acids were also formed preferentially from the C1 position, with about 50% of the acidic carboxyl groups associated with formic acid, 25% with acetic acid, and 25% with higher acids. The main path for furfural formation involved explusion of the C6 atom; however, C1 expulsion also occurred. Interestingly, at 225°C, 50% of the glucose evaporated and was recovered.

Heyns and his colleagues[271,272] described the results of a detailed GC-MS study of the products of D-glucose pyrolysis at 300°C and the pyrolysis of a wide variety of carbohydrates between 300° and 500°C. For D-glucose, the chief pyrolysis product was III (Fig. 19). However, the yield of the anhydrosugar products was very low, and 55 additional compounds were also identified. Recognizing that some of the volatile furan compounds contained more than 6 carbon atoms, Heyns et al.[271] posited that the furans were not formed from D-glucose itself, but by fragmentation of a glucan polymer. Because the same volatile degradation products were observed from 300° to 500°C for D-erythrose, D-xylose, D-ribiose, D-arabinose, D-glucurono-6,3-lactone, cellobiose, maltose, lactose, sucrose, raffinose, amylose, amylopectin, and cellulose, Heyns and Klier[272] concluded that by dehydration, degradation, and condensation reactions, all these carbohydrates form similar *polymeric* intermediates, which then undergo secondary thermal degradation.

• *Polysaccharides*

The research of Bryce and Greenwood[273] has shown the relative thermal stability of various polysaccharides in vacuum to be amylose < starch < amylopectin < cellulose. The products of starch pyrolysis consist of a condensable sirup composed primarily of levoglucosan, a variety of permanant gases, and char. In agreement with the work of Heyns and Klier,[272] Bryce and Greenwood[274] report that the volatile products of potato starch are qualitatively identical to those evolved from various starches, their amylose and amylopectin components, cellulose, simple sugars such as D-glucose, maltose, and maltotriose. Table 11 displays the yields of volatile materials produced from starch and various other saccharides[275] held for 18 hours at 300°C. In addition to the compounds displayed in Table 11, water vapor, CO_2, and CO are major products of starch pyrolysis.

Shafizadeh and his colleagues[276,277] have described the pyrolysis of xylan and related model compounds. Yields of 31% char, 16% tar, 31% liquid condensate, and 8% CO_2 were obtained from xylan. The tar fraction contained a mixture of oligosaccharides with average DP of 6-8 and some other compounds. The aqueous pyrolyzate contained mainly 2-furaldehyde and water.

Probably the most interesting research reported in the literature on hemicellulose pyrol-

Volatile product	Saccharide Pyrolyzed					
	Starch	Amylopectin	Amylose	Maltose	Isomaltose	D-Glucose
Acetyldehyde	400	460	240	430	480	250
Furan	395	210	225	350	260	230
Acetone	230	335	90	255	395	130
Acrolein	15	15	20	30	30	20
2-Methylfuran	230	185	135	290	375	135
Butyraldehyde	10	15	15	10	10	5
2-Propanone	85	60	15	50	110	15
2,5-Dimethylfuran	35	20	10	50	110	15
Formaldehyde	1300	1400	1700	n.d.[b]	n.d.	n.d.
Formic acid	2600	n.d.	n.d.	n.d.	n.d.	n.d.
Acetic acid	300	n.d.	n.d.	n.d.	n.d.	n.d.
2-Furaldehyde	3000	2500	3500	n.d.	n.d.	n.d.
Pyrolytic residue, %	60	30	10	30	40	20
Total sirup, %	25	35	75	50	30	70

[a] Amount expressed as: (moles of compound $\times 10^7$)/g. of saccharide. [b] n.d. = not determined.

Table 11 Amounts of Volatile Materials[a] Produced from Starch and Related Materials after Pyrolysis at 300° for 18 Hours[275]

ysis involved the production of 1,6-anhydro-β-D-mannopyranose from ivory nutmeal.[278] In these experiments, Furneaux and Shafizadeh were able to increase the yield of the anhydrosugar from <5% to 27% by boiling the nutmeal in hydrochloric acid for 10 min followed by a distilled water wash. This acid pretreatment reduced the ash content of the nutmeal from 1.29% to 0.12%. Following pretreatment, the pyrolytic char yield was also reduced from 33% to 27% at 500°C, and DSC studies revealed a dramatic change from exothermicity to endothermicity during pyrolysis. These results appear to indicate the catalytic influence of ash on char formation and suggest ways to improve the anhydrosugar yields from hemicellulose materials.

Mechanisms and Kinetics

• *Anhydrosugars and Related Monomers*

Although the anhydrosugars are generally stable in the vapor phase to temperatures of 500°C, most discussions of their degradation chemistry have been presented in the context of moderate temperatures (<500°C). In keeping with this tradition, the mechanisms of anhydrosugar, vapor-phase pyrolysis are reviewed in this section, although some recent work has been included in a following section.

Berkowitz-Mattuck and Noguchi[70] proposed the free radical mechanism depicted in Fig. 22 to explain the variety of carbonyl compounds observed as products of cellulose pyrolysis. Although multiple pathways for further

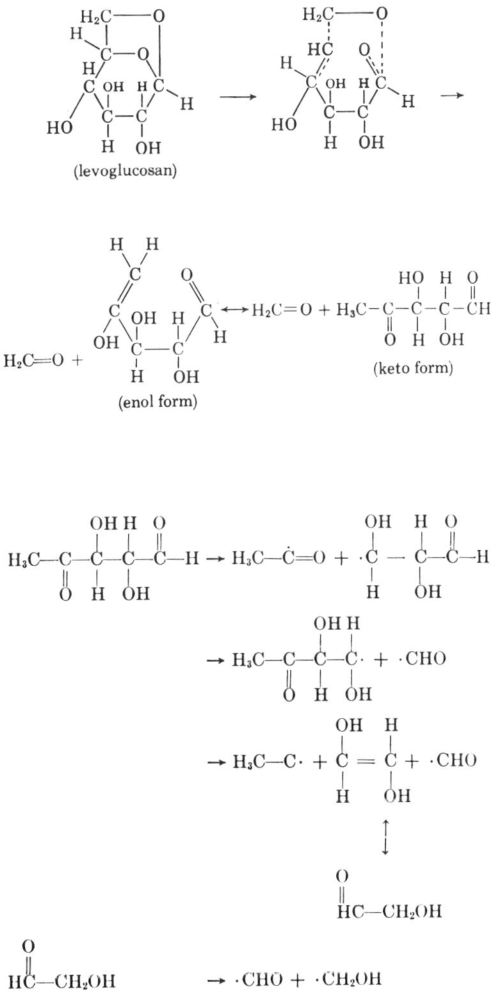

Fig. 22. Mechanism of levoglucosan pyrolysis.[70]

degradation of the carbonyl compounds are clearly available, one pathway which could predominate at higher temperatures has the overall stoichiometry:

$$C_6H_{10}O_5 \rightarrow 4CO + \text{other products.}$$

This pathway represents the ultimate degradation of carbonyl groups to CO, which typically occurs at high temperatures.

Based on the variety of pyrolysis products obtained from 3-deoxy-D-erythro-hexosulose, Shafizadeh and Lai[234] proposed the degradation mechanism displayed in Fig. 23. The variety of condensation, decarbonylation, decarboxylation, and disproportionation reactions displayed in Fig. 23 ultimately results in the formation of condensed phase products (sirups, tars, and char) or a variety of low molecular weight compounds. The latter reactions are favored by high temperatures and low pressures, as well as by certain catalytic additives.

• *Mono and Disaccharides*

In 1968, Shafizadeh[120] reviewed the acid-catalyzed degradation of D-glucose to levulinic acid at temperatures of 160° to 240°C. Figure 24 summarizes the mechanism, which Shafizadeh speculates could be operative at higher temperatures in the absence of catalysts. It is interesting to note that the mechanism evidences competing

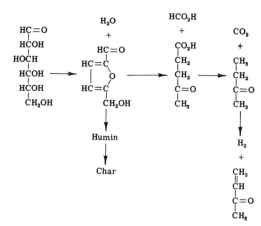

Fig. 24. Acid-catalyzed Degradation of D-Glucose.[120]

reactions, resulting in the formation of condensed materials (humin and char) or lower molecular weight products. If the rate of the condensation reaction is slow relative to the formation of levulinic acid, and if the temperature is sufficiently high to affect the decomposition of the 3-buten-2-one, the overall stoichiometry of the mechanism is:

$$C_6H_{12}O_6 \rightarrow 2CO_2 + H_2O + H_2$$
$$+ CO + CH_4 + C_2H_4.$$

The significance of this result will be discussed further.

The elegant work of Hounimer and Patai[269,270] on the pyrolysis of D-glucose established the main source of anhydrosugar formation to be depolymerization of the glucan created from the D-glucose at lower temperatures. Dehydration reactions were shown to play an essential role in (1) the formation of the glucan by intermolecular dehydration, (2) the formation of an enol by intramolecular dehydration which rearranges to a carbonyl function, and (3) secondary reactions. Hounimer and Patai[270] found that 18% of the H_2O collected had resulted from the formation of carbon-carbon double bonds. Concurrent with the dehydration reactions was the fragmentation of the carbon skeleton, leading to the formation of furans and carbonyl compounds (See Figs. 25–27). The fragmentation reactions are facilitated by the intramolecular de-

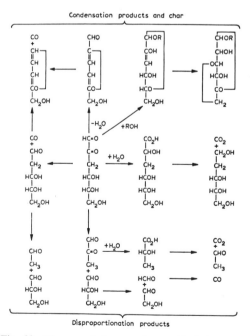

Fig. 23. The pyrolytic reactions of 3-deoxy-D-*erythro*-hexosulose.[234]

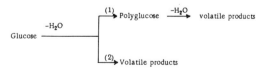

Fig. 25. Competitive mechanisms of glucose pyrolysis.[270]

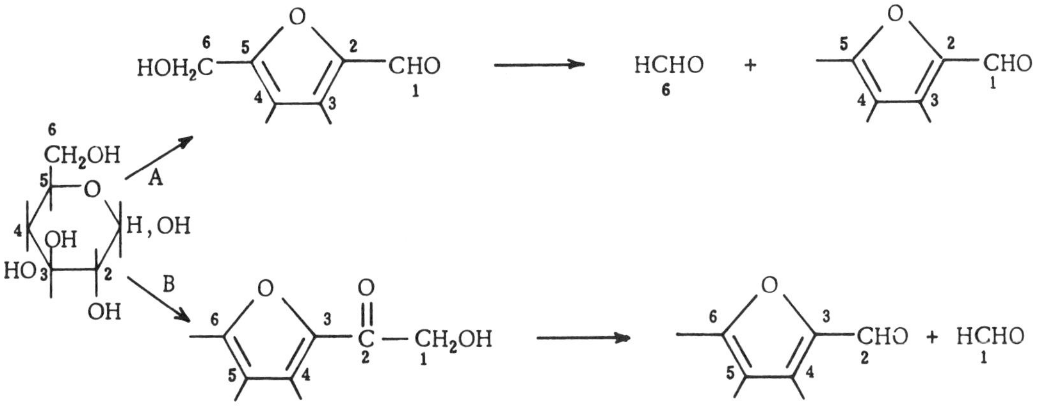

Fig. 26. Competitive pathways for furfural formation from glucose.[270]

hydration reactions. These form methylene and carbonyl groups which weaken the carbon skeleton. The rate of the fragmentation reactions was observed to increase more rapidly with increasing temperature than the rate of the dehydration reactions, leading to the expectation that fragmentation predominates at high temperatures.

Hounimer and Patai[279] also reported studies on the pyrolysis of various glucosides at 250°C. They found that all simple glucosides undergo cleavage with the subsequent release of the aglycone component. In some cases, the component could be quantitatively recovered. Because no glucose was detected in the products, Hounimer and Patai concluded that the cleavage of the glycosidic link in simple aglycones is not hydrolytic but involves immediate polymerization to higher oligomers. The authors also note the effects of the α and β configurations on levoglucosan formation.

Shafizadeh and his colleagues have also described extensive studies of the pyrolysis of a variety of phenyl glucosides and related model compounds.[276,277,280-284] In addition to confirming the earlier work of Hounimer and Patai,[279] this research provided further evidence for the role of heterolytic processes in the formation of randomly linked oligosaccharides by polymerization reactions. It was observed that (1) the pyrolysis reactions of the phenyl glucosides were strongly catalyzed by addition of a Lewis acid, (2) the phenolic aglycones were quantitatively

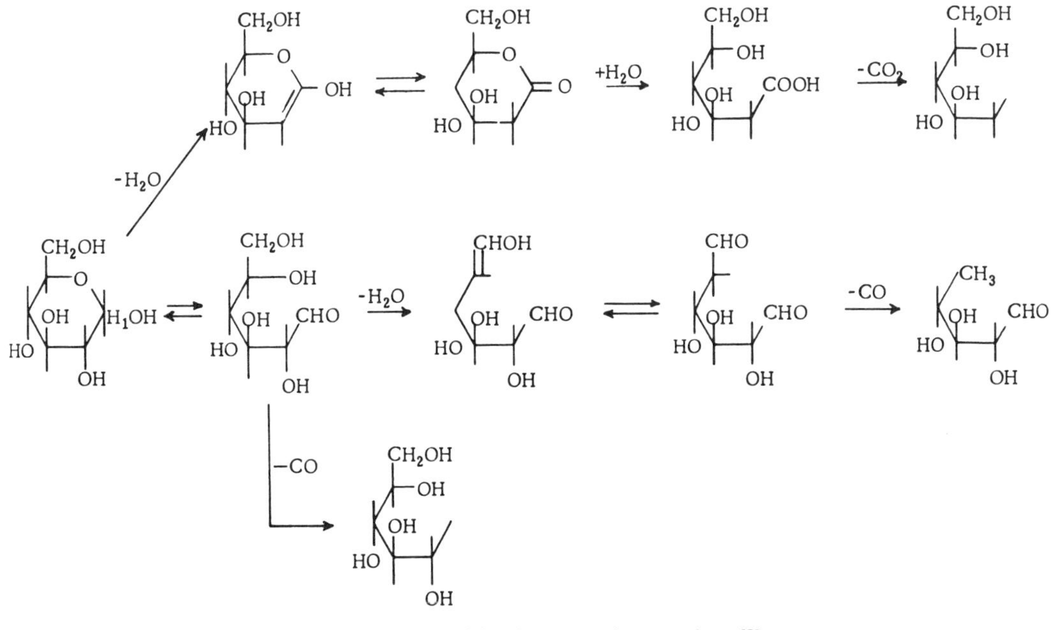

Fig. 27. Competitive fragmentation reactions.[270]

99

recovered as free phenols and not as a variety of products derived from condensation of aryloxy free radicals, (3) the measured rate of formation of free radicals was more consistent with the electron density of glycosidic oxygen than resonance and stability of phenolic free radicals, (4) the measured stability, concentration, and physical characteristics of the free radicals indicated their association with char, (5) the substituents on the aglycone or glycosyl moiety produced an inductive effect on the cleavage of the glycosidic group, thereby influencing the reaction products, and (6) the free hydroxyl groups played a significant role in the cleavage of the glycosic group and the formation of pyrolysis products.[282] Therefore, Shafizadeh and his colleagues concluded that the transglycosylation process proceeds through the heterolytic cleavage of existing glycosidic bonds involving nucleophilic displacement of the glycosidic group by one of the free hydroxyl groups of the sugar molecule. A mechanistic picture of the results of these studies is given in Fig. 28.

The effects of configuration on the pyrolysis reactions of α-and β-D-glucopyranosides were described by Shafizadeh, Lai, and Sussott.[281] Little difference was noted between the untreated α-and β-D-glucopyranoside; however, pyrolysis of the alkali-treated β-D-anomer gave a quantitative yield of levoglucosan. This latter result strongly contrasted with the low yield of levoglucosan obtained from the α-D-anomer and offers some insight into the generally low yields of levoglucosan obtained from starch relative to those derived from cellulose. The increased thermal stability of the β-glycosidic bond was also noted by other workers.[272,285]

• *Polysaccharides*
The permanent gases, CO_2 and CO, as well as H_2O and various anhydrosugars, are the major volatile products of starch pyrolysis at moderate temperatures. Puddington[285] first reported that the formation of CO and CO_2 follows a first-order reaction. The work was expanded by

Bryce and Greenwood[273] who observed an initial rapid first-order reaction followed by a slower first-order reaction. Rapid decomposition of starch, amylose, and amylopectin was observed at 219°C, whereas an equivalent decomposition rate was not obtained with cellulose until temperatures of 270° to 290°C were reached.[273]

For temperatures above about 270°C, Bryce and Greenwood[273] found that the molar ratio $CO_2:CO$ was independent of temperature, with a value between 3:1 and 4:1 for starch, amylose, amylopectin, and cellulose. Similarly, the molar ratio $CO:H_2O$ was observed to fall between 3:1 and 4:1 for the three starches but reached 5:1 for cellulose. At lower temperatures, no linear correlation was found between the production of water and the permanent gases, CO_2 and CO. At higher temperatures ($>350°C$), Cerniani[286] observed the formation of methane and unsaturated hydrocarbons in addition to CO_2, CO, and H_2O.

Because the production versus time curves for CO_2, CO, and H_2O were all nonsigmoidal (that is, the curves manifested a continuously decreasing first derivative), Bryce and Greenwood[273] concluded that there was no induction period, autocatalysis, or liquid phase present during the decomposition of the three starches and cellulose studied. This conclusion was in agreement with the earlier work of Puddington[285] and Murphy.[187] Bryce and Greenwood[273] found that the decomposition reactions of starch, amylose, amylopectin, and cellulose all exhibited an apparent activation energy lying between 29 and 30 kcal/g mol.

The role of competitive pyrolysis reactions in the thermal degradation of xylan and related model compounds has been described by Shafizadeh and his colleagues.[276,277] Depending upon local conditions, the authors found pathways to be available for (1) random condensation of the glycosyl unit formed by thermal cleavage of the glycosidic bond or (2) direct fragmentation of the unit. The competitive reactions involved a complex series of dehydration, disproportiona-

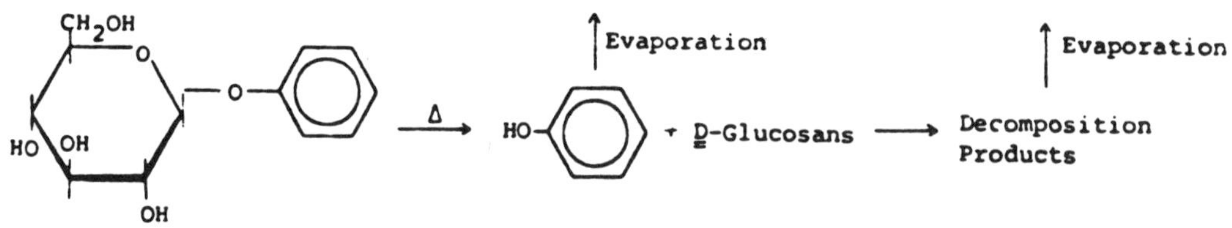

Fig. 28. Pyrolytic reactions of phenyl β-D-glucopyranoside.[122]

tion, and fragmentation mechanisms, some of which were influenced by substituents, as well as the presence of acidic or basic catalysts.

A study of the kinetics of xylan pyrolysis has been reported by Bar-Gadda.[287] In the temperature range of 200° to 235°C, a first-order nucleation mechanism best described the DSC data; from 240° to 255°C, a Carter-Valensi model was followed; and from 270° to 290°C, an Avrami-Erofeyev equation best fit the data. Although these findings are of interest in light of the similar findings by Dollimore and Holt[190] for cellulose pyrolysis, the role of competitive reactions in the posited three temperature range mechanism is not clear.

High Temperature Phenomena

Because the high temperature phenomena of noncellulosic carbohydrates involve both solid- and gas-phase chemistry, these phenomena are considerably more complex than those of the high temperature pyrolysis of cellulosic carbohydrates. A preliminary discussion of the gas-phase pyrolysis chemistry was given earlier, and will be expanded in this section. The solid-phase pyrolysis phenomenon closely resembles that of cellulose and will be discussed prior to the gas-phase chemistry.

Products

When noncellulosic carbohydrates are subject to very rapid heating, they undergo pyrolysis at temperatures exceeding 500°C. Table 12 lists the major products of the very rapid pyrolysis of a variety of biomass materials.[74] Although it is believed that all biomass materials can be completely volatilized by sufficiently rapid heating, the results given in Table 12 indicate the formation of some char from many of the materials tested.

The high temperature, vapor-phase pyrolysis of volatile matter derived from cellulose and hemicellulose typically results in the formation of a hydrocarbon-rich synthesis gas[73,76] and a refractory tar. Intermediate products, which are highly unstable at elevated temperatures, have been observed by Milne[114-116] (see Fig. 16) and Schulten.[197,198]

Mechanisms and Kinetics

As with cellulose, the high yields of CO obtained from the flash pyrolysis of several carbohydrates (see Tables 10 and 12) are very difficult to explain unless a new high temperature pyrolysis pathway is presumed to exist. The low temperature pyrolysis of both dextrose and cellobiose was reported by Puddington[285] to result in CO/CO_2 mass ratios of 0.14 and 0.12, respectively. The mass ratios obtained by Hopkins et al.[74] (see Table 10) were greater than 5. Thus, the high yield of CO cannot be explained by the low temperature pathway discussed above. Similarly, the vapor-phase pyrolysis of volatile matter (largely levoglucosan) derived from cellulose pyrolysis is known (see Table 10) to result in CO/CH_4 and CO/C_2H_4 logarithmic product ratios of 1.9 and 2.2, respectively. The values of these ratios decline with increasing temperature.[77] Thus, vapor-phase pyrolysis is also unable to account for the high yields of CO evi-

Sample	CO_2	CO	H_2	C_2H_2	CH_4	C_2H_4	HC	GAS	LIQ	SOL*	TOTAL
Lignin	2.1	31.6	4.0	3.9	0.2	0.2	0.0	42.0	13.2	20.0 S	75.2
Lignin	1.1	26.9	3.3	3.9	0.3	0.2	0.0	35.7	18.2	26.1 S	80.0
Redwood	4.2	42.3	3.3	5.9	0.6	0.5	0.0	56.8	15.1	6.7 SP	78.6
Redwood	2.3	26.2	1.8	4.7	0.7	0.5	0.0	36.2	16.3	31.7 SP	84.2
Dextrose	3.6	24.8	1.9	4.1	0.7	0.0	0.0	35.1	54.2	10.1 U	99.4
Dextrose	2.5	14.6	0.9	2.2	0.9	0.4	0.0	21.5	62.7	16.5 U	100.7
D-cellobiose	4.5	23.9	1.5	3.2	1.0	1.6	0.0	35.7	49.4	8.9 U	94.0
D-cellobiose	4.3	22.1	1.5	3.3	1.1	1.6	0.1	34.0	52.0	38.7 U	124.7
Kraft paper	2.4	22.8	1.3	1.8	0.7	0.5	0.1	29.6	10.0	43.1 PA	82.7
Kraft paper	2.1	14.9	0.7	1.3	0.6	0.4	0.1	20.1	12.3	52.0 PA	84.4
Leucaena	3.2	49.3	3.3	5.5	0.7	0.5	0.1	62.6	10.1	8.2 P	80.9
Corn Cob	4.0	4.3	0.3	0.7	0.2	0.0	0.0	9.5	11.2	73.4 U	94.1
Corn Cob	1.5	4.3	0.4	0.7	0.2	0.3	0.3	7.7	7.6	83.0 U	98.3
Calatropis	3.1	20.8	1.6	2.1	0.5	0.2	0.1	28.4	26.0	34.4 P	88.8
Newsprint	1.9	20.3	1.1	2.5	0.7	0.5	0.5	27.5	17.1	42.3 PA	86.9
Cow Manure	3.0	37.9	2.8	3.0	0.5	0.0	0.0	47.2	8.9	21.4 P	77.5

* Note: The form of the solid left in the reactor after the flash varied with the substrate. The letters to the right of the solid mass yield indicate the type of solid found in the reactor cell after the experiment. They are as follows: P-primary char, S-secondary char, A-ash, U-unpyrolyzed substrate.

Table 12   Products of the Flash Pyrolysis of Various Biomass Materials[74] (Mass % Yields of Original Sample)

denced in Tables 10 and 12. The only apparent explanation for this behavior is the existence of a high temperature, fragmentation pathway which results in the catastrophic fission of the carbohydrate ring structure and the immediate formation of CO. With so much energy available for the cleavage of molecular bonds at these high temperatures, it is not surprising to find similar behavior among all solid phase carbohydrate materials.

In 1972, Shafizadeh and Lai[288] reported an elegant series of experiments involving the vapor-phase pyrolysis of synthetic levoglucosan labeled at the C1, C2, and C6 positions. By tracing the labeled carbon atoms of the major pyrolysis products back to their original positions in the anhydrosugar molecule, Shafizadeh and Lai were able to deduce the pyrolysis pathways. The results, summarized in Table 13, indicate that the major products are each formed by a variety of pathways. Carbon dioxide and carbon monoxide are primarily derivatives of C1 and C2. The observed formation of CO_2 from the C6 and C2 carbon atoms indicates that extraordinary rearrangements must occur during vapor-phase pyrolysis. 2-Furaldehyde results primarily from C1, C2, C3, C4, and C5, whereas acetaldehyde includes predominantly C6 and C2 atoms. The results given in Table 13 indicate that the uncatalyzed pyrolysis of levoglucosan more closely resembles the acid-catalyzed than the alkaline-catalyzed results. In general, these results appear to be in good agreement with the earlier work of Hounimer and Patai[269,270] on the moderate temperature pyrolysis of D-glucose.

Unfortunately the results of Shafizadeh and Lai[288] offer no insight into the effects of temperature, residence time, and concentration on the products or the mechanisms of the gas-phase-pyrolysis reactions. More recently, research in my laboratory[77] on the vapor-phase pyrolysis of cellulose-derived volatile matter has described these effects using a semibatch, tubular, laminar

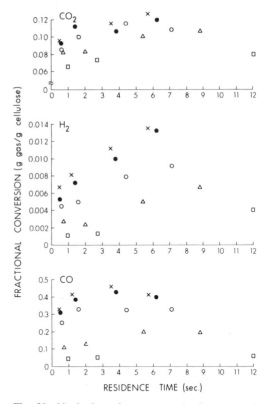

Fig. 29. Nonhydrocarbon gas production vs. residence time for various gas-phase temperatures:[76] (□) 500°, (△) 600°, (○) 650°, (●) 700°, (×) 750°C.

flow reactor. Volatile matter, consisting primarily of levoglucosan and its furanose isomer,[289] evolved from rapidly pyrolyzing cellulose, was carried by flowing superheated steam into the higher temperature, gas-phase cracking zone of the reactor. Variations in the cracking zone's temperature and steam flow rate permitted an evaluation of the effects of temperature and residence time parameters on the product distribution.

Results of these experiments are summarized in Figs. 29–32. The temperature-dependent asymptotic yields evidenced by all the major gaseous products (except perhaps CO_2) could not be

Compound	Neat			5% NaOH			5% ZnCl$_2$		
	1-^{14}C	2-^{14}C	6-^{14}C	1-^{14}C	2-^{14}C	6-^{14}C	1-^{14}C	2-^{14}C	6-^{14}C
2-Furaldehyde	60.8	103.4	35.8	30.2	100.7	73.0	86.0	95.8	16.6
2,3-butanedione	24.8	54.6	31.3	16.5	31.0	56.5	64.8	57.7	26.9
Pyruvaldehyde	27.3	26.3	19.1	23.3	42.0	30.7	49.7	46.7	29.6
Acetaldehyde	10.1	30.5	36.0	6.3	29.2	55.1	4.4	7.3	29.8
Glyoxal	15.4	19.2	6.9	25.5	48.3	36.0	29.5	28.2	28.2
Carbon dioxide	34.3	24.5	6.3	31.2	17.7	8.9	43.7	33.3	9.5
Carbon monoxide	21.1	18.9	16.8	38.4	18.0	13.7	36.7	27.6	11.4

Table 13 Percentage of the pyrolysis products traced to the labeled carbons of 1,6-anhydro-β-D-glucopyranose.[288]

102

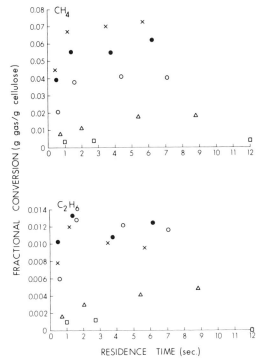

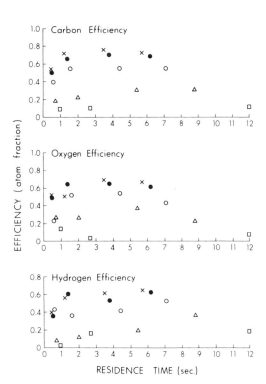

Fig. 30. Paraffinic hydrocarbon gas production vs. residence time for various gas-phase temperatures: [76] (□) 500°, (△) 600°, (○) 650°, (●) 700°, (×) 750°C.

Fig. 32. Carbon, hydrogen, and oxygen efficiency vs. residence time for various gas-phase temperatures: [76] (□) 500°, (△) 600°, (○) 650°, (●) 700°, (×) 750°C.

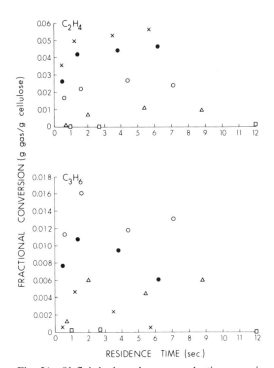

Fig. 31. Olefinic hydrocarbon gas production vs. residence time for various gas-phase temperatures: [76] (□) 500°, (△) 600°, (○) 650°, (●) 700°, (×) 750°C.

correlated to conditions within the reactor using the customary kinetic severity factor.[77] Instead, a competitive mechanism (see Fig. 33) was shown to account for the distribution of gaseous and condensable products. The mechanism, which involves (1) the gas phase cracking of volatiles to permanent gas species with activation energy $E_1 = 49$ kcal/gmol, and (2) the gas phase polymerization of volatiles to a refractory condensable (tar) product with $E_2 = 15$ kcal/gmol, is similar to the solid phase pyrolysis chemistry discussed earlier. The rate law and constants given in Reference 77 correlated the data for carbon efficiency given in Figure 32 with a correlation coefficient R=0.999, giving credence to the mechanism and rate law.

As discussed in Ref. 77, competitive reactions give rise to the following functional dependence of product yields on temperature:

$$\ln (\text{yield of species i/yield of species j}) \propto T^{-1}. \quad (1)$$

Figure 34 displays the yield data for various gas species plotted according to Eq. 1. In Fig. 34, the yield ΔCH_4 of methane (for example) represents the asymptotic (long residence time) formation of methane by the vapor-phase pyrolysis of

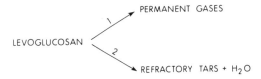

Fig. 33. Vapor-phase, global pyrolysis mechanism of levoglucosan.[77]

cellulose-derived volatile matter (believed to be primarily levoglucosan). Thus, ΔCH_4 = (observed asymptotic yield of methane exiting the gas-phase reactor) − (yield of methane resulting from the primary pyrolysis reactions). The linearity of the data displayed in Fig. 34 provides evidence for the role of competitive reactions in the formation of the various permanent gas species.

The variable $\Delta\eta$, indicated in Figure 34, represents the asymptotic (long residence time) increase in the carbon efficiency due to the vapor-phase pyrolysis reactions. Thus, $\Delta\eta$ = (carbon contained in the asymptotic yield of permanent gases exiting the gas-phase reactor ÷ carbon in the cellulose substrate) − (carbon contained in

the permanent gases entering the gas phase reactor ÷ carbon in the cellulose substrate). The value of $\Delta\eta$ can be viewed to be a reaction coordinate for the gas-phase cracking reactions which lead to the formation of permanent gases. The constant value of $\ln(\Delta CO/\Delta\eta)$, evidenced in Fig. 34, indicates that as $\Delta\eta$ increased with increasing temperature (see Fig. 32), ΔCO increased equally rapidly. Apparently, a fraction of the carbon atoms composing the volatile matter are *dedicated* to CO formation by the gas-phase cracking reaction, irrespective of the prevalent gas-phase temperature.

A thoughtful examination of the trends evidenced in Fig. 34 suggests that for cellulose-derived volatile matter, the gas-phase cracking reaction involves competition between dehydration reactions, resultng in methane and ethylene formation and decarboxylation reactions. The role of dehydration reactions may not be immediately evident from Fig. 34, but can be deduced from the declining yields of CO_2 and the constant relative yield of CO. Recognizing (1) an oxygen balance must be maintained, (2) the pres-

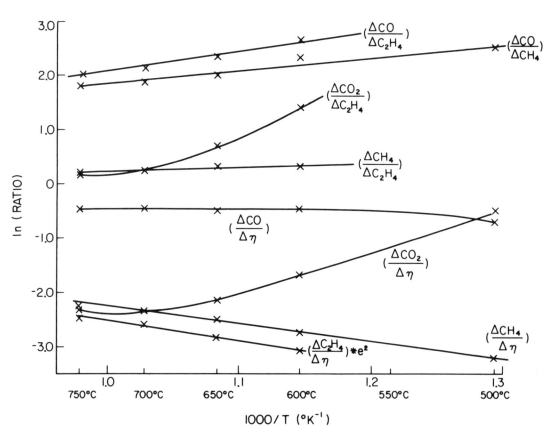

Fig. 34. Ln of the ratios of various gas-phase yields, and the gas-phase carbon efficiency, vs T^{-1} for cellulose-derived volatile matter.[77]

ence of oxygen in the permanent gas species declined with increasing temperature, and (3) water was the only major species undetected in the experiments, it seems necessary to conclude that higher, gas-phase temperatures favorably influence the formation of water over CO_2.

No clear mechanistic explanation for these results presently exists; however, the mechanisms discussed may offer some hints which could lead to the final explanation.

The constant value of $\Delta CH_4/\Delta C_2H_4$ displayed in Fig. 34 strongly suggests that methane and ethylene are products of the same pathway. Moreover, gas-phase pyrolysis data indicate that the molar ratio $CH_4:C_2H_4$ is 1:1 for the vapor-phase pyrolysis reactions. This ratio is in agreement with the mechanism due to Shafizadeh given in Fig. 24. As shown in Fig. 34, the increased formation of CH_4 and C_2H_4 at higher temperatures comes at the expense of CO_2 formation, suggesting the existence of a second low temperature pathway involving decarboxylation chemistry. Although the scheme given in Figure 24 could be modified to better account for CO_2 formation, its real weakness is the low yield of CO projected by the mechanism. Actual yields of CO are about 3.6 moles CO per mole of levoglucosan decomposed. This high yield of CO closely resembles the scheme of Berkowitz, Mattuck, and Noguchi (see Fig. 22); however, their scheme does not appear to provide for the concurrent formation of C_2H_4 and CH_4. A suitable combination of the two mechanisms would overcome the problems evidenced by each one individually; however, no combination of these mechanisms easily accounts for the apparent dedication of carbon atoms to CO formation. The free radical reaction

$$CO_2 + H\cdot \rightarrow CO + OH\cdot$$

could act to enhance the formation of CO by consuming the CO_2 formed by the acid-catalyzed mechanism. Moreover, subsequent reactions of the resultant $OH\cdot$ radicals should lead to the formation of H_2O, thus giving the impression of competition between decarboxylation and dehydration pathways. However, the ability of this complex series of free radical reactions, connecting free radical and concerted displacement or acid-catalyzed, carbonium-ion pathways, to maintain the constant formation of CO over a wide range of temperatures, seems remote to this reviewer.

Effects of Various Parameters

Few quantitative studies of the effects of heating rate, pressure, particle size, and gaseous environment on the pyrolysis of noncellulosic carbohydrates exist in the literature. High heating rates are thought to favor the transglycosylation and fragmentation reactions, thereby reducing char formation.[74] Conversely, high pressures favor the char forming, exothermic condensation reactions.[290] No studies of the effects of particle size or gaseous environment on the pyrolysis of noncellulosic carbohydrates are known to this reviewer.

As is the case with cellulose, extensive studies have been described on the effects of additives on the pyrolysis of noncellulosic carbohydrates. At low temperatures, acids are used commercially to enhance the formation of white and yellow dextrins from starch.[210] Alkaline catalysts are sometimes used in the preparation of British gums from starch. Both dextrin and gum formation involve complex transglycosylation and repolymerization reactions,[210] which apparently are accelerated by the presence of either acid or alkali catalysts. At higher temperatures, the addition of simple salts favorably influence the degradation of starch and the formation of CO_2 and CO.[210]

Acid catalysts enhance the rates of the condensation reactions of levoglucosan[229,263] and glucose[235] at low temperatures, and the gas-phase dehydration reactions of levoglucosan[261,262,288] and related molecules[266,267] at higher temperatures. Alkaline catalysts enhance the fission reactions of various sugars,[272] levoglucosan,[288] and related anhydrosugars,[268,272] resulting in the formation of various carbonyl compounds and CO_2. No catalysts have been reported which enhance the formation of anhydrosugars from noncellulosic carbohydrate materials.

Summary and Critique

Below 200°C, all noncellulosic carbohydrates undergo condensation reactions, forming complex, highly branched dextrins. The mechanism of dextrinization is thought by some workers to involve a carbonium-ion intermediate, but convincing evidence for the heterolytic mechanism is lacking. In view of the wealth of evidence supporting the role of free radical reactions in cellulose pyrolysis, their absence here would be surprising. Extensive studies involving ESR, free radical sensitizers, and other techniques are needed to gain insight into this question.

At higher temperatures, the condensed polymers formed at lower temperatures undergo transglycosylation reactions to produce anhydrosugars. The extensive research of Shafizadeh indicates that these transglycosylation reactions involve a heterolytic mechanism. In addition to the production of anhydrosugars, the transglycosylation reactions can also result in further condensation of the branched polymer. Adding further complexity to the picture, fragmentation reactions produce a variety of carbonyl compounds in this temperature range. With the possible exception of the condensation pathways, none of these competing reactions are strongly favored by any known set of pyrolysis conditions.

Above 500°C, two competing global reactions in the vapor phase consume the reactive anhydrosugars and related solid phase pyrolysis products. A polymerization reaction results in the formation of a refractory tar, while cracking reactions produce a variety of light hydrocarbons, CO, CO_2, H_2, and H_2O from the volatile, solid-phase pyrolysis products. Curiously, a fraction of the carbon atoms which participate in the vapor-phase cracking reactions are *dedicated* to CO formation. Further competition is evidenced in the cracking pathway between dehydration reactions which form CH_4 and C_2H_4, and decarboxylation reactions. No mechanistic explanation of these results is presently available.

At very high heating rates, the fragmentation reactions play an increasingly important role in the solid-phase pyrolysis reactions. Extraordinary yields of CO are generated by these reactions with little dependence on the material undergoing pyrolysis.

The nonspecificity of the noncellulosic carbohydrate pyrolysis reactions is a troublesome reality. Conditions available for manipulation by an engineer usually do not cause pyrolysis to occur via a single pathway. The favorable influence of acid washing on the formation of anhydrosugars from ivory nutmeal is a noteworthy exception to this rule. Conditions which favor char formation are also well known. Future efforts aimed at directing the pyrolysis reactions towards the formation of high value products merit greater emphasis by the research community.

CONCLUSIONS

The global mechanism of carbohydrate pyrolysis involves three competing reactions. Condensation (polymerization) reactions, resulting in the formation of char, CO_2, CO, and H_2O, dominate the pyrolysis phenomena below 250°C. At somewhat higher temperatures, depolymerization reactions generate volatile anhydrosugars and related monomeric compounds. When subjected to very rapid heating, cellulosic materials undergo catastrophic fragmentation, producing H_2, CO, and other simple carbonyl compounds. Noncellulosic carbohydrates seem to undergo similar fragmentation at very high heating rates. These fragmentation reactions are observed only under extraordinary conditions and do not play a role in most pyrolysis experiments. Little agreement exists in the literature on the mechanistic details of these reactions.

Two of the three competing pyrolysis pathways result in the formation of low value products: char, CO, CO_2, H_2O, and H_2. These competing parasitic reactions reduce the yields of higher value anhydrosugar and other monomers at all temperatures. Moreover, the parasitic reactions are easily catalyzed. Consequently, high yields of high value products are difficult to effect. This is particularly true for the hemicellulose component of biomass materials. Secondary reactions also reduce the yields of high value products. Responding to these obstacles, Shafizadeh[291] has proposed a commercial process, based on dilute acid hydrolysis of the biomass, producing simple sugars from the hemicellulose component prior to vacuum pyrolysis of the lignocellulosic residue. Russian literature[175] suggests that the Soviet Union has already begun to implement sirup-producing commercial processes based on these concepts, and similar technologies are now being developed in the U.S.[117] The implications of these developments will be more fully discussed in Part 2 of this review to appear in Vol. 3 of this review series.

ACKNOWLEDGMENTS

The preparation of this review was supported by the National Science Foundation under Grant No. PFR-8008690, the U.S. Department of Energy under Sub-Contract No. B-C5822-A-Q, and the Coral Industries Endowment. The author thanks O. Zaborsky (NSF), B. Berger, S. Friederick, M. Gutstein (DOE), D. Stephens (BPNL), and D. Chalmers (Coral Industries) for their interest in this work. The author also thanks M. Seibert for his assistance with the preparation of this manuscript.

REFERENCES AND NOTES

1. M. King Hubbert. 1956. *Am. Petroleum Inst.*
2. M. King Hubbert. 1962. *Nat. Acad. Sci.-Nat. Res. Counc. Publ. 1000-D.*
3. U.S., Congress, Senate, Committee on Interior and Insular Affairs, Ser. no. 93-40(92-75). 93d Cong., 2d session, 1974.
4. M. King Hubbert. 1978. *McGraw Hill enclycopedia of energy.* New York: McGraw Hill Book Co.
5. T.A. Siddiqi. In *Critical energy issues in Asia and the Pacific.* F. Fesharaki, H. Brown, C. Siddiyao, T. Siddiqi, K.R. Smith, and K. Woodard, eds. 1982. Boulder, Colo: Westview Press.
6. L.E. Rodin. 1975. In *Proceedings of NAS Symposium.* Washington, D.C.: National Academy of Sciences.
7. A.D. Poole and R. H. Williams. 1976. *Bull. At. Sci.* 32:48.
8. R.E. Inman. 1977. Mitre Technical Report No. 7347:1
9. J.A. Alich and J. G. Witwer. 1977. *Solar Energy* 19:625.
10. L.L. Anderson. 1977. In *Fuels from waste.* L.L. Anderson and D.A. Tillman, eds. New York: Academic Press, 2–16.
11. C.C. Burwell. 1978. *Science* 199:1041.
12. A.D. Poole. 1974. In *The energy conservation papers.* Edited by R.H. Williams. Cambridge, Mass.: Ballinger Publishing Co., 219–309.
13. Office of Technology Assessment. Energy from biological processes. 1 OTA-E-123.
14. D. Pimentel et al. 1981. *Science* 212:1110–1115.
15. D.G. Wilson, ed. 1977. *Handbook of solid waste management.* New York: Van Nostrand Reinhold.
16. G. Tchobanoglous; H. Theisen; and R. Eliassen. 1977. *Solid wastes.* New York: McGraw Hill Book Co.
17. P.N. Cheremisinoff and A.C. Moresh. 1976. *Energy from solid wastes.* New York: Marcel Dekker.
18. E.M. Wilson et al. 1978. EPA-600/7-78-086, NTIS. Springfield, Va.
19. S.L. Law. 1977 (Dec.). *J. Wat. Poll. Cont.*
20. B.W. Haynes; J.C. McConnell; and S.L. Law. 1978. Bureau of Mines Rept. of Inves., 8293.
21. B.W. Haynes; S.L. Law; and W.J. Campbell. 1977. Bureau of Mines Rept. of Inves., 8244.
22. R.C. Feber and M.J. Antal. 1977 (Sept.). Environmental Protection Technology Series, EPA-600/2-77-147.
23. J. Shelton and A.B. Shapiro. 1977. *The woodburners encyclopedia.* Waitsfield, Vermont: Vermont Crossroads Press.
24. H.F.J. Wenzl. 1970. *The chemical technology of wood.* New York: Academic Press.
25. J.S. Goldstein. 1977. ACS Symposium Series 43, Am. Chem. Soc.
26. L.E. Wise and E.C. Jahn. 1952. *Wood chemistry.* New York: Reinhold Publishing Co.
27. B.L. Browning. 1963. *The chemistry of wood.* New York: Interscience Publishers.
28. C. Doree. 1947. *The methods of cellulose chemistry. New York: Van Nostrand Co.*
29. R.F. Gould, ed. 1966. Adv. in Chem. Series 59, Am. Chem. Soc.
30. J.A. Perl. 1967. *The chemistry of lignin.* New York: Marcel Dekker, Inc.
31. F.E. Brauns. 1952. *The chemistry of lignin.* New York: Academic Press.
32. E. Adler. 1977. *Wood Sci. Technol.* 11:169

33. K.J. Freudenberg and A.C. Neish. 1968. *Constitution and Biosynthesis of Lignin.* New York: Springer.
34. H. Tarkow, et al. Wood. 1970. In *The encyclopedia of chemical technology.* 2nd ed. Edited by A. Standen. New York: Interscience Publishers.
35. A.B. Booth and J.S. Autenrieth. 1970. Terpenes and terpenoids. In *The encyclopedia of chemical technology.* 2nd ed. Edited by A. Standen. New York: Interscience Publishers.
36. H.I. Enos et al. 1970. Rosin and rosin derivatives. In *The encyclopedia of chemical technology.* 2nd ed. Edited by A. Standen. New York: Interscience Publishers.
37. M.J. Antal. 1978 (July). Biomass energy enhancement—a report to the president's council of environmental quality. NTIS PB 296-624/OGA.
38. Courtesy of S. Bosdech. DeKalb AgResearch, Inc., DeKalb, Ill.
39. J.E. Halligan, K.L. Herzog, and H.W. Parker. 1975. *Ind. Eng. Chem., Process Des. Develop.* 14:64.
40. I.L. Bogert and D.S. Green. 1977. *Prog. in Wat. Tech.* 9:467.
41. M.J. Antal, Jr. 1980. NTIS DE 081904166 (PB-81-134793).
42. SERI, Retrofit '79. *Proceedings of a workshop on air gasification.* SERI/TP-49-183.
43. N.J. Weinstein and R.F. Toro. 1976. *Ann Arbor Sci.* Ann Arbor, Mich.
44. F. Shafizadeh; K.V. Sarkanen; and D.A. Tillman, eds. 1976. *Thermal uses and properties of carbohydrates and lignins.* New York: Academic Press.
45. L.L. Anderson and P.A. Tillman, eds. 1977. *Fuels from waste.* New York: Academic Press.
46. D.A. Tillman. 1979. *Progress in biomass conversion.* New York: Academic Press.
47. P.N. Cheremisinoff. 1980. *Biomass: applications, technology and production.* New York: M. Dekker.
48. S. Sofer and O. Zaborsky. 1981. *Biomass conversion processes for energy and fuels.* New York: Plenum Press.
49. T. Reed. 1981. *A survey of biomass gasification.* Park Ridge, N.J.: Noyes Data Corp.
50. J. Coombs and D.O. Hall, eds. *Biomass, an international journal.* Essex, England: Applied Science Publishers.
51. *Proceedings, specialists workshop on fast pyrolysis of biomass.* 1980. SERI/CP-622-1096.
52. J. Diebold and J. Scahill. 1982 (Mar.). SERI/PR-234-1456.
53. J.L. Kuester. 1980. Olefins from cellulose pyrolysis. Preprint. Alternate Feedstocks for Petrochemicals Symposium. Division of Petroleum Chemistry. American Chemical Society National Meeting. Las Vegas; Aug. 24–29, 1980.
54. F. Shafizadeh and A.G.W. Bradbury. 1979. *J. Appl. Polym. Sci.* 23:1431–42.
55. S.L. Madorsky, V.E. Hart, and S. Straus. 1956. *J. Res. Nat. Bur. Stand.* 56:343–54.
56. S.L. Madorsky. 1958. *J. Res. Nat. Bur. Stand.* 60:343
57. S.L. Madorsky. 1964. *Thermal degradation of organic polymers.* New York: Interscience Publishers, J. Wiley and Sons, Inc.
58. Y. Halpern and S. Patai. 1969. *Israel J. of Chem.* 7:673–83.
59. M. Weinstein and A. Broido. 1970. *Combus. Sci. & Tech.* 1:287–92.

60. A. Basch and M. Lewin. 1973. *J. Polym. Sci.* 11:3071–93.
61. A. Basch and M. Lewin. 1973. *J. Polym. Sci.* 11:3095–101.
62. A. Basch and M. Lewin. 1974. *J. Polym. Sci.* 12:2053–63.
63. A.G.W. Bradbury; Y. Sakai; and F. Shafizadeh. 1979. *J. Appl. Polym. Sci.* 23:3271–80.
64. A.E. Lipska and W.J. Parker. 1966. *J. Appl. Polym. Sci.* 10:1439–53.
65. P.C. Lewellen; W.A. Peters; and J.B. Howard. 1976. Presented at the Sixteenth International Symposium on combustion. MIT.
66. M.R. Hajaligol; J.B. Howard; and J.P. Longwell; W.A. Peters. 1982. *Ind. Eng. Chem. Process Des. Dev.* 21:457.
67. B. Iatridis and G.R. Gavalas. 1979. *Ind. Eng. Chem. Prod. Res. Dev.* 18:127–30.
68. J.B. Howard; W.A. Peters; and M.A. Serio. 1981. Coal devolatilization information for reactor modeling. AP1803 Research Project 986–5.
69. S.B. Martin. 1965. Presented at Tenth Symposium (International) on Combustion. *The Comb. Inst.* 877–896.
70. J.B. Berkowitz-Mattuck and T. Noguchi. 1963. *J. Appl. Polym. Sci.* 7:709–25.
71. K.A. Lincoln. 1965. *Pyrodynamics* 2:133–143.
72. K.A. Lincoln. 1980. In *Proceedings of the Specialists' Workshop on Fast Pyrolysis of Biomass.* Copper Mountain, Colo. SERI/CP-622-1096.
73. J. Lede et al. 1980. *Revue Phys. Appl.* 15:545–552.
74. M. Hopkins; M.J. Antal, Jr.; and J. Kay. *J. Applied Polym. Sci.* Submitted.
75. M.J. Antal, Jr. 1978. In *Energy from biomass and wastes.* Edited by D.E. Klass. Washington, D.C.
76. M.J. Antal, Jr. 1981. In *Biomass as a nonfossil fuel source.* Edited by D.E. Klass. American Chemical Society. Washington, D.C.
77. M.J. Antal, Jr. *Ind. Eng. Chem. Prod. Res. Dev.* In press.
78. T.W. Mattocks. 1981. MSE thesis. Princeton University.
79. W. Mok and M.J. Antal, Jr. *Thermochim. Acta.* Submitted.
80. R.J. McCarter. 1973. *J. Appl. Polym. Sci.* 17:1833–46.
81. R.J. McCarter. 1972. *Textile Res. J.* 42:709–19.
82. R.R. Baker. 1975. *J. Thermal Analysis* 8:163–73.
83. A. Cutler. 1983. Ph.D. thesis. Princeton University.
84. F.J. Kilzer and A. Broido. 1965. *Pyrodynamics* 2:151–63.
85. A. Broido and M. Weinstein. In *Thermal Analysis Proceedings 3rd ICTA.* 1971. Edited by Hans G. Wiedemann. Basel: Birkhauer, pp. 285–296.
86. A. Broido. 1976. In *Thermal uses and properties of carbohydrates and lignins.* By F. Shafizadeh; K.V. Sarkanen; and D.A. Tillman, eds. New York: Academic Press, pp. 19–35.
87. P.D. Garn. 1965. *Thermoanalytic methods of investigation.* New York: Academic Press, Inc.
88. W.J. Smothers and Y. Chiang. 1966. *Handbook of differential thermal analysis.* New York: Chemical Publishing Co.
89. E.M. Barrall, II and J.F. Johnson. 1970. In *Techniques and methods of polymer evaluation.* P.E. Slade, Jr. and L.T. Jenkins, eds. New York: Marcel Dekker, Chap. 1.
90. R.C. Mackenzie. 1970. *Differential thermal analysis.* London: Academic Press, Inc.
91. T. Daniels. 1973. *Thermal analysis.* New York: J. Wiley and Sons, Inc.
92. E.M. Barrall, II. 1972. In *Guide to modern methods of instrumental analysis.* Edited by T.H. Gouw. New York: Wiley, Chap. 12.
93. W.W. Wendlandt. 1974. *Thermal methods of analysis.* 2nd ed. New York: J. Wiley and Sons, Inc.
94. H. Roderig; A. Basch; and M. Lewin. 1975. *J. Polym. Sci.* 13:1921–32.
95. M. Lewin; A. Basch; and C. Roderig. 1975. In *Proceedings of the International Symposium on Macromolecules.* Edited by E.B. Mano. Rio De Janeiro, Brazil; July 26–31, 1975. Elsevier, Amsterdam.
96. Y. Halpern and S. Patai. 1969. *Israel J. Chem.* 7:691–6.
97. W.K. Tang. 1967. *Forest Prod. Laboratory Paper* 71. Madison, Wisc.
98. W.K. Tang and W.K. Neil. 1964. In *Thermal analysis of high polymers: J. Polym. Sci. C.* Edited by B. Ke. New York: Interscience, p. 65.
99. D.F. Arseneau. 1971. *Can. J. Chem.* 49:632–638.
100. D.F. Arseneau. 1971. In *Proceedings of the Third ICTA.* Davos.
101. M.J. Antal, Jr.; H.L. Friedman; and F.E. Rogers. 1980. *Comb. Sci. and Tech.* 21:141–52.
102. W. Mok and M.J. Antal, Jr. *Thermochim. Acta.* Submitted.
103. J.H. Flynn. 1980. *Thermochim. Acta* 37:225.
104. M.J.Antal, Jr. 1982. In *Thermal Analysis Proceedings of the 7th ICTA.* Kingston, Canada.
105. J.C. Arthur, Jr. and O. Hinojosa. 1966. *Textile Res. J.* 36:385–7.
106. J.C. Arthur, Jr.; T. Mares; and O. Hinojosa. 1966. *Textile Res. J.* 36:630.
107. P.J. Baugh; O. Hinojosa; and J.C. Arthur, Jr. 1967. *J. Appl. Polym. Sci.* 11:1139.
108. T. Maresad and J.C. Arthur, Jr. 1969. *J. Polym. Sci.* 7:419–25.
109. H. Hatakeyama et al. 1969. *Tappi* 52:1724–8.
110. H. Hatakeyama and J. Nakano. 1970. *Tappi* 53:472–5.
111. F. Shafizadeh; P.P.S. Chin; and W.F. DeGroot. 1977. *Forest Sci.* 23:81–9.
112. F. Shafizadeh and W.F. DeGroot. 1977. In *Fuels and energy from renewable resources.* New York: Academic Press, Inc.
113. R.A. Susott; F. Shafizadeh; and T.W. Aanerud. 1979. *J. Fire & Flammability* 10.
114. T.A. Milne and M.N. Soltys. 1981. Fundamental pyrolysis studies. Annual Report for Fiscal Year 1981. SERI/PR-234-1454.
115. T.A. Milne and M.N. Soltys. 1982. Fundamental pyrolysis studies. Quarterly Report. SERI/PR-234-1537.
116. T.A. Milne and M.N. Soltys. 1982. Fundamental pyrolysis studies. Quarterly Report. SERI/PR-234-1617.
117. M.W. Hopkins; C.I. DeJenga; and M.J. Antal, Jr. To appear in *Prog. Sol. Energy.*
118. G. David and M. MacKay. 1967. Forestry Branch Dept. Pub. 1201, ODC 813.4.
119. J.R. Welker. 1970. *J. Fire and Flammability* 1:12.
120. F. Shafizadeh. 1968. *Adv. in Carbohydrate Chem.* 23:419–74.
121. F. Shafizadeh. 1975. In *Appl. Polym. Symp. No. 28.* New York: J. Wiley and Sons, pp. 153–74.
122. F. Shafizadeh. 1980. In *Proceedings of the Specialists Workshop on Fast Pyrolysis of Biomass.* SERI/CP-622-1096.
123. P.M. Molton and T.F. Demmitt. 1977. BNWL-2297 UC. Richland, Wash.: Battelle Pacific Northwest Lab.

124. T. Milne. 1981. In *Biomass gasification: principles and technology*. Edited by T. Reed Park Ridge, N.J.: Noyes Data Corp.

125. F.J. Kilzer. 1971. In *Cellulose and cellulose derivatives*. Vol. 5, 2nd Ed. New York: Wiley-Interscience, 1015–46.

126. Sirups are commonly referred to as "tars" in the literature, but this use of the word tars is misleading. Tars are composed of hydrocarbons, whereas the sirups are composed of anhydrosugars and related carbohydrates. The spelling of the "sirups" was chosen to differentiate the pyrolysis products from edible syrups.

127. M. Lewin and A. Basch. Cellulose. In Fire retardancy. In *The encyclopedia of polymer science and technology. Suppl. Vol. 2.* 1977. New York: J. Wiley & Sons, Inc., pp. 340–62.

128. W.D. Major. 1958. *Tappi* 41:530.

129. S.T. Kosiewicz. 1980. *Thermochimica Acta* 40: 322–6.

130. O.P. Golova and R.G. Krylova. 1957. *Dokl. Akad. Nauk, SSSR* 116:419.

131. S. Patai and Y. Halpern. 1970. *Israel J. Chem.* 8:655–62.

132. T.V. Gatovskaya et al. 1959. *Zh. Fiz. Khim.* 33:1418; *Chem. Abs.* 54:8212. (1960).

133. A. Paucault and G. Sauret. 1958. *Compt. Rend. Acad. Sci.* 246:608–11.

134. D.P.C. Fung. 1969. *Tappi* 52:319.

135. A. Broido and F.J. Kilzer. 1963. *Fire Res. Abstr. Rev.* 5:157–61.

136. A. Broido et al. 1973. *J. Appl. Polym. Sci.* 17:3627–35.

137. A. Broido and H. Yow. 1977. *J. Appl. Polym. Sci.* 21:1677–85.

138. A. Broido and M. Weinstein. 1970. *Comb. Sci. Tech.* 1:279–85.

139. F. Shimazu and C. Sterling. 1966. *J. Food Sci.* 31:548.

140. P.K. Chatterjee and C.M. Conrad. 1966. *Text. Res. J.* 36:487–94.

141. P.K. Chatterjee. 1968. *J. Appl. Polym. Sci.* 12:1859–64.

142. K. Kato and H. Komorita. 1968. *Agr. Biol. Chem.* 32:21–6.

143. A. Basch and M. Lewin. 1975. *Polymer Letters Edition* 13:493–9.

144. E.L. Back; M.T. Htun; and F. Johanson. 1967. *Text. Res. J.* 37:432–3.

145. E.L. Back. 1967. *Tappi* 50:542–7.

146. N.S. Hon. 1975. *J. Polym. Sci., Polym. Chem. Ed.* 13:1347–61.

147. N.S. Hon. 1975. *J. Polym. Sci., Polym. Chem. Ed.* 13:955–9.

148. W.A. Reeves. 1962. *J. Text. Inst.* 53:22–36.

149. H. Mehta. 1969. *Text. Res. J.* 39:387–90.

150. R.S. Parikh. 1967. *Text. Res. J.* 37:538.

151. R.H. Barker and S.L. Vail. 1967. *Text. Res. J.* 37:1077–8.

152. A. Pictet and J. Sarasin. 1918. *Compt. Rend.* 166:38. *Helv. Chim. Act.* 1:87 (1918).

153. R.F. Schwenker, Jr. and L.R. Beck, Jr. 1963. *J. Polym. Sci.* Part C(2):331–40.

154. S. Glassner and A.R. Pierce, III. 1965. *Anal. Chem.* 37:525.

155. S.B. Martin and R.W. Ramstad. 1961. *Anal. Chem.* 33:982.

156. C.T. Greenwood; J.H. Knox; and E. Milne. 1961. *Chem. Ind.* (London) 46:1878.

157. D. Gardiner. 1966. *J. Chem. Soc.* (C):1473–6.

158. G.A. Byrne; D. Gardiner; and F.H. Holmes. 1966. *J. Appl. Chem.* 16:81–8.

159. A.E. Lipska and F.A. Wodley. 1969. *J. Appl. Polym. Sci.* 13:851–65.

160. F.A. Wodley. 1971. *J. Appl. Polym. Sci.* 15: 835–51.

161. W.E. Franklin. 1979. *Anal. Chem.* 51:992–6.

162. H.R. Schulten; U. Bahr; and W. Gortz. 1981. *J. Anal. Appl. Pyrolysis* 3:229–41.

163. A. Broido and M.A. Nelson. 1975. *Comb. & Flame* 24:263–8.

164. R.C. Smith and H.C. Howard. 1937. *J. Amer. Chem. Soc.* 59:234–6.

165. F. Shafizadeh; A.G.W. Bradbury; and W.F. DeGroot. To appear.

166. E. Rensfelt et al. 1978. In *Energy from biomass and wastes*. Edited by D.E. Klass, Washington, D.C.

167. O.P. Golova; R.G. Krylova; and I.I. Nikolaeva. 1959. *Vysokomol. Soedin* 1:1235.

168. A.M. Pakhomov; O.P. Golova; and I.I. Nikolaeva. 1957. *Izv. Akad. Nauk. SSSR, Otdel. Khim. Nauk.* 521. *Chem. Abs.* 52:5811. (1958).

169. O.P. Golova et al. 1957. *Dokl. Akad. Nauk. SSSR* 115:1122. *Chem. Abs.* 52:4165. (1958).

170. Some reviewers[120,123] associate Madorsky with a different point of view, however, this author is unable to find the attributed support for the carbonium ion mechanism in his writings.

171. A.N. Kislitsyn et al. 1971. *Zh. Prikl. Khim.* (Leningrad) 44:2587.

172. A.N. Kislitsyn and Z.M. Rodionova. 1969. *Tr. Tsentr. Nauchn.-Issled. Inst. Lesokhim. Prom.* 20.

173. A.M. Pakhomov. 1957. *Izv. Akad. Nauk. SSSR Otdel Khim. Nauk.* 1497. *Chem. Abst.* 52:5811. (1958).

174. O.P. Golova; A.M. Pakhomov; and E.A. Andrievskaya. 1957. *Izv. Akad. Nauk. SSSR, Otdel. Khim. Nauk.* 1499. *Chem. Abstr.* 52:5811. (1958).

175. O.P. Golova. 1975. *Russian Chemical Reviews* 44:687–97.

176. F. Shafizadeh and Y.L. Fu. 1973. *Carbohydrate Res.* 29:113–22.

177. F. Shafizadeh et al. 1979. *J. Appl. Polym. Sci.* 23:3525–39.

178. F. Shafizadeh; T. Cochran; and Y. Sakai. 1979. *AIChE Symposium Series 184* 75:24–34.

179. R.D. Cardwell and P. Luner. 1976. *Wood Sci. Tech.* 10:131–47.

180. R.D. Cardwell and P. Luner. 1976. *Wood Sci. Tech.* 10:183–98.

181. R.D. Cardwell and P. Luner. In *Preservation of paper and textiles*, 362–81.

182. A.F. Roberts. 1970. *J. Appl. Polym. Sci.* 14:244–7.

183. K. Akita and M. Kase. 1967. *J. Polym. Sci. A* 5:833.

184. M.V. Ramiah. 1970. *J. Appl. Polym. Sci.* 14:1323.

185. F.L. Browne and W.K. Tang. 1962. *Fire Res. Abst. Rev.* 4:76.

186. C.H. Mack and D.J. Donaldson. 1967. *Textile Res. J.* 37:1063.

187. E.J. Murphy. 1962. *J. Polym. Sci.* 58:649.

188. V.K. Shivadev and H.W. Emmons. 1974. *Comb. Flame* 22:223.

189. D.L. Urban and M.J. Antal, Jr. 1982. *FUEL* 61:799–806.

190. C. Dollimore and B. Holt. 1973. *J. Polym. Sci. A* 11:1703.

191. C. Fairbridge; R.A. Ross; and S.P. Sood. 1978. *J. Appl. Polym. Sci.* 22:497–510.

192. C. Kala; I.S. Gur; and H.L. Bhatnagar. 1980. *Indian J. Chem.* 19A:641–5.

193. E. Chornet and C. Roy. 1980. *Thermochimica Acta* 35:389–93.
194. R.D. Cardwell and P. Luner. 1978. *Tappi* 61: 81–4.
195. C. Fairbridge; R.A. Ross; and P. Spooner. 1975. *Wood Sci. Tech.* 9:257.
196. I.L. Eventova et al. 1974. *Khim. Volokra* 4: 29–31.
197. H.R. Schulten and W. Gortz. 1978. *Anal. Chem.* 50:428–33.
198. H.R. Schulten; U. Bahr; and W. Gortz. 1981. J. Anal. Appl. Pyrolysis 3:137.
199. J.G. Wiegerink. 1940. *J. Res. Natl. Bur. Stand.* 25:435.
200. R.C. Waller; K.C. Bass; and W.E. Roseveare. 1948. *Ind. Eng. Chem.* 40:138.
201. A.J. Stamm. 1956. *Ind. Eng. Chem.* 48:413.
202. M.J. Antal, Jr.; L. Hofmann; J.R. Moreira; C.T. Brown; and R. Steenblick. *Solar Energy.* In press.
203. L.K. Mudge; L.J. Sealock, Jr., and S.L. Weber. 1974. *J. Anal. Appl. Pyrolysis* 1:165.
204. X. Deglise; C. Richard; A. Rolin; and H. Francois. Personal communication.
205. J.E. Halligan; K.L. Herzog; and H.W. Parker. 1975. *Ind. Eng. Chem. Process Des. Dev.* 14:64.
206. O. Wolf. 1950. *Staerke* 2:273.
207. J.A. Radley. 1953. *Starch and its derivatives*. 3rd ed. vol. 2. London: Chapman Hall Ltd., p. 107.
208. C.C. Gaper and D.M. Rathman. 1959. In *Industrial gums*. Edited by R.L. Whistler. New York: Academic Press, Inc., p. 699.
209. G.V. Caesar. 1950. In *Chemistry and industry of starch*. 2nd ed. Edited by R.W. Kerr. New York: Academic Press, Inc., p. 345.
210. C.T. Greenwood. 1967. *Adv. Carbohyd. Chem.* 22:483.
211. Olof Theander. In *Carbohydrate sweeteners in foods and nutrition*. P. Koivistoinen and C. Hyvonen, eds. 1980. London: Academic Press, pp. 185–199.
212. S. Peat et al. 1958. *J. Chem. Soc.* 586.
213. M. Cerny and J. Stanek. 1970. *Fortschr. Chem. Forsch.* 14:526–55.
214. R.J. Dimler. 1952. *Adv. Carbohy. Chem.* 7:37.
215. M. Cerny and J. Stanek, Jr. 1977. *Adv. Carbohyd. Chem.* 34:23.
216. F. Shafizadeh et al. 1970. *Carbohyd. Res.* 13:184–6.
217. F. Shafizadeh et al. 1970. *Carbohyd. Res.* 15:165–78.
218. F. Shafizadeh; R.A. Susott; and C.R. McIntyre. 1975. *Carbohyd. Res.* 41:351.
219. F. Shafizadeh and R.A. Susott. 1973. *J. Org. Chem.* 38:3710–15.
220. Y.J. Park; H.S. Kim; and G.A. Jeffrey. 1971. *Acta Cryst.* B27:220–7.
221. W. Sandermann and H. Augustin. 1964. *Holz Roh Werkst.* 22:377–386. *Chem. Abstr.* 66:11994. (1967).
222. M.S. Bains. 1979. *Carbohydr. Res.* 34:169–73.
223. *International critical tables*. Vol. 5. 1929. New York: McGraw-Hill, p. 166.
224. D.R. Stull; E.F. Westrum; and G.C. Sinke. 1969. *The chemical thermodynamics of organic compounds*. New York: Wiley.
225. See Ref. 215, pp. 50–63 and references therein.
226. A. Pictet. 1918. *Helv. Chim. Acta* 1:276.
227. H. Pringsheim and K. Schmalz. 1922. *Ber.* 55B:3001–7.
228. J.C. Irvine and J.W.H. Oldham. 1925. *J. Chem. Soc.* 127:2903.
229. J. daS Carvalho; W. Prins; and C. Schuerch. 1959. *J. Am. Chem. Soc.* 81:4054–8.
230. M.L. Wolfrom; A. Thompson; and R.B. Ward. 1959. *J. Am. Chem. Soc.* 81:4623.
231. M.L. Wolfrom et al. 1961. *J. Org. Chem.* 26:4617.
232. H. Abe and W. Prins. 1961. *Makromol. Chem.* 42:216.
233. A. Bhattecharya and C. Shuerch. 1961. *J. Org. Chem.* 26:3101.
234. F. Shafizadeh and Y.Z. Lai. 1975. *Carbohyd. Res.* 40:263–74.
235. P.T. Mora and J.W. Wood. 1958. *J. Am. Chem. Soc.* 80:685.
236. H.W. Durand; M.F. Dull; and R.S. Tipson. 1958. *J. Am. Chem. Soc.* 80:3691–7.
237. P.T. Mora et al. 1958. *J. Am. Chem. Soc.* 80:693.
238. H. Sugisawa and H. Edo. 1964. *Chem. Ind. (London)* 892.
239. G.G.S. Dutton and A.M. Unrau. 1962. *Can. J. Chem.* 40:1196-1200.
240. G.G.S. Dutton and A.M. Unrau. 1963. *Can. J. Chem.* 41:2439–46.
241. G.G.S. Dutton and A.M. Unrau. 1964. *Can. J. Chem.* 42:2048–55.
242. H. Sugisawa. 1966. *J. Food Sci.* 31:381.
243. K. Kato and H. Komorita. 1968. *Agr. Biol. Chem. (Tokyo)* 32:715–20.
244. G.N. Richards and F. Shafizadeh. 1978. *Aust. J. Chem.* 31:1825–32.
245. J.R. Katz. 1934. *Rec. Trav. Chim.* 53:554.
246. J.R. Katz and A. Weidinger. 1939. *Z. Physik, Chem.* A184:100–22.
247. D. Costa and G. Costa. 1951. *Chim. Ind. (Milan)* 33:71–6.
248. A.T. Perkins and H.L. Mitchell. 1957. *Trans. Kansas Acad. Sci.* 60:437.
249. H. Morita. 1956. 1957. *Anal. Chem.* 28:64. 29:1095.
250. M.C.P. Varma. 1958. *J. Appl. Chem. (London)* 8:117.
251. B. Brimhall. 1944. *Ind. Eng. Chem.* 36:72–5.
252. R.W. Kerr and F.C. Cleveland. 1953. *Staerke* 5:261–6.
253. A. Thompson and M.L. Wolfrom. 1958. *J. Am. Chem. Soc.* 80:6618.
254. M.L. Wolfrom; A. Thomson; and R.B. Ward. 1961. *Ind. Eng. Chem.* 53:217.
255. G.V. Caesar; N.S. Gruenhut; and M.L. Cushing. 1947. *J. Am. Chem. Soc.* 69:617–21.
256. H. Rüggeberg. 1952. *Staerke* 4:78.
257. J.D. Geerdes; B.A. Lewis; and F. Smith. 1957. *J. Am. Chem. Soc.* 79:4209.
258. G.M. Christensen and F. Smith. 1957. *J. Am. Chem. Soc.* 79:4492–5.
259. F. Orsi. 1973. *J. Thermal Anal.* 5:329–35.
260. C.E. Weill; B. Carroll; and J.W. Liskowitz. 1980. *Thermochimica Acta* 37:65–9.
261. A. Broido; M. Evett; and C.C. Hodges. 1975. *Carbohyd. Res.* 44:267–74.
262. Y. Halpern; R. Riffen; and A. Broido. 1973. *J. Org. Chem.* 38:204–9.
263. F. Shafizadeh; C.W. Philpot; and N. Ostojic. 1971. *Carbohyd. Res.* 16:279–87.
264. F. Shafizadeh; R.H. Furneaux; and T.T. Stevenson. 1979. *Carbohyd. Res.* 71:169–91.
265. G.R. Bedford and D. Gardiner. 1965. *Chem. Commun.* 13:287–8.
266. F. Shafizadeh et al. 1978. *Carbohyd. Res.* 61: 519–28.
267. F. Shafizadeh and P.P.S. Chin. 1976. *Carbohyd. Res.* 46:149–54.

268. F. Shafizadeh et al. 1978. *Carbohyd. Res.* 67:433–47.
269. Y. Hounimer and S. Patai. 1967. *Tetrahedron Letters* 14:1297–300.
270. Y. Hounimer and S. Patai. 1969 *Israel J. Chem.* 7:513–24.
271. K. Heyns; R. Stute; and H. Paulsen. 1966. *Carbohyd. Res.* 2:132–49.
272. K. Heyns and M. Klier. 1968. *Carbohyd. Res.* 6:436.
273. D.J. Bryce and C.T. Greenwood. 1966. *Appl. Polym. Symp.* 2:149–58.
274. D.J. Bryce and C.T. Greenwood. 1963. *Staerke* 15:285–90.
275. D.J. Bryce and C.T. Greenwood. 1963. *Staerke* 15:359–63.
276. F. Shafizadeh et al. 1971. *J. Org. Chem.* 36:2813–18.
277. F. Shafizadeh; G.D. McGinnis; and C.W. Philpot. 1972. *Carbohyd. Res.* 25:23–33.
278. R.H. Furneaux and F. Shafizadeh. 1979. *Carbohyd. Res.* 74:354–60.
279. Y. Hounimer and S. Patai. 1969. *Israel J. Chem.* 7:535–46.
280. F. Shafizadeh; R.A. Susott; and G.D. McGinnis. 1972. *Carbohyd. Res.* 22:63–73.
281. F. Shafizadeh; Y.Z. Lai; and R.A. Susott. 1972. *Carbohyd. Res.* 25:387–94.
282. F. Shafizadeh; M.H. Meshreki; and R.A. Susott. 1973. *J. Org. Chem.* 38:1190–4.
283. F. Shafizadeh and Y.Z. Lai. 1973. *Carbohyd. Res.* 31:57.
284. Y.Z. Lai and F. Shafizadeh. 1974. *Carbohyd. Res.* 38:177–87.
285. I.A. Puddington. 1948. *Can. J. Res.* B26:415.
286. A. Cerniani. 1951. *Ann. Chim. (Rome)* 41:293–308.
287. R. Bar-Gadda. 1980. *Thermochimica Acta* 42:153–63.
288. F. Shafizadeh and Y.Z. Lai. 1972. *J. Org. Chem.* 37:278–284.
289. C.I. DeJenga; M.J. Antal; and M. Jones. 1982. *J. Appl. Polym. Sci.* 27:4313–22.
290. W. Mok. 1983. M.S.E. thesis. Princeton University.
291. F. Shafizadeh. 1980. *AS/ISES 1980, Proc. of the 1980 Ann. Mtg.* 31:122–125.

Advances in Solar Energy © 1983 American Solar Energy Society, Inc.

RECOMBINANT GENETIC APPROACHES FOR EFFICIENT ETHANOL PRODUCTION

H. W. STOKES, S. K. PICATAGGIO, AND D. E. EVELEIGH

Department of Biochemistry and Microbiology, Cook College, New Jersey Agricultural Experiment Station, Rutgers University, New Brunswick, New Jersey 08903

Abstract

Ethanol has potential use as an alternative liquid transportation fuel and as a chemical feedstock. The recent restrictions on the availability of oil focused attention on ethanol production through the microbial fermentation of biomass. The economic incentives to develop such fermentation processes are borderline. This paper discusses the manner in which recombinant genetic approaches may potentially relieve certain of the economic bottlenecks and put the fermentative production of ethanol on a firmer basis. It then reviews the genetics of bacteria and yeasts. The manner in which these processes may be used is illustrated with protocols for the development of new microbial strains that can utilize a range of less expensive substrates (cellulose and starch) and that are more tolerant of high ethanol concentrations.

INTRODUCTION

The recent OPEC oil embargo focused attention on the dependence of the United States on foreign oil supplies to meet current energy needs. The finite nature of fossil oil reserves was also sharply brought into focus. A wide variety of solutions were proposed during this period to conserve petroleum reserves. One of these was the fermentative conversion of biomass to chemical feedstocks. In this scenario biomass, including grain (corn), trees, and agricultural and urban wastes, would be fermented through the action of microorganisms to industrially useful chemicals, (for example, ethanol, butanol, acetone, glycerol, and acetic acid). The production of these chemicals was envisaged in itself to be energy sparing and could be considered from both immediate and long-term perspectives. Ethanol was a particularly attractive chemical to produce because it could be used directly as a liquid transportation fuel and implemented as a 10% blend in gasoline called "gasohol." Furthermore, the general principles governing ethanol fermentations were well known.[1] A range of federal government incentives were introduced to promote fuel alcohol production. For example, a four cents per gallon federal excise tax, the

Crude Oil Windfall Profit Tax, was passed in 1980. Fuel alcohol production in the U.S. rapidly reached 185,000 tons/annum by 1980. Although, gasohol did not gain complete market acceptance, today it still has considerable use as both a liquid transportation fuel and as an octane booster. The economics of fuel alcohol production in the U.S. result in marginal profits, but there is potential to increase the efficiency of ethanol production, and hence, considerable worldwide research efforts are being directed in this area. In Brazil, a major gasohol program has been launched based on sucrose (from sugar cane) and starch (from cassava) as fermentation

Harold (Hatch) W. Stokes is currently a Research Associate at the University of Connecticut. Born and educated in Australia, he received his Ph.D. from Monash University in Microbial Genetics (1980). He explored the genetics of Zymomonas *while a Research Associate at Cook College, Rutgers University.*

Stephen Picataggio obtained his B.S. degree (1975) from Wagner College and then worked at Best Foods, N.J. He subsequently enrolled as a Doctoral Candidate in the Graduate Microbiology Program at Rutgers University.

Douglas E. Eveleigh is a Professor of Microbiology at Rutgers University. Trained in Britain he has worked in Canada and the U.S. on various topics in applied microbiology.

substrates. Further details on large-scale fermentative production of alcohol are available from recent reviews.[2-7]

During these developments, two major questions were raised. First, does the production of fuel alcohol result in a positive energy balance?[8] This question can be answered affirmatively provided that there is integration of the several operational phases and there is a near-by market.[9] Second, how profitable is the large-scale production of fuel alcohol? This question is extremely complex, but the answer depends principally on the type and availability of substrate (which also governs the process design), the by-product credits, and the energy expended in recovery of the ethanol from the dilute aqueous fermentation broth.[9] In the U.S., the process of using corn as substrate is profitable.[6] However, as the world price of oil continues to escalate, it is attractive to consider new strategies to improve the overall fermentation. Some general limitations and potential solutions are outlined in Table 1. It is apparent from this table that several economic bottlenecks potentially can be relieved through the genetic improvement of microbial strains. This review focuses on this latter aspect and discusses:

(1) the general type of genetic approach that may be taken to increase ethanol productivity;

(2) the fundamentals of microbial genetics with emphasis on recombinant DNA methodologies; and

(3) specific genetic strategies by which the range of substrates used by a microorganism can be increased (cellulose and starch).

GENETIC APPROACHES TO INCREASE ETHANOL PRODUCTIVITY

Ethanol Fermentations [10, 11]

Yeast *(Saccharomyces cerevisiae)* is well known for its production of ethanol in the brewing and wine industries. Glucose is metabolized via the Embden-Meyerhof pathway (see Fig. 1A). A key enzyme is pyruvate decarboxylase for the conversion of pyruvate to acetaldehyde, and thence to ethanol. Two moles of ethanol are theoretically produced per mole of glucose, and the *in vivo* efficiency is generally well over 90%. Pyruvate decarboxylase is rarely found in bacteria but occurs in *Sarcina ventriculi* and *Erwinia amylovora*. These bacteria also employ the Embden-Meyerhof pathway and principally produce ethanol. However, they also produce small amounts of acetate and lactate, respectively. Another efficient bacterial ethanol producer is *Zymomonas mobilis*. It also has pyruvate decarboxylase, but employs the Entner-Doudoroff pathway for the catabolism of glucose to pyruvate (see Fig. 1B). The efficiency of the conversion of glucose to ethanol is equivalent to that of yeast, but this pathway yields 1 mole ATP/mole glucose, in contrast to the Embden-Meyerhof pathway which yields 2 moles ATP/mole glucose. Many other bacteria, including enterobacteria, clostridia, and lactic acid bacteria, produce ethanol, but the pathway is via acetyl-CoA and acetaldehyde dehydrogenase. This is illustrated in Fig. 1C. However, as each of these bacterial groups produces a unique spectrum of products, the quantitative

	Solution
I. *Productivity of the Microbe*	
A. Low yield	Strains with greater tolerance to high substrate and ethanol concentrations
B. Low Fermentation Rates	Relieve the regulatory controls of the ethanol pathway
II. *More Effective Processing*	
A. Reduce the costs of operation and of capital equipment	(1) Utilization of cheaper substrates (e.g., cellulose)
	(2) Use of "high density" cultures in combination with vacuum fermentation
	(3) Development of continuous culture and immobilized cell systems.
B. Lower the cost of distillation	(1) Use of strains with greater ethanol tolerance
	(2) Application of thermophilic cultures

Table 1. Limitations to the fermentative production of ethanol

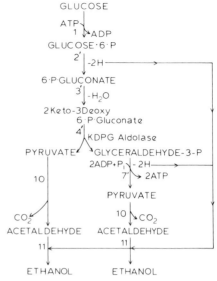

A. The Embden-Meyerhof Pathway

GLUCOSE
↑ 1 (ATP → ADP)
GLUCOSE·6·P
↑ 2
FRUCTOSE·6·P
↑ 3 (ATP → ADP)
FRUCTOSE·1·6·diP
↑ 4
2 GLYCERALDEHYDE
↓ 5 (NAD⁺ → NADH)
P_i 2 DIPHOSPHOGLYCERATE
↑ 6 (ADP → ATP)
2 PHOSPHOGLYCERATE(3)
↑ 7
2 PHOSPHOGLYCERATE(2)
↑ 8 (→ H_2O)
2 PHOSPHOENOLPYRUVATE
↑ 9 (ADP → ATP)
2 PYRUVATE
↓ 10
2 ACETALDEHYDE + $2CO_2$ (NADH → NAD⁺)
↓ 11
2 ETHANOL

Net reaction: Glucose + 2ADP → 2 Ethanol + $2CO_2$ + 2ATP

B. The Entner-Doudoroff Pathway

GLUCOSE
ATP ↓ 1 ADP
GLUCOSE·6·P
2′ ↓ −2H →
6·P·GLUCONATE
3′ ↓ −H_2O
2 Keto-3 Deoxy 6·P·Gluconate
4′ ↓ KDPG Aldolase
PYRUVATE ← GLYCERALDEHYDE-3-P
2ADP+P_i ↓ −2H →
10 7′ ↓ 2ATP
CO_2 PYRUVATE
ACETALDEHYDE 10 ↓ CO_2
ACETALDEHYDE
11 ↓ 11 ↓
ETHANOL ETHANOL

Net reaction: Glucose + P_i + ADP → 2 Ethanol + $2CO_2$ + ATP + H_2O

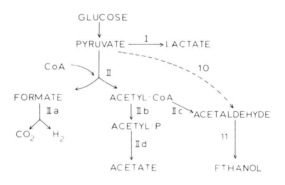

C. Enterobacteria — Mixed Acid Pathway

GLUCOSE
↓
PYRUVATE — 1 → LACTATE
CoA ↓ II
FORMATE ... ACETYL·CoA
IIa ↓ IIb ↓ IIc → ACETALDEHYDE
CO_2 H_2 ACETYL·P 11 ↓
IId ↓
ACETATE ETHANOL

Fig. 1. Major fermentation pathways leading to ethanol production.

A. Embden-Meyerhof pathway
1. Hexokinase
2. Glucose phosphate isomerase
3. Phosphofructokinase
4. Aldolase
5. Glyceraldehyde 3-phosphate dehydrogenase
6. 3-Phosphoglycerate kinase
7. Phosphoglycerate mutase
8. Enolase
9. Pyruvate kinase
10. Pyruvate decarboxylase
11. Alcohol dehydrogenase

B. Entner-Doudoroff Pathway
2′. Glucose-6-phosphate (G-6-P) dehydrogenase
3′. 6-Phosphogluconate (6-P-G) dehydratase
4′. 2-Keto-3-deoxy-6-phosphogluconate (2-K-3-D-6PG) aldolase
7′. See Embden-Meyerhof 7 & 8.

C. Mixed Acid Pathway
I. Lactate dehydrogenase
II. Pyruvate formate lyase
IIa. Formate hydrogen lyase
IIb. Phosphotransacetylase
IIc. Acetaldehyde dehydrogenase
IId. Acetate kinase

115

yield of ethanol is relatively low. The major emphasis for fuel alcohol production has been in the application of yeast and *Z. mobilis*. However, *S. ventriculi* has the attribute of growing in very acidic media which reduces the chances of contamination. In addition, the production of acetate can be removed via mutational selection to yield more efficient production of ethanol. Our detailed understanding of the genetics of *Escherichia coli* also makes this bacterium a potential producer of fuel ethanol.

INCREASED PRODUCTIVITY THROUGH SYNTHESIS IN NEWLY CONSTRUCTED STRAINS (TABLE 2)

Since microbes that possess pyruvate decarboxylase show efficient conversion of glucose to ethanol, how can their fermentations be improved? The use of high density cultures,[12-14] flocculent strains,[15-17] the application of continuous culture[18-19] and immobilized cell techniques,[20-25] and their combined use in vacuum fermentations[26-28] are engineering approaches to this end. A range of physiological approaches are also well worth considering and are outlined in the following section.

Enhanced Production of Enzymes in the Fermentation Pathways

This approach is potentially possible since respiratory deficient (ρ^-) yeast mutants have been shown in general to produce greater levels of several enzymes of the glycolytic pathway, and such mutants gave more efficient yields of ethanol compared to the wild strains.[29,30] Amplification of the glycolytic enzymes should be possible through recombinant DNA (rDNA) techniques to increase gene dosage. However, such an approach could be extremely complex if applied to all of the enzymes of the glycolytic pathway. Increasing the yields of key enzymes could be more profitable. Thus, increased levels of hexokinase[29] are of particular interest since the initial entry and catabolism of glucose have been indicated as limitations on the fermentation rate. It must be cautioned that any approaches to enhance the uptake and phosphorylation of sugars must be considered in relation to the individual microbe under study. Thus, yeast phosphorylates glucose via a specific glucokinase and by two nonspecific hexokinases. In contrast, *Z. mobilis* has two distinct constitutive enzymes, a glucokinase and a fructokinase.[31] The former enzyme shows inhibition in the presence of high concentrations of nucleotide triphosphates, whereas the latter, uninhibited by nucleotides, is repressed in the presence of glucose. From these results, and with the knowledge that glucose uptake rates do not increase in sugar concentrations of greater than 10%, it was concluded "that glucose phosphorylation and possibly uptake are dependent on the energy status of the cell."[31] In the Enterobacteriaceae and many other bacteria, glucose transport into the cell is mediated by a phosphoenol pyruvate/glucose phosphotransferase system.[32,33,34] Clearly enhancing the uptake of available sugars such as glucose, fructose, and sucrose are complex propositions. Finally, it is pertinent to note that as one set of rate-limiting steps is selectively removed, others will become apparent. The latter will have to be assessed on a continuing basis.

Enhanced Fermentation Through:

1. Increased production of enzymes of the fermentative pathways.
2. Removal of the regulatory controls of the fermentative pathway.
3. Channeling of metabolic pathways into a homofermentative mode.
4. Extending the use of substrates to include inexpensive substrates—cellulose, xylose and starch.
5. Mutation and selection of osmophilic, thermotolerant, and molasses tolerant strains.
6. Increased tolerance to ethanol: cloning of fatty acid synthetase genes to yield more ethanol permeable membranes.
7. Increasing the production of alcohol dehydrogenase in order to maintain high fermentative rates.
8. Preparation of highly flocculent strains that permit high cell densities and facilitate harvesting the culture.

Table 2. Modification of the microorganism

Removal of Regulatory Controls of the Fermentative Pathway

A more definitive approach to gaining more rapid fermentation involves removal of the regulatory control of glycolysis. Phosphofructokinase (PFK) (see Fig. 1.A.1) is probably the rate-limiting step in the Embden-Meyerhof pathway. It is inhibited by high concentrations of ATP in an allosteric manner. However, as PFK from lactic acid bacteria[35] is insensitive to ATP, it should be possible to clone the gene for this enzyme into an ethanol producer and enhance the fermentation rate. It is cautioned that the indiscriminate insertion of nonregulated enzymes into a new host could also result in untoward events, but these can be assessed in due course.

Channeling of Metabolic Pathways

Complex genetic manipulations are being envisaged, and therefore the use of *E. coli* for ethanol production can also be entertained. It makes relatively small amounts of ethanol (see Fig. 1C) if its production of formate, acetate, lactate, and succinate can be limited through mutational inactivation, and the gene for pyruvate decarboxylase (for instance, from *Z. mobilis*) is incorporated, the metabolism of pyruvate can be channeled directly into ethanol production (see Fig. 1.C.10).

Extending the Substrate Range to Include Inexpensive Materials

A direct approach to reduce the cost of ethanol fermentation is the use of cheaper substrates. Two major, readily available supplies include cellulose and xylan from biomass. Unfortunately, there are no homofermentative, ethanol-producing microbes which can use cellulose or the pentose sugars. Mutational inactivation of superfluous pathways (lactic and acetic acid production) of the thermophilic, cellulolytic *Clostridium thermocellum* has yielded strains that produce up to 5% ethanol.[36] Enhancement of ethanol yield has also been obtained through co-culture of *C. thermocellum* with *C. thermosaccharolyticum*, and of *C. thermohydrosulfuricum* with *Thermoanaerobium brockii*.[37] Genetic approaches using these rapidly growing thermophiles have only recently been initiated. However, there have been steps taken to broaden the substrate utilization pattern of homofermentative, ethanol-producing yeast and *Zymomonas* to include xylose and cellulose. One approach is to promote the direct use of xylose by yeast. *Saccharomyces cerevisiae* will use xylulose, but not xylose. In fact, fungi in general, if they metabolize xylose, do so via xylitol (see Fig. 2). It has been shown that a few yeasts, including *Pachysolen tannophilus*,[38,39] *Candida tropicalis* ATCC 1369,[40] and *Candida* sp. XF 217,[41] can convert xylose directly to ethanol. Although these yeasts produce alcohol efficiently, they are intolerant of 4 to 5% alcohol and also require the continued uptake of small amounts of oxygen. In contrast, the ability of *Saccharomyces cerevisiae* to yield 10 to 12% alcohol from glucose gives it a major competitive

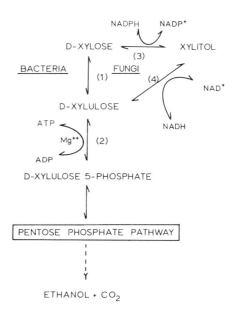

Fig. 2. Pathways for D-Xylose catabolism.
1. D-Xylose isomerase
2. D-Xylulokinase
3. Aldoreductase
4. D-Xylulose reductase

edge, and genetic engineering approaches have focused on modifying its metabolism to permit the use of xylose. As noted above, *Saccharomyces* species can use D-xylulose. Thus, efforts have emphasized cloning xylose isomerase into the yeast to facilitate the direct use of xylose (see Fig. 2). Several laboratories are working on this approach. To date, published reports are limited to the isolation of a Col E1 hybrid plasmid with genes for xylose utilization that complement xylose-negative strains of *E. coli* and *Salmonella typhimurium*.[42] Genes expressed include a regulatory element and structural genes for D-xylose isomerase and D-xylulose kinase. Enhancement of D-xylulose kinase activity in strains still carrying the chromosomal gene for this enzyme was attributed to a gene-dosage effect of the multicopy Col E1 hybrid plasmids. These results are most encouraging. Developments for gaining the use of cellulose include the successful cloning of cellobiase genes from *Escherichia adecarboxylata*[43] and cellulase genes from *Cellulomonas fimi*[44] into *E. coli*. The concept is to subsequently transfer them to efficient ethanol producing species. Similar approaches include development of amylolytic *S. cerevisiae* strains in order to facilitate the direct fermentation of starch-based media. Both of these approaches are considered later.

Osmo- and Thermo-Tolerance and Resistance to Molasses

In order to obtain a highly efficient process, it is advantageous to achieve 10 to 12% ethanol concentrations. However, this implies the use of osmotolerant strains because it would be necessary to initiate the fermentation with 20 to 24% sugar solutions. An engineering approach to circumvent this problem might be the continuous addition of substrate to maintain a 5 to 10% concentration. Osmophilic strains tolerant of much higher concentrations are well known, but they are not highly efficient ethanol fermentors. It would be advantageous to have strains that combine osmotolerance with a high efficiency of ethanol production. In an analogous fashion, it would be advantageous to use microbial strains that can tolerate high molasses concentrations. Here, yeast shows greater tolerance of molasses than does Z. mobilis.[45] The bacterium exhibits considerably reduced conversion efficiencies with 10% glucose in the presence of molasses, while the yeast maintains 90% efficiency for conversion even with 20% sugars. However, the bases for osmophily and tolerance of molasses are not understood. Selection of strains with greater tolerance has therefore been approached through the simple expediency of selecting resistant mutants. Slight success has been achieved in this manner with Z. mobilis mutants capable of growth in 20% molasses (v/v).[16]

Temperature is an important factor in ethanol production. The use of thermotolerant strains results in lower process costs since fermentors need less cooling, and also less energy is expended in the distillation of the product. More rapid fermentations may be a further benefit. However, tolerance to alcohol is markedly reduced at higher temperatures and thermotolerant strains do not necessarily maintain efficient ethanol productivity. Strains showing greater thermotolerance have been isolated by selection of resistant strains at high temperatures, including *Zymomonas* strains that grow at 40°C compared to routine cultivation at 30°C.[16,46] Their productivities are not increased markedly over that of the wild strains. An additional point is that fermentation by *Saccharomyces uvarum* has a higher optimum temperature than that for growth.[47] This should be noted in the future selection of thermotolerant strains in relation to their fermentative mode. Selection of mesophilic strains adapted to higher temperatures results in limited advantages. In the long term, one can expect the development and application of thermophilic bacteria growing at 60° to 70°C.[37,48]

Increasing Ethanol Tolerance

A major aspect regarding enhancement of ethanol productivity (see Table 2) is tolerance to ethanol. Ethanol markedly affects the rates of growth and fermentation, the viability of the cell, and also the efficiency of substrate conversion.[1,49]

Microorganisms show considerable variation in tolerance to ethanol. Maximum tolerance is around 15% wt/vol (19.6% v/v),[50] or even 25% wt/vol under optimal conditions. In yeast fermentations, yields of 4 to 9% wt/vol[5] are routinely obtained. In saké production, 14% wt/vol is reached. This enhanced tolerance is due in part to the slow growth rate at low temperature, and also to the nutritional effects of a proteolipid formed by *Aspergillus oryzae* in the prior koji fermentation of the rice.[51] Strains of yeast[52] and Z. mobilis,[16,51] showing greater tolerance to externally suplied ethanol, have been readily isolated. It was assumed that such ethanol tolerant strains would show more efficient ethanol production. This was not found to be true for Z. mobilis strains.[16,52] The reasons for this apparent discrepancy are gradually being clarified. Certain workers now consider that intracellular ethanol concentrations can be considerably higher than the external concentration,[53,54,55] even though efflux across membranes is very rapid. At high fermentation rates, the rates of ethanol efflux from the cell are impaired due to increased product concentration, and the final effect is that intracellular concentrations may reach as high as 30%.[54] This concept has been supported by kinetic analysis of continuous culture of yeast in which high ethanol concentrations were achieved by use of high glucose concentration.[56] Furthermore, it is now clear that the inhibition of growth rate, of fermentation rate, and of viability are all discrete phenomena.[56] Thus, simple plate screening for ethanol-tolerant mutants need not reflect selection of cultures with enhanced ethanol productivity. Selection in continuous culture is more suitable for isolation of mutants with rapid fermentation rates at high ethanol concentrations.[56]

High ethanol concentrations also inhibit RNA and protein synthesis in yeast,[57] but the primary inhibitory action of ethanol is still undefined. In *E. coli*, exposure to ethanol results in an increased proportion of unsaturated fatty acids in the cell membrane, while the synthesis of phosphatidyl ethanolamine is inhibited. On this basis, Ingram[58] has suggested gaining greater ethanol tolerance of *Z. mobilis* by cloning fatty acid synthetase genes from extremely ethanol-tolerant lactobacilli into the *Zymomonas* to yield a cell with more fluid membranes. This could facilitate the efflux of ethanol from the cell.

Alcohol Dehydrogenase

As a result of high internal ethanol concentration, it has been suggested that the fermentative rate of yeast is reduced due to the inactivation of alcohol dehydrogenase.[53] Increasing the synthesis of this enzyme through gene amplification is one approach to circumvent this problem. In this instance, there is detailed information available regarding alcohol dehydrogenase (ADH). Yeast possesses three,[59] and *Z. mobilis* two,[60] ADH isozymes. The yeast ADH I and ADH II genes have been cloned. The latter was selected by use of a recipient yeast that lacked ADH, and yet allowed constitutive expression of the cloned gene. The constitutive function was a result of a mutation in an unlinked regulatory gene.[61] It should be possible to test the hypothesis that limiting levels of ADH reduce the fermentation rate at high ethanol concentration through the preparation of strains with enhanced enzyme levels mediated by gene amplification techniques.

Flocculent Strains

The engineering aspects of dense cell cultures have already been referred to. The isolation and high productivities of highly flocculent mutants are readily accomplished by selection of sedimenting flocs in continuous culture. Flocculation in *S. cerevisiae* RSRJM80, and also in *S. diastaticus* G, is controlled by single genes in each species. Thus, it should be quite possible to produce flocculent strains through rDNA techniques.

From this brief review, it is apparent that rDNA techniques can be used in a variety of ways to enhance the rate of fermentation, to produce new strains, and to develop microbes that facilitate fermentor operation (see Table 2).

DEVELOPMENT OF GENETICS OF *ZYMOMONAS* AND YEAST

Z. mobilis *Genetics—Introduction*

One method of enhancing the ethanol-producing properties of *Z. mobilis* is by introducing genes from other microorganisms to increase the substrate range of the organism. There are several genetic techniques which are potentially useful for the introduction of novel genes into *Z. mobilis*. The advent of rDNA in particular has provided strong impetus for the development of new strains of *Z. mobilis*. However, fundamental to the success of any cloning strategy is at least a rudimentary knowledge of the genetics of the organism of interest. In the case of *Z. mobilis*, the question of genetic characterization is only just beginning to be addressed. We will briefly discuss some of the techniques used in bacterial genetics and how they relate to gene cloning in *Z. mobilis*.

The three classic methods of genetic analysis in bacteria are transformation, transduction, and conjugation. A fourth method of genetic recombination, protoplast fusion, is being increasingly used in the study of prokaryotes. However, since this technique is more commonly used in the study of eukaryotes, it will be discussed in the yeast section.

Transformation

Transformation is a process by which naked DNA is taken up by a cell,[62] and may result in genetic recombination. While foreign DNA is normally cloned into a plasmid which does not usually recombine with the host chromosome, it is by transformation that recombinant DNA molecules are most commonly introduced into a cell. Mechanisms of transformation are not well defined. For efficient transformation to occur, the bacterial cells have to be made "competent" such that they can irreversibly bind DNA to the cell membrane. Genetic analysis using transformation has been most successfully employed in *Bacillus subtilis* where the induction of competence is relatively easy.[63] In other organisms, however, transformation is dependent on more exacting conditions. Such conditions normally include suspension of cells in a 0.01 to 0.2 M solution of a divalent cation—Ca^{2+} in the case of *E. coli*,[64] or Mg^{2+} in the case of *Pseudomonas aeruginosa*[65]—, at 0°C followed by a "heat shock." Attempts to establish conditions for transformation in *Z. mobilis* in our laboratory, using a variety of conditions, have so far been unsuccessful.

Transduction

Another widely used method of genetic transfer is transduction. This is a process whereby chromosomal or plasmid DNA is passively packaged into a bacteriophage head during the phage's infective cycle.[66] First discovered in *Salmonella typhimurium*,[67] it has been used extensively for genetic studies in many organisms, including *E. coli*[68] and *P. aeruginosa*.[69] In addition, bacteriophage cloning vectors have been successfully developed in *E. coli*.[70] Although bacteriophages are normally host specific, we have screened *P. aeruginosa* bacteriophages for infection of *Z. mobilis*, since the latter is believed to be closely related to such aerobic soil bacteria as *Pseudomonas*.[71] However, of the several screened, none showed evidence of infection of *Z. mobilis*.[72] Similarly, direct screening for wild-type bacteriophages capable of infecting *Z. mobilis* has also been unsuccessful.

Conjugation

The only method of genetic transfer to be employed successfully in *Z. mobilis* is conjugation. In conjugation, gene transfer from a donor cell to a recipient is promoted by a class of extrachromosomal elements known as sex plasmids, or sex factors.[73] Sex plasmids, like bacteriophage, are usually host specific. While sex plasmids have been found in several microorganisms, the development of genetic maps in many species, including *Z. mobilis*, has been hindered by a lack of such plasmids. Recently, however, it has become possible to begin to develop genetic maps in some species of gram-negative bacteria which do not possess native sex plasmids. In 1968, a group of R (drug resistant) plasmids were isolated that were atypical of previously identified plasmids because they could transfer themselves conjugally to a large number of gram-negative bacteria.[74] While these plasmids do not possess the ability to promote chromosomal recombination, it was possible to isolate variants that did.[75] It was subsequently found that such variants were capable of promoting gene exchange in several gram-negative bacteria.[76] Recently, it has been shown that one such variant, R68-45, will promote genetic recombination in *Z. mobilis*.[77] Using R68-45 in conjunction with auxotrophic mutants,[16] the construction of a genetic map, including those genes related to ethanol production, should become possible in *Z. mobilis*.

Application of Transposons

Another useful tool in genetic analysis is the transposon. Transposons are discrete genetic elements capable of moving, physically and functionally intact, from one segment of DNA to another.[78] Transposable elements possess genes which code for readily selectable phenotypes. As such, they are useful for a number of genetic manipulations.[79] Most commonly, they can be used for generating mutants[80] since their insertion into a structural gene results in the inactivation of that gene. Transposons could be useful in *Z. mobilis*. The expression of at least one transposon, the tetracycline resistant transposon Tn10, in *Z. mobilis* has been demonstrated.[81] More directly, the transposon Tn951 is known to encode for genes for the catabolism of lactose.[82] Since *Z. mobilis* does not utilize lactose, the introduction of this transposon into *Z. mobilis* may result in the use of this sugar for ethanol production. Integration of Tn951 into the *Z. mobilis* chromosome could be achieved by the use of a suicidal plasmid carrier.[83] The subsequent incorporation of genes coding for the catabolism of galactose into such *Z. mobilis* strains will increase the efficiency of lactose use.

Apart from their utility for gene mapping studies, plasmids are essential tools for genetic engineering. As such, the identification and characterization of plasmids suitable for cloning in *Z. mobilis* are essential. Since plasmids can replicate independently of the host chromosome, they can provide a vector into which foreign DNA can be inserted. Insertion of foreign DNA into a vector is carried out after both species of DNA are cut with a restriction endonuclease, as outlined in Fig. 3. After insertion of foreign DNA and religation of the vector, the plasmid is reintroduced into a cell by transformation. The mechanics of genetic engineering have been reviewed.[84] A number of cloning vectors have been developed for such organisms as *E. coli* and *B. subtilis*.[85] As shown in Table 3, the host range of most of these vectors is limited.

Development of Z. mobilis *Cloning Vectors*

The development of plasmid vectors suitable for introducing foreign DNA into *Z. mobilis* can be approached in one of two ways: (1) develop native *Z. mobilis* plasmids as vectors, and (2) develop a plasmid from another source with

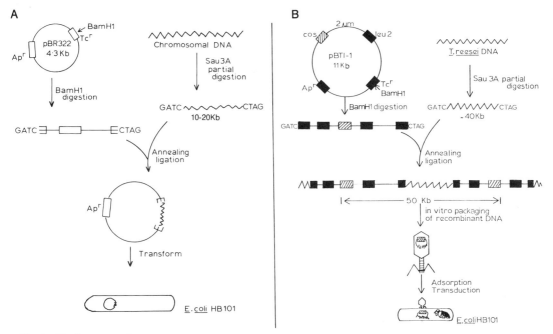

Fig. 3. Cloning protocols.
    A. Bacterial plasmid cloning
    B. Cosmid cloning

   Ap^r = Ampicillin resistance gene
   Tc^r = Tetracycline resistance gene
 Leu 2 = β-isopropylmalate dehydrogenase gene. Confers prototrophy on appropriate Leu$^-$ host.
   cos = Cohesive ends of bacteriophage λ (see text).
Sau 3A = Restriction endonuclease Sau 3A
Bam H1 = Restriction endonuclease Bam H1. Insertion of foreign DNA at this site inactivates the tetracycline
           resistance gene.

the ability to replicate in *Z. mobilis*. One particularly promising approach is the use of derivatives of broad host range plasmids as cloning vehicles. As discussed previously, these plasmids are capable of replication in many gram-negative bacteria. Genetically engineered derivatives of such plasmids have recently been developed which incorporate many of the properties necessary in cloning vehicles.[85] One such plasmid, pRK290,[86] is capable of replication in *Z. mobilis*.[87] pRK290 possesses a number of properties useful for cloning in *Z. mobilis* and other organisms which are not well defined, genetically. First, the plasmid is capable of replication in *E. coli*, making it possible to clone and study foreign DNA in this genetically well-characterized host. Second, although being conjugally nontransmissible, pRK290 can be mobilized into *Z. mobilis* from *E. coli* using a helper plasmid.[86] This is particularly useful since a transformation system has not yet been developed for *Z. mobilis*.

The alternative cloning strategy is to develop plasmids native to *Z. mobilis* as vectors for this host. It is known that strains of *Z. mobilis*

possess plasmids.[16, 87] Indeed, plasmids in *Z. mobilis* appear to be quite widespread.[88] What, if anything, these plasmids code for is not known. It has been suggested that an alcohol dehydrogenase in *Z. mobilis* may be plasmid encoded.[60] The evidence for this is indirect and is based on the irreversible loss of an alcohol dehydrogenase isozyme under certain culture conditions. Strains of *Z. mobilis* are resistant to several antimicrobial agents,[89] but it is not known if these are plasmid encoded. Since *Z. mobilis* shows greatest resistance to the aminoglycoside antibiotics, it is likely that the observed resistance is a nonspecific effect. Most anaerobic bacteria are resistant to aminoglycosides as a result of the inability to transport the antibiotic across the cell membrane.[90]

Since antibiotic resistance determinants, or some similar selectable marker, are essential for a cloning vehicle, a useful strategy may be to identify and clone replication genes from *Z. mobilis* plasmids into cloning vehicles developed in other hosts. Such an approach would provide many of the advantages of pRK290. In addition, it would allow the direct selection of

Vector	Original Host	Host Range
Gram-Negative bacteria		
Col E1[85]	*Escherichia coli*	*E. coli*
pBR322[85]	*E. coli*	*E. coli*
RSF1010[91]	*E. coli*	*E. coli*, some *Pseudomonas* spp.
pRK290[86]	*Klebsiella aerogenes*	Most Gram-Negative bacteria
RP1[74]	*Pseudomonas aeruginosa*	Most Gram-Negative bacteria
Gram-Positive bacteria		
pUB110[85]	*Bacillus*	*Bacillus*
SCP–2[85]	*Streptomyces*	Some *Streptomyces* spp.
Fungi		
YIp[117]		
YEp[110]		
YRp[119] (ars)	Hybrids of *E. coli*/yeast DNA	*E. coli, Saccharomyces cerevisiae*
(Cen)		
pBTI-1[121] (yeast cosmid)	Hybrid of λ/*E. coli*/yeast DNA	*E. coli, S. cerevisiae* Packagable in λ capsids

Table 3. Plasmid vector systems.

inserted foreign DNA by insertional inactivation.[85] Since pRK290 has only one drug resistance marker, recombinant plasmids cannot be screened by this method. Several multiply-resistant *Pseudomonas* plasmids have recently been developed as cloning vehicles.[91] Like pRK290, some of these plasmids, although non-transmissible, are also capable of being mobilized by "helper" plasmids. Insertion of a *Z. mobilis* replication gene into such a plasmid may provide a useful cloning vehicle which can be readily transferred into, and easily maintained in, *Zymomonas*.

Yeast Genetics—Introduction

Yeasts have been used for the preparation of beverages for thousands of years. During this period, strains were inadvertently selected through ritualistic fermentation practices. The importance of specific strains, however, was not appreciated until Hansen introduced the pure culture concept to the brewing industry in 1881. Since then, man has attempted to improve the performance of ethanol-producing yeast strains with regard to such characteristics as ethanol yield, efficiency of substrate utilization, rate of fermentation, strain viability, and reproductive vigor. Although classical breeding of strains has been employed, many improvements continue to be accomplished by unconscious selection and fortuitous manipulation of fermentation conditions.

The yeast genetic system has now been well defined and the existence of literally hundreds of mutant loci allows the construction and purposeful breeding of industrial strains by classical breeding techniques. With the development of the techniques for the transformation of yeast with exogenous DNA, it is now possible to specifically alter the genotype of host cells. In combination with the techniques of recombinant DNA, specific sequences of cloned DNA can be introduced into yeast to bring about the desired phenotypic change without the complex intermingling of parental genotypes that occurs in conventional breeding programs. These techniques greatly expand the potential for genetic manipulation of industrially important yeast strains.

The following section will give a brief overview of the yeast genetic system and the advances leading to the development of yeast as a host for cloned DNA. For a more comprehensive examination of yeast genetics and molecular biology, the reader is referred to recent reviews.[92-95]

Classical Yeast Genetics

Saccharomyces cerevisiae belongs to that diverse group of microorganisms known as the yeasts, which are limited to the fungi in which the predominant form is unicellular.[96] This rather vague grouping of microorganisms is comprised of some 39 genera and 350 species, the main criteria used for classification being morphological and physiological characteristics. The life cycle of *S. cerevisiae* normally alternates between the haploid and diploid states (see Fig. 4). Both ploidies can exist stably on simple, well-defined media. In heterothallic strains, the haploid cultures express one of two stable mating

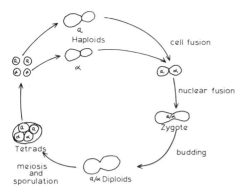

Fig. 4. Life cycle of heterothallic strains of *S. cerevisiae*.

The a and α refer to the yeast mating-type locus (MAT), each being determined by the specific allele present at this locus. The yeast mating type, as well as other heterozygous chromosomal genetic markers, segregate 2:2 among the ascospores following meiosis and sporulation. Extrachromosomal elements, in contrast, segregate either $4^+:0^-$ or $0^+:4^-$ among the meiotic progeny.

types, called a and α, depending upon the mating-type allele at the mating-type locus (MAT). The genetics of mating type (and mating-type interconversion in homothallic strains) has been reviewed.[97] In the absence of cells of the opposite mating type, the haploid strains divide asexually by a multilateral budding process. A daughter cell is initiated as an outgrowth from the mother cell followed by nuclear division, cross-wall formation, and finally cell separation. The timing of the cell cycle is under strict genetic control.[93]

Haploids of both mating types produce distinct diffusible sex factors. In the presence of cells of opposite mating type, the sex factors arrest the cell cycle at the stage immediately preceding bud initiation and DNA synthesis in cells of the opposite mating type.[98,99] In the arrested condition, cells of opposite mating type become competent to agglutinate, and this results in the fusion of the cells to form a zygote. Nuclear fusion follows and the first bud produced by the zygote contains a diploid nucleus. The a/α diploids that result are unable to mate, are stable, and can divide mitotically in a cell cycle similar to that of the haploid strains. They can be induced to undergo meiosis through nitrogen starvation to yield four haploid ascospores.

Tetrad analysis is one of the primary meiotic mapping techniques by which genes are located on the yeast chromosomes. By analysis of the ascospore segregation patterns of various heterozygous genetic markers, gene to gene and gene to centromere linkages can be established. Genes associated with the yeast chromosome usually segregate 2:2 among the ascospores (barring chiasma and chromatid interference or gene conversion events), while extrachromosomal elements segregate in a non-Mendelian ($4^+:0^-$ or $0^+:4^-$) manner. Thus, one can readily determine whether a recombinant phenotype is of chromosomal or extrachromosomal origin.

Other techniques by which genes are mapped on the yeast chromosome, including random spore analysis, trisomic analysis, mitotic crossing over, and chromosome loss and transfer, are reviewed elsewhere.[100] By the combined use of these techiques, extensive linkage maps of the yeast genome have been developed.[101] The genetic map consists of 17 linkage groups attached to independently segregating centromeres. This is interpreted to mean that there is a minimum of 17 chromosomes per haploid genome. To date, mapping studies have located 312 genes on the 17 chromosomes and 3 fragments (which most likely belong to the established chromosomes).

An interesting feature revealed by the linkage map is that the different chromosomes vary in size from 90 centiMorgans (cM) to over 450 cM.[102] Since in yeast there is an average of 3.0 kilobases (Kb)/cM, the yeast genome is relatively small and at 13,800 Kb (or 9×10^9 daltons), the genome complexity is only about three times that of *E. coli*. Unlike the bacterial genome, however, functionally related genes are, in general, not clustered. Exceptions to this rule exist with genes such as ARG5 and ARG6 on chromosome V and the GAL 1, 7, 10 cluster on chromosome II.[103] In general, eukaryotic genes do not seem to be organized in the operons characteristic of procaryotes.

The genetic analysis of industrial yeast strains has been complicated by the fact that they are often polyploid and unable to mate like their haploid ancestors. However, in some cases hybridization of strains by mating has resulted in improved fermentation rates and increased alcohol yields in the fermentation of molasses.[104] The application of classical yeast genetics to the brewing industry has been reviewed.[105] Others have applied the technique of protoplast fusion[106] for the improvement of industrial yeast strains. A diagramatic representation of this protocol is presented in Fig. 5. By removal of the

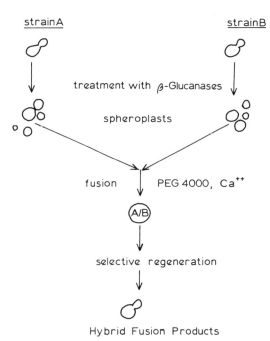

strainA   strainB

treatment with β-Glucanases

spheroplasts

fusion   PEG 4000, Ca⁺⁺

A/B

selective regeneration

Hybrid Fusion Products

Fig. 5. Spheroplast fusion in *S. cerevisiae.*

A and B refer to strains with distinguishable characteristics (i.e., heterozygous genetic markers). Treatment of cells with β-glucanases, which degrade cell wall polymers (glucans), results in the formation of osmotically sensitive spheroplasts. Spheroplasts of both strains are mixed and fused in the presence of polyethylene glycol (PEG, molecular weight 4000-6000) and calcium ions. The spheroplasts are then plated in a hypertonic medium which permits the regeneration of cell wall material and growth of the fusion products.

yeast cell wall with any of a number of lytic enzyme preparations and subsequent treatment of protoplasts with polyethylene glycol and Ca⁺⁺, protoplasts derived from different strains can be fused regardless of mating type or ploidy, thus circumventing the normal mating process. Although Russell and Stewart[107] were able to construct diploids from brewing yeasts by this technique, it was not specific enough to introduce selected traits without the loss of the strain's integrity. Apparently, the complex intermingling of parental genomes results in a fusion product which is very different from either parent.[107] The more recent advances in yeast transformation and recombinant DNA allow the introduction of specific DNA sequences into the host strain and should allow elimination of these problems. However, before we consider these techniques, the extrachromosomal genetic elements of yeast deserve mention.

Extrachromosomal Elements

The extrachromosomal genetic elements of yeast include the 2 μm plasmid circle, the mitochondrial genome, and a double-stranded RNA species. This last genetic element, present in some strains of *S. cerevisiae,* encodes a diffusible protein toxin that is lethal to strains lacking this trait by inducing the release of ATP and potassium into the medium.[108] The strains harboring these elements are appropriately named "killer strains." The double-stranded RNA is encapsulated in intracellular, virus-like particles, and segregates in a non-Mendelian fashion. The mitochondrial genome of *S. cerevisiae* carries the information for some of the enzymes involved in respiration and oxidative phosphorylation, as well as for the mitochondrial rRNAs and tRNAs. An interesting feature of this genome is the complex mosaic structure of some mitochondrial genes.[109]

Sinclair[110] first observed circular plasmid DNA molecules in yeast that measured 1.95 μm in contour length. Since then, the yeast 2 μm plasmid circle has been extensively studied both for its potential as a cloning vector and as a model for eukaryotic DNA replication. The majority of laboratory strains of *S. cerevisiae* harbor 2 μm plasmid DNA in approximately 50 to 100 copies/cell.[111, 112] They are stably maintained and, although termed cytoplasmic elements on the basis of their non-Mendelian segregation pattern, are believed to reside in the nucleoplasm. Like chromosomal DNA, the 2 μm plasmid is also arranged in a nucleosome structure by a normal complement of histones.[113] Recently, Falco et al.[114] reported the presence of 2 μm DNA sequences integrated in the yeast chromosome. Some top and bottom fermenting brewing yeasts were the first natural strains found to lack this plasmid.[115, 116] A number of hybrid yeast *E. coli* cloning vectors, utilizing the 2 μm plasmid origin of replication, have been constructed for use as a cloning vehicle in yeast and are discussed below.

Gene Cloning in **Saccharomyces cerevisiae**

The ability to study the fundamental nature of gene structure, function, and expression in yeast has been greatly enhanced by the advances made in the development of recombinant DNA techniques and the yeast transformation system. The yeast transformation protocol is based on the coprecipitation of yeast spheroplasts and the transforming DNA in the presence of polyethylene glycol and Ca^{++}. The genetic basis for the ability of a strain to be transformed, or the mechanism by which transformation occurs, is not known. However, some factors which effect the transformation frequency are the cloning vector employed and the regenerative ability of the particular strain in use.

The transformation of yeast with exogenous DNA was first demonstrated conclusively by Hinnen et al.[117] They succeeded in transforming a nonrevertable *leu* 2 strain to LEU$^+$ by using a derivative of the bacterial plasmid, ColE1, containing the yeast *Leu* 2 gene. Although the transformation frequency with this hybrid plasmid (pYeleu 10) was low (10^{-6} 10^{-7} transformants/ viable cell), the transforming plasmid DNA was found to be stably integrated into the yeast genome at the *leu* 2 locus, as well as at several other genetic sites. The cloned DNA demonstrated the 2:2 segregation pattern characteristic of a chromosomal genetic element. In no case have these recombinant molecules been detected as autonomously replicating plasmids.

Transformation frequencies two to three orders of magnitude greater have been obtained[118] using a hybrid plasmid containing the *E. coli* vector pMB9 fused to yeast 2 μm DNA sequences and the yeast *Leu* 2 gene as a selectable marker. This vector demonstrates the characteristic of autonomous replication in both yeast and *E. coli*. Struhl et al.[119] have also described high frequency transformation of yeast by using bacterial plasmid vectors containing segments of yeast 2 μm plasmid DNA. The autonomous nature of these hybrid plasmids was demonstrated by the 4$^+$:0$^-$ segregation pattern characteristic of an extrachromosomal element in tetrads as well as by detection of covalently closed, circular plasmid molecules in agarose gels. However, transformants of this type are highly unstable and the transformed phenotype is rapidly lost under nonselective growth conditions. This segregational instability, the frequency of which varies with the particular vector, is due to complete loss of the replicating plasmids and is associated with the copy number of the hybrid in yeast transformants. In contrast to resident 2 μm plasmid DNA which is present in approximately 50 copies/cell, pYep 6 is maintained at only 5 to 10 copies/cell and is lost at the rate of 40% segregation per cell division under nonselective conditions.[119] In addition, segregational instability is increased in hybrid plasmids containing large DNA inserts. This instability

can be reduced by using a 2 μm plasmid-free host, as well as by maintaining the transformants under selective pressure.

A second type of vector, the hybrid yeast/*E. coli* plasmid (YRp7), also transforms yeast at a high frequency (10^3 to 10^4 transformants/μg).[119] The yeast DNA sequences in this hybrid are believed to contain a chromosomal origin of replication (autonomously replicating sequences, ars). YRp7 is maintained as a replicating plasmid in yeast transformants and is present in only 1 to 5 copies/cell. In no case is the plasmid found integrated into the yeast chromosome. The segregation of YRp7 among meiotic progeny was 0$^+$:4$^-$ as the replicon was lost from germinating ascospores. Although these plasmids demonstrate extreme mitotic and meiotic instability, Clark and Carbon[120] have constructed stable derivatives by cloning the centromere from chromosome III (CEN 3) onto an ars plasmid to essentially create a minichromosome. The centromere stabilizes the ars plasmid in yeast transformants, and the hybrid plasmid is maintained in a single copy/cell. Upon meiosis, the vector segregates in a Mendelian manner. Undoubtedly, these will prove to be useful and reliable vectors for gene cloning in yeast.

Morris et al.[121] have constructed a series of yeast-cosmid vectors suitable for the cloning of large DNA fragments into yeast. These vectors contain bacterial plasmid DNA sequences for replication and selection in *E. coli*, in addition to yeast 2 μm DNA sequences or chromosomal replicators (ars) and yeast markers necessary for replication and selection in yeast. In addition, these vectors contain the cohesive ends of bacteriophage lambda which allow the packaging of recombinant molecules into lambda phage heads. The use of the *in-vitro* packaging system selects for recombinant molecules containing 23 to 40 Kb of cloned DNA (see Fig. 3). This feature is particularly useful for construction of genome libraries and significantly reduces the number of recombinants that need to be screened for a desired sequence as compared to shotgun cloning using plasmid vectors. Recently, centromere sequences were introduced into these vectors to further stabilize the large recombinant plasmids in yeast transformants.[122]

In summary, two basic types of transformation events occur in yeasts: an integrative type in which the transforming DNA becomes stably associated with the yeast genome, and a nonintegrative type, in which the transforming DNA is maintained as a plasmid. The type of transformation event is determined solely by the specific yeast sequences contained on the vector.[123] The plasmid vectors used in yeast transformation are described according to their mechanism of maintenance in yeast transformants.[124] Vectors resulting in the stable addition of the entire plasmid to the yeast genome are referred to as yeast integrating plasmids (YIp).[117] Vectors which are maintained episomally by inclusion of all, or part, of the yeast 2 μm plasmid sequence are termed YEp, for yeast episomal plasmids.[110] Vectors which include a yeast sequence allowing autonomous replication, presumably due to a yeast chromosomal replicon, are called YRp, for yeast replicating plasmids.[119] All of these plasmids have several characteristics in common. They all contain a bacterial replicon which makes possible their amplification in *E. coli*. The DNA of procaryotic origin is derived from a certified EK2 vector, pBR322.[125] They also contain markers which allow for their selection in *E. coli*. Some of these markers are drug-resistance genes derived from pBR322; however, others are yeast DNA sequences which are expressed in *E. coli*. These include the *Leu* 2, *His* 3, *Ura* 3, and *Trp* 1 genes, all of which complement the mutations which inactivate the analogous bacterial functions,[126] and can also be used to select for the presence of the vector in yeast transformants. A list of these vectors is presented in Table 3.

As mentioned, these vectors serve as the vehicles by which cloned DNA can be introduced, via transformation, into yeast. The techniques by which recombinant molecules are constructed have been reviewed.[127] The cloning vectors described above, like the bacterial vectors, contain unique restriction sites for a variety of restriction endonucleases. These sites are suitable for cloning foreign DNA and result in the insertional inactivation of drug-resistance genes. A diagramatic representation of plasmid and cosmid cloning protocols is given in Fig. 3. The full cosmid cloning procedure has been described.[128, 129]

As the yeast vectors also contain the yeast sequences necessary for their replication and selection in yeast, the recombinant molecules isolated in *E. coli* can be amplified, recovered by standard plasmid extraction techniques, and directly introduced into appropriate yeast strains by transformation.

Often, the most difficult aspect of a cloning strategy entails the selection of those transformants containing the gene or genes of interest. One of the most common methods for analyzing a collection of recombinants for a specific DNA sequence is the technique of colony hybridization.[130] The advantage of this technique is that it does not require expression of the cloned DNA squence. A number of yeast ribosomal RNA and transfer RNA genes have been identified by this approach.[131-133] In another case,[134] the iso-l-cytochrome C gene of yeast was identified and cloned using a synthetic oligodeoxynucleotide hybridization probe. Hybridization approaches are most suitable in cases where one does not expect expression of the cloned DNA or where specific hybridization probes can be readily obtained.

Perhaps the most direct means of selection involves the detection of transformants which display the desired phenotypic change. This selection obviously requires expression of the cloned gene and is subject to the many complexities of accurate transcription and translation, as well as, in some cases, secretion. However, in a number of cases the functional expression of cloned eukaryotic genes have been demonstrated in *E. coli,* and this has led to the identification of genes from yeast as well as other eukaryotic microorganisms.[135-137] In these cases, a bacteriophage lambda, or hybrid plasmid containing a eukaryotic DNA sequence, was isolated by the complementation of a non-revertible *E. coli* mutant lacking a defined enzymatic activity. The usual assumption has been that the cloned eukaryotic DNA sequence, which specifically complements mutations of a defined *E. coli* structural gene, contains the analogous structural gene from the eukaryote. This approach has been surprisingly successful and has led to the isolation of a number of yeast genes most notably the *Leu* 2 gene,[136] the *His* 3 gene,[138] the *Ura* 3 gene,[126] the *Trp* 1 gene,[119] and the *Arg* 4 gene.[139] The isolation of these genes was essential to the development of the yeast transformation system. Other genes isolated by this approach include the catabolic dehydroquinase encoded *qa*-2 gene of *Neurospora crassa,*[137] and the gene encoding β-galactosidase of *Kluyveromyces lactis.*[140] Since *E. coli* lacks the post-transcriptional and post-translational machinery characteristic of eukaryotes, it seems likely that the eukaryotic structural genes which are expressed in this host lack intervening sequences, and that their products do not require further modification (e.g., glycosylation) for activity. Undoubtedly, there are a number of problems limiting the expression of eukaryotic genes in *E. coli,* and the recent development of complementation cloning in yeast could make *E. coli* selections obsolete.[141]

The advances made in the development of the yeast transformation system make possible the identification and isolation of any yeast gene by direct complementation of the analogous yeast mutation. The only limitation of this approach is the availability of the specific yeast mutant. Since the yeast genome is relatively small, a few thousand independent recombinant molecules are sufficient to screen for a desired sequence among the entire genome. For the isolation of genes which complement nutritional markers or temperature-sensitive mutants, it is possible, in most cases, to simply plate the primary yeast transformants under conditions which are restrictive for the growth of the mutants.[141] However, in other cases, it is often necessary to screen the primary transformants for the desired trait after they have been selected with the markers on the yeast vector by replica-plating techniques. By this approach, a number of yeast genes which could not have been similarly identified in *E. coli* were isolated. These include the *His* 4 gene[142] which encodes a complex, multifunctional enzyme involved in histidine biosynthesis; some of the CDC genes controlling the yeast-cell-division cycle;[120, 143] the genes controlling mating-type interconversion in homothallic strains,[144] and the structural gene (ADH1) for the constitutive yeast alcohol dehydrogenase.[145] Many other yeast genes have now been cloned and are described in reviews by Petes,[92] Olsen,[141] and Davis.[146] In addition, some procaryotic genes are also expressed in yeast, and include the *E. coli lac* Z gene[147] and the gene encoding resistance to a 2-deoxystreptamine antibiotic, G418.[148]

The results obtained with the yeast transformation system demonstrate the excellent potential for the cloning and expression of other eukaryotic genes in yeasts. As a lower eukaryote, one would expect this cloning system to be compatible with the expression of other eukaryotic genes at the posttranscriptional and posttranslational level. However, nontranslated sequences within structural genes (introns) have so far been found in relatively few yeast genes. Hence, the extent to which yeasts are capable of accurately processing genetic information from other eu-

karyotic sources is still uncertain. Beggs et al.[149] studied the expression of the chromosomal rabbit β-globin gene in yeast, and found that, compared to the mature β-globin mRNA, the transcripts produced in yeast lacked 20 to 40 nucleotides from the 5′ end, and contained all of the small intron and part of the large intron. These results suggest improper splicing of the β-globin mRNA transcripts in yeast. However, there are cases in which the expression in yeast of cloned DNA from other eukaryotic sources has been demonstrated. Dickson[150] demonstrated the expression of the β-galactosidase gene (*Lac* 4) from the yeast *Kluyveromyces lactis* in *S. cerevisiae.* Although these transformants could not grow on lactose (most likely due to the lack of a specific permease), intracellular enzyme activity was demonstrated. Minerleau-Puijalon et al.[151] demonstrated that a copy DNA sequence encoding chicken ovalbumin was successfully expressed in *S. cerevisae,* and produced approximately 1,000 to 5,000 molecules per cell. Its expression apparently was under the control of the *E. coli* regulatory sequence contained on the hybrid plasmid. More recently, the expression of a copy DNA sequence, encoding mature human leukocytic interferon-D, was cloned into an autonomously replicating yeast plasmid under the control of the yeast alcohol dehydrogenase (ADH 1) promoter.[152] Yeast transformed with this plasmid synthesized up to one million molecules of biologically active interferon per cell.

Accurate processing of genetic information from other eukaryotic organisms would make yeasts the organisms of choice for cloning almost any eukaryotic gene. Their rapid growth and reproduction, as well as their ease of culture on an industrial scale, afford them the ability to bring about chemical conversions rapidly and efficiently. A great potential now exists for the manipulation of industrially useful yeast strains, particularly with regard to ethanol production from biomass.

SPECIFIC CLONING STRATEGIES FOR IMPROVING ETHANOL PRODUCTION

Starch Utilization

In brewing, the cereal substrates are precooked to liquefy and gelatinize the starch grains. The subsequent combined synergistic action of α- and β-amylases yields a fermentation substrate composed of glucose and maltose. The amylases are obtained from germinated barley and are added after the cooking step. The process can be accelerated by the addition of thermostable bacterial α-amylase to the hot starch mash, and also by subsequent use of fungal glucoamylase, an enzyme that acts by unzippering dextrins directly to yield glucose. Thus, to gain effective amylolysis, an α-amylase plus a combination of β-amylase, maltase (α-glucosidase), or glucomylase is required.

It would be most advantageous to have a yeast that has complete amylolytic activity, since this would reduce the need for addition of α-amylase to the starch mash prior to actual fermentation. One yeast, *S. diastaticus*, has limited amylolytic properties and is able to ferment dextrins, but not starch per se. Indeed, attempts have been made to use this yeast for dextrin utilization in the production of low carbohydrate beers, but the product was designated as having a nonacceptable taste. Hybridization of *S. uvarum* and *S. diastaticus* resulted in strains with lower dextrinizing activity.[153] However, for fuel alcohol production, the dextrinizing properties of *S. diastaticus* are of potential application. Three dextrinase genes have been implicated and are located in three distinct linkage groups.[154]

In contrast, the brewing yeasts *S. cerevisiae* and *S. uvarum* lack amylase genes, although they can hydrolyze maltose (maltase or α-glucosidase production). The structural maltase genes are under the control of up to seven genes.[155, 156] Thus, maltase genes have been cloned, but in this instance regulation may be through any of these other genes that occur in unlinked loci. There is considerable interest in the genetics of the yeast amylase and maltase systems. This should result in rational approaches to the construction of truly amylolytic, ethanol-producing yeasts. However, the revived interest in the less-well-known amylolytic strains, *Schwanniomyces castelli*[157] and *Schwanniomyces alluvius*,[158] may lead to simpler applications of starch-using yeasts for fuel alcohol production. The advantages here for gasohol production are high alcohol concentration (10 to 12%), and problems regarding flavor that are irrelevant. Yeast that can use starch, without the necessity of prior swelling of the substrate by boiling, would also be of considerable economic benefit.

An additional approach to starch use is the prehydrolysis of the substrate by the addition of α-amylase and glucoamylase, and subsequent culture of an ethanol-producing microorganism on the glucose broth. In this regard, rDNA methodology can be used to facilitate hyperamylase production. This would involve gene insertion into an appropriate host, gene amplification, and hyper-enzyme production. In this approach, further insight into the enzyme-synthesis regulatory system could be gained in the *Bacillus* system. It has already been manipulated by mutation and transformation techniques to give α-amylase yields 1,500 times greater than the wild strain.[159] Also of direct interest is the report on the cloning of the genes for a thermostable α-amylase in *Bacillus subtilis*.[160] The large-scale production of thermostable α-amylase would be of considerable benefit in starch hydrolysis, since following gelation at 100°C, such thermostable enzymes can be added directly to the hot starch mash.

Cellulose Utilization

Since a high proportion of biomass is cellulose,[161] the prospects for the production of alcohol from this substance has received particular interest. Cellulolytic organisms break down cellulose by converting it to cellobiose. Cellobiose, in turn, is converted to glucose by a β-glucosidase.[162] The cloning of a β-glucosidase gene from *Escherichia adecarboxylata* into the *E. coli* plasmid vector pBR322 has been reported.[43] It was possible to identify the gene by the ability of *E. coli* to use cellobiose.

By a slightly more sophisticated technique, a cellulase gene from the Gram-positive organism *Cellulomonas fimi* was cloned into *E. coli*.[44] Selection of the appropriate clone in this case was achieved by an immunological screening method using anti-cellulase antibodies. A recombinant, identified in this way, was subsequently shown to possess cellulolytic activity in a colorimetric assay. Low intracellular levels of endoglucanase were expressed. While this endoglucanase, like the *E. adecarboxylata* β-glucosidase, was cloned into pBR322, and thus is confined to *E. coli*, development of new vectors should allow the subcloning of these genes into organisms more immediately useful for the production of ethanol.

Little is known about the genetic organization of bacterial cellulases. However, the cloning of cellulase genes in *E. coli* will provide useful information on the method of expression and organization of these genes.

Increasing Ethanol Tolerance by rDNA

Tolerance of high ethanol concentrations is critical to cell growth, fermentation, and viability.[57] As mentioned before, one approach proposed by Ingram[58] to gain greater ethanol tolerance of Zymomonads is to clone into them the fatty acid synthetase genes of *Lactobacillus homohiochi* and *L. heterohiochi*. These bacterial contaminants of sake can grow at alcohol concentrations greater than 16% wt/v. It is assumed that their alcohol tolerance is due, in part, to the exceptionally long monounsaturated fatty acids in their membranes. Initial cloning of these genes can be accomplished in *E. coli* using broad, host-range plasmids, e.g., pRK290. Transfer of the chimeric plasmids could then be mediated to *Z. mobilis* via conjugation with the aid of a helper plasmid. It should also be possible to accomplish this step by direct incorporation of the *Lactobacillus* genes into *Z. mobilis* plasmids. The latter are under study in our laboratory and should yield chimeric plasmids stable in *Z. mobilis* since they will retain their replication function as discussed above. Details of the Ingram approach are yet to be published.

SUMMARY

The production of alcohol by biotechnological methods is by no means a new idea. Certainly current techniques for the production of alcoholic beverages by fermentation have been very successful. However, the economic incentives to produce lower priced fuel alcohol by fermentation is borderline. Nonetheless, the increasing awareness of the scarcity of fossil fuels may bring about an expansion of the applications of fermentation alcohol in the future. For this to come about, it is clear that new technologies must be adopted. The use of recombinant DNA technology to produce novel strains of *Zymomonas* and yeast may represent one way in which the cost of fermentation fuel alcohol can be reduced. This, coupled with improved techniques of large scale fermentation, may represent a significant step in the production of fuel alcohol and chemical feedstocks.

ACKNOWLEDGMENTS

New Jersey Agricultural Experiment Station Publication No. F–01111–1–82, supported by state funds, and a Department of Energy grant.

REFERENCES

1. A H. Rose. 1977. In *Economic microbiology: Alcoholic beverages*. Vol. 1. Edited by A. H. Rose. New York: Academic Press, pp. 1–44.
2. J. Bu'Lock. 1978. In *Microbial technology: Current state, future prospects*. Twenty-ninth Symposium of the Society for General Microbiology, U.K. Edited by A. T. Bull, D. C. Ellwood and C. Ratledge. Cambridge: Cambridge University Press, pp. 309–325.
3. N. Kosaric; D. C. M. Ng; I. Russell; and G. S. Stewart. 1980. *Adv. Appl. Microbiol.* 26:147–227.
4. D. M. Munnecke. 1981. In *Biomass conversion processes for energy and fuels*. Edited by S. S. Sofer and O. R. Zaborsky. New York: Plenum, pp. 339–355.
5. R. C. Righelato. 1980. *Phil. Trans. R. Soc. Lond.* B 290:303–312.
6. K. Venkatasubramanian and C. R. Keim. 1981. *Ann. New York Acad. Sci.* 369:187–204.
7. National Alcohol Fuels Commission. 1981. *Fuel alcohol, an alternative for the 1980s*. Final report GPO. Washington, D.C.
8. R. S. Chambers; R. A. Herendeen; J. J. Joyce; and P. S. Penner. 1979. *Science* 206:789–795.
9. D. Brandt. 1981. In *Biomass conversion processes for energy and fuels*. Edited by S. S. Sofer and O. R. Zaborsky. New York: Plenum, pp. 357–373.
10. H. W. Doelle. 1975. *Bacterial metabolism*. 2d ed. New York: Academic Press.
11. G. Gottschalk. 1978. *Bacterial metabolism*. 1st ed. New York: Springer-Verlag.
12. A. Margaritis and C. R. Wilke. 1978. *Biotech. Bioeng.* 20:709.
13. K. J. Lee; M. Lefebvre; D. E. Tribe; and P. L. Rogers. 1980. *Biotech. Letters* 2:487–492.
14. R. Maleszka; I. A. Veliky; and H. Schneider. 1981. *Biotech. Letters* 3:415–420.
15. I. G. Prince and J. P. Barford. 1982. *Biotech. Letters.* 4:263–268.
16. M. L. Skotnicki; K. J. Lee; D. E. Tribe; and P. L. Rogers. 1982. "Genetic engineering of microorganisms for chemicals." *Basic Life Sciences* 19:271–290.
17. Alcon Biotechnology Ltd. 1982. Portsmouth, U.K. Brochure.
18. T. K. Ghose and R. D. Tyagi. 1979. *Biotech. Bioeng.* 21:1387–1400.
19. P. L. Rogers; K. J. Lee; M. L. Skotnicki; and D. E. Tribe. 1981. In *Advances in biotechnology*. Vol. 2. Edited by M. Moo-Young and C. W. Robinson. Toronto: Pergamon Press, pp. 189–194.
20. M. Wada; J. Kato; and I. Chibata. 1981. *Europ. J. Appl. Microbiol. Biotechnol.* 11:67–71.
21. Y-Y. Linko; H. Jalanka; and P. Linko. 1981. *Biotech. Letters* 3:263–268.
22. W. Grote; K. J. Lee; and P. L. Rogers. 1980. *Biotech. Letters* 2:481–486.
23. E. J. Arcuri; R. M. Worden; and S. E. Shumate II. 1980. *Biotech. Letters* 2:499–504.
24. A. Margaritis; P. K. Bajpai; and J. B. Wallace. 1981. *Biotech. Letters* 3:613–618.
25. G. Amin and H. Verachtert. 1982. *Europ. J. Appl. Microbiol. Biotechnol.* 14:59–63.
26. A. Ramalingham and R. K. Finn. 1977. *Biotechnol. Bioeng.* 19:583–589.
27. B. Maiorell and C. R. Wilke. 1980. *Biotechnol. Bioeng.* 22:1749–1751.

28. J. H. Lee; J. C. Woodward; R. J. Pagan; and P. L. Rogers. 1981. *Biotech. Letters* 3:177–182.
29. M. Bacila and J. Horii. 1979. *TIBS.* 4:59–61.
30. G. Moulin; H. Boze; and P. Galzy. 1982. *J. Ferment. Technol.* 60:25–29.
31. H. W. Doelle. 1982. *Europ. J. Appl. Microbiol. Biotechnol.* 14:241–246.
32. S. Cromie and H. W. Doelle. 1982. *Europ. J. Appl. Microbiol. Biotechnol.* 14:69–73.
33. S. S. Dills; A. Apperson; M. R. Schmidt; and M. H. Daier, Jr. 1980. *Microbiol. Revs.* 44:385–418.
34. A. H. Romano; J. D. Trifone; and M. Brustolon. 1979. *J. Bacteriol.* 139:93–97.
35. H. W. Doelle. 1972. *Biochim. Biophys. Acta.* 258:404.
36. D. I. C. Wang; I. Biocic; H. S. Fang; and J. D. Wang. 1979. Proceedings of the Third Annual Biomass Energy Systems Conference. Springfield. Va.: National Technical Information Service, and personal communication.
37. J. G. Zeikus and T. K. Ng. 1982. Ann. Reports Ferment. Processes 5:263–289.
38. H. Schneider; P. Y. Wang; Y. K. Chan; and R. Maleska. 1981. *Biotech. Letters* 3:89–92.
39. P. J. Slininger; R. J. Bothast; J. E. Van Cauwenberg; and C. P. Kurzman. 1982. *Biotech. Bioeng.* 24:371–384.
40. T. W. Jeffries. 1981. *Biotech. Letters* 3:213–218.
41. C. S. Gong; L. D. McCracken; and G. T. Tsao. 1981. *Biotech. Letters* 3:245–250.
42. R. Maleszka; P. Y. Yang; and H. Schneider. 1982. *Can. J. Biochem.* 60:144–151.
43. R. W. Armentrout and R. D. Brown. 1981. *Appl. Environment. Microbiol.* 41:1355–1362.
44. D. J. Whittle; D. G. Kilburn; R. A. J. Warren; and R. C. Miller, Jr. 1982. *Gene* 17:139–145.
45. H. J. J. Van Vuuren and L. Meyer. 1982. *Biotech. Letters* 4:253–256.
46. K. Holland. 1980. Unpublished observations.
47. S. W. Brown and S. G. Oliver. 1982. *Biotech. Letters* 4:269–274.
48. L. G. Ljundahl; F. Bryant; L. Carreira; T. Saki; and J. Weigel. 1981. *Trends in the biology of fermentation for fuels and chemicals.* Edited by A. Hollander. New York: Plenum Press.
49. S. Aiba; M. Shoda; and M. Nagatani. 1968. *Biotech. Bioeng.* 10:845–864.
50. S. L. Chen. 1981. *Biotech. Bioeng.* 23:1827–1836.
51. S. Hayashida; D. D. Feng; and M. Hongo. 1974. *Agric. Biol. Chem.* 38:2001–2006.
52. J. Burrell. 1981. Unpublished observations.
53. T. W. Nagodawithana and K. H. Steinkraus. 1976. *Appl. Microbiol.* 31:158–162.
54. J. M. Navarro and G. Durand. 1978. *Ann. Microbiol. Inst. Pasteur.* 129B:215–225.
55. D. S. Thomas and A. H. Rose. *Arch. Mikrobiol.* 1979. 122:49–55.
56. G. K. Hoppe and G. S. Hansford. 1982. *Biotech. Letters* 4:39–44.
57. S. W. Brown; S. G. Oliver; D. E. F. Harrison; and R. C. Righelato. 1981. *Europ. J. Appl. Microbiol. Biotechnol.* 11:151–155.
58. L. O. Ingram. 1982. Seminar.
59. M. Ciriacy. 1975. *Mutation Res.* 29:315–326.
60. C. Wills; P. Kratofil; D. Londo; and T. Martin. 1981. *Arch. Biochem. Biophys.* 210:775–785.
61. V. M. Williamson; E. T. Young; and M. Ciriacy. 1981. *Cell* 23:605–614.
62. O. T. Avery; C. M. McLeod; and M. McCarty. 1944. *J. Exp. Med.* 79:137–158.
63. F. E. Young and J. Spizizen. 1961. *J. Bact.* 81:823–829.
64. S. N. Cohen; A. C. Y. Chang; and L. Hsu. 1972. *Proc. Nat. Acad. Sci.* (USA) 69:2210–2214.
65. M. J. Sinclair and A. F. Morgan. 1978. *Aust. J. Biol. Sci.* 31:769–688.
66. M. Susskind and D. Botstein. 1978. *Microbiol. Rev.* 42:385.
67. N. D. Zinder and J. Lederberg. 1952. *J. Bact.* 64:679.
68. M. L. Morse; E. M. Lederberg; and J. Lederberg. 1956. *Genetics* 41:142.
69. B. W. Holloway. 1969. *Bact. Rev.* 33:419–443.
70. E. M. Lederberg and S. N. Cohen. 1974. *J. Bact.* 119:1072.
71. J. Swings and J. DeLey. 1977. *Bact. Rev.* 41:1–46.
72. H. W. Stokes and E. L. Dally. 1982. Unpublished observation.
73. F. Jacob and E. L. Wollman. 1961. Sexuality and the Genetics of Bacteria. New York: Academic Press.
74. P. M. Chandler and V. Krishnapillai. 1974. *Genet. Res.* 23:239–250.
75. D. Haas and B. W. Holloway. 1976. *Molec. Gen. Genet.* 144:243–251.
76. B. W. Holloway. 1979. *Plasmid* 2:1–19.
77. M. L. Skotnicki; D. E. Tribe: and P. L. Rogers. 1980. *Appl. Env. Microbiol.* 40:7–12.
78. P. Starlinger. 1980. *Plasmid* 3:241–259.
79. N. Kleckner; J. Roth; and D. Botstein. 1977. *J. Mol. Biol.* 116:125–159.
80. S. Harayama; T. Masataka; and T. Lino. 1981. *Mol. Gen. Genet.* 184:52–55.
81. E. L. Dally. 1982. M. S. Thesis. Rutgers University.
82. S. Baumberg; G. Cornelius; M. Panagiotakopoulos; and M. Roberts. 1980. *J. Gen. Microbiol.* 119:257–262.
83. M. Sato; B. J. Staskamicz; N. J. Panopoulos; S. Peters; and M. Honma. 1981. *Plasmid* 6:325–331.
84. R. L. Sinsheimer. *Ann. Rev. Biochem.* 1977. 46: 415–438.
85. H. V. Bernard and D. R. Helinski. 1980. *Genetic engineering.* Vol 2. New York: Plenum Press, pp. 133–167.
86. G. Ditta; S. Stanfield; D. Corbin; and D. R. Helinski. 1980. *Proc. Natl. Acad. Sci.* (USA) 77:7347–7351.
87. E. L. Dally; H. W. Stokes; and D. E. Eveleigh. 1982. *Biotech. Letters* 4:91–96.
88. H. W. Stokes and E. L. Dally. 1982. Unpublished observations.
89. H. W. Stokes; E. L. Dally; R. L. Williams; B. S. Montenecourt; and D. E. Eveleigh. 1980. *Chemistry in energy production.* A.C.S. Symp. pp. 115–121.
90. L. E. Bryan and S. Kwan. 1981. *J. Antimicrob. Chemother.* 8 Suppl. D:1–8.
91. M. Bagdasarian; R. Lurz; B. Ruckert; F. C. H. Franklin; M. M. Bagdasarian; J. Frey; and K. N. Timmis. 1981. *Gene* 16:237–247.
92. T. D. Petes. 1980. *Ann. Rev. Biochem.* 49:845–876.
93. L. H. Hartwell. 1974. *Bacteriol. Rev.* 38:164–198.
94. M. Guerineau. 1979. In *Viruses and plasmids in fungi,* Edited P. A. Lemke. New York: Marcel Dekker, pp. 155–181.
95. *The molecular biology of the yeast Saccharomyces.* 1981. J. N. Strathern; E. W. Jones; J. R. Broach, eds. Cold Spring Harbor Laboratory.

96. J. Lodder. 1970. In *The yeasts: A taxonomic study.* New York: Interscience.

97. J. Herskowitz and J. Oshima. 1981. In *The molecular biology of the yeast Saccharomyces.* J. Strathern; E. W. Jones; J. R. Broach, eds. Cold Spring Harbor Laboratory, pp. 181–210.

98. E. Bücking-Throm; W. Duntze; L. H. Hartwell; and J. R. Manney. 1973. *Expl. Cell Res.* 76:99–110.

99. E. Bücking-Throm and W. Duntze. 1970. *J. Bacteriol.* 104:1388–1390.

100. R. K. Mortimer and D. Schild. 1981. In *The molecular biology of the Yeast Saccharomyces.* J. Strathern; E. W. Jones; J. R. Broach, eds. Cold Spring Harbor Laboratory, pp. 11–26.

101. R. K. Mortimer and D. Schild. *Microbiol. Rev.* 1980. 44:519–537.

102. J. N. Strathern; C. S. Newton; J. Herskowitz; and J. B. Hicks. 1979. *Cell* 18:309–315.

103. R. Bigilas; J. Keesey; and T. R. Ferik. 1977. *JCN-UCLA Symp.* 8:179–189.

104. D. H. Clayton; T. A. Howard; and P. A. Martin. 1972. *Amer. Soc. Brew. Chem. Proc.* 30:78–81.

105. R. S. Tubb. 1979. *J. Inst. Brew.* 85:286–289.

106. P. van Solnigen and J. B. van der Platt. 1977. *J. Bacteriol.* 130:946–947.

107. J. Russell and G. G. Stewart. 1979. *J. Inst. Brew.* 85:95–98.

108. R. B. Wickner. 1981. In *The molecular biology of the yeast Saccharomyces.* J. N. Strathern; E. W. Jones; J. R. Broach, eds. Cold Spring Harbor Laboratory, pp. 415–444.

109. B. Dujon. In *The molecular biology of the yeast Saccharomyces.* 1981. J. N. Strathern; E. W. Jones; J. R. Broach, eds. Cold Spring Harbor Laboratory, pp. 505–635.

110. J. H. Sinclair; B. J. Stephens; P. Sanghavi; and M. Rabinowitz. 1967. *Science* 156:1234–1237.

111. C. P. Hollenberg; P. Borst; and E. F. J. van Bruggen. 1970. *Biochem. Biophys. Acta.* 209:1–15.

112. G. D. Clark-Walker and G. G. Miklos. 1974. *Eur. J. Biochem.* 41:359–365.

113. J. R. Broach. 1982. *Cell* 28:203–204.

114. S. C. Falco; Y. Li; J. R. Broach; and D. Botstein. 1982. *All.* 29:573–584.

115. D. M. Livingston. 1977. *Genetics* 86:73–84.

116. G. G. Stewart; J. Russell; C. J. Panchal. 1980. *Abstr. VI in Int. Symp. on Yeasts.* London (Ontario), Canada, pp. 212–221.

117. A. Hinnen; J. B. Hicks; and G. R. Fink. 1978. *Proc. Natl. Acad. Sci.* (USA) 75:1929–1933.

118. J. D. Beggs. 1978. *Nature* 275:104–109.

119. K. Struhl; D. J. Stinchcomb; S. Scherer; and R. W. Davis. 1979. *Proc. Nat. Acad. Sci.* (USA) 76:1035–1039.

120. L. Clark and J. Carbon. 1980. *Nature* 287:504–509.

121. D. Morris; J. Noti; F. K. Osborne; and A. Szalay. 1981. *DNA* 1:27–35.

122. D. Morris. Personal Communication.

123. C. Segen; P. J. Farabraugh; A. Hinnen; J. M. Walsh; G. R. Fink. 1979. In *Genetic engineering principles and methods.* Vol 1. J. K. Setlow and A. Hollander, eds. New York: Plenum Press, pp. 117–132.

124. D. Botstein; S. C. Falco; S. E. Stewart; M. Brennan; S. Scherer; D. T. Stinchcomb; K. Struhl; and R. W. Davis. 1979. *Gene* 8:17–24.

125. F. Bolivar; R. Rodriquez; P. J. Greene; M. C. Betlack; H. Heyneker; J. H. Cresa; S. Falfon; and H. W. Boyer. 1977. *Gene* 2:95–113.

126. M. L. Bach; F. Lacrante; and D. Botstein. 1979. *Proc. Natl. Acad. Sci.* (USA) 76:386–390.

127. *Methods in enzymology.* Vol. 68. 1979. Edited by R. Wu. New York: Academic Press.

128. G. Hohn. In *Methods in enzymology.* Vol. 68. 1979. Edited by R. Wu. New York: Academic Press, pp. 299–308.

129. J. Collins. In *Methods in enzymology.* Vol. 68. 1979. Edited by R. Wu. New York: Academic Press, pp. 309–325.

130. M. Grunstein and D. S. Hogness. 1975. *Proc. Natl. Acad. Sci.* (USA) 72:3961–3965.

131. R. A. Kramer; J. R. Cameron; and R. W. Davis. 1976. *Cell* 8:227–232.

132. J. D. Petes; J. R. Broach; P. Wensink; L. M. Hereford; G. R. Fink; and D. Botstein. 1978. *Gene* 4:37–49.

133. M. V. Olson; B. D. Hall; J. R. Cameron; and R. W. Davis. 1979. *J. Molec. Biol.* 127:285–295.

134. D. L. Montgomery; B. D. Hall; S. Gillam; and M. S. Smith. 1978. *Cell* 14:673–680.

135. K. Struhl and R. W. Davis. 1977. *Proc. Natl. Acad. Sci.* (USA) 74:5255–5259.

136. B. Ratzkin and J. Carbon. 1977. *Proc. Natl. Acad. Sci.* (USA) 74:487–491.

137. D. Vapnek. 1977. *Proc. Natl. Acad. Sci.* (USA) 74:3508–3512.

138. K. Struhl; J. R. Cameron; and R. W. Davis. 1976. *Proc. Natl. Acad. Sci.* (USA) 74:1471–1475.

139. L. Clarke and J. Carbon. 1978. *J. Mol. Biol.* 120: 517–532.

140. R. C. Dickson and J. S. Markin. 1978. *Cell* 15: 123–130.

141. M. V. Olson. 1918. In *Genetic engineering principles and methods.* Vol. 3. J. K. Setlow and A. Hollander, eds. New York: Academic Press, pp. 57–88.

142. A. Hinnen; P. J. Farabraugh; C. Ilgen; and G. R. Fink. 1979. In *Eukaryotic gene regulation.* Vol. 14. R. Axel; T. Maniatis; M. Fox, eds. New York: Academic Press, pp. 43–51.

143. K. A. Nasmyth and S. S. Reed. 1980. *Proc. Natl. Acad. Sci.* (USA) 77:2119–2123.

144. K. A. Nasmyth and K. Thatchell. 1980. *Cell* 19: 753–764.

145. V. M. Williamson; J. Bennetzen; E. J. Young; K. Nasmyth; and B. D. Hall. 1980. *Nature* 283:214–216.

146. R. W. Davis; G. R. Fink; and D. Botstein. 1981. *Ann. Rev. Biochem.* 50:112–158.

147. J. J. Pantkier; P. Fournier; H. Heslot; and A. Ramback. 1980. *Curr. Genetics* 2:109–113.

148. A. Jiminez and J. Davies. 1980. *Nature* 280: 869–871.

149. J. D. Beggs; J. van den Berg; A. van Ooyen; and C. Weissman. 1980. *Nature* 283:835–840.

150. R. C. Dickson. 1980. *Gene* 10:347–356.

151. O. Mercereau-Puijalen; F. Lacroute; and P. Kourilsky. 1980. *Gene* 11:163–167.

152. R. A. Hitzeman; F. E. Hagie; H. L. Levine; D. V. Goeddel; G. Ammerer; and B. D. Hall. 1981. *Nature* 293:717–722.

153. C. C. Emeis. 1971. *Amer. Soc. Brew. Chem. Prac.* 29:58–62.

154. H. Tamaki. 1978. *Molecular Gen. Genet.* 164: 205–209.

155. F. K. Zimmerman; N. A. Khan; and N. R. Eaton. 1973. *Molec. Gen. Genet.* 123:29–41.

156. D. B. Mowshowitz. 1979. *J. Bact.* 137:1200–1207.

157. G. G. Stewart. Personal Communication.

158. G. B. Calleja; S. L. Rick; A. Nasim; C. V. Lusena; C. C. Champagne; I. A. Veliky; and F. Morannelli. 1982. In *The XIII Proceedings of the International Congress of Microbiology*. Boston, Mass.; August 8–13, 1982.

159. K. Yamane and B. Maruo. 1980. In *Molecular breeding and genetics of applied microorganisms*. Edited by K. Sakaguchi and M. Okanish:

Tokyo, Japan: Kodansha-Academic Press, pp. 117–123.

160. S. Shinomiya; K. Yamane; T. Mizakami; F. Kawamura; and H. Saito. 1981. *Agric. Biol. Chem.* 45:1733–1735.

161. J. A. Bassham. 1975. *Biotech. Bioeng. Symposium*. No. 5. Sept. 19, 1975.

162. C. S. Gong and G. T. Tsao. 1979. *Ann. Rep. Ferment. Processes* 3:111–140.

Advances in Solar Energy © 1983 American Solar Energy Society, Inc.

CRYSTALLINE SILICON AS A MATERIAL FOR SOLAR CELLS

M. RODOT

Centre National de la Recherche Scientifique, 1 Pl. A. Briand, F-92190, Meudon, France

Abstract

As a material for solar cells, hyperpure single crystalline silicon is progressively replaced by less pure, polycrystalline materials. The various techniques of silicon purification and of ingot, ribbon and sheet growth are reviewed. The physical effects of impurities as well as those of grain boundaries and other crystalline defects are discussed, in order to define the material requirements to obtain efficient solar cells; it is shown that grain boundaries can getter impurities and that impurities can passivate grain boundaries, so that more impure and imperfect crystals, i.e. cheaper materials than presently used, can be envisaged in future. From a detailed comparison of the various proposed techniques, it is concluded that the present "chemical route" consisting of halogenide purification and ingot growth may be challenged soon by a "metallurgical route" in which the active layer is epitaxially grown on a wafer recrystallized from upgraded metallurgical grade silicon.

INTRODUCTION

In the three decades following their discovery,[1] solar cells used mostly one semiconductor material: monocrystalline, ultra-pure, electronic grade silicon. For space applications this material allowed to obtain very high efficiencies[2] reaching 15% (AMO) and even 20% under air mass 1 (AM1). For terrestrial applications where a compromise on cost and efficiency has to be made, this material still accounts for more than 95% of the commercial cells. However, recent developments show the possibility of using silicon under a less pure or less perfect form, or both, opening the way to a larger extension of the solar cell market. For instance, in 1980 was published an efficiency of 12.9% for a cell made from metallurgical grade silicon containing as much as 70 ppm (parts per million) iron[3] and in 1981 was announced 17% efficiency for a polycrystalline cell of 4 cm² area.[4] Since these trends tend to modify the whole industrial and commercial status of this field, it is useful to review them, which is the object of this paper.

The Impact of Silicon Material Quality

Figure 1 shows the price decomposition of a typical photovoltaic (PV) generator, as seen by a french producer.[5] The major recent and future decrease is that of the silicon material, which accounts for 40% of the total price in 1982. According to the largest silicon producer, Wacker,[6] the solar cell material will cost 50 to 90¢/W in 1990, allowing a world production of 100 MW/yr (instead of 4¢/W and 5 MW/yr in 1980). More optimistic prospects have even been presented.[7] This progress on the material is accompanied by gains in the cell technology and encapsulation, which are in part due to improved automatized processes, made possible by the production increase.

The evolution of crystalline silicon as a material for solar cells is dominated by three main issues:

(1) A new solar grade (SG) silicon, with a purity specification less stringent than that of electronic grade (EG) material is being created. This adaptation of the chemical industry to the needs of PV industry was not possible when the consumption of the latter was so low (60 t/yr in

Michel Rodot, Ph.D., is Director of Research of the French National Center for Scientific Research (CNRS) at Meudon, France. Dr. Rodot had previous responsibility for a laboratory devoted to semiconducting materials, in which the first French high efficiency solar cells were made in 1960, as well as other works on infrared and luminescent materials. From 1976–81 he led the CNRS program for solar energy research and is now working in connection with industry in the field of polycrystalline silicon for solar cells.

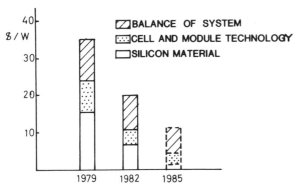

Fig. 1. Components of solar cell prices, in 1980 $ per peak W.[5]

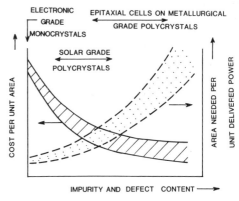

Fig. 2. Scheme showing the main effects of silicon imperfections on solar cell efficiency and cost.

1980, as compared to 3,000 t used by microelectronics); it becomes possible and necessary with the expected PV market growth. In 1990 several new production units of normal size (at least 500 t/yr) will have to be devoted specifically to PV needs. Several conceptions of such units are being actively studied now.

(2) Czochralski (CZ)-grown crystals are being replaced by ingots, ribbons, or sheets made by one of the numerous, potentially cheaper, processes used to crystallize silicon melts. Some of these processes, like the edge-defined film-fed growth (EFG) and the web-dendrite processes, have been studied for 10 to 15 years; a larger number have been invented and tried in the last few years. Most processes give polycrystals of various qualities. The competition between them is very open.

(3) Eventually, the solar cell will be built, not inside the polycrystalline solar grade material, but in a purer silicon layer deposited on the top of it. This would allow to admit still more impurities and defects in the material used as substrate. This concept has been developing quickly since 1980; it impulses new studies of cheaper substrates and may even merge with the concept of "sheets," which implies using non-silicon substrates.

When the purity and perfection requirements of the material are made less stringent, its cost may be decreased, but at the same time the admittance of impurities and physical defects has adverse effects on the cell efficiency (see Fig. 2). The best material, in terms of $/W, should be the result of a compromise. Knowing the behavior of impurities and defects in silicon solar cells is a prerequisite to assess the best technological process. This matter will be treated later. No mention will be made of a spe-

cial point defect problem, that of the radiation damage experienced by space solar cells.

With these fundamental data in mind, we shall be able to describe more precisely the state of the different technologies for silicon purification, crystal growth, and cell fabrication. In our conclusion we shall present some information on the economic and social achievements which may result from present and future improvements of the silicon material for solar cells.

Two excellent reviews[8,9] have recently treated some topics overlapping with the present paper.

Other Materials Problems with Solar Cells

The problems thus sketched are by no means the only ones concerning materials raised by the development of photovoltaics.

Once the basic material has been defined, the fabrication of PV modules using crystalline silicon implies the formation of p-n junctions (or other types of junctions), of contacts, of an anti-reflective layer, and also the encapsulation of individual cells in modules. The junction formation is essentially a silicon material problem and will be quoted, though not exhaustively, later. For the sake of homogeneity, this paper will not deal with contact metals, antireflecting materials, and plastic encapsulants.

Of course there also exist decisive materials problems in the development of other solar cells. Among the numerous solar cell types,[10] those using Si-H amorphous alloy perform remarkably well, and other semiconductor compounds are also promising. Each material raises the same questions—essentially cost-quality optimization—as does silicon, plus, very often, problems of instabilities which do not exist in silicon terrestrial PV cells. These questions are more complex than for crystalline silicon because of the larger number of possible defects in compounds and alloys. Though very interesting, they are out of the scope of this review. It may be possible that, in the long range, one of these materials performs better than crystalline silicon, either because of cheaper fabrication (in the thin film option) or because of smaller needed area (in the option of sophisticated cells under concentrated solar radiation). They may also emerge as solutions for limited markets (e.g., small Si-H cells for pocket computers). But the potential improvements of crystalline Si described in this review seem sufficient to ensure its commercial vitality in the next 10 years.

PATHS FOR SILICON PURIFICATION AND CRYSTALLIZATION

Silicon is extracted from the raw material silica, SiO_2, which is in the form of quartzite rock or quartz sand. Metallurgical grade (MG) Si, of purity 98 to 99% and cost of $1.20/kg, is a product of silica reduction by carbon in an arc furnace.[11] From this first step, two general routes are possible.

The Chemical Route

This route is similar to that of EG Si, except for simplifications.

From MG Si, one usually gets refined Si by a series of steps, including the formation, distillation, and reduction of halogenides. Two of these compounds, $SiCl_4$ and SiF_4, are implied in other technologies (famed silica and fertilizers, respectively) which use ten times as much of them as does the electronics industry: an oppor-

---

A. *CHEMICAL PURIFICATION*[a]
    Zinc reduction of $SiCl_4$[b] (the Dupont process)
    Hydrogen reduction of $SiHCl_3$ and $SiCl_4$[c] (the Siemens process)
    Conversion of $SiCl_4$ to $SiHCl_3$, SiH_2Cl_2, and finally SiH_4, and pyrolysis of SiH_4[b,d]
    Hydrogen reduction of SiH_2Cl_2[b]
    Sodium reduction of $SiCl_4$
    Sodium reduction of SiF_4

B. *CRYSTAL GROWTH TECHNIQUES FROM THE MELT*[f]

B.1. *Ingot growing and sawing*
    CZ = Czochralski growth[g]
    HEM : heat exchange method[h]
    Bridgman growth[i]
    Vacuum casting[j]
B.2. *Self-supported ribbons*
    HRG[k] = horizontal ribbon growth, and
    LASS = low angle silicon sheet
    RFC[l] = ribbon from cylinder
    RFP[m] = ribbon from powder
    RTR[n] = ribbon to ribbon
    WD[o] = web dendrite
    ESP[p] = edge-supported pulling
    EFG[q] = Edge-defined film fed growth and similar techniques derived from the Stepanov technique : CAST, inverted Stepanov
    ICC[r] = interface controlled crystallization
    Ultra-fast quenching[s]
    Plasma torch technique[t]
B.3. *Sheets grown on low-cost substrates*
    SOC = silicon on ceramics[u]
    RAD = ribbon against drop[v]

[a] Reviews have been published by S. Pizzini,[8] E. Sirtl,[66] J. Dietl[9]
[b] R. Lutwack[68]
[c] E. Sirtl[67]
[d] W. C. Breneman[68]; R. Lay[69]
[e] W. H. Reed[71]
[f] Some general information on crystallization processes is given by M. C. Fleming[15] and, for ribbons and sheets only, by a special issue of *J. Crystal Growth*.[129] Crystal growth of Si for solar cells has also been reviewed by J. Dietl,[52] T. F. Ciszek,[83] and E. Fabre.[130]
[g] G. Fiegl[131]
[h] F. Schmid, C. P. Khattak[84]
[i] cf. note[85]
[j] J. Dietl,[9] D. Helmreich[95]
[k] B. Kudo, D. Jewett[97]
[l] The theory of this method is described by H. Rodot[87]
[m] D. Casenave[88]
[n] A. Baghdadi, K. R. Sarma[98]
[o] R. G. Seidensticker, C. S. Duncan[80]
[p] T. F. Ciszek,[73] J. L. Hurd and T. F. Ciszek, J. *Crystal Growth* (to be published)
[q] cf. note[103]
[r] Information on this method is given by J. Dietl[9]
[s] K. I. Arai[110]
[t] M. S. Alaee[111]
[u] J. D. Zook, B. G. Koepke, B. L. Grung, M. H. Leipold (Ref. 129 p 260); S. B. Schuldt[113]
[v] C. Belouet.[115]

Table 1. Use of Solar Grade Silicon (The Chemical Route)

tunity for economic interactions and savings which is largely utilized. These compounds, as well as other ones playing the role of intermediate products ($SiHCl_3$, SiH_2Cl_2, SiH_4, and so on, may be used in various reaction schemes (see Table 1A). The principles of this type of chemistry have been discussed in several papers.[12,8] Some of these routes which appear to be the most promising[13] will be more fully described later. The resulting silicon cost is estimated between \$10 and 20/kg, instead of \$50 and 75/kg for electronic grade silicon.[14] Its impurity content, although little publicized up to now, seems to be hardly higher than that of EG Si.

In a second step, this refined silicon is melted and solidified in a controlled way, so that it is finally transformed into a thin slice of mono- or polycrystalline Si. The various known processes[9] (see Table 1B) can be classified under three main headings. In the first class of processes, large ingots are grown, and then slices are sawed from the ingots. In the second class, self-supported ribbons are directly obtained from the melt, with the evident advantage of by-passing the sawing step. A third class uses either temporary or permanent substrates, onto which sheets are grown which may be made very thin. Modeling of crystallization phenomena is a complex matter.[15] In all cases, some kind of progressive crystallization takes place which allows the formation of a hole-free and crack-free crystal. Supplementary purification may also occur, but this effect is often masked by a contrary one, that is, the introduction of impurities from crucibles, dies, or other solids which are generally in contact with the melt. The defects incorporated in the crystal include dislocations, stacking faults, twinning, and grain boundaries; some techniques produce small size grains, but many of them deliver large crystals, much larger than 0.1 to 1 cm. The best performing techniques and their results will be described later. Economic performance is largely dependent on the growth speed, that is, the area of silicon which is produced per unit time; best results are lower than \$2/W and will fall below \$1/W using \$20/kg raw material. The gain over the present situation (\$4 to 7/W) is sufficient to open large opportunities to the new generation of solar cells which will be made by these techniques.

The Metallurgical Route

This second route (see Table 2) involves a further drastic simplification of chemical purifi-

A. *UPGRADED METALLURGICAL GRADE (UMG) SILICON*
    Direct arc reactor technique (DAR)[a]
    Pyrometallurgical treatments[b, c, j]

B. *CRYSTALLIZATION AND EPITAXY STEPS*
    UMG-SI + zone melting + CVD + H − passivation[d]
    UMG-Si + CZ + CVD[e, c]
    MG-Si + CZ or HEM + gettering + CVD[f, g]
    DAR-Si + ESP + CVD[h]
    DAR-Si + HEM[i]
    UMG-Si + progressive crystallization on graphite + grain boundary segregation + CVD[j]

[a] V. D. Dosaj[72]
[b] J. R. McCormick,[3], S. Pizzini[8], J. Dietl[9]
[c] F. Secco D'Aragona[74]
[d] T. L. Chu[75]
[e] P. H. Robinson[76]
[f] T. Warabisako[77]
[g] R. G. Wolfson[78]
[h] T. F. Ciszek[73]
[i] C. P. Khattak[16]
[j] T. L. Chu[56]

Table 2. Use of Metallurgical Grade Silicon (The Metallurgical Route)

cation, compensated by a crystallization step which has a highly efficient purifying function. A third step, epitaxial deposition of a pure silicon layer on the substrate thus obtained, seems in general necessary to get good reproducible results; this third step may perhaps, in some cases, be bypassed. Whereas in the chemical route impurities and defects were minimized separately, this one is a more global and ambitious approach, trying to take the best advantage of defect-impurity interactions which are exemplified by grain-boundary passivation and segregation phenomena.

Step one[8] rejects the halogenide intermediates as too expensive. In a first form, there is a radical improvement of the initial silica reduction, which is conducted in very pure conditions with specially chosen silica and carbon raw materials (the direct arc reactor (DAR) or submerged electrode arc furnace). In a second form, full use is made of acid leaching, slagging, gas blowing, and other pyrometallurgical treatments which are common in the metallurgy of refractory metals. The product of step one may be called upgraded metallurgical grade silicon (UMG).

UMG silicon is submitted, in step two, to melting and progressive crystallization in conditions maximizing impurity segregation.[9] The crystal quality and purity thus obtained are seldom sufficient to obtain directly efficient solar cells.[16] This is the reason for step three which consists of depositing on this material, by chemical vapor deposition (CVD) or other means, a thin layer which will be the active absorber of the solar cell. While this third step may be an economic burden able to destroy the advantage gained in step one, it may also be viewed as a part of the cell processing which in any case is necessary after the silicon slice production. To match this goal, cell technology may have to be reconsidered.

In spite of much recent work on this route, it is further from industrial development than the classical chemical route is.

PHYSICAL BACKGROUND

Silicon solar cells are simple electronic devices, implying no gradients of bandgap or electron affinity, but only p-n junctions and, eventually, carrier concentration gradients. Alternative solutions like Schottky barriers or MIS-devices, which are less performant than p-n junctions in the case of perfect silicon, remain so, at least provisionally, when impurities and grain boundaries, or both, are present, and will not be considered here. Cell modeling shows that the physical parameter which is mostly affected by defects is the carrier recombination length in the region where radiation is absorbed; however, other factors also may play a role. This chapter will discuss these points in the frame of a simple cell modeling scheme; we refer to several books for a more refined theoretical description of solar cells.[17-20] We will also quote those properties of impurities and defects which are of importance for solar cell fabrication.

The Criterion of Base Recombination Length

The essential part of a silicon cell is illustrated by Fig. 3. Its total thickness may be of the order of the absorption length for solar radiation $L_\lambda = 1/\alpha_\lambda$ (α_λ = absorption coefficient): fifty μm of crystalline Si absorb 85% of solar radiation; it may also be much larger due to mechanical strength requirements. The photo-emf, V_{oc}, originates mainly from the barrier, or space-charge region (SCR), present at the p-n junction and is shown, on very general bases,[20] to be an increasing function of both the barrier height in the dark and the radiation intensity, ϕ_o. At least for moderate ϕ_o's, it is found that J is the sum of two terms:

$$J = J_{sc}(\phi_o) - J_{BK}(V). \qquad (1)$$

The first term is the short-circuit current and integrates the contributions of top, barrier and base regions; this term is influenced by the recombination length, L, in each region, since carrier recombination causes a net decrease of J_{sc} for a given flux, Φ_o, and also by the recombination speed, s, on the top and bottom faces of the cell. The second term is the obscurity current, comprising also three contributions: those of top and base regions, proportional to $[\exp(eV/kT) -$

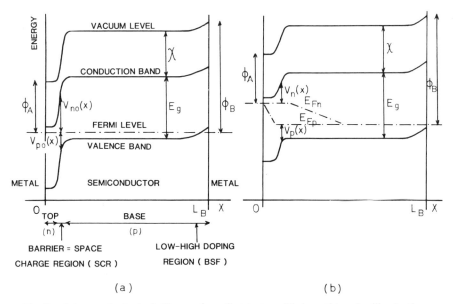

Fig. 3. Scheme of a typical silicon solar cell: (a) at equilibrium, (b) under illumination.

1], and that of SCR, proportional to [exp $(eV/n_{scr} kT) - 1$], where $n_{scr} = 1.5$ to 2.

For a simple absorption and recombination modeling, it follows that the cell is optimized if:

(1) the top region is very thin and highly doped, and the front surface has low s, and

(2) the base width L is larger than L_λ and the base recombination length, L_n, is larger than L, the base doping is as high as is compatible with $L_n \gg L$, L_λ, and the back surface has a low s. A back surface field (BSF) produced by a doping gradient may help to reduce both surface and volume recombination in this region.

We assume here, to conform to the usual cell conformation, that the base is p-type and the top region n-type.

Under these conditions, the photoelectrons created in the base and able to reach the junction are responsible for the light current, J_{sc}. In each layer (x, x + dx) the number of absorbed photons is:

$$dN_\lambda = N_\lambda (1 - R_\lambda) \frac{1}{L_\lambda} \exp\left(-\frac{x}{L_\lambda}\right) dx$$

$(R_\lambda = \text{reflectivity})$

and their contribution to the current is:

$$dJ_\lambda (x) = eN_\lambda (1 - R_\lambda) \frac{1}{L_\lambda} \exp$$

$$-\left[\frac{x}{L_\lambda} + \frac{x}{L_n}\right] dx.$$

Integrating with respect to base thickness and wavelength gives:

$$J_{sc} = \int_\lambda J_\lambda; \quad J_\lambda = eN_\lambda (1-R_\lambda) \frac{1}{1 + \dfrac{L_\lambda}{L_n}}$$

$$\left[1 - \exp - L\left(\frac{1}{L_\lambda} + \frac{1}{L_n}\right)\right]. \quad (2)$$

Equation (2) shows explicitly that high recombination affects the value of J_{sc}.

To express the open-circuit voltage V_{oc}, the term J_{BK} (V) may be put under the form:

$$J_{BK} = J_o \left(\frac{\exp V}{nkT} - 1\right) \quad (3)$$

where n is ideally equal to 1 (Shockley's model) but may be larger if space-charge region recombination is important. The saturation current, J_o, is a function of $1/L_n$ and $1/L_p$, so that the open-circuit voltage:

$$V_{oc} = n \left(\frac{kT}{e}\right) Ln \left(\frac{J_{sc}}{J_o}\right) \quad (4)$$

also decreases when recombination is increased.

The model sketched here has been developed by Davis[21] who drew the curves of Fig. 4 to illustrate the effect of L_n on J_{sc} and V_{oc}. Figure 4a shows the variation of J_{sc} with L_n according to Eq. (2); Fig. 4b shows the variation of V_{oc} with

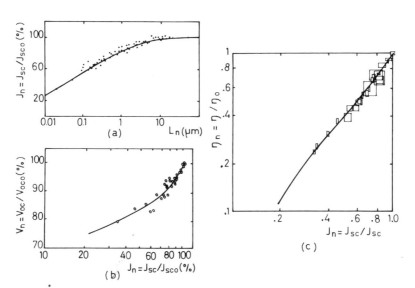

Fig. 4. Variations of solar cell performance with lifetime.[21] (a) Short-circuit current vs. base lifetime. (b) Open-circuit voltage vs. short-circuit current. (c) Efficiency vs. short-circuit current. Curves describe a simplified model (see text) in which L_λ has been given the mean value 0.17 μm.

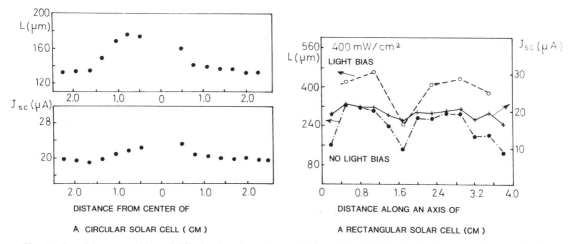

Fig. 5. Local homogeneities of diffusion length and correlated variations of short-circuit current for two single crystalline solar cells (light flux 2mW/cm,[2] 1 μm wavelength) (from Chu[24]).

J_{sc} according to Eq. (4), with J_{sc} given by Eq. (2) and J_0 proportional to $1/L_n$.

If, for a constant doping, L_n is reduced from 10 to 1 μm, J_{sc} is multiplied by a factor 0.81, V_{oc} by 0.92, and the solar cell efficiency by 0.7.

Besides their effect on recombination losses just described (bulk effect), impurities and defects may also bring about a junction effect, that is, an increase of the nonideality coefficient n of Eq. (3). Complete measurement of solar cell characteristics is the most accurate way of assessing the effects of defects and has been often used to this purpose (see e.g. Ref. 21 for impurities and Ref. 22 for grain boundaries). To get more physical insight, it may be useful, however, to measure the electron recombination length, L_n, or lifetime τ_n in the base material. It is known that $L_n = (\tau_n \mu_n kT/e)^{1/2}$ (μ_n = mobility). Both μ_n and τ_n can be altered by imperfections. Several measurement techniques have been discussed by individuals[23] and in standard textbooks; results are generally technique-sensitive and may be largely influenced by light intensity or other parameters. Some of these techniques can detect local heterogeneities, as is shown in Fig. 5 for the case of an apparently perfect single crystal (using a sophisticated technique).[24] It is often enlightening to observe the effect of extended defects on recombination by using mobile light-probe (LBIC) or electron-probe (EBIC), and detecting the photocurrent through a p-n junction or Schottky barrier placed at a fixed distance from the probe, while the probe is scanned on the Si sample.[25]

IMPURITY EFFECTS

The main inpurity properties which are relevant to our problem are: the solubility of impurities in Si, their segregation coefficients and diffusion constants, their electronic activity and effect on diffusion length, and finally their influence on solar cell characteristics.

Impurity Solubility and Segregation Coefficient

Only Ge is soluble in all proportions in Si. Other impurities have solubility limits which are higher for atoms somewhat similar to silicon, lower for very different atoms; they are correlated with impurity tetrahedral radii and heats of sublimation from silicon.[26] Except at the highest temperatures (retrograde solubility), most impurities have solubilities which decrease exponentially with T, so that impurity precipitation occurs by low temperature annealing (see Fig. 6). A closely related property is the segregation coefficient, k, which describes the purifying effect of unidirectional freezing (see Table 3).

It has been pointed out recently that dissolved impurity content can be 10 to 100 times more than the equilibrium solubility limit in crystals quenched at high rates after laser- or electron-beam annealing[27] (see Fig. 30).

A microscopic-scale description of solutions may be very useful, but is not always available. Carbon, group III, and group V atoms dissolve substitutionally in Si; Au, Fe, Cu, and Ni dissolve mostly interstitially, although an appreciable atomic fraction is located on lattice sites (see Fig. 6 for Au; this eventuality has been contested for Fe.[28, 29]). Oxygen, which is interstitial

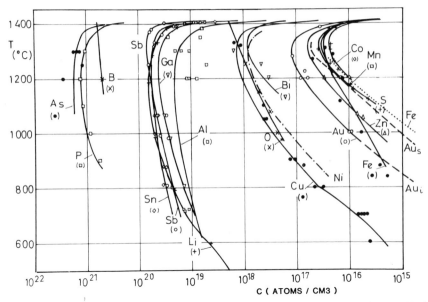

Fig. 6. Solid solubilities of impurity elements in silicon _____ from[26]; _._._. (Ni) from[138]; ___ (Au in interstitial and substitutional sites) from[139]; .... (Fe) from Feichtinger.[29]

in low concentrations, can also form various Si-O complexes,[30] with compositions ranging from Si_2O to SiO_6.

If several impurities are simultaneously present, they may interact in a number of ways. The presence of acceptors can reciprocally increase the solubility of donors. In other cases, pairs of impurities or more complex defects have been found, for example, Fe-O complexes[31] or Fe-B complexes.[32]

Impurity Diffusion

Atomic diffusion data have been collected for various impurities and are reviewed in several

papers.[33, 34, 35] The diffusion coefficient varies with temperature according to an Arrhenius law (see Fig. 7):

$$D = D_o \exp - \frac{Q}{kT} (D, D_o \text{ in cm}^2/\text{s}, Q \text{ in eV}).$$

The usual donors (P, As, Sb, Bi) and acceptors (B, Al, Ga, In) have been extensively studied; they diffuse substitutionally, that is, slowly; their activation energy, Q, is of order 3 to 4 eV. All interstitial impurities diffuse at a much higher rate, with Q's of order 0.5 to 1.0 eV; intermediate behaviors may sometimes be met (Ti) (see Fig. 7).

Impurities	Solubility at 1,100° C (cm^{-3})	Segregation coefficient	Impurities	Solubility at 1,100° C	Segregation coefficient
Ag		4.10^{-5} [a]	Li	$6 .10^{19}$	$.10^{-2}$
Al	2.10^{19}	$2.8.10^{-3}$ [b]	Mg		$3.2.10^{-6}$ [b]
As	$1.5.10^{21}$	0.3	Mn	$1.5.10^{16}$ [c]	$1.3.10^{-5}$ [b]
Au	$4 .10^{16}$	$2.5.10^{-5}$	Mo		$4.5.10^{-8}$ [b]
B	$5 .10^{20}$ [c]	0.8	Na		2.10^{-3}
Be		$1.3.10^{-4}$	Ni		$3.2.10^{-5}$ [b]
Bi	$1.5.10^{17}$	$7 .10^{-4}$	O	$4 .10^{17}$	0.5
C	$.10^{16}$	0.07 [b]	P	$1.5.10^{21}$	0.35
Ca		1	S	$3 .10^{15}$	$.10^{-5}$
Co	$.10^{16}$ [c]	8.10^{-6}	Sb	$5 .10^{19}$	$2.3.10^{-2}$
Cr		$1.1.10^{-5}$ [b]	Sn	$5 .10^{19}$	$1.6.10^{-2}$
Cu	$6 .10^{17}$	$6.9.10^{-4}$ [b]	Ta		$.10^{-7}$
Fe	$9 .10^{15}$	$6.4.10^{-6}$ [b]	Ti		$3.6.10^{-6}$ [b]
Ga	$3.5.10^{19}$	8.10^{-3}	V		4.10^{-6} [b]
Ge	no limit	0.33	Zn	$1. 10^{16}$	1.10^{-5} [b]
In		4.10^{-6}	Zr		$1.5.10^{-8}$

[a] J. A. Van Vechten.[132]
[b] S. Pizzini[8];
[c] A. R. Bean[133] (at 1200° C); other data from F. A. Trumbore[26]

Table 3. Solubilities and Segregation Coefficients of Impurities in Silicon.

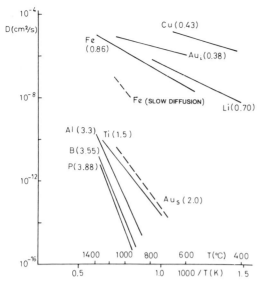

Fig. 7. Diffusion coefficients of impurities in monocrystalline Si (Ti from Ref. 135, all others from Ref. 34.)

Interaction between the diffusions of simultaneously present impurities is frequently observed. However, most studies have been devoted to usual donors and acceptors, which favor the diffusion of one another. How the complex defects migrate is not known in general; in the particular case of the Fe-B complex, the activation energy for diffusion was found to be close to that of interstitial iron.[36]

Electronic Activity of Impurities

Among the different impurities which have been studied in silicon,[35] the elements of columns III and V are the best known and most simple; each substitional atom is, respectively, a shallow acceptor and a shallow donor, so that they are currently used for doping. An exception is Al, which is incorprated only partly (1/10) in this way, while most Al atoms are incorporated in another way without producing doping.

Most transition metals bring in deep electronic levels which can be detected, for instance, by deep-level transient spectroscopy (DLTS)[37] Table 4 shows, for some impurities, the energy of the trap level and the capture rate of the trap for electrons and holes. However, very different results may be obtained for a given impurity, depending on the temperature and speed of sample quenching and on the presence of other impurities.

If N_x (cm^{-3}) atoms of impurity, x, produce $N_t = a_x N_x$ traps with a capture cross section for electrons, C_n, the impurity causes a specific recombination probability:

$$\frac{1}{\tau_n} = C_n a_x N_x. \tag{5}$$

From τ_n we can deduce the recombination length L_n:

$$L_n = \left(\frac{D_n}{C_n a_x N_x}\right)^{\frac{1}{2}}. \tag{6}$$

Impurity	Behavior	Trap Level Energy (eV)	Capture Rates (cm^2/s) C_n	C_p
Au[a, c]	acceptor	Ec − 0.55	2.10^{-8}	1.10^{-7}
	donor	Ev + 0.35	6.10^{-7}	$2.4.10^{-7}$
Co[c]	acceptor	Ec − 0.53	2.10^{-8}	
	donor	Ev + 0.35		$> 1.10^{-8}$
Cr[c]	acceptor	Ec − 0.57	4.10^{-9}	
	donor	Ev + 0.36		$> 5.10^{-7}$
	donor	Ev + 0.38		$> 5.10^{-7}$
Cu[a, c]	donor	Ev + 0.24	−	$2. \ 10^{-7}$
	donor	Ev + 0.41		$8. \ 10^{-6}$
Fe[a]	donor	Ev + 0.40	$1.5.10^{-8}$	$3. \ 10^{-9}$
Mo[b]	donor	Ev + 0.30		
Ni[c]	acceptor	Ec − 0.43	$1. \ 10^{-8}$	
	donor	Ev + 0.14		$> 4.10^{-7}$
Ti[b]	donor	Ev + 0.30		
	acceptor	Ec − 0.26		
Ti[c]	donor	Ev + 0.35		$> 3.10^{-7}$
	acceptor	Ec − 0.55	3.10^{-8}	
V[c]	donor	Ev + 0.31		$> 4.10^{-7}$
	acceptor	Ec − 0.58	3.10^{-8}	

[a] H. G. Grimmeiss.[41]
[b] A Rohatgi.[40]
[c] L. C. Kimerling.[140]

Table 4. Deep Levels in Silicon

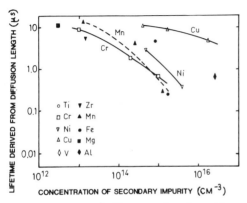

Fig. 8. Lifetime τ of silicon samples of equal boron doping (3 Ωcm) containing a secondary impurity x, as a function of concentration N_x (derived from measurements of solar cell short-circuit current) (from Othmer and Chen.[23])

Some systematic measurements have been made of L_n or τ_n in crystals containing one impurity. We quote, in Figs. 8 and 9, measurements of L_n by the penetrating light or surface photovoltage method.[23,38] The recombination is found to increase steeply due to traces of V or Ti, a little less with Fe, Cr, Mn, Ni, and moderately with Cu and Al. For Al, the results are changed after cell processing, which may bring in other impurities or change the nature of the defects. Results for Fe are different in these two studies, indicating the possible existence of different defects (due to other impurities like O or C?).

Through DLTS measurements it is found that the trap density is generally lower than the impurity concentration,[39] determined by mass spectrometry or neutron activation; the factor a_x is of order 0.1 to 0.2 for Ti, 0.006 for Fe. Gettering by other impurities may be responsible for this fact.

Effect of Impurities on Solar Cell Efficiency

From Eqs. (2) and (6) one can see how the presence of an impurity in concentration N_x affects the short-circuit current of a solar cell. A series of experiments[21] has given the results reported on Fig. 4a; Figs. 4b and 4c illustrate the corresponding measurements of the open-circuit voltage, V_{oc}, and the cell efficiency, η. Most results fit well with the simple model developed above. How the efficiency decrease is a function of the added impurity is shown by Fig. 10, which confirms the trends shown by Figs. 8 and 9 about minority carrier lifetime in the base material, and also shows that n-base devices are affected less by impurities than p-base ones. Another observation is that the N_x dependence of η does not fit well the calculated curve for Fe, Cu, and Ni; furthermore, with these impurities it is found that the dark characteristic of solar cells is dominated by space-charge region recombination instead of base recombination,[39,40] that is, the factor n of Eq. (3) changes from 1 to 2 by these additions; this is a second degradation mechanism, added to the main one.

Precipitates which occur when the impurity content overpasses the solubility limit are by themselves another cause of cell degradation. They are active mainly in the space-charge region where they produce a junction excess current (increase of J_o).[41]

Carbon and oxygen are important impurities not considered in Fig. 10. Their influence has been discussed by Pizzini,[8] who concludes that their concentration should not be increased above its value in electronic grade Czochralski crystals, that is, 2.10^{16} cm^{-3} for carbon, 5.10^{17} to 10^{18} cm^{-3} for oxygen.

When several impurities are simultaneously

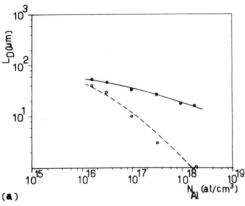

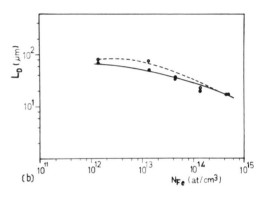

Fig. 9. Diffusion length, L_n, of electrons in B-doped ($6.3.10^{16}$ cm^{-3}), P-compensated (4.10^{16} cm^{-3}) crystals as a function of (a) Al-doping, (b) Fe-doping, (——— before cell processing; _ _ _ _ _ after cell processing) (from Pizzini[38]).

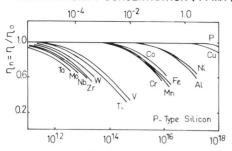

METAL IMPURITY CONCENTRATION (PPMA)

P- Type Silicon

METAL IMPURITY CONCENTRATION (ATOMS/CM3)

(a)

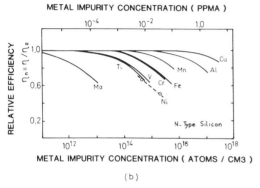

METAL IMPURITY CONCENTRATION (PPMA)

N- Type Silicon

METAL IMPURITY CONCENTRATION (ATOMS / CM3)

(b)

Fig. 10.   Solar cell efficiency vs. impurity concentration:[21] (a) p-base cells; (b) n-base cells. The baseline cell, without impurity, is used as a reference (efficiency = 1).

present, Eq. (6) is modified to express the additivity of recombination probabilities:

$$\frac{1}{L_n^2} = \frac{1}{L_{no}^2}$$

$$+ K_x N_x + K_y N_y + K_z N_z + \dots \quad (7)$$

This additivity is roughly verified,[21] although the resulting efficiency tends to be lower than that indicated by Eq. (7). However, there are exceptions due to interactions between impurities. A typical case is that of Cu and Ti which, if present together, give better cell efficiency than for Ti only at the same concentration.[42] Similarly, oxygen is known to getter transition metal atoms into complexes which may be less detrimental to the base recombination length than these atoms alone.[31]

Interactions between impurities are yet poorly understood. Interaction between impurities and defects (especially grain boundaries) is a major problem which will be touched upon later.

To summarize, it has been established that solar cell performances are degraded by impurities through two main mechanisms. One is the decrease of bulk recombination length, which plays a major role in decreasing J_{sc} and, consequently, V_{oc} and the efficiency, η. The second mechanism is the increase of dark current J_0 due to increased SCR recombination, which affects V_{oc} and the fill factor, and thus the efficiency, η. The worst poisons are Ti, Mo, V, Fe, and other transition elements, which should be avoided in solar grade Si in spite of their abundance in metallurgical grade Si. Other current impurities like Cu, Al, O, and C are less damaging and may be tolerated at a moderate level. It can be added that the most mobile impurities (Fe, Cu) may affect the cell properties because of their diffusion during the cell fabrication steps involving annealings, or even may produce degradation of solar cells with time.[43]

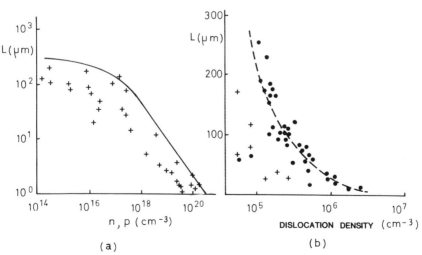

(a)   (b)

Fig. 11.   (a) Trends of diffusion length in Si single crystals, as a function of doping level: _____ Shockley-Read mechanism for perfect crystals; + experimental points. (b) Diffusion length in Si ribbons, as a function of dislocation density, measured: ● away from grain boundary, x on grain boundaries (from Baghdadi[136]).

143

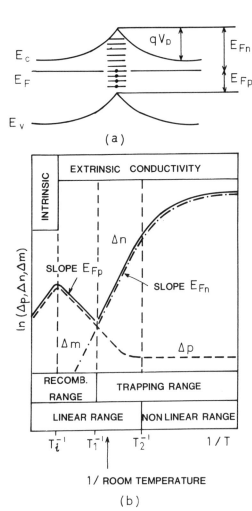

(a)

(b)

Fig. 12. (a) Band structure near a dislocation.
(b) Dependences of excess charge concentrations on reciprocal temperature (n, p = free carriers, m = charge trapped on dislocations) (from T. Figielski[44]).

Dislocations and Grain Boundaries

Even with single crystals, the recombination length is largely affected by physical defects; the dislocation density is the most important parameter together with doping level (see Fig. 11a). In polycrystals, dislocations and grain boundaries both play a significant role (see Fig. 11b).

A general feature is that physical defects produce chemical heterogeneities. For instance, impurities accumulate in grain boundaries. In experimental work, it is difficult to distinguish between the effect of defects and that of impurities accumulated near defects. However, for clarity we shall try to treat first the cases of purely physical defects.

Both edge and screw dislocations imply dangling bonds, which give birth to an electric activity.[44] The band structure is perturbed (see Fig. 12a) and the excess charge induced by illumination is given by Fig. 12b where:

$\Delta m = \Delta n - \Delta p$ is the induced charge fixed on dislocations,

Δn, Δp are the induced free electron and hole concentrations.

Usually at room temperature Δm is small (no long-lived trapping) and Δn and Δp are proportional to the generation rate, G, that is, electron and hole lifetimes can be defined ($\tau_n = \Delta n/G$, $\tau_p = \Delta p/G$).

Dislocations appear in the crystals as a means to relieve stresses; they are often controlled by temperature gradients during cooling. Their density, as revealed by etch-pit density measurements, may often be kept below 10^5 cm^{-2}, in which case they only influence a little the carrier lifetime. When it reaches 10^6 cm^{-2}, the lifetime is reduced and solar cell properties are degraded accordingly (see Fig. 13).

A simple treatment of grain boundaries, g.b.'s, has been published by Card and Yang.[45] A concentration N_{is} of interface states is present along the grain boundaries, with approximately:[46]

$N_{is} < 10^{11}$ cm^{-2} eV^{-1} for a low angle g.b,
10^{11} cm^{-2} eV$^{-1} < N_{is} < 10^{13}$ cm^{-2} eV^{-1} for a medium angle g.b.,
$N_{is} > 10^{13}$ cm^{-2} eV^{-1} for a high angle g.b.

The resulting diffusion potential, V_d, is similar to that of Fig. 12a. Its sign is opposite for n- and p-type polycrystals, and its magnitude under a given illumination depends on N_{is} and the doping level as shown by Fig. 14a. Recombination at g.b.'s can be defined by a recombination velocity, S, analogous to that defined for a surface (see Fig. 14b). Considering all the g.b.'s existing in a given volume, as a function of grain size, d_G, one can define an effective volume lifetime, τ_{eff}, which depends on doping level, interface state density, and grain size as shown by Fig. 14c. Finally, Fig. 14d shows the variations of J_{sc}, J_o, and V_{co} with τ_{eff}, illustrating two main degradation mechanisms occurring when $\tau < 10^{-7}$s: a bulk effect (recombination losses, affecting J_{sc} and V_{oc}) and a junction effect (nonideality due to SCR recombination), affecting V_{oc} and fill factor. In this model, τ_{eff} is proportional to grain size, d_G, (the shape of the grains playing a minor role). In a survey of experiments on solar cells with

144

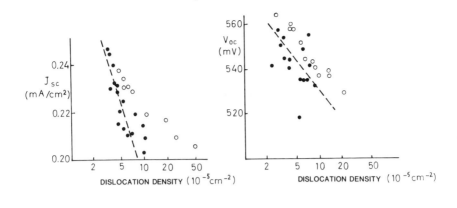

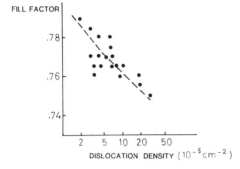

Fig. 13. Plots of I_{cc}, V_{cc}, and fill factor vs. dislocation density for RTR ribbon (from B. L. Sopori[137]; see also Ref. 51).

very small grain sizes,[47] this variation was confirmed and used, together with the mobility versus grain size variation, to account for the degradation of solar cell characteristics (see Fig. 15). Zehaf[48] also found a $\tau \propto d_G$ law in Bridgman cast crystals.

Experimentally, Fig. 16a shows the decrease of diffusion length from its intragrain value, L_b, as measured with a scanned electron beam; it occurs at a distance L_b on each side of the g.b.[49] This observation led to a model of polycrystalline solar cell in which the short-circuit current is a function of both L_b and d_G (see Fig. 16b).

A more refined approach has been proposed by Fossum,[50] who considers columnar grains, calculates the g.b. recombination velocity S, then works out a model of the solar cell as a three-dimensional device (see Fig. 17a). For high values of S, the effective lifetime varies according to the law:

$$\frac{1}{\tau_n \text{(eff)}} \cong \frac{1}{\tau_n \text{(bulk)}} \left[1 + 2\pi^2 \frac{L_n^2 \text{(bulk)}}{d_G^2} \right] \quad (8)$$

as shown by Fig. 17b, and the dark current is nonideal (coefficient n equal to 2 instead of 1 for ideality). A comparison with experiment shows

that this latter finding is well verified (other papers confirm this conclusion) and that $S \cong 10^5$ cm.s^{-1} but no comparison is made between Eq. (8) and experimental results. However a variation law similar to Eq. (8) has been found by Sopori[51] both in SILSO slices and RTR ribbons. This author also measured the short-circuit current of solar cells made on materials having different kinds of defects and found, approximately, a universal curve (see Fig. 18) independent of the nature of the defect.

The fact that low-angle grain boundaries are less damageable than high-angle ones has been rarely verified,[52] except in the case of parallel twins which are consistently found to cause no degradation. EBIC images and spectral photoresponse to a scanned electron beam are beautiful tools to specify the electric activity of each type of defect (see Fig. 19 and Ref. 9). But detailed understanding of this activity implies to consider the interaction of impurities with physical defects.

Impurity Segregation Towards Defects

It is known that dislocations can attract impurities, for example, Al, Cu, O, C, in single crystal

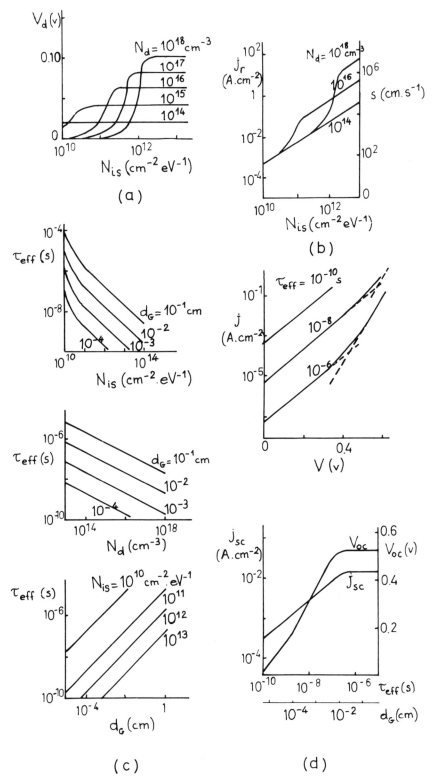

Fig. 14. (a) Diffusion potential at grain boundary under solar illumination, as a function of interface state density, N_{is}, and donor concentration N_d. (b) Recombination current density and recombination velocity S at g.b. (c) dependence of minority carrier lifetime on grain size, interface state density and donor concentration. (d) dark current (forward bias) and solar cell properties of a p-n junction of polycrystalline silicon (from Card[45]).

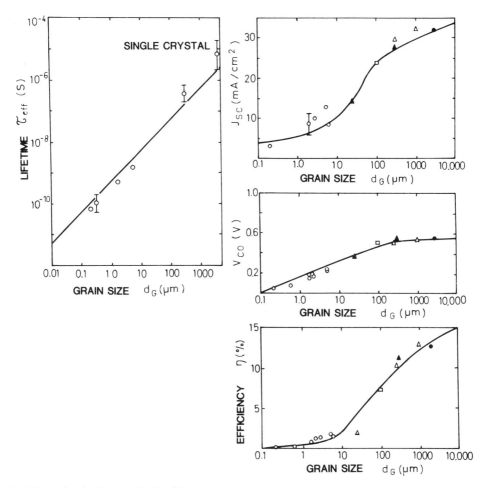

Fig. 15. Effect of grain size on effective lifetime and solar cell characteristics: —— calculated; ○△●▲□ experimental results (from Ghosh[47]).

Si.[53] The same is true for grain boundaries; in a preliminary study of Si bicrystals[54] it was shown that g.b.'s are composed of several types of dislocations, some of which are decorated with impurities.

Direct evidence of impurity accumulation at g.b.'s has been obtained repeatedly. Let us quote some examples.

Kazmerski[55] has examined, by Auger and SIMS techniques, Si surfaces freshly fractured under ultravacuum. Many of these fractures follow grain boundaries. He found that, while inside the grains only Si and B (doping agent) were visible, regions of thicknesses < 80 Å near g.b.'s contained C, O, Al, Ni, B, and other impurities in large contents (10^{11}-10^{14} cm^{-2}).

Helmreich[52] has demonstrated the accumulation of Al and Mg in g.b.'s for UMG Si, by use of an energy dispersive X-ray spectrometer.

Using high resolution SIMS, Chu[56] could prove the segregation of many impurities (C,

Mg, Ti, Fe) toward g.b.'s by recrystallization of MG Si. A further segregation is observed if the polycrystal is heat treated in a helium atmosphere at 700°C for 12 h.

Another observation concerns a more complex defect, called "grit structure," in SILSO cast polysilicon, that is, a region of the material with numerous defects and a very low photoresponse; it was found that C accumulated in this region.[9]

Interactions between defects can change greatly the electrical activity of grain boundaries. It has been argued[9, 57] that g.b.'s act as dislocation sinks, reducing the dislocation density inside the grains of a polycrystal below its usual level in single crystals. But the most spectacular effects are due to impurities. For Bridgman-cast SEMIX polycrystals, the scanning photoresponse pattern of a shallow p-n junction showed that g.b.'s caused either reduced or enhanced photoresponse[57, 58]; on the whole, little

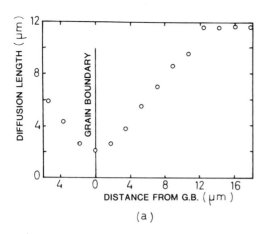

(a)

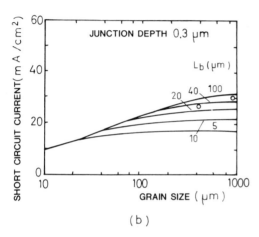

(b)

Fig. 16. (a) Diffusion length variation near a grain boundary. (b) Effect of grain size on the short-circuit current for different intragrain diffusion lengths L_b (from Koliwad[49]).

difference was seen between diodes containing grain boundaries and diodes with no grain boundaries. This behavior is explained by dopant penetration along g.b.'s, which increases the p-n junction area and causes additional collection of light generated carriers. So the final behavior of g.b.'s may be determined, not only by the material itself, but also by the heat treatments performed to complete the solar cell. Similar observations are due to Redfield,[59] who creates potential barriers by O diffusion from the grains to the g.b.'s, and to Turner,[60] who saw enhanced photoresponse at g.b.'s after the so-called GILD p-n junction formation process.

Passivation
This leads us to the techniques of passivation, by which one tries to improve polycrystalline solar cells by introducing impurities into grain boundaries.

Two main ideas have been developed: passivation by hydrogen, and penetration of donors along the g.b.'s. But other proposals will also be mentioned.

The introduction of hydrogen is reminiscent of the role of hydrogen in saturating dangling bonds of amorphous silicon. Seager[61] showed that diffusion of hydrogen provides a significant reduction in both the state density and the accompanying grain-boundary potential barrier, while oxygen, nitrogen, and sulfur hexafluoride caused an increased density of states. Further experimenters[62] exploited this possibility in different ways. An example of these results is given

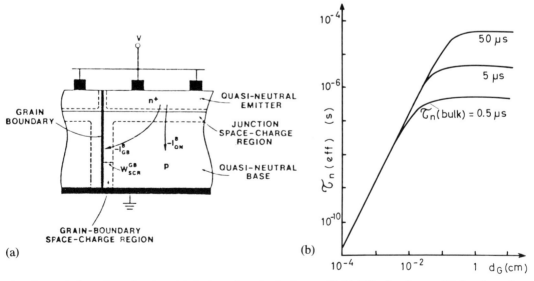

(a)

(b)

Fig. 17. (a) A three-dimensional model of solar cell made of columnar Si. (b) Effective electron lifetime for various grain sizes, d_{G_2}, and intragrain lifetimes, τ_n (diffusion coefficient $D_n = 25$ cm^2/s) (from Fossum[50]).

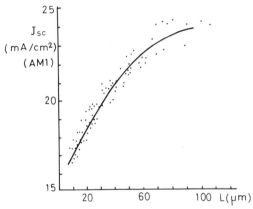

Fig. 18. Short-circuit current vs. bulk lifetime for polycrystals containing different types of physical defects (from Sopori[51]).

by Fig. 20, in which H is seen to improve both the photoresponse at g.b.'s and the dark current characteristics.

The introduction of dopant into the grain boundaries has been suggested by Di Stefano[63] and invoked by Storti[58] and others[64] to account for improvements in both lifetime and short-circuit current. Thus, p-n junction formation is a step which, if properly conducted, may lead to a better photocarrier collection (see Fig. 21).

Other examples of passivation are also instructive. Koliwad[49] has improved I_{cc} and V_{co} of fine-grained polysilicon cells by adding some copper (Cu reduces the g.b.'s barrier) while the same addition degrades coarse-grained cells (Cu

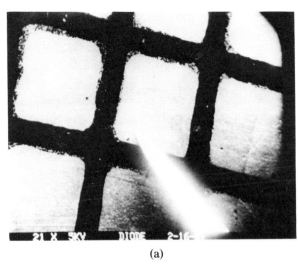

(a)

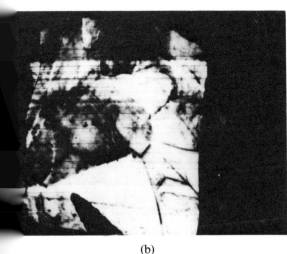

(b)

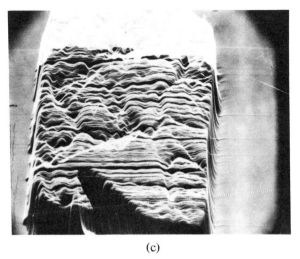

(c)

Fig. 19. (a) SEM image (7 x) of a Schottky diode pattern. (b) and (c) EBIC images (16 x) of the central Schottky diode shown in (a); the individual scans are shown in contour representation in (b) and in image representation in (c) where black means reduced current. SEMIX material studied by S. M. Johnson, C. D. Wang, and R. W. Armstrong (private communication).

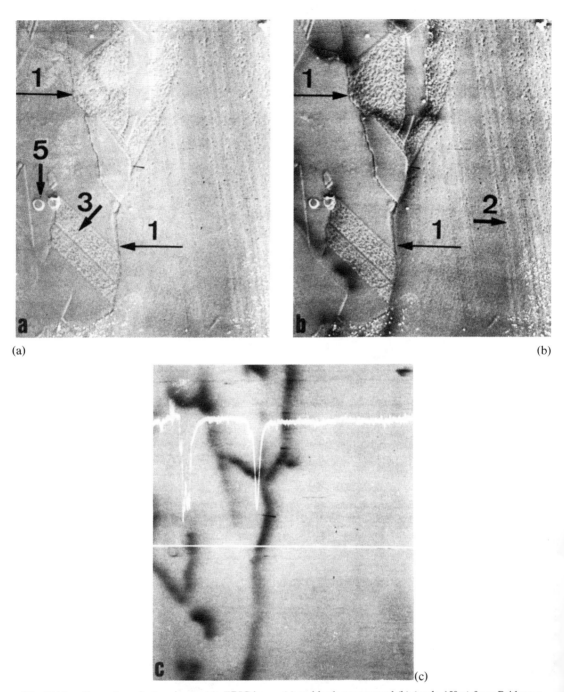

(a)

(b)

(c)

Fig. 19 bis.   Secondary electron image (a), EBIC image (c) and both superposed (b) (scale 150 x) for a Bridgman-grown polycrystal (from C. Belouet, private communication). On (c) is superposed the photoelectric signal obtained along line L.

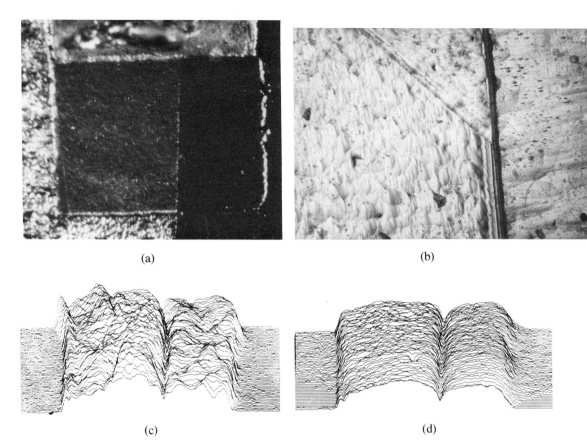

(a)   (b)

(c)   (d)

Fig. 19 ter.   (a) and (b): Optical photographs of a mesa-diode in CGE polycrystal, before and after etching by SIRTL-etch. (c) and (d): LBIC scanning of the same diode at low and high light bias respectively.

The two grains shown by the photographs have different mean photoresponse and important photoelectric accidents at zero light bias (from J. P. Crest, private communication).

decreases the bulk lifetime). Janssens[65] proposes to sinter an Al-SiO₂-Si structure over the Si polycrystal.

Conclusions on Material Requirements

Starting from a simple description of defects and impurities, and of their influence on recombination and solar cell performances, we have reached conclusions which are far from being so simple.

First, we know which impurities (the transition elements) are the most harmful, and which impurities can be tolerated to a limited extent, like Al, Cu, O, and C. We also know that coarse-grained (1 mm or so) polycrystalline Si is acceptable, especially if the grains are columnar and the p-n junction perpendicular to the grain boundaries, and that small-grained (1 μm or so) polycrystals are not acceptable. Precipitates are very harmful, twins are not; dislocations should be kept at a density lower than 10^5 cm^{-3} inside the grains.

But the first difficulty arises when one tries to define properly the defects and impurities present in a given material. While the mean impurity content can be measured accurately, and the physical defects can be defined and counted, the microscopic nature of defects, impurity coupling, detailed nature of grain boundaries, segregation of impurities toward g.b.'s, is hardly known at all. Much new work is required in this field.

Mainly for this reason, the influence of defects on carrier lifetime is only partly accounted for: the case of one impurity is simple, but real crystals have multiple impurities. In polycrystals, the (largely unknown) impurities in g.b.'s are more important than the grain misorientation. Furthermore, the measurement of τ_{eff} or L_{eff} is in itself a difficult problem (all methods are not necessarily convergent). Local photoelectric response measurements are a very powerful tool to characterize polycrystals.

Finally, the degradation of solar cell per-

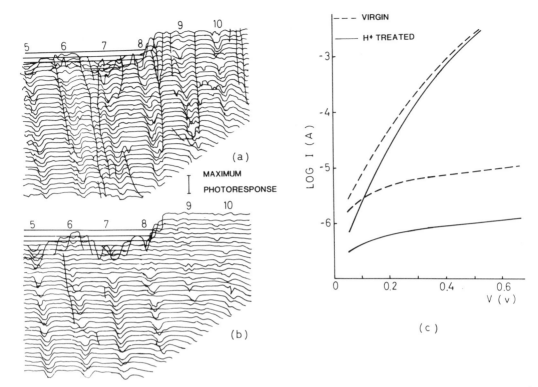

Fig. 20. (a) EBIC map of a section of an n/p polycrystalline solar cell. Dips in the current response correspond to metal-top contact finger or to grain boundaries (the latter dips are connected by hand-drawn lines). (b) The same section after grain-boundary hydrogenation (16 h at 350 torr hydrogen plasma). (c) Dark characteristic of a roughly 1 cm² n/p polycrystal solar cell, before and after the heat treatment (from Seager[62]).

formances due to defects and impurities is essentially related to: (1) bulk lifetime decrease, and (2) junction effects. But its detailed prediction is not yet possible (compare the discrepancies between Figs. 14 and 15, 15a and 17b). One very positive observation emerges from the last two years' studies: impurities and grain boundaries can help prevent one another from being harmful for solar cells. Avoiding invalid generalizations, one can say that in some cases, grain boundaries can attract impurities and consequently decrease

their degrading action, and on the other hand, impurities can passivate grain boundaries and improve the properties of polycrystals with small size grains. This is a very important asset, hardly exploited yet, of polycrystalline impure silicon as a solar cell material.

CHEMICAL PURIFICATION TECHNOLOGIES

The preparation of crystalline Si for solar cells begins with chemical purification steps. We have

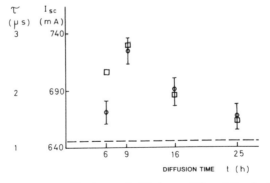

Fig. 21. Effect of a double phosphorous diffusion on the lifetime, τ ($\square$), and the short-circuit current, I_{sc} ($\bigcirc$), of a solar cell. First diffusion (750 °C during time, t) pushes the dopant along the g.b.'s of columnar SILSO polycrystal, before a second diffusion, in usual conditions, produces the junction. Efficiency has been raised to 12.5% (from Ferraris[64]).

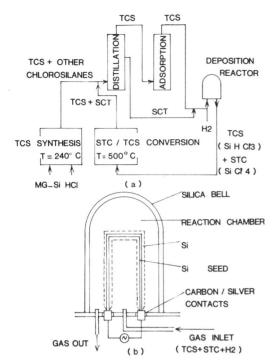

Fig. 22. The Siemens process: (a) flow chart; (b) the $SiHCl_3$ pyrolysis reactor (from J. Dietl[9]).

seen that the purity requirement is not exactly known. Reaching the highest possible purity at moderate cost is the goal of the methods described in the following section. To be ready for solar cell production, the material thus obtained has only to be remelted and crystallized without reintroducing impurities; this way has been called the chemical route. An alternative way is the metallurgical route, in which a chemical purification aiming at a moderate purity at the lowest cost is followed by a crystallization which produces a further purification.

Halogenide Processes [8, 9, 66]

Almost all of the silicon used by the electronic and solar industries is presently prepared by the Siemens process. The process is schematized in Fig. 22a: MG Si (previously prepared by reduction of SiO_2 in an arc furnace) is converted to $SiHCl_3$, which is then reduced by H_2 in a specially designed reactor[67] (see Fig. 22b). In this process, all impurities which are transition metals or dopants, including boron, are removed; their final concentration, much smaller than 1 ppb (part per billion) is perfectly adapted to microelectronics. This process is expensive ($70/kg), mainly because of its high energy consumption (300 kWh/kg) and low production rate (3.10^{-3} g/h per cm^2 of Si seed in the decomposition reactor).

Several other processes (see Table 1A) have been studied in order to adapt the price to the needs of solar cells. We shall describe the three main ones,[13] the flow charts of which are presented in Fig. 23.

The Union Carbide Process

In this process, MG Si reacts with $SiCl_4$ and H_2, and the resulting $SiHCl_3$ is converted to SiH_4; metal impurities are removed as solid chlorides. SiH_4 is then distilled, so as to remove boron (to a level of 0.13 ppb), and finally pyrolyzed in a free-space reaction[68, 69] to recover the Si (other systems have been proposed[70]). Fe, Cr, Ni, Mn, Cu, and Zn are found at a level lower than 0.6 ppm (part per million). The energy use of the process is only 91 kWh/kg, the overall Si yield is 85%. The price of Si granules is estimated by Union Carbide to be $11/kg, while an independent estimation gave $14/kg (in 1980 dollars), for a plant producing 1,000 t/yr.

The Hemlock Process

This process is similar to the Siemens process, except that the compound which is produced and then decomposed is SiH_2Cl_2, instead of $SiHCl_3$. Compared to the Siemens process, it was found that the rate of Si deposition was doubled and that the energy consumption was cut in half. The expected price is $20/kg.

The Battelle Process

This process includes the conversion of MG Si into $SiCl_4$, the reduction of $SiCl_4$ by Zn in a fluidized bed reactor, and the electrolytic regeneration of Zn from $ZnCl_2$. The obtained Si granules are very pure except for a Zn content of order 100 ppm and their price might be $17/kg. This process failed several years ago, when it was studied as the Du Pont process, because of boron incorporation in the Si; however, it seems that now $SiCl_4$ raw material is commercially available with only traces of boron. Table 1A also quotes two processes based on the same principle: the reduction of $SiCl_4$ by Na,[71] which seems to be abandoned now, and that of SiF_4 by Na.[9]

Only in the case of the Union Carbide process have the laboratory studies been completed and a 100 t/yr pilot plant been built. However, it seems that the proposed free-space reactor has not yet been tested at full scale. Owing to its low cost, this technology may bring an important contribution to the solar cell industry in the next five years.

153

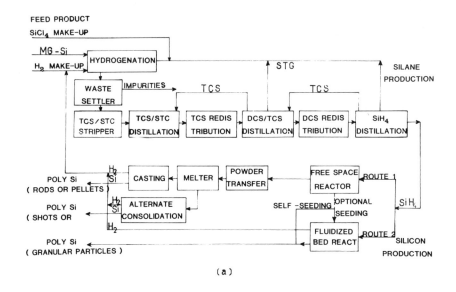

(a)

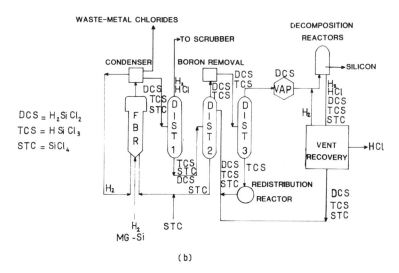

(b)

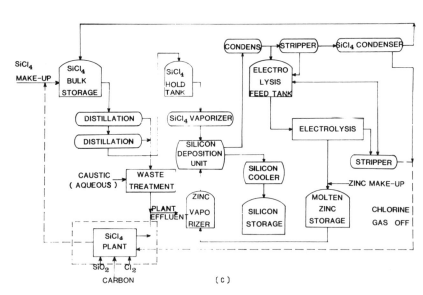

(c)

Fig. 23. Flow charts of Si purification processes: (a) Union Carbide; (b) Hemlock; (c) Battelle (from R. Lutwack[13]).

154

Pyrometallurgical Processes

In a longer range perspective, taking in account the tolerance of solar cells to somewhat higher impurity contents, several teams have proposed to simplify drastically the Si purification. Two main ways may be followed (see Table 2A).

The Direct Arc Reactor (DAR) Technique[72]

An idea first experimented by Dow-Corning was to abandon any purification of MG Si. For this purpose, the reduction of silica by carbon in the arc furnace has to be specially conducted. Very pure silica sources can be selected (for example, rock crystal) and eventually, pretreated before the reduction step (purified sands). Carbon black powder formed into pellets by special binders can be used as reductants. The arc furnace itself can be improved (the most recent version is called Submerged Electrode Arc Furnace). The resulting purity of MG Si can be raised to 99.99% (50 to 100 ppm Al or Fe, 10 ppm B or P). In at least two cases,[16, 73] this silicon has been used to get solar cell efficiencies better than 10%. However, one weakness of the process is the unavoidedly high carbon content of the Si.

Partial Refining of Metallurgical Grade Si[3, 8, 9]

In another concept, it is admitted that a purification of MG Si is necessary, but this purification is limited to simple and cheap steps. Several classical metallurgical processes can be used.

Acid leaching is a selective removal of impurities based on the observation that impurity content is maximum at the grain boundaries of MG Si, often in the form of precipitated silicides or metal eutectics. When MG Si is crushed and the resulting powder is attacked by acids, metal impurity content can be decreased by a factor 2 to 100. Aqua regia seems to be one of the best acids to be used.[56, 74]

Blowing a reactive gas through molten Si may remove those impurities, which have an affinity with the gas larger than that of Si. Al can be removed by chlorine, Al and Ca by oxygen, B by H_2/H_2O mixtures. However, these reactions should be conducted with due consideration of thermodynamic equilibria and chemical engineering principles, which seems to be a delicate task at the present level of knowledge.

Liquid-liquid extraction (or slagging) is a process by which impurities are pumped from melted Si into an oxide melt (slag) in close contact. Alkaline earth metal silicates are generally used as slags. Although efficient for removing

Al, this process brings in new impurities (Ca, Mg) and seems unable to lead to very high Si purity.

A combination of these processes is able to give a silicon which is much improved compared to MG Si, and has been called UMG (upgraded metallurgical grade) Si. The Dallas University[75] and RCA[76] teams were the first ones to obtain 10% efficiency cells using this material, and many others continued this research.[16, 56, 73, 77, 78] but in order to get this result, UMG Si (as well as DAR Si) has to be further purified by progressive crystallization, as shown by Table 2B.

It is impossible to report on the Si purity reached by all these methods. However, Table 5 has been prepared as an illustration of typical results. More detailed results may be found in Pizzini's review[8] and in original papers.

Purification by Progressive Freezing of a Melt

In principle, the segregation coefficient, k (see Table 3), describes the ratio of impurity contents in the solid and liquid phases when the latter is progressively crystallized. The purifying effect of this process is very large for transition elements. However, these values of k are true for equilibrium conditions and perfect mixing of the liquid, which implies a very slow crystallization. At fast freezing rates, impurities may accumulate in the liquid near the interface, the effective segregation coefficient may be smaller than k, and impurities may also precipitate as a second phase in front of the freezing interface. This phenomenon, known as constitutional supercooling,[79] occurs for low k's, high impurity concentrations, small temperature gradients in the furnace, and high freezing rates. In practice, it limits the latter parameter to some cm/h (see, for example, Ref. 74).

The methods currently used to grow Si ingots (see Table 1, B1) are capable of a significant purifying action, in particular Czochralski[3] and HEM (see Table 6). The same is true for some, but not all of the ribbon-growth techniques (see Table 1, B2), in particular for the WD[80] and ESP (see Table 6) ribbons.

Purification during crystal growth may be also improved by removing impurities accumulating on the surface of the melt; for instance, a SiO_2 slag may be used to transform carbon into volatile compounds. Impurity evaporation has been shown to help purification in the case of zone melting under plasma[81] (see results of this technique in Table 6).

155

Content (ppma)[a]	Metallurgical Grade (MG)[b]	Direct Arc Reactor (DAR)[b]	Slagged and Leached (UMG)[c]	MG-Si+HEM Purification[d]	Solar Grade (Union Carbide Process)[e]	Electronic Grade (Siemens Process)[f]
Al	2200	70-320	75-150	< 2-13		0.5
B	40		20-40	< 10		
C	700-1000					0.4[b]
Cr	16	5	< 5-15	< 1	0.05	0.0015
Cu	13	2-25	5-30	< 1-6	0.1	
Fe	2000	25-850	60-80	7-19	0.3	< 0.025
Mn	140	5-75	< 5-40	< 10	0.05	< 0.003
Mo		3	< 5	< 10		< 0.0006
Ni	24	5-25	< 5-30	< 1	0.15	< 0.01
O						10-20[b]
P	27		40-50	< 50		0.0025
Ti	105	5-40	5-15	< 1		
V		5		< 3		< 0.003
Zn					0.2	
Zr		3				< 0.02

[a] ppma = atomic parts per million.
[b] From S. Pizzini.[8]
[c] From J. R. McCormick.[3]
[d] From C. P. Khattak.[134]
[e] From R. Lay.[69]
[f] Neutron activation analyses of G. Revel (private communications).

Table 5. Typical Impurities in Si of Different Grades

The combination of pyrometallurgical processes and gradient freezing is able to transform metallurgical grade Si into a material which is suited to the production of good epitaxial solar cells and even, in some cases, of good diffused solar cells.

Conclusions on Purification Techniques

Although the problem of impurities should not be arbitrarily separated from the whole material elaboration process, we shall try, at this stage, to point out the assets and liabilities of the chemical and metallurgical routes.

(ppma)	From D. Morvan[81] MG Si	From D. Morvan[81] Idem Zone Melted[a]	From C. P. Khattak[134] MG Si	From C. P. Khattak[134] Idem + HEM	From C. P. Khattak[134] UMG Si	From C. P. Khattak[134] Idem + HEM[b]	From T. F. Ciszek[73] MG Si	From T. F. Ciszek[73] Idem + ESP[c]	From D. Casenave[88] MG Si	From D. Casenave[88] Idem + RFP
Al	2900	33	880	13	70-100	0.6-1	> 3400	240	220	1-2
B			10	10	3	8	17	34	100	100
Ba	2-23	0.2					10	< 0.02		
C							180	4		
Ca	1500		34	3-6	28-35		290	< 0.06		
Co	2.5	0.01				< 0.01				
Cr	22-40	0.09	90	1	5	< 0.005-0.01	40	0.4		
Cu	18		12		2	0.02				
Fe	1900	3.5	2000	15-19	25-40	< 0.03-0.8	> 2500	2.5	2100	0.4
Mg	120-540	< 2	2	1		< 0.005-0.05	85	0.2		
Mn	40-70	0.9	7	9	5	< 0.005-0.4	550	2.2	20	0.2
Mo			4	10	3	< 0.01-0.05	1.5	0.03		
Ni			21	1	5	< 0.03	40	0.4		
P			50	50	3-7	3	14	2.8		
Ti	95		510	1	6	0.005-0.08	290	< 0.6	30	0.1
V	47	1	30	3	5	< 0.003-0.03	250	< 0.5	0.5	0.2
Zn	17	< 0.1				< 0.005				
Zr	0.1	0.03			3	< 0.02	13	< 0.03		

[a] Self-supported Si twice zone melted under plasma.
[b] Slagging with SiO_2 is used to decrease the number of SiC particles. Single crystals were obtained. Diffused solar cells were made, of which half had efficiencies between 8 and 12.3% and half were short-circuited by residual SiC particles.
[c] Diffused solar cells had efficiencies up to 3.4%, and epitaxial solar cells up to 8%.

Table 6. Purification of Si by Progressive Freezing

The main technical advantage of the halogenide processes (the chemical route) is to remove impurities to such a low level that they need not be considered in the further crystal growth and cell technology steps. The price to be paid to reach this goal is illustrated by the figure of $20/kg, which is hoped to be reached with full-size (500 to 1000 t/yr) plants. If we consider standard ratios (20 tons are needed for 10,000 m² of solar cells producing 1 peak megawatt) and moderately improved technologies for crystal growth and cell technology, this Si price will represent, in 1985, about one-eighth of the total cell cost (instead of one-fourth today). It then can be argued that a further reduction of the Si cost would be only of marginal interest, and that improving the other steps would be more beneficial. This way allows important progress without too many risks.

The main technical advantage of the pyrometallurgical processes (the metallurgical route) is to produce useable silicon at the lowest possible cost, which could represent in 1985 only one-twentieth of the total cell cost. It can be hoped that this initial advantage will not be destroyed by increased requirements of the further steps. This hope seems reasonable as far as the epitaxial layer deposition is concerned, since high-speed production lines are conceivable for this process, which is at the same time a substitute for diffusion in the p-n junction elaboration step. It is more questionable that the crystal growth step may remain as cheap and simple if impurity contents are higher, since the requirement of slow growth needed for purification is in direct contradiction with that of fast growth needed for cost effectiveness. However, if deeper insight shows a possible conciliation there, the metallurgical route may very well be the best one.

In order to forecast the winner with safety, some data are presently lacking, principally about impurity behavior in polycrystals and impurity effects on crystal growth. The author inclines to optimism about the metallurgical route, but only in the long range. The size of the solar cell business is an important factor here: the silicon produced by the chemical route may be pure enough to allow other applications than solar cells, which may help to amortize the initial investment when the solar cell market is still small.

CRYSTAL GROWTH TECHNOLOGIES

Silicon powder or chinks purified, as explained above, have to be converted into slices on which the solar cells will be built. They are melted and the melt is then freezed in a controlled way. Progressive solidification is necessary to obtain a massive crystal, free of holes or cracks; in the metallurgical route it is also a purifying step as mentioned above.

There are a great variety of melting and solidification techniques. A first classification is possible according to the nature of the obtained product, which may be an ingot, a ribbon, or a silicon sheet deposited on a substrate. After describing these techniques and products, we shall close this part by mentioning the epitaxial layer deposition which is an almost necessary step of the metallurgical route.

Each technique is characterized by the production cost of a unit area Si wafer and by the efficiency of the solar cells which may be produced on large-area wafers. The former parameter is heavily dependent on the production rate, expressed in cm²/min (in the case of ingots, different rates have to be considered for the ingot growing and cutting). The latter is in close relation with the crystal quality (nature and number of defects). These parameters have been accurately defined for several important techniques (compare Tables 9 and 10 below); Koliwad[82] predicts their expected progress up to 1986. But it is clear that other techniques are still in a less advanced, but hopeful, stage of development.

There are some other features which are common to most of the techniques. One of these is the problem of the container materials for molten Si. As summarized by Dietl,[9] these materials are essentially high-density graphite, quartz or quartz ceramics, and mullite. High-purity graphite and quartz are available, as is graphite with adjustable thermal expansion coefficient[83] or quartz coated with mechanically unstable quartz,[84] both allowing to avoid cracks during Si cooling. Sometimes reuseable graphite crucibles, or molds, were used.[85, 86] In a few cases, silicon is self-supported, without any crucible or substrate.[87, 88] The contact of molten Si with crucibles, substrates, or dies may cause the introduction of impurities (in particular, formation of SiC with graphite). In the case of ribbons, Dietl[9] has very clearly classified the different techniques from the point of view of the strength of the silicon die or silicon substrate interaction. In some cases (RTR, RAD techniques), the ribbon may be first built on a substrate, then separated from the latter: the distinction between ribbon and sheet is then somewhat arbitrary. Finally, here again each technique has specific characters and systematic comparisons are not always directly possible.

Of course, the thickness of the Si wafer is an important element of the cost-effectiveness of each technique. It is of order 500 μm for sliced ingots (limited by the high-speed sawing possibilities), 200 μm for ribbons (limited by edge and meniscus effects), 50 μm for sheets. Other important technical parameters include the possibility of continuous operation, energy consumption, ease of thermal control, and so on.

The fundamental mechanisms of crystal growth, in the specific geometrical and thermal conditions of each technique, are of paramount importance. Some general rules can be stated at this stage. Let us recall, first, that the speed of growth is directly correlated with the speed at which latent heat is extracted from the solid-liquid interface; it is expected to be higher when the interface area is larger and when the cooling crystal is in contact with large thermal masses.[89] The heat transport regime affects the shape of the liquid zone which, especially for narrow zones, is also dependent on the forces (interfacial tensions, hydrostatic pressure) exerted on the liquid; in particular the natural "growth angle" between crystal and liquid surfaces is a fundamental parameter to discuss the stability of crystal shape in the case of Czochralski, ribbon and sheet growth.[90] On the other hand, a curved

shape of the interface, as well as a steep thermal gradient in the region of the growing solid, may be the cause of stress-induced dislocations, grain boundaries, and even irregularities of the shape of the solid.[91] Some fundamental considerations of this kind are indeed necessary to optimize each technique, but the detailed picture is specific of the considered technique. Finally, impurities are known to be a source of physical defects, although most past work was devoted to ultrapure silicon.

Ingots

Czochralski Pulling

According to Fiegl,[78] this classical technique tends to produce larger size crystals (15-cm diameter) and to incorporate continuous feeding of the melt (see Fig. 24a) in order to reach larger pulling rates (up to 18 cm/h) and silicon yields (85%). Added-on costs as low as $8 to 14/m^2 are expected in 1986 for single crystals weighing 160 kg. Crystal quality is very high, though perhaps less homogeneous in very large crystals. A difficulty inherent to this technique is the low packing factor of circular solar cells.

Heat Exchanger Method

This technique[84] avoids any moving part and uses a controlled heat extraction (by a helium jet stream) from the bottom of the crucible to pro-

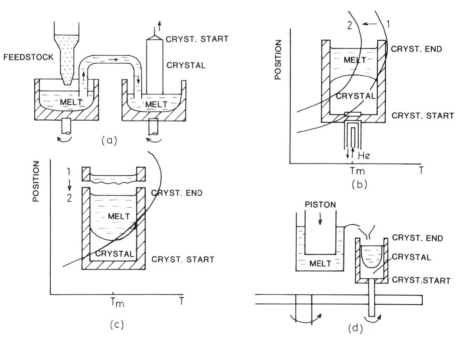

Fig. 24. Schemes illustrating the ingot growing techniques: (a) continuously fed Czochralski;[131] (b) heat exchanger[9] method; (c) Bridgman technique;[9] (d) mold casting.[9]

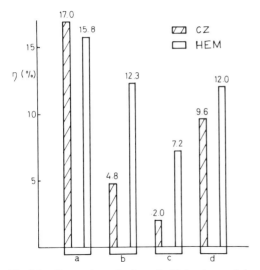

Fig. 25. Comparison of solar cell efficiencies made by Czochralski and Heat exchanger method: (a) diffused cells, from electronic grade meltstocks, (b) diffused cells, from solar grade meltstocks, (c) diffused cells, from metallurgical grade meltstocks, (d) epitaxial cells, from metallurgical grade meltstocks (from F. Schmid[93]).

gressively freeze the melt (see Fig. 24b). Freezing rate may reach 1 cm/hr. for square cross sections up to 30 × 30 cm (50 kg ingots). Single crystals can be obtained if starting from a seed. Impurities tend to segregate to the melt top, far from the solid-liquid interface, so that this technique is highly purifying (see Table 6). In a recent study,[92] the oxygen content was found to be 5 to 25 ppm, and the carbon content 6 ppm, apart from some SiC precipitates (attributed to the oil of the vacuum pump); the dislocation density was 10^6 cm^{-2} and the average diffusion length 30 μm; 90% of the ingot was useable to make efficient solar cells from a pure charge. Figure 25 illustrates the ability of the HEM technique to use rather impure Si[93].

Bridgman Growth [83, 85]

This is perhaps the simpler technique of Si growth, in which the melt and its container are moved within a fixed temperature profile, causing crystallization to proceed from bottom to top (see Fig. 24c). The polycrystal thus obtained has been found to be mechanically as strong as a single crystal and to give substantially similar solar cells. The industrial goal is to grow cubic ingots of size 30 cm in 3 h (that is, about 20 kg/h), with silicon yields better than 95%. In practical growth conditions, segregation hardly occurs when impurities are added,[94] which may limit the technique to the use of pure meltstocks; however, internal segregation toward grain boundaries was mentioned already[57, 58] to occur and account for grain-boundary passivation.

Mold Casting [9, 86, 95]

During mold casting, the melt is poured into (reuseable) containers in graphite, which are preheated and put into rotation. A homogeneous layer of chill solid silicon is instantly developed and prevents the molten silicon to react with the mold walls. The thermal regime is arranged so that heat is extracted from the bottom and the melt surface remains liquid up to the end of crystallization. For a 16 × 16 cm^2 cross section, the solidification rate was found to be 5 kg/h. A semicontinuous installation has been conceived for large-scale industrial production (see Fig. 24d). The obtained material led to cell efficiencies better than 10%.[96]

Slice Cutting

To transform an ingot into slices is not a cheap operation. By the usual ID technique (internal diameter saw), the cutting edge is 0.03 to 0.04 cm, the cutting rate 5 cm/min, and the larger number of slices is 20 to 22 per cm. The sheet-added value is of order $45/m^2, and may reach $16/m^2 in 1986[82], which would be still more expensive than the added value of Czochralski ($14/m^2) or HEM ($11/m^2) ingot growing. Other techniques are also considered:[9, 82] multiple-blade slurry technique (MBS) using steel blades in conjunction with abrasive particles, multiple-wire slurry technique (MWS), and fixed abrasive slicing technique (FAST) using a series of wires for simultaneous slicing of a number of wafers.

Ribbons

Horizontal Ribbon Growth

Kudo[97] and other authors have succeeded in pulling a ribbon from a silicon melt without any shaping die (see Fig. 26a). A solid wedge is generated at the surface by helium blowing and this solid is then pulled into a ribbon forming a small angle with the surface melt; a natural meniscus joins the ribbon and the crucible containing the melt. The solid is cooled by its large horizontal surface, which allows an easy heat dissipation and a high pulling speed. This speed reached 85 cm/min for polycrystal growth, and 41.5 cm/min for single-crystal growth. Ribbon thickness is 0.02 to 0.2 cm, maximum ribbon width is 5 cm. A high level of automation guarantees a good stability of operation for ribbons several meters

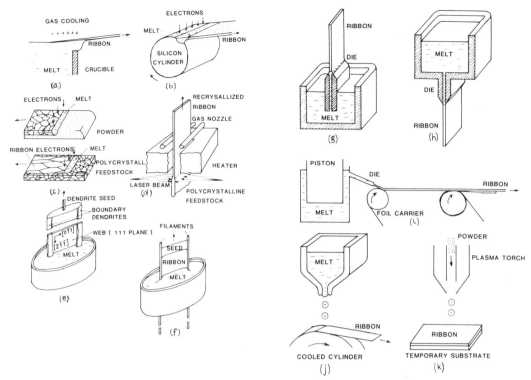

Fig. 26. Schemes illustrating the ribbon growing techniques: (a) horizontal ribbon growth; (b) ribbon from cylinder; (c) ribbon from powder, with its two steps: formation of the ribbon and recrystallization; (d) ribbon-to-ribbon (recrystallization step only); (e) web dendrite; (f) edge-supported pulling; (g) edge-defined, film-fed growth; (h) inverted Stepanov technique; (i) interface controlled crystallization; (j) ultrafast quenching; (k) plasma-torch technique.

long. Single crystals of purity and diffusion length comparable to CZ crystals have been obtained (dislocation density 10^5 cm^{-2}) as well as solar cells with 10% efficiency (AM1, without antireflection coating).

Ribbon from Cylinder

Figure 26b shows a somewhat similar case in which the liquid zone is formed by electron bombardment on a generating line of a rotating cylinder.[87] The thermal and hydrodynamic equilibria have been calculated, but no constant thickness ribbons have yet been obtained.

Ribbon from Powder

This method is one of the simplest possible, based on the concept of recrystallization. The same process, that is, moving a small liquid zone created by electron bombardment, is used first to convert silicon powder into a ribbon, and then to increase the grain size of this ribbon[88] (see Fig. 26c). Starting from Si powder, this method may be well fitted to certain techniques of UMG Si production, and it has indeed been proved to be efficient for Si purification. Ribbon thick-

nesses of order 0.04 cm have been obtained, and 0.02 cm may be possible (the faster the pulling rate, the thinner the ribbon). Growth rate is typically 20 cm^2/min, and can be improved. Most of the obtained silicon consists of very large columnar crystals (see Fig. 27), so that minority carrier lifetime, on the first tests, was 25 to 35 μm. No solar cell has yet been made with this very promising material.

Ribbon-to-Ribbon Technique

The recrystallization of a preformed ribbon is also the principle of the RTR-technique.[98, 99] A Si film is first built on a molybdenum substrate by chemical vapor deposition (from a SiHCl$_3$ + H$_2$ plasma). The Mo substrate is easily separated from the Si film, due to their different expansion coefficients, and is reuseable. After removing, by acid etch, the Mo-contaminated surface layer, the Si film is then recrystallized with the help of two CO$_2$ lasers, as schematized in Fig. 26d. Typical ribbon thickness is 75 to 400 μm, width 2 to 7.5 cm, length 22 cm. A continuous ribbon production is conceivable with this technique. The maximum growth rate is 55 cm^2/min in a system which probably can be improved; nondendritic growth and dendritic growth both can be obtained, according to the chosen temperature profile, as discussed by Baghdadi.[99] A detailed study of the defects in RTR ribbons was published by Sopori.[100] Dislocations ($10^4 - 10^6$ cm^{-2}), stacking faults, and twinning were found, but the crystal quality was still sufficient for the minority carrier diffusion length to reach 20 to 100 μm and the solar cells built on these ribbons to have efficiencies up to 11 to 12%.

Web Dendrites

Though used previously for other semiconductors, the method illustrated by Fig. 26e was first applied to silicon in 1971.[101] Its present state is described by papers of Seidensticker and co-workers.[80] Initially a [2 $\bar{1}$ $\bar{1}$] dendrite is plunged

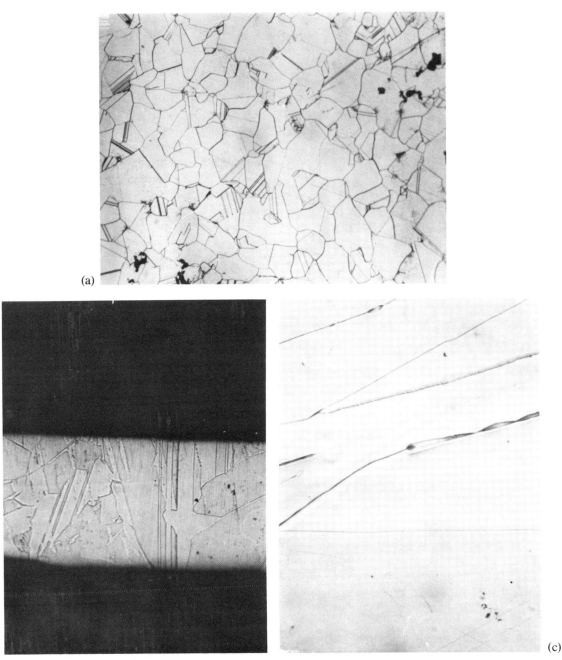

(a)

(c)

Fig. 27.   Ribbons grown from Si powder using local electron-beam melting (RFP method): (a) after ribbon formation from powder (30 μm large grain); (b) and (c) after ribbon recrystallization: views across ribbon and of ribbon surface (scale 100 x) (from D. Casenave[88]).

in a melt which is supercooled by a few degrees around this seed. The structure shown by Fig. 26e, with a web between two dendrites, is then spontaneously developing. The web is very thin (0.01 to 0.02 cm). The growth velocity is about 2 to 5 cm/min and the area growth rate has reached 27 cm²/min. The obtained ribbons are single crystals with twins as only notable defects, so that very efficient (15.5%) solar cells have been made. The effects of impurities added in the melt on web growth and purity have been extensively studied by Seidensticker.[80] Adding impurities in the 20 to 50 ppm range does not prevent ribbon growth, but higher contents cause interface instabilities. The effective segregation coefficients were found to be dependent on growth velocity and ribbon thickness; for transition elements they were about three times larger than the equilibrium coefficients given by Table 3, which allows a very substantial purification to occur.

Edge-Supported Pulling

This more recent technique is similar to web dendrite, except that the bounding dendrites of Fig. 26e are replaced by two filaments emerging from the melt (see Fig. 26f). The filaments may be in quartz, graphite, mullite, or other materials. Initially, a single-crystal seed inserted between the filaments is plunged in the melt; the meniscus is then stabilized, without need of supercooling the melt.[102] Ribbons 0.01 to 0.2 cm thick, 3 to 4 cm wide, and up to 25 cm long were pulled at a growth speed of 0 to 3 cm/min. They are formed by grains of width 0.1 to 4 mm, extending several centimeters along the growth direction, in spite of the nucleation process at the filaments. These large grains are favorable for the production of high quality solar cells (efficiencies up to 13.8%).

Edge-Defined, Film-Fed Growth

The preceding ribbon techniques are being developed specifically for Si and other semiconductors, and use more or less spontaneous shaping of the ribbon. In the following techniques, a die is used to give the proper shape to the liquid just before crystallization occurs. The EFG method was invented in 1969 by La Belle (see Ref. 103) as a modification of Stepanov's technique and was used for Si growth for more than 10 years. A complete review on this technique has been written by Wald.[104] The shape of the meniscus shown by Fig. 26g is essentially controlled by surface tension of the liquid and wetting angle at the die-liquid interface. Its stability, which is minimal at the ribbon edges, depends on the rate at which heat is extracted by radiation and conduction from the freezing crystal and on the heat input from the die, which is increased at the die edges to compensate for increased losses. Good stability can then be obtained for ribbons of some cm wide and 0.01 to 0.03 cm thick, grown at speeds of the order of 3 to 5 cm/min. The edge problem is removed if the die is given a tubular shape. Silicon tubes have been grown and even tubes of polygonal section which are then easily cut into ribbons.[105] The modelization of heat and mass transport, as a function of die geometry and temperature distribution, growth speed, ribbon thickness, and cooling profile, is still being improved to include special features like asymmetric geometry of the die, or transport of carbon and oxygen impurities from the die and atmosphere into the ribbon, or enhancement of growth speed by cooling elements placed near the growth interface. Incorporating these improvements into EFG apparatus now allows growth to 10 cm wide, 200-μm-thick ribbons, at speed 4 cm/min, from which 11 to 12% efficiency solar cells have been made.[106] Record efficiencies are beyond 14%.

Crystallographic defects in EFG crystals[107] depend on growth rate and include parallel arrays of dislocations, twins, and stacking faults, and also large angle grain boundaries. Carbon originating from the graphite die is an important impurity, the concentration of which can surpass 10^{18} cm^{-3}. SiC precipitates (which can trap other impurities) do occur in these ribbons; their density is minimized, at least in half of the ribbon thickness, by the use of asymmetric dies. Ribbons grown from quartz crucibles are better than those grown from graphite ones, but cell processing is also important to determine the final quality of the cell;[108] PH$_3$ diffusion has given the best results, with post-processing diffusion lengths over 50 μm and cell efficiencies over 12.5%.

Machines currently produce ribbons of a length from 1.25 to 50 m. Their energy consumption is low (120 kWh/m²). Their cost efficiency can be increased if several ribbons are grown simultaneously from the same melt. A 10-ribbon machine, including a melt replenishment system, has been built; the already-mentioned tube growth is another approach, which led to growth rates reaching 146 cm²/min. Such improvements, as well as those of ribbon quality, account for the expected cost decrease illustrated by Table 10.

A similar technique, now less developed, is the inverted Stepanov technique (see Fig. 26h). According to Ricard,[109] high quality 0.1-cm-thick ribbons of regular shape are readily obtained and efforts are engaged towards reduction of film thickness.

The so-called interface-controlled crystallization[9] (see Fig. 26i) technique is also a kind of shaping technique designed to produce continuous ribbons, well suited for large-scale industrial production.

Ultra-Fast Quenching Technique

A technique well in use for the production of amorphous metals (see Fig. 26j) has been applied to the production of Si ribbons;[110] it uses the projection of Si droplets on a rotating cylinder. This method is unique by its production rate, which is potentially several thousands of cm^2/min (and even 300,000 cm^2/min); it is the only method yet proposed to produce km^2 of solar cells which may be needed in the next decades. However the counterpart of this production speed is a bad quality of the Si ribbon; very fine grains cause the solar cell efficiency to be limited to 4%. Of course recrystallization might be attempted on these ribbons, but at the price of reducing the production speed to usual values.

Plasma-Torch Technique

The same is true, probably, of the technique which consists of projecting Si powder, by a plasma torch,[111] on a cold substrate from which the Si ribbon can be easily detached (see Fig. 26k). A recent study[112] showed that grains as large as 100 μm and doping control could be obtained, but no solar cell production has yet been attempted.

Sheets

Silicon-On-Ceramics

The SOC process[113] has been designed to deposit melted Si onto a substrate and obtain a thin layer, by just dipping the substrate in a Si melt for a short time. In a recent version called SCIM (silicon coating by inverted meniscus), the substrate is moved across a trough where molten Si is poured (see Fig. 28a), a process which is reminiscent of classical metallurgy. Cheap substrates

Ref.	Starting Material[g]	Crystal Growth Technique[g]	Active Layer Deposition[g]	Results[h]	
				L_n (μm)	η (%)
a	UMG (Fe = 200)	Slow ZM (1 cm/min)	CVD 25 μm (Fe = 200) + H_2 passivation + SJ	8-12	9.75
b	UMG (Fe = 60-80)	2 CZ	CVD 50 μm + SJ		8.2-12.9[i]
c	UMG (Fe < 100)	2 CZ (Fe < 2)	Gettering[j] + CVD (30-50μm) + SJ		10.5-10.8[k]
c	MG (Fe = 6300)	2 CZ (Fe = 3)	CVD 30-50 μm + SJ		9.9
d	MG	HEM	Gettering[l] + BSF + CVD 18 μm + SJ	13	11
e	MG/DAR (Fe = 1400)	Slow ESP	CVD + SJ		7.9-10.5[m]
f	UMG (Fe = 350-500)	Bridgman + heat treatment[n]	Gettering[l] + CVD 20-25 μm + SJ + heat treatment[n]		8.95

MG   = metallurgical grade
DAR = direct arc reactor
CZ   = Czochralski pulling
ESP = edge-supported pulling
SJ   = shallow junction

UMG = upgraded metallurgical grade
ZM   = zone melting
HEM = heat exchanger methode
CVD = chemical vapour deposition
BSF = back surface field

[a] From T. L. Chu[75]
[b] From P. H. Robinson[76]
[c] From T. Warabisako[77]
[d] From R. G. Wolfson[78]
[e] From T. F. Ciszek[73]
[f] From T. L. Chu[56]
[g] Indications on Fe content in ppm are given in parentheses.
[h] Typical values of minority carrier diffusion length, L_n, in the epitaxial layer and solar cell efficiency η.
[i] Best diffused (instead of epitaxial) cells: 4.5%.
[j] Gettering by back surface grinding.
[k] Best diffused cells: 8.4%.
[l] Gettering by HCl etching at 1175° C.
[m] Best diffused cells: 4%.
[n] 12 h at 700°C, to enhance impurity segregation towards grain boundaries.

Table 7.   Properties of Epitaxial Solar Cells

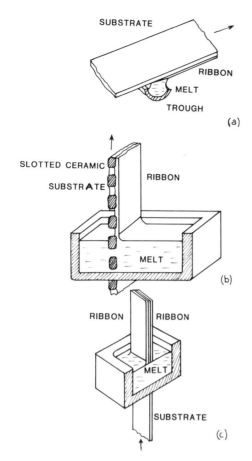

Fig. 28. Schemes illustrating: (a) the SCIM-process; (b) the SOC-process with a slotted substrate; (c) the RAD process, for the deposition of Si sheets on a substrate.

which can lead to crack-free Si sheets include carbon-coated mullite or fused quartz. Mullite can be made to have the same expansion coefficient as silicon; on the other hand, slotted substrates (see Fig. 28b) allow for back electrical contacts to the Si ribbon. The Si-ceramic interface is irregular and the ribbon has a high density of defects and impurities (grain boundaries, oxide inclusions, and so on). Nevertheless, interesting values of minority carrier diffusion lengths have been found[114] and solar cell average efficiency is in the 7 to 10% range. Growth conditions of these supported ribbons are very stable, so that high growth rates are possible for thin sheets, for example, 15 cm/min; wide sheets are readily obtained, so that high throughput rates seem to be feasible (several hundreds of cm²/min).

Ribbon-Against-Drop

The RAD process[115] (see Fig. 28c) was developed along the same lines with the intention to obviate the discontinuous nature of dip-coating techniques. Several carbon ribbon substrates are simultaneously pulled vertically either around or inside a molten zone which is maintained on top of a polycrystalline Si rod. It is a kind of crucible-less, zone-floating, technique, in which the meniscus is stabilized by the carbon ribbon; the thermal and hydrodynamic regimes have been discussed by Belouet.[115] The production rate is currently 7.5 to 10 cm/min, the diffusion length may reach 60 μm in as-grown layers, and solar cell efficiency 9%.[116] A SiC layer, some μm thickness, is formed at the carbon-silicon interface, but a recent paper[117] mentions the possibility to dissolve the carbon substrate and the SiC layer, so as to produce self-supported ribbons rather than sheets by the RAD process.

Epitaxial Layers

We have seen that, in the metallurgical route, the crystallization process conducted using one of the preceding techniques has to be completed by an epitaxial layer deposition. This layer is made of ultrapure Si and will serve as the active layer when the substrate is too impure to have a good photoelectric yield.

Chemical vapor deposition is generally used for this purpose; at deposition temperatures of order 1100 to 1200° C, the layer grows epitaxially on the substrate. The gas used for the silicon transport may be SiH_4, $SiHCl_3$, SiH_2Cl_2, or $SiCl_4$, mixed with hydrogen. The growth rate is 2 to 10 μm/min for usual gas flows at the deposition temperature of 1150°C. The needed layer of thickness 30 to 50 μm, is built in 10 to 20 minutes, which is short enough to allow a cheap, large-scale production.

Impurity diffusion from the substrate into the epitaxial layer is the principal limit on the quality of this layer and on the efficiency of the solar cell which will be built on it. It would be useful to measure the impurity profile and the minority carrier diffusion length in the layer as a function of the substrate impurity content and CVD conditions. Only scarce results appear in the literature on this point; Table 7 is a compilation of some of them. In order to improve the layer quality, an interesting possibility is to purify the substrate's superficial layer prior to the active layer deposition. Such a gettering treatment can be performed by several techniques[118] which are all based on the outdiffusion of impurities. For this reason, they are most efficient for the fast diffusing species, like Fe. Assuming the use of this technique, we can consider that all harmful impurities have diffusion coefficients in polycrystalline layers which are smaller than 10^{-8} cm^2 s^{-1} (compare Fig. 7) and conclude, accordingly, that the active layer surface is of more than 10 times higher purity than the substrate. It would be useful to verify quantitatively the value of this ratio.

At this point we shall close the discussion on impurities in silicon solar cells by estimating the allowed impurity concentration at different steps of the metallurgical route:

• in the active layer itself, using the discussion above, and specifically the results illustrated by Fig. 10,
• in the substrate, using the factor 10 just mentioned,
• in the UMG Si, from which the substrate is made by one of the techniques of crystal growth mentioned above and allowing a good segregation. We shall take the conservative option that the effective segregation coefficient is 3 to 10 times less favorable than the equilibrium coefficient, to allow for a crystal growth speed large enough to keep the process cheap. This is quite arbitrary, of course, but probably representative of many crystal growth techniques.

The results of these estimations are given by Table 8. Confronted with Table 5, they lead to the conclusion that the upper grades of UMG and DAR materials are quite convenient to ob-

tain 10% efficiency solar cells, as has been verified by different experimenters quoted in Table 7.

Conclusions on Crystallization Techniques

We shall confine these concluding remarks to the growth of highly purified silicon (the chemical route), which is the main problem for the next few years. It is clear that all the techniques described above are still in evolution, with two main directions of improvement: improvement of the material quality and technological improvement.

A dozen techniques are already able to produce 10% efficiency solar cells, in spite of the defects present in the material. Seven of them are listed in Table 9, due to Yoo,[119] who compared different materials available in 1981. Future progress is expected. For instance, all techniques are not yet optimized from the point of view of post-growth-controlled cooling: too rapid cooling is an important source of stress-induced dislocations. The techniques giving slightly inferior quality are essentially those which use a temporary or permanent substrate; they may need a second step of recrystallization (like the RTR process) if they are handicapped by small grain size, or of epi-layer deposition if they are incorporating impurities.

On several technological points great progress is possible: temperature control, in relation

(Contents in ppma)	Active Layer[a]	Substrate[b]	UMG Si	
			If $k_{eff} = 10\ k^c$	If $K_{eff} = 3\ k^c$
Ag	0.08	0.8	2000	6000
Al	0.01	0.1	3	10
Au	0.0002	0.002	10	30
B[c]	2	20	20	20
C	1	10	15	50
Co	0.002	0.02	250	750
Cr	0.002	0.02	175	550
Cu	0.8	8	330	1000
Fe	0.0005	0.005	75	250
Mn	0.002	0.02	150	500
Mo	0.000002	0.00002	50	150
Ni	0.008	0.08	250	800
O[c]	20	200	200	400
P[c]	1	10	10	20
Ta	0.000001	0.00001	10	30
Ti	0.00001	0.0001	3	10
V	0.00001	0.0001	3	10
Zr	0.000002	0.00002	125	400

[a] Conservative estimation of impurity content allowable near the p-n junction.
[b] Assuming that the 50 μm epitaxial layer is produced by 20 min CVD at 1100°C and that the diffusion coefficient is of order 10^{-8} cm^2 s^{-1} at this temperature.
[c] For B, k_{eff} is taken equal to 1, and for O and P, between 0.5 and 1.0.
[d] Each impurity is assumed to behave independently.

Table 8.   Allowed Impurity Contents, Estimated at Different Steps of the Metallurgical Route[d]

| Technique | Unwanted Impurities | Structural Defects | | Solar Cell Efficiency | | |
		Grain Boundaries (g.b.)	Dislocations (cm^{-3})	Process [e]	Average (%)	Best (%)
INGOTS						
CZ[b]				SJ+BSF	15.5	16.2
CZ[c]				SJ+BSF	14.1	15.5
HEM	Inclusions	Very large grain if any	Typically 10^5	GET+SJ+BSF	15.0	15.7
SILSO[d]	Inclusions (SiC)	Columnar structure Grain size $\simeq$ mm	Typically 10^5	SJ+BSF	12.1	12.6
RIBBONS						
RTR	Mo	Columnar structure large grain (cm size)	Typically 10^5	SJ	12.2	13.1
WEB	Mo		Typically 10^5 Twins associated with dislocations	SJ+BSF+BSR	14.3	15.5
EFG	Inclusions (SiC)	Low and high angle g.b. Mostly columnar with few intersecting g.b.	Typically 10^6 Twins associated with dislocations	SJ+BSR	12.1	13.6
SHEETS						
SOC	Al and transition metals	Mainly columnar structure Few intersecting g.b.	Typically $> 10^6$ Dislocation clusters Stacking faults associated with dislocations	SJ	8.8	9.6

[a] From H.I. Yoo.[119]

[b] Control samples.

[c] Advanced Czochralski.

[d] Mold casting.

[e] GET = gettering by $POCl_3$; SJ = shallow junction by $POCl_3$ diffusion at 875°C; BSF = back surface field, BSR = back surface reflector. Finally, evaporated contacts and multilayer antireflective coating.

Table 9. Si Quality Reached by Several Growth Techniques [a]

with growth speed (and with thickness of ribbons and sheets), multiribbon machines, and automation. The resulting costs reached today and expected for 1986 are listed in Table 10 (from Koliwad[82]) for five techniques. The expected progress is considerable, and the preceding discussion helps to understand why and how such progress may be reasonably expected. Once again, other techniques not included in Table 10 may have just as bright prospects. Among them,

| | Throughput Rate (Individual Machine) | Maximum Module Efficiency | Sheet-Added Price [d] (1981) | | At Module Efficiency | Sheet-Added Price (1986 Goal) ($/Wp) | At Module Efficiency |
			$/kg	$/Wp			
INGOTS							
CZ[e]	1.0 kg/h[b]	> 15%	96.5		9.6%	0.19	15%
				0.69 to 1.[c]			
HEM	3.1 kg/h[b]	> 15%	31.5		12.3% ?	0.26	15%
RIBBONS							
EFG[f]	42 cm²/min	13.2%	92.6	1.15	8.1%	0.20	12%
WD[g]	27 cm²/min	14.5%	196.6	0.82	11 %	0.29	14%
SHEETS							
SOC[h]	5 cm²/min	10.5%	—	—	—	0.19	11%

[a] From K. M. Koliwad.[82]

[b] These figures can be transformed into 150 cm²/min (for CZ) and 450 cm²/min (for HEM), for wafers of 10-cm size, 0.1-cm total thickness (half for the wafer, half for the kerf loss). Cutting machines are assumed numerous enough for the growing process to be rate limiting.

[c] Depending on the chosen cutting system.

[d] This price is for scaling-up of presently demonstrated technologies.

[e] Diameter 100 mm (it is possible, though not usual, to grow 150 mm diameter, 150 kg crystals).

[f] Ribbon width 100 mm, 1 ribbon only (multiribbon machines do exist already).

[g] Ribbon width 50 mm.

[h] Sheet width 12.5 mm.

Table 10. Technical and Economic Comparison of Five Si Growth Techniques [a]

certainly Bridgman (SEMIX), mold casting (SILSO), HRG, ESP, and RAD at least have to be seriously considered.

SOME PROBLEMS OF CELL TECHNOLOGY

To make a p-n junction in a crystalline Si wafer, the standard process is diffusion at 900° C, either from gaseous sources like PH_3, or from solid phases.[120] This diffusion process has to be preceded and followed by auxiliary steps (etching, cleaning). The back-surface field is often made by Al-alloying; however, in the screen-printing option, BSF and p-n junction can be made both in a single step. Resulting efficiencies are lower if L_n in the wafer is lower, that is, in the imperfect materials. This is illustrated by Fig. 29, which also shows that cell improvement due to BSF is larger when L_n itself is larger. The early results shown by Fig. 29 have been improved later on (see Table 9).

Cold junction processes are developing very fast as an alternative to diffusion.[121] They consist essentially of two steps: impurity deposition and pulsed annealing. Impurity deposition can be made by ion implantation or glow discharge in a doping gas, but also by simpler methods like evaporation, spray, sputtering, or spin-on deposition.[122] Pulsed annealing may use electron beams, laser beams or microwaves.[123] In some cases, these two steps are merging in only one step.[60] In general cold junction processes have several advantages:
- lower energy consumption,
- process amenable to full automation,
- bulk kept at low temperature,

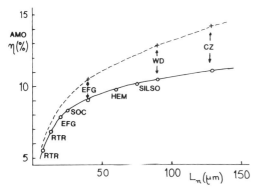

Fig. 29. Air-mass-zero efficiencies of simple cells and cells with back-surface field (BSF), made from materials of different base diffusion lengths (from H. Yoo[96]). Efficiencies under AMO are currently several points lower than under AM1 which is a better measurement for terrestrial cells.

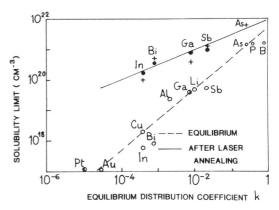

Fig. 30. Correlation of the solubility limits at thermal equilibrium and after laser treatment with equilibrium distribution coefficients (from E. Fogarassy[126]).

- minimization of preannealing and postannealing steps, which should lead to decreased costs.

An accurate cost comparison[124] has shown this potential advantage of the cold junction processes. Several types of high throughput machines (10 MW_p/yr) are being developed. On the other hand, in all cases when these techniques were applied to polycrystalline Si, they were found to be satisfactory (see for example, Wolfson,[78] Bell[125]).

An interesting material property related to pulsed annealing is the anomalously high impurity solubility resulting from the high freezing rates (of order 10 m/s) following the formation of the superficial melt (see Fig. 30).[126]

We have mentioned the possibility of making p-n junction by epitaxy, a process yet rarely employed.

As a conclusion, it can be said that the junction formation will become cheaper through increased automation, possible use of cold junction processes, and higher throughput rates. The use of more imperfect silicon may require other process changes. A cost decrease by a factor of 4 is a short-term perspective[124] which will accompany the major cost decreases due to the use of new purification and crystal growth techniques, studied in the preceding parts. Another effective, though slower, cost decrease will affect the other technological steps, not studied in this review (contacts, AR layer, and encapsulation).

CONCLUSIONS

The technical evolution just described proves why and how the price decrease of solar cell panels will follow the short-term favorable trends of Fig. 1, which are also stated, in more detail, in Table 11. In 1985, polycrystalline ingots will be used; techniques for the production and use of solar grade and UMG Si will be developed. If they would leave the pilot plant step and enter that of full-size production in this transitory situation, the price might be divided by a factor 2 or 3 (only), as compared to the price of 1982. Full use of the techniques just mentioned, and perhaps that of ribbon growing, are expected in the second half of our decade; this will bring about a new cost decrease, but no efficiency increase, since the use of more and more imperfect Si is anticipated in this period.

In Table 12, we try to make clear the complete price structure of solar photovoltaic electricity, resulting both from the progress in solar cells themselves and—not discussed here—anticipated progress in "balance-of-system," that is, power conditioning and storage. The method used was previously presented by several papers[127, 128]; the latter one, by the American Physical Society study group headed by H. Ehrenreich, also gives the end-of-century prices which are quoted in the last column of Table 12 and shows a good continuity with our predicted prices for 1990. It is hardly reasonable to anticipate that such a long-term forecast may be accurate; however, we shall try to draw some conclusions from the data of Tables 11 and 12, taken as illustrating the probable evolution.

The competitivity of PV generators, relative to diesel generators in isolated sites, will be reached between 1985 and 1990 according to the generator size and site. A representative price for 50 kW diesel electricity is 30¢/kWh in 1980, 35 to 40¢/kWh in 1985. As a consequence, the period 1985–90 should see a considerable market development, whereas the period 1982–85 may be more difficult for PV industry. The development of PV generators in tropical isolated sites has a great social as well as economic importance, as shown by the value of almost maintenance-free sources useable for refrigerators, TV receptors, water pumps, house lighting, and village workshops. However, local uses of PV generators in the USA, and even more in Europe, will develop more slowly; the largest applications are the residential ones, which require in most places an electricity price of 10¢/kWh or less.

In the period 1985-90 an interesting competition between crystalline and amorphous Si solar

| Step | Diffused Cells | | Epitaxial Cells |
	1982	1984-85	1985-86
Raw material	3.0	1.0	0.4
Crystallization	2.0	1.0	1.1
Epi-layer	—	—	}1.0
Cell technology[b]	2.5	1.0	
Encapsulation	1.5	1.0	1.0
Total Price[a]	9.0	4.0	3.5

[a] Prices in 1982 \$/peak W. Cell efficiencies considered to be 10% in all this period.
[b] Includes formation of p-n junction, contacts and antireflective coating.

Table 11. Probable Short-Range Evolution of the Price Components of PV Panels Made of Crystalline Si[a]

cells will perhaps take place. The latter are up to now limited to very small power applications, but their efficiency increase has been fast recently (8% reported at the recent Stresa Conference of the European Communities, and 10% at the IEEE Conference in San Diego). If new progress can result from the systematic use of tandem cells of amorphous alloys and the elimination of cell instabilities, their potential cheapness makes them a serious contender for kW-size generators in some years. However, one of the main purposes of the present review was to show that the crystalline silicon solar cells also have good perspectives for improvements, which will make competition difficult for new contenders.

We feel that the race towards high efficiencies will be resumed for crystalline Si as soon as the new cheap materials will be firmly established in production. It might be the decisive factor which will determine the choice of the winning ingot, ribbon, or sheet technique. It might be the main line of progress leading to the end-of-century results of Table 12, which will establish the PV generators as a major contributor to the energy needs of the next century.

In this perspective, it is urgent to solve the problems related to the physics of defects and impurities which have been raised. These problems are of a very fundamental nature and need the work of goal-oriented teams of physicists. Of course, they are also very important for shorter-range progress related to the use of polycrystalline Si, UMG Si, epitaxial and cold junction cells, or both. Schematically, these problems, which remain to be solved, may be summarized as follows:

• multi-impurity effects in single crystals, in par-

168

	1982[1] (kw-size Generator, 6-h Storage)	1990[1] (100 kw-Size, 6-h Storage)	2000[1] (Network-Coupled Plant, No Storage)
C_m ($/m²) [b]	800.	250.	160.
η_m [b]	0.09	0.10	0.13
η_p [c]	0.80	0.85	0.90
C_s ($/m²) [d]	50.	40.	20.
C_{PC} ($/kWp) [e]	4000.	1200.	140.
$K_s H_s$ ($/kWp) [f]	4000.	600.	0
C_I ($/kWp) [g]	26700.	5800.	2100.
H_p (hr/yr) [h]	3000.	3000.	2500.
$\overline{OM}$ ($/kWp) [h]	0.01	0.01	0.004
$\overline{FL}$ ($/kWh) [j]	0	0	0.009
C_E ($/kWh) [k]	1.345	0.30	0.118

[a] Prices in 1982 dollars.
[b] C_m = cost, η_m = efficiency of crystalline Si panels (or modules). Years 1982-90 are assumed to be mainly devoted to price decrease and 1990-2000 to efficiency increase.
[c] η_p = efficiency of the non-panel parts of the plant.
[d] C_s = area-proportional costs, except panels, that is: land, panel supports, and connections. The area considered is the cell area.
[e] C_{PC} = cost of the power conditioning components.
[f] H_s = storage time; $K_s H_s$ = storage cost.
[g] C_I = investment price, given by the formula[128]:

$$C_I = f_{ind} \left(\frac{C_s + C_m}{\eta_p \eta_m I_p} + C_{PC} + K_s H_s \right)$$

where I_p is the peak solar radiation intensity (I_p = 1 kW/m²) and F_{ind} is a financial factor taking account of engineering and commercial costs (f_{ind} = 1.35).
[h] H_p = nominal yearly time of energy production at peak power.

[i] $\overline{OM}$ = cost of the operation and maintenance.

[j] $\overline{FL}$ = cost of the fuel-powered back-up system.

[k] C_E = energy price, given by the formula[128]:

$$\overline{C_E} = \frac{A}{H_p} = C_I + \overline{OM} + \overline{FL}$$

where A is a financial factor describing the amortization rate (A = 0.15).
[1] For 1982 and 1990 we have considered the typical autonomous systems mainly produced at these dates. For 2000 we have inserted figures proposed by H. Ehrenreich[128] for a network-coupled plant typical of this date. The PV electricity will then be only 2 to 3 times more expensive than that produced by coal-powered thermal plants.

Table 12. Probable Long-Range Evolution of the Price of PV Generators and Photovoltaic Energy Price [a]

ticular trap density as a function of transition metal and oxygen content,
• effect of precipitates, in particular carbide precipitates, on junction characteristics,
• chemical nature of the grain boundaries,
• passivation of the grain boundaries,
• influence of impurities on crystal growth,
• effective impurity segregation by cheap crystal growth techniques,
• diffusion barriers and gettering treatments, or both, for epitaxial cells,
• junction technologies for imperfect silicon.

Obviously, the development of the techniques able to produce high quality solar grade Si ingots, ribbons, and sheets is of high priority. A number of possible paths between MG Si and high efficiency solar cells are in close competition; it is important to qualify not only the individual techniques of Si purification, crystal growth, and junction growth, but also the whole process of getting a good final product from the original raw material. Among these different paths, those which use a partial chemical purification, a complementary purification during crystal growth, and an epi-layer deposition seem to warrant high interest.

ACKNOWLEDGEMENTS

It is a pleasure to thank W. Armstrong, E. Bates, J. Calzia, T. L. Chu, T. F. Ciszek, E. Fabre, S. J. Fonash, D. Helmreich, C. P. Khattak, A.

Neugroschel, R. Revel, H. Rodot, B. L. Sopori, N. Tsuya, F. V. Wald, J. D. Zook for useful information. Special thanks are due to W. Armstrong, C. Belouet, J. P. Crest and P. Pinard for sending original photographs.

REFERENCES

1. D. M. Chapin; C. J. Fuller; and G. L. Pearson. 1961. *J. Appl. Phys.* 252:246.
2. J. Lindmayer and J. F. Allison. 1972. *9th. IEEE Photovoltaic Spec. Conf. Proc.* New York: IEEE, 83.
3. J. R. McCormick. 1980. *14th IEEE Photovoltaic Spec. Conf. Proc.* New York: IEEE, 298.
4. G. M. Storti. 1981. *15th IEEE Photovoltaic Spec. Conf. Proc.* New York: IEEE, 442.
5. J. Grenier. Private communication. These figures are prices, and not costs; the 1985 figure is typical of the upper limit of European expectations and may be considered as conservative. However, U.S. Department of Energy expectations for 1982 ($6 to 13/W system price, including $2.80/W module price) have not been confirmed in practice.
6. W. Freiesleben. 1981. *3rd Photovoltaic Solar Energy Conf.* Dordrecht: Reidel, 166.
7. According to Y. Hamakawa. 1981. *3rd Photovoltaic Solar Energy Conf.* Dordrecht: Reidel, 22, solar cell production may reach 50 MW in 1985 and 3000 MW in 1990.
8. S. Pizzini. 1982. *Solar Energy Mat.* 6:253.
9. J. Dietl; D. Helmreich; and E. Sirtl. 1981. *Crystals: growth, properties and applications.* Vol. 5. Berlin: Springer-Verlag, 43.
10. E. Bucher. 1978. *Appl. Phys.* 17:1.
11. Alternative processes have been proposed, but are not in use: the aluminothermic treatment of SiO₂ (E. Vigouroux. 1899. *C. R. Acad. Sci.* 129:334) and the electrolysis of fluosilicates (R. Boen and J. Bouteillon. 1978. *1st Photovoltaic Solar Energy Conf.* Dordrecht: Reidel, 860. D. Elwell and R. S. Feigelson. 1982. *Solar Energy Mat.* 6:123).
12. L. P. Hunt. 1975. *11th IEEE Photovoltaic Spec. Conf. Proc.* New York: IEEE, 259; J. Amouroux; P. Fauchais; D. Morvan; D. Rocher. 1979. *Ann. Chim.* 4:231.
13. R. Lutwack. 1981. *3rd. Photovoltaic Solar Energy Conf.* Dordrecht: Reidel, 220.
14. The energy consumption is also reduced. This was imperative to obtain valuable solar cells. Using electronic grade Si, the energy payback time is 12 yrs, which is unacceptable. Although not exactly known, the corresponding figure for solar grade Si should be no more than 3 yrs, and perhaps .5 yr. See L. P. Hunt. 1976. *12th IEEE Photovoltaic Spec. Conf. Proc.* New York: IEEE.
15. See e.g., M. C. Flemings. 1974. *Solidification processing.* New York: McGraw-Hill; B. Chalmers. 1964. *Principles of solidification.* New York: J. Wiley; J. C. Brice. 1973. *The growth of crystals from liquids.* Amsterdam: North Holland.
16. See, however, C. P. Khattak; F. Schmid; and L. P. Hunt. 1980. Comm. to the Electrochem. Soc. St. Louis, May 1980, who obtained 12.3% efficiency after a simple Heat-Exchanger Method growth of silicon produced by Submerged Electrode Arc Furnace. Best efficiency was only 7.2% for HEM growth of MG silicon (F. Schmid; M. Basaban; and C. P. Khattak. 1981. *3rd Photovoltaic Solar Energy Conf.* Dordrecht: D. Reidel, 252); 8.5% for CZ growth of MG silicon (J. I. Hanoka; H. E. Strock; and P. S. Kotval. 1978. *13th Photovoltaic Spec. Conf. Proc.* New York; IEEE, 485).
17. H. J. Hovel. 1975. "Solar cells." In *Semiconductors and semimetals.* Vol. 3. New York: Academic Press.
18. R. C. Neville. 1978. *Solar energy conversion: the solar cell.* Elsevier.
19. A. Laugier and A. J. Roger. 1981. *Les photopiles solaires.* Technique et documentation.
20. S. J. Fonash. 1980. "Photovoltaic devices." *Critical reviews in solid state and materials sciences* 110.
21. J. R. Davis; A. Rohatgi; R. H. Hopkins; P. D. Blais; P. Rai-Choudhury; J. R. McCormick; and H. C. Mollenkopf. 1980. *IEEE Trans. Electron Dev.* ED 27:677.
22. G. M. Storti; S. M. Johnson; H. C. Lin; C. D. Wang; *14th IEEE Photovoltaic Spec. Conf. Proc.* New York: IEEE, 191.
23. J. M. Reynolds and A. Meulenberg Jr. 1974. *J. Appl. Phys.* 45:2582; S. Othmer and S. C. Chen. 1980. *13th IEEE Photovoltaic Spec. Conf. Proc.* New York: IEEE, 1238. The latter paper shows how the presence of long-lived traps may affect the interpretation of photoconductivity measurements in terms of lifetime.
24. T. L. Chu and E. D. Stokes. 1980. *13th IEEE Photovoltaic Spec. Conf. Proc.* New York: IEEE, 95.
25. The use of EBIC technique is described by J. I. Hanoka and R. O. Bell. 1980. *Ann. Rev. Mater. Sci.* 11:353 and that of LBIC by D. L. Lile and N. M. Davis. 1975. *Solid State Electron* 80:699 and also B. L. Sopori and A. Baghdadi. 1979. *Solar Cells* 1:237.
26. F. A. Trumbore. 1960. *Bell Syst. Tech. J.* 39:205.
27. K. A. Jackson and M. J. Leamy. 1978. *AIP Proc. Conf. on Laser Solid Interaction and Laser Processing.* Edited by S. D. Ferris; M. J. Leamy; and J. M. Poate. Boston, 102; C. W. White; J. Narayan; and R. T. Young. *Ibid.* 275; R. Stuck; E. Fogarassy; J. C. Muller; A. Grob; J. J. Grob; and P. Siffert. 1979. "Laser and electron beam processing of materials." *Materials Research Soc. Symp.* Los Angeles.
28. W. Frank; V. Gösele; and A. Seeger. 1978. *Inst. Phys. Conf. Ser.* 46:514.
29. H. Feichtinger. 1978. *Inst. Phys. Conf. Ser.* 46: 528; E. Weber and H. G. Riotte. 1980. *J. Appl. Phys.* 51:1484.
30. P. Capper; A. W. Jones; E. J. Wellhouse; and J. G. Wilkes. 1977. *J. Appl. Phys.* 48:1646.
31. G. A. Rozgony and C. W. Pearce. 1978. *Appl. Phys. Letters* 37:747.
32. K. Wunstel and P. Wagner. 1982. *Appl. Phys.* A 27:207.
33. B. I. Boltaks. 1963. *Diffusion and point defects in semiconductors.* London: Inforsearch; D. Shaw *Atomic diffusion in semiconductors.* London: Plenum, 1973.
34. A. Seeger and K. P. Chik. 1968. *Phys. Stat. Sol.* 29:455.
35. H. Wolf. 1969. *Silicon semiconductor data.* New York. Pergamon.
36. J. D. Gerson; L. J. Cheng; and J. W. Corbett. 1977. *J. Appl. Phys.* 48:4821.

37. H. G. Grimmeiss. 1977. *Ann. Rev. Mater. Sci.* 7:341 gives a description of DLTS measurements and some results on capture coefficients C_n, C_p. The relation between lifetime and the coefficients C_n, C_p is discussed by H. J. Queisser. 1978. *Solid State Electron* 21:1495.

38. S. Pizzini; L. Giarda; A. Parisi; A. Somi; and G. Soncini. 1980. *15th IEEE Photovoltaic Spec. Conf. Proc.* New York: IEEE, 902.

39. A. Rohatgi; J. R. Davis; R. H. Hopkins; P. Rai-Choudhury; P. G. McMullin; and J. R. McCormick. 1980. *Solid State Electr.* 23:415.

40. A. Rohatgi; R. B. Campbell; J. R. Davis; R. H. Hopkins; P. Rai-Choudhury; H. Mollenkopf; and J. R. McCormick. 1980. *14th IEEE Photovoltaic Spec. Conf. Proc.* New York: IEEE, 908.

41. W. Shockley. 1949. *Bell Syst. Tech. J.* 28:435; J. E. Lawrence. 1965. *J. Electr. Chem. Soc.* 112:796.

42. A. M. Salama. 1978. *13th IEEE Photovoltaic Spec. Conf. Proc.* New York: IEEE, 496.

43. R. B. Campbell; M. H. Hanes; R. H. Hopkins; A. Rohatgi; P. Rai-Choudhury; and H. C. Mollenkopf. 1981. *15th IEEE Photovoltaic Spec. Conf. Proc.* New York: IEEE, 530.

44. T. Figielski. 1978. *Solid State Electron.* 21:1403.

45. H. C. Card; E. S. Yang. 1977. *IEEE Trans. on Electron Devices.* ED 24:397.

46. L. L. Kazmerski. 1978. *Solid State Electron* 21:1545.

47. A. G. Ghosh; C. Fishman; and T. Feng. 1980. *J. Appl. Phys.* 51:446.

48. M. Zehaf; H. Amzil; J. P. Crest; G. Mathian; S. Martinuzzi; J. Oualid. 1982. "Comm. Electrochem. Soc. Conf." Montreal. To be published.

49. K. M. Koliwad; T. Daud; and J. K. Liu. 1979. *2nd Photovoltaic Solar Energy Conf.* Dordrecht: D. Reidel, 710; K. M. Koliwad and T. Daud. 1980. *14th Photovoltaic Spec. Conf. Proc.* New York: IEEE, 1204.

50. J. G. Fossum; A. Neugroschel; F. A. Lindholm; and J. A. Mazer. 1980. *14th Photovoltaic Spec. Conf. Proc.* New York: IEEE, 184.

51. B. L. Sopori. 1981. "Comm. intern. workshop on the physics of semiconductor devices (Delhi).

52. D. Helmreich and H. Seiter. *2nd Photovoltaic Solar Energy Conf.* Dordrecht: D. Reidel, 742.

53. R. Bullough and R. C. Newman. 1963. *Progr. in Semicond.* Vol. 7. Edited by A. F. Gibson and R. E. Burgess. New York: John Wiley.

54. J. J. Aubert; J. J. Bacmann; A. Bourret; J. Daval; J. Dessaux; and A. Rocher. 1980. *3rd Photovoltaic Solar Energy Conf.* Dordrecht: D. Reidel, 589.

55. L. L. Kazmerski; P. J. Ireland; and T. F. Ciszek. 1980. *J. Vacuum Sci. Technol.* 17:34.

56. T. L. Chu; S. S. Chu; L. L. Kazmerski; R. Whitney; C. L. Lin; and R. M. Davis. *Solar Cells* 5:29.

57. R. G. Rosemeier; R. W. Armstrong; S. M. Johnson; G. M. Storti; and C. C. Wu. 1981. *15th IEEE Photovoltaic Spec. Conf. Proc.* New York: IEEE, 1331.

58. G. M. Storti; S. M. Johnson; H. C. Lin; and C. D. Wang. 1980. *14th IEEE Photovoltaic Spec. Conf. Proc.* New York: IEEE, 191.

59. D. Redfield. 1981. *15th IEEE Photovoltaic Spec. Conf. Proc.* New York: IEEE, 1179. In this paper it is concluded that the crystal disorientations play only a minor role, if any, on the electrical activity of grain boundaries.

60. J. C. C. Fan; T. F. Deutsch; G. W. Turner; D. J. Ehrlich; R. L. Chapman; and R. M. Osgood Jr.

15th Photovoltaic Spec. Conf. Proc. 1981. New York: IEEE, 432 and G. B. Turner; D. Tarrant; D. Aldrich; R. Pressley; and R. Press. 1982. *4th Photovoltaic Solar Energy Conf.* Dordrecht: Reidel to be published, both propose laser-induced annealing of a silicon wafer in contact with a doping gas. It is interesting to note that Turner observed a passivation of grain boundaries in a SILSO wafer, by this technique.

61. C. H. Seager and D. S. Ginley. 1979. *Appl. Phys. Lett.* 34:337.

62. J. I. Pankove; M. A. Lampert; and M. L. Tarng. 1978. *Appl. Phys. Lett.* 32:439; C. H. Seager; D. S. Ginley; and J. D. Zook. 1980. *Appl. Phys. Lett.* 36:831; J. Mimila-Arroyo; F. Dueñas-Santos; and J. L. Delvalle. 1981. *15th IEEE Photovoltaic Spec. Conf. Proc.* New York: IEEE, 259; P. H. Robinson and R. V. D'Aiello. 1981. *Appl. Phys. Lett.* 39:63.

63. T. H. Di Stefano and J. J. Cuomo. 1977. *Appl. Phys. Letters* 30:351.

64. W. A. Orr and F. P. Califano. 1980. *14th IEEE Photovoltaic Spec. Conf. Proc.* New York: IEEE, 1198; F. Ferraris; F. C. Matacotta; S. Pidatella; and L. Sardi. 1980. *3rd Photovoltaic Solar Energy Conf.* Dordrecht: D. Reidel, 625.

65. R. Janssens; R. Mertens; and R. Van Overstraeten. 1980. *3rd Photovoltaic Solar Energy Conf.* Dordrecht: D. Reidel, 620.

66. E. Sirtl. 1979. *2nd Photovoltaic Solar Energy Conf.* Dordrecht: D. Reidel, 84.

67. E. Sirtl; L. P. Hunt; and R. Sawyer. 1974. *J. Electrochem. Soc.* 121:919.

68. W. C. Breneman; E. G. Farrier; and H. Morihara. 1978. *13th Photovoltaic Spec. Conf. Proc.* New York: IEEE, 339.

69. R. Lay and S. K. Iya. 1978. *15th Photovoltaic Spec. Conf. Proc.* New York: IEEE, 565.

70. S. J. Solomon. 1978. *15th Photovoltaic Spec. Conf. Proc.* New York: IEEE, 569.

71. W. H. Reed; R. N. Meyer; M. G. Fey; F. J. Harvey; and F. G. Arcella. 1978. *13th Photovoltaic Spec. Conf. Proc.* New York: IEEE, 370.

72. V. D. Dosaj, L. P. Hunt; and A. Schei. 1978. *J. Metals* 30:8.

73. T. F. Ciszek; M. Schietzelt; L. L. Kazmerski; J. L. Hurd; and B. Fernelius. 1981. *15th Photovoltaic Spec. Conf. Proc.* New York: IEEE, 581.

74. F. Secco d'Aragona; H. M. Liaw, and D. M. Heminger. 1981. *Proc. Symp. on Materials and New Processing Techn. for Photovoltaics.* Princeton: Electrochem. Soc., 119.

75. T. L. Chu; S. S. Chu; E. D. Stokes; C. L. Lin; and R. Abderrassoul. 1979. *13th Photovoltaic Spec. Conf. Proc.* New York: IEEE, 1106; T. L. Chu. 1979-1980. *Solar Energy Mat.* 2:265.

76. P. H. Robinson; R. V. D'Aiello; and D. Richman. 1980. *14th Photovoltaic Spec. Conf. Proc.* New York: IEEE, 54. See also J. I. Hanoka; H. B. Strock; and P. S. Kotval. 1981. *J. Appl. Phys.* 52:5829.

77. T. Warabisako; T. Saitoh; E. Kuroda; H. Itoh; N. Nakamura; and T. Tokuyama. 1980. *Jap. J. Appl. Phys.* 19 (suppl. 1): 538.

78. R. G. Wolfson and R. G. Little. 1981. *15th Photovoltaic Spec. Conf. Proc.* New York: IEEE, 595.

79. D. T. J. Hurle. 1961. *Solid State Electronics* 3:37; W. A. Tiller. 1963. In *The art and science of growing crystals.* New York: J. Wiley. Constitutional supercooling is also associated with a "cellular growth" of Si crystals which gives them an

irregular, "lamellar" aspect. Such a material is improper for solar cell production.

80. C. S. Duncan; R. G. Seidensticker; J. P. McHugh; R. H. Hopkins; M. E. Skutch; J. M. Driggers; and F. E. Hill. 1980. *14th Photovoltaic Spec. Conf. Proc.* New York: IEEE, 25; R. G. Seidensticker and R. H. Hopkins. 1980. *J. Crystal Growth* 50:221.

81. D. Morvan; J. Amouroux; and G. Revel. 1980. *Rev. Phys. Appl.* 15:1229.

82. K. M. Koliwad and M. H. Leipold. 1981. *3rd Photovoltaic Solar Energy Conf.* Dordrecht: D. Reidel, 228.

83. T. F. Ciszek; G. H. Schwuttke; and K. H. Yang. 1979. *J. Crystal Growth* 46:527.

84. F. Schmid. 1976. *12th Photovoltaic Spec. Conf. Proc.* New York: IEEE, 146; C. P. Khattak and F. Schmid. 1978. *14th Photovoltaic Spec. Conf. Proc.* New York: IEEE, 484.

85. Bridgman growth is a long-used technique to get Si polycrystals. It seems to be used for the so-called SEMIX material for solar cells. See J. Lindmayer, Z.C. Putney. 1978. *14th Photovoltaic Spec. Conf. Proc.* New York: IEEE, 208. A similar technique is used by J. Fally and C. Guenel. 1981. *3rd Photovoltaic Solar Energy Conf.* Dordrecht: D. Reidel, 598. The conditions of crystal growth are not very explicit in these papers.

86. D. Helmreich. *Proc. Symp. Electr. and Opt. Prop. of Polycrystalline or Impure Semiconductors and Novel Silicon Growth Methods.* 1980. Edited by Ravi and O'Hara. The Electrochem. Soc., 184.

87. H. Rodot; P. Cassagne; and M. Hamidi. 1981. *J. Electronic Mat.* 10:481.

88. D. Casenave; R. Gauthier; L. Vandekerkove; and P. Pinard. 1982. *Appl. Phys. Lett.* 40:698. The thermal equilibrium is described by the same authors in *Solar Cells* 5:367 (1982) and the purifying effect in *Solar Energy Mat.* 5:417 (1981).

89. As an example of exact treatment of the heat transport problem, see e.g., the work of J. D. Zook and S. B. Schyldt. 1980. *J. Crystal Growth* 50:51.

90. The growth angle of Si, ϕ_0, has been measured by several experimenters. Values between 8 and 12° have been found by T. Surek; S. R. Coriell; and B. Chalmers. 1980. *J. Crystal Growth* 50:21. By a new technique, M. Hamidi and H. Rodot *Rev. Phys. Appl.* (to be published) found 11° for (100) and (111) monocrystals and 8-10° for polycrystals. The crystal shape stability in the case of Czochralski growth, zone-floating and meniscus-controlled ribbon growth is discussed in the former paper by T. Surek, et al.

91. See e.g., a good discussion of stresses in RTR ribbons, by R. W. Gurtler. 1980. *J. Crystal Growth* 50:69. Post-heating of the crystal was found to be important in this case, and also for the RFP ribbons (see D. Casenave[88]) and other techniques.

92. K. A. Dumas; C. P. Khattak; and F. Schmid. 1981. *15th Photovoltaic Spec. Conf. Proc.* New York: IEEE, 954.

93. F. Schmid and C. P. Khattak. *Optical Spectra* (May 1981).

94. S. M. Johnson; G. M. Storti; R. W. Armstrong; R. G. Rosemeier; M. E. Taylor; and W. F. Regnault. 1981. *15th Photovoltaic Spec. Conf. Proc.* New York: IEEE, 949.

95. B. Authier. 1978. *Adv. Solid State Phys.* 18:1.

96. H. Yoo; P. Iles; D. Tanner; G. Pollock; and F. Uno. 1981. *15th Photovoltaic Spec. Conf. Proc.* New York: IEEE, 312.

97. B. Kudo. 1980. *J. Crystal Growth* 50:247; D. W. Jewett and H. E. Bates. 1980. *14th Photovoltaic Spec. Conf. Proc.* New York: IEEE, 1404. The first suggestion of HRG technique is from C. E. Bleil. 1969. *J. Cryst. Growth* 5:99.

98. K. R. Sarma; R. N. Legge; and R. W. Gurtler. 1980. *J. Electron. Mat.* 9:841.

99. A. Baghdadi and R. W. Gurtler. 1980. *J. Crystal Growth* 50:236.

100. B. L. Sopori. 1981. *J. Electronic Mat.* 10:517.

101. D. L. Barrett; D. R. Hamilton; and E. H. Meyers. 1971. *J. Electrochem. Soc.* 118:952.

102. J. L. Hurd and T. F. Ciszek. *J. Crystal Growth.* To be published.

103. The Stepanov method was originally described by A. V. Stepanov. 1959. *Zh. Tekh. Fiz.* 29:382, and used for Si later on: L. M. Zatulovskii and P. M. Chaikin 1969. *Izv. Akad. Nauk SSSR, Ser. Fiz.* 33:1998. The introduction of a wettable die gave birth to the EFG process, first published by H. E. La Belle and A. I. Mlavsky. 1971. *Mater. Res. Bull.* 6:581, and fully described by T. Surek; B. Chalmers; and A. I. Mlavsky. 1977. *J. Cryst. Growth* 42:453; K. V. Ravi. 1977. *Ibid.* 39:1; F. V. Wald. In *Crystals: growth, properties and applications.* Vol. 5. 1981. Berlin: Springer-Verlag, 149. CAST (Capillary Action Shaping Technique) is very similar to EFG. See T. F. Ciszek and G. H. Schwuttke. 1975. *Phys. Stat. Sol. a,* 27: 231. The inverted Stepanov, or PDG (Pending Droplet Growth), technique is a variant described by K. M. Kim; S. Berkman; H. E. Temple; and G. W. Cullen. 1980. *J. Crystal Growth* 50:212, and J. Ricard. 1978. *1st Photovoltaic Solar Energy Conf.* Dordrecht: D. Reidel, 882.

104. F. V. Wald. 1981. In *Crystals: growth, properties and applications.* Vol. 5. Berlin: Springer-Verlag.

105. A. S. Taylor; R. W. Stormont; C. C. Chao; and E. J. Henderson. 1981. *15th Photovoltaic Spec. Conf. Proc.* New York: IEEE, 589.

106. J. P. Kalejs and F. V. Wald (Tyco C°). Private communication.

107. C. V. H. N. Rao; M. C. Cretella; F. V. Wald; and K. V. Ravi. 1980. *J. Crystal Growth* 50:311.

108. K. V. Ravi; R. C. Gonsiorawski; A. R. Chaudhuri; C. V. Hari Rao; C. T. Ho; J. I. Hanoka; and B. R. Bathey. 1981. *15th Photovoltaic Spec. Conf. Proc.* New York: IEEE, 928.

109. J. Ricard (Pechiney). Private communication.

110. K. I. Arai; N. Tsuya; and T. Takeuchi. *14th Photovoltaic Spec. Conf. Proc.* New York: IEEE, 31.

111. M. S. Alaee and T. Hashem. 1978. *3rd Photovoltaic Solar Energy Conf.* Dordrecht: D. Reidel, 584.

112. R. Suryanarayanan and G. Zribi. "Comm. int. conf. on polycrystalline semiconductors." Sept. 1982, Perpignan. *J. Phys.* 43:C1-375.

113. J. D. Zook; B. G. Koepke; B. L. Grund; and M. H. Leipold. 1980. *J. Crystal Growth* 50:26; S. B. Schuldt; J. D. Heaps; F. M. Schmit; J. D. Zook; and B. L. Grung. 1981. *15th Photovolt. Spec. Conf. Proc.* New York: IEEE, 934.

114. According to J. D. Zook. 1981. *3rd Photovoltaic Solar Energy Conf.* 1981. Dordrecht: D. Reidel 589, a diffusion length of 39 μm has been found within a grain, while the grain-boundary recombination velocity was s = 10^5 cm/s.

115. C. Belouet; J. J. Brissot; R. Martres; and Ngo-Tich Phuoc. 1977. *1st Photovolt. Spec. Conf.* Dordrecht: D. Reidel, 164; C. Belouet. 1980. *J.*

Crystal Growth 50:279; C. Belouet; J. Schneider; C. Belin; C. Texier; and R. Martres. 1980. *14th Photovoltaic Spec. Conf. Proc.* New York: IEEE, 49.

116. C. Belouet; C. Belin; J. Schneider, and J. Paulin. 1981. *3rd Photovoltaic Solar Energy Conf.* Amsterdam: D. Reidel, 558.

117. C. Texier-Hervo; M. Mautref; and C. Belouet. 1982. *4th Photovoltaic Solar Energy Conf.* To be published.

118. Gettering by $POCl_3$ (a glass used for n-doping) is described by A. Rohatgi; R. B. Campbell; J. R. Davis; R. H. Hopkins; P. Rai-Choudury; H. Mollenkopf; and J. R. McCormick. 1980. *14th Photovoltaic Spec. Conf. Proc.* New York: IEEE, 908; annealing for 1 h at 1100°C is found to have no effect on Mo, a partial effect on Ti (which is removed on about 20 μm thickness), a complete removing action on Fe. T. Saitoh; T. Warabisako; E. Kurado; H. Itoh; S. Matsubara; and T. Tokuyama. *Ibid.* 912, improve by more than 50% the efficiency of solar cells made from MG-Si, by grinding the wafer back surface and subsequent annealing (damage gettering). T. C. Chu[56] and R. G. Wolfson[78] anneal the substrate with HCl + H_2, before the CVD deposition.

119. H. I. Yoo; P. A. Iles; D. C. Leung; and S. Hyland. 1981. *15th Photovoltaic Spec. Conf. Proc.* New York: IEEE, 598.

120. These solid phases may be CVD-grown oxides, sprayed-on layers on screen-printable pastes. The use of screen printing for this step, as well as for metallization, has been proposed by L. Frisson; M. Honoré; R. Mertens; R. Govaerts; and R. Van Overstraeten. 1980. *14th Photovoltaic Spec. Conf. Proc.* New York: IEEE, 941.

121. See papers by P. Siffert and by E. Fabre, to be published in *4th Photovoltaic Solar Energy Conf.* Dordrecht: D. Reidel.

122. J. Michel; C. Fages; J. C. Muller; P. Siffert; D. Hoonhout; T. de Jong; and F. W. Saris. 1981. *3rd Photovoltaic Solar Energy Conf.* Dordrecht: D. Reidel, 713 have compared implantation and glow discharge. E. Fogarassy; R. Stuck; J. C. Muller; M. Hodeau; A. Wattiaux; M. Toulemonde; and P. Siffert. *Ibid.*, 639 have compared the methods using a deposited solid.

123. See numerous references in *Laser and electron beam processing of electronic materials.* 1979. Vol. 80-1. Proc. Conf. Los Angeles: The Electrochem. Soc., and also *Laser and electron beam processing of materials.* 1980. New York: Academic Press.

124. H. Goldman and M. Wolf. 1980. *14th Photovoltaic Spec. Conf. Proc.* New York: IEEE, 923.

125. R. O. Bell; C. T. Ho; K. V. Ravi; F. V. Wald; J. C. Muller; and P. Siffert. 1979. *2nd Photovoltaic Solar Energy Conf.* Dordrecht: D. Reidel, 153.

126. E. Fogarassy; R. Stuck; J. J. and A. Grob; and P. Siffert. 1980. In *Laser and electron beam processing of materials.* New York: Academic Press, 117.

127. R. Chabbal and M. Rodot. 1977. *1st Photovoltaic Solar Energy Conf.* Dordrecht: D. Reidel, 29.

128. *Solar photovoltaic energy conversion.* 1979. American Physical Society report of a study group headed by H. Ehrenreich.

129. *Shaped crystal growth* in *J. Crystal Growth* 50 (1980).

130. E. Fabre; C. Belouet, *3rd Photovoltaic Solar Energy Conf.* (Dordrecht: D. Reidel, 1981) 244.

131. Si monocrystals have been grown by Czochralski technique for three decades. Recent developments are presented by G. Fiegl, A. C. Bonora, *14th Photovoltaic Spec. Conf. Proc.* (IEEE, New York 1978) 303; G. Fiegl, *3rd Photovoltaic Solar Energy Conf.* (Dordrecht: D. Reidel, 1981) 209.

132. J. A. Van Vechten. 1980. *Handbook on semiconductors.* Vol. 3. Edited by T. S. Moss, S. Keller. North Holland, 1.

133. A. R. Bean and R. C. Newman. 1971. *J. Phys. Chemical Solids* 32:1211.

134. C. P. Khattak; M. Basaran; F. Schmid; R. V. D'Aiello; P. H. Robinson; and A. H. Firester. 1981. *15th Photovoltaic Spec. Conf. Proc.* New York: IEEE, 1432.

135. V. P. Boldyrev; I. I. Pokrovski; S. G. Romanovskaia; A. V. Tkatch; and I. E. Chimanovitch. 1977. *Fiz. Tekh. Poluprovod.* 11:1199.

136. A. Baghdadi; R. W. Gurtler; R. N. Legge; R. J. Ellis; and B. L. Sopori. 1978. *13th Photovoltaic Spec. Conf. Proc.* New York: IEEE, 363.

137. B. L. Sopori. 1981. *Proc. Symp. on Electronic and Optical Properties of Polycrystalline and Impure Silicon.* Electrochem. Soc., 66.

138. J. H. Aalbrets and M. L. Verheijke. 1962. *Appl. Phys. letters* 1:19.

139. W. P. Wilcox and T. J. La Chapelle. 1964. *J. Appl. Phys.* 35:240.

140. L. C. Kimerling; J. L. Benton; and J. J. Rubin. 1981. *Inst. Phys. Conf. Ser.* 5:217.

Advances in Solar Energy © 1983 American Solar Energy Society, Inc.

A REVIEW OF LARGE WIND TURBINE SYSTEMS

JAMES I. LERNER

Wind Energy Consultant, 420 Santa Ynez Way, Sacramento, CA 95816

and

HORST SELZER

ERNO-Raumfahrttechnik GmbH, Hünefeldstr. 1-5, D-2800 Bremen, W. Germany

INTRODUCTION

The objective of this article is to review the technical and economic status of large wind turbine systems which are nearing the stage of readiness for commercial operation. Owing to the comprehensive nature of this subject, an article of this length does not permit the elaboration required for in-depth coverage of both theory and results. Consequently, theoretical coverage will be provided only to the extent necessary to provide background to the subject of large wind turbines.

Initial emphasis is placed on some of the major findings to date and on the design trade-offs which illustrate that large wind turbines are designed on the basis of compromise. Economics of these systems is covered, and the major determinants are discussed: energy capture, production costs, and operations and maintenance costs.

As the programs in Europe and in the U.S. developed rather independently, this article tries to respond to the differences in the findings and topics of interest by two combined papers. Although the topics covered are the same, the presentation stresses different points of view. Contradicting values—like costs for maintenance and repair—indicate that there is still a great need for more test data in order to find the optimum solution.

JAMES I. LERNER: LARGE WIND TURBINE SYSTEMS IN THE UNITED STATES

THEORY

Aerodynamics

Two goals of the wind turbine designer are to select aerodynamically efficient airfoils and to be able to accurately predict their performance. During the past several years advances have been made in both of these goals. The latest airfoils which have been developed for general aviation show promise for wind turbine applications. A recent comparison of these airfoils with a more conventional airfoil used on the NASA MOD-0 100 kW turbine indicated that significant improvement in power output is possible at higher wind speeds.[1] Performance improvements in turbine output ranging from 17% at 25 miles per hour to 70% at 32 miles per hour were calculated for a 200-ft-diameter wind turbine. These improvements would result in increased annual electrical energy capture of approximately 5% for a site with highly energetic wind resources (mean annual wind speed equals 8 meters per second at 10 meters height). Of perhaps greater significance is the fact that at 32 miles per hour the high performance rotor is stalled over only 16% of its span, while the

Dr. James I. Lerner was educated at M.I.T., where he received a B.S. degree, and at Stanford where he received M.S. and Ph.D. degrees in aeronautical and astronautical engineering. From 1975 to 1981 he was senior technical advisor at the California Energy Commission, where he established and managed the California Wind Energy Program. In 1981 he established a full-time wind energy consulting practice. One of his clients, the City of Fairfield, is planning a 100 megawatt wind farm project with construction to commence in 1984 and full-scale operation in late 1985.

NASA 230 series airfoil is completely stalled. One possible problem which results with such efficient airfoils is the necessity to limit the power output so as not to exceed the generator rating. This is accomplished by regulating the blade pitch angle to "shed" power above rated power. For stall-regulated, fixed pitch designs, a problem arises with the need to provide additional means to control the rotor.

Attempts have been made to improve the prediction methods for estimating rotor performance. Classical aerodynamic theory is inadequate at higher wind speeds when the airfoils are at high angles of attack. This is of critical importance for fixed pitch and tip-controlled rotor performance, since the maximum power production is a critical design parameter. A semiempirical method, which has provided better agreement with measured values of power output, has provided theoretical power estimates within 10% of measured values for two large turbines with rotor diameter of 38 m and 61 m.[2]

Related to the question of rotor aerodynamic performance is the aerodynamic interference which occurs when wind turbines are sited in closely spaced arrays or clusters. This wake interference is an important factor in the design of turbine arrays. Numerical modeling recently performed by Lissaman et al., yielded the result that for a 36-unit square array of 91-m diameter, 2.5 MW rated, MOD-2 turbines at 10 diameter spacing, a loss of about 5% in annual energy production occurs for a site with a representative ambient turbulence intensity of 10% in a unidirectional wind.[3] The effect of reduced crosswind spacing for a rectangular array further increases the energy loss. Field tests to assess the accuracy of the model by measurement of the growth rate of the wake are currently being conducted by DOE on the three-unit MOD-2 array located at Goodnoe Hills, Wash.

Structural Dynamics

A fundamental requirement is the avoidance of resonance which occurs when the disturbing forces created by the rotation of the rotor excite the natural frequencies of the system. These natural frequencies are associated with the fundamental modes of vibration of the rotor, the drive train, and the tower. These phenomena have been successfully analyzed and predicted for designs such as the MOD-0 and MOD-1 turbines, with stiff tower and rigid hub, and for a soft design, such as the MOD-2, in which the machine passes through a 2/rev tower resonance going up to and coming down from operating speed. Analysis and field experience with the NASA MOD-0 and MOD-1 turbines is reviewed in a recent paper by Sullivan.[4]

The trend in large wind turbine design is reduction in cost through reductions in weight and dynamic loads. This trend will result in larger turbines with greater flexibility and greater number of rigid body degrees of freedom such as teetered rotors, free yaw, and softer towers. According to Thresher,[5] this gives cause for concern since the computer codes have not yet been completely validated for these designs in which greater motion is involved. In addition, atmospheric excitation, such as gusts, is more important for these more flexible structures.

Control

Horizontal-axis wind turbines require control for two primary purposes: to keep the rotor aligned perpendicular to the wind direction, and to maintain the proper power input to the generator. This can be accomplished either by active or passive means. Heading control for downwind rotors can be accomplished passively, that is, the rotor realigns itself whenever the wind direction changes. This means of control is utilized on the Hamilton Standard WTS-4, a 78-m (256-ft) diameter, 4-MW, wind turbine. Both analytical studies and tests performed on the NASA MOD-0 wind turbine confirm the effectiveness of this means of passive free yaw control.[6, 7]

Machines with upwind rotor, such as the Boeing MOD-2, require an active system to sense and maintain the proper heading. This requires the use of sensors to monitor wind direction and a hydraulically powered drive system to change the heading whenever average wind direction changes, resulting in parasitic power loss.

Excessive loads in high winds must be controlled for safe operation. This can be accomplished passively by a fixed pitch rotor, which is designed to stall at high wind speeds; however, the peak power results in very substantial rotor and drive train loads which add to weight and cost. In the conceptual design of the Boeing MOD-5B, a 128-m (420-ft) diameter, 7.2-MW turbine, a fixed pitch design was found to be less cost-effective than a variable pitch system even though the fixed pitch system was found to produce 5% more energy capture.[8]

Active pitch control is achieved by means of full-span or partial-span control. WTS-4 employs

Fig. 1. Hamilton Standard WTS-4.

full-span, while MOD-2 employs partial-span control.

Safety

Safety can be assured through redundancy, safe life design, and quality assurance during each phase of the project. Special attention was devoted to safety in the MOD-2. A failure modes and effects analysis was conducted in which over 750 failure modes were analyzed, and cor-

rective actions were implemented.[9] Loss of control of the rotor is perhaps the most obvious potential failure. The MOD-2 pitch control surfaces are independent, and either one of the two surfaces is capable of slowing the rotor.

Another probable failure mode of the MOD-2 is the rapid progression of a fatigue crack in the steel blade which could result in the catastrophic loss of the rotor. This potential failure is prevented by means of a crack detection system

Fig. 2.   DOE/NASA 2500 kW Experimental Wind Turbine, Goodnoe Hills, Wash.

capable of detecting a 5-cm crack 30 days prior to the time the crack reaches a 12-cm critical length.

The importance of quality control cannot be overestimated. The failure of a hydraulic actuator due to hydraulic fluid contamination resulted in a costly failure of a MOD-2 turbine in June 1981. Extensive damage to the rotor quill shaft

and generator required that these items be replaced and necessitated a thorough design review and modifications to prevent similar failures.

WIND CHARACTERISTICS
For large wind turbines, the average wind energy available over the rotor disc is the key deter-

minant of the suitability of a candidate turbine site from the standpoint of annual energy production. Predictions of annual energy capture based on measurements made at a 10-m height and an assumed one-seventh power law variation of wind speed with elevation could seriously overestimate the energy capture. For example, a turbine such as the MOD-2 with 61-m hub height would experience approximately two-thirds the annual incident energy for a wind shear exponent of .07 compared to the energy for a one-seventh power law. These concerns are highlighted by Simon in connection with measurements performed by Pacific Gas and Electric Company in Northern California in which an abundance of hours during the energetic summer wind season experienced a decrease of wind speed with increasing height.[10]

Atmospheric turbulence results in cyclic loads which limit the lifetime of wind turbines. These turbulence inputs must be considered in the structural design of the wind turbine. Thresher has determined that for a simplified wind turbine dynamic system model the inclusion of the turbulent gradients across the rotor result in a significant increase in the number of peaks per unit time which would tend to decrease the predicted tower fatigue life compared to predictions in which gradient terms are excluded.[11]

DESIGN

Design Criteria

The most important design goal is cost of energy. All other objectives are secondary. To illustrate, consider the first large wind turbine that was designed with economics as the key design goal, the MOD-2, 2.5-MW, horizontal-axis wind turbine designed and fabricated for DOE and NASA by Boeing Engineering and Construction Company. The MOD-2 was designed with a cost goal of 4¢/kWh (1977 U.S. dollars), which translates to about 6¢/kWh in current 1983 dollars. The following assumptions were made:

- 100th unit turnkey cost
- dedicated production facility producing 20 turbines per month
- utility economics with fixed charge rate = 0.18
- 25-unit wind farm
- cost includes 69KV transformer, excludes land
- site mean annual wind speed at 9.1-m height, 6-¼ m/s (14 mph).

Fig. 3. Boeing MOD-2 2.5 MW wind turbines located at Goldendale, Wash. The tip controls are rotated to the "feather" position while the turbines are parked awaiting the next period of steady winds. (Photo by Dr. James Lerner, October 31, 1981)

Other design criteria for MOD-2 were:
- 30-year service life
- maximum survival wind speed at 9.1-m height, 53.6 m/s (120 mph)
- minimum on-line availability = 0.90
- ability to withstand environmental stress including lightning strikes, seismic disturbances, extreme temperatures, hail, snow, and ice
- ability to operate remote and unattended.

Based on these criteria, an in-depth analysis was conducted of design trade-offs with the primary objective of obtaining a design for least cost of electricity. Detailed results of these trade studies are presented in the MOD-2 preliminary design report.[12]

Similar exercises were performed by Hamilton-Standard in the design of the WTS-4, and are summarized by Doman.[13] The design of a large wind turbine is a very complex process due primarily to the many choices which must be made. Some of the more important choices are listed in the following table, which represents only a portion of the choices which must be made:

Large Horizontal-Axis Wind Turbine
Design Trade-Offs

Rigid vs. Teetered Rotor
Fixed vs. Variable Pitch Rotor
Partial vs. Full-Span Rotor Control

Upwind vs. Downwind Rotor
Metal vs. Fiber glass Rotor Material
Free Yaw vs. Yaw-Driven
Tilted vs. Non-Tilted Rotor
Epicyclic vs. Parallel Shaft Gearbox
Single Speed vs. Multi-Speed Generator
Synchronous vs. Induction Generator
Soft vs. Stiff Tower
Ground Computer vs. Nacelle Microprocessor
Optimum Machine Size

Rotor Blades (Materials)

A variety of materials have been utilized for blades ranging in size from 19 to 46 m. Blade design is important since the rotor is currently the major contribution to the cost of electricity. For the MOD-2 the rotor represents approximately 25% of the cost. Substantial improvements are anticipated for the larger, third generation MOD-5 turbines currently under development.

Aluminum blades using aircraft technology for construction were utilized on the 38-m diameter MOD-0 and -0A machines developed and tested by NASA. Since the cost in high-rate production was not expected to drop below $40 per pound, NASA began to develop low-cost blade technology. These developments have focused on a variety of sizes, materials, and manufacturing methods and are described in Refs. 14 and 15. The leading candidates are summarized in Table 1.

Material	Description	Estimated Cost
Steel	Boeing MOD-2, 91-m diameter rotor consisting of five sections bolted together. Welded construction. Sturdy, low cost.	$5/kg ($2.25/lb)
	No major problems after 2,000 h of operation on the first three prototypes. Boeing has selected steel for the fixed portion of the MOD-5B rotor (128-m diameter.	
Laminated Wood	Gougeon Brothers developed for NASA 19-m laminated, wood/epoxy composite blades for three of the MOD-0A turbines. Over 14,000 h of operation with no visible fatigue damage. GE has selected this concept for MOD-5A (122-m diameter).	$12/kg ($5.50/lb)
Fiber glass	Hamilton-Standard developed a 19-m filament-wound fiber glass blade for NASA in 1976. The company has invested heavily in a highly automated filament winding blade fabrication facility in conjunction with the WTS-4 program.	Not Available

* Manufacturing cost, quantity production from Ref. 14 escalated to 1983 dollars.

Table 1 Large Wind Turbine Blade Concepts

Rotor Blades (Design Concepts)

Other aspects of blade design include number of blades, placement, pitch control, and connection to the hub.

Number of Blades (Two versus Three)

A three-bladed rotor is more efficient than a two-bladed rotor; however, the increased weight of the rotor and drive train result in a substantial increase in system cost. For the MOD-2, the three-bladed rotor results in an approximately 8% increase in cost of electricity.[12]

Placement Relative to Tower

The choice of an upwind or downwind rotor involves many trade-offs. The MOD-2 was designed as an upwind machine because the annual energy output was 2.6% greater than for a downwind design.[12] An unexpected benefit of this decision is reduced noise emissions relative to the downwind design. In contrast, the WTS-4 is a downwind design which allows free yaw. The discovery of disturbing noise emissions from the NASA MOD-1, a 2-MW downwind machine may dictate that greater emphasis should be given to upwind rotor placement in order to avoid the acoustic problems associated with tower shadow.

Pitch Control

Fixed pitch versus variable pitch has already been discussed with respect to control. A disadvantage of fixed pitch is the lack of both starting and over-speed control; hence, all large turbines in the U.S. have variable pitch. Partial span pitch appears to offer many advantages compared to full span in that the system is simplified and results in significant weight reduction. For MOD-2, the choice of a 30% span tip control results in a nearly 10% reduction in cost of electricity over full-span control.[12]

Blade Articulation (Teetered versus Rigid Rotor)

"Teetering" refers to the addition of one degree of freedom to the rotor so that the rotor is free to rotate in and out of the plane of the rotor disc, thereby reducing flap-wise cyclic bending moments from gusts. For the MOD-2, selection of a teetering versus fixed rotor resulted in a system weight reduction of 61,000 lb (28,000 kg) or about 10% of total system weight.[12] This results in a 4% reduction in cost of electricity. A similar choice was made for the WTS-4. In addition, for the downwind design with free yaw, the teeter hinge can be modified so that the teeter axis is inclined. Proper selection of this angle corrects the heading trim error and causes the rotor to face the wind.[6]

Tower

The primary choice to be made is soft versus stiff tower. A soft tower is characterized by having a fundamental bending mode between one and two cycles per revolution for a two-bladed rotor, while a stiff tower is characterized by having a fundamental frequency that is greater than two times the rotor operating frequency. The soft tower is designed so that its natural frequency is low enough so that the tower can attenuate the two per revolution rotor loads. The soft tower selected for MOD-2 is a tubular steel shell of variable thickness which weighs about one-half the weight a stiff truss and costs about one-third as much as the truss. This weight reduction results in a 23% reduction in cost of electricity for MOD-2.[12] The start-up control system for a soft tower design requires prevention of continuous idling at or near tower resonance, which might occur under certain wind conditions, thereby causing large tower oscillations to build up. A secondary benefit of the soft tower design is improved aesthetics of a slender cylindrical tower compared to a four-legged truss tower.

Drive Train

The drive train assembly consists of a low-speed shaft, a quill shaft, a gearbox, and a high speed shaft leading to the generator. After the rotor, the drive train is the second largest contributor to the cost of electricity.

Gearbox

A primary decision is the selection of a parallel shaft, epicyclic, or a hybrid design. For the MOD-2, a trade study revealed that a lightweight, compact epicyclic gearbox manufactured by STAL LAVAL (Sweden) weighing 35,000 lb would be more suitable than a 110,000 lb parallel shaft gearbox. The parallel shaft gearbox would increase the cost of electricity by 16%.[12] For the GE MOD-5A, a hybrid design is being developed in which the first stage is an epicyclic, and the succeeding stages are parallel shaft gears.[16] A thorough review of problems and design approaches associated with gears for large wind turbines is found in Ref. 17.

Quill Shaft

For the MOD-2, rotor torque is transmitted to the gearbox through a soft quill shaft. This soft shaft is designed to reduce drive train torque fatigue and improve power quality of the generator by isolating the gearbox and generator from the torsional oscillations of the rotor. This quill shaft acts as a soft spring in torsion and absorbs moments caused by wind gustiness and changes in wind direction.

Generator

Two major choices are variable versus constant speed and synchronous versus induction generator.

Variable versus Constant Speed (or Multi-Speed)

Since rotor aerodynamic performance is optimized by operating at an optimum ratio of tip speed to wind speed, a variable speed rotor, by adjusting its rotational speed as wind speed changes, captures more energy than a constant speed rotor. In order to take advantage of this, the generator must be modified to accept a variable speed input and convert this to a regulated, constant frequency electrical output. This option was considered and rejected for the MOD-2 owing to the poor efficiency and higher cost of the electrical equipment which could not overcome the slight improvement in annual energy capture compared to a fixed speed system. The variable speed system results in a 17% increase in cost of electricity for MOD-2.[12] For the Boeing MOD-5B, a variable speed generator was selected over a constant speed generator because the increase in annual energy production more than compensates for the additional cost of the variable speed system.[8] One reason for this selection was the discovery that the variable speed system was so effective at reducing the two-per-revolution torque oscillations, which permitted a deletion of the quill shaft and a reduction in the gearbox size.

Synchronous versus Induction Generator

There do not appear to be overwhelming advantages of either system. Both types have advantages and disadvantages. The induction generator is considerably simpler, while the synchronous generator requires complex controls. In addition, although induction generators are less costly, they require additional equipment to control power factor, thus negating any cost advantages. For MOD-5B, Boeing has selected a variable speed induction generator.[8]

Lifetime

Large wind turbines must be designed and built to operate reliably and economically for a sufficient length of time to recover the initial capital investment and operation and maintenance costs. A design goal of 30 years' life, typical of conventional utility electric power plants, has been adapted for large wind turbines. During this period of time, the dynamically loaded structures of the wind turbine will experience approximately 200 million cycles of loading, thus, fatigue is a major factor in the design of large wind turbines. This requires the determination of allowable stress to insure 30-year life. Stress analyses or component tests must be performed on the various structural elements to determine safe allowable stress levels. An example of the exhaustive work that must be performed is the development tests for the MOD-2, in which material and component tests were performed to extend the fatigue data base for materials (steel) and to test key components in order to verify their static strength, fatigue characteristics, and operational characteristics.[18]

A specially designed, crack detection system for MOD-2 was developed to detect cracks in the rotor blade and shut the system down prior to catastrophic failure of the rotor. The development of this system required tests to determine flow rate through a crack and fracture toughness in order to evaluate the ability of the system to detect cracks prior to reaching a critical length. For example, a crack at a critical portion of the blade will grow to critical length in 10 days. The system has the capability to detect a change in flow rate at least 10 days prior to the crack length growing to critical size.

The various tests that were performed on the MOD-2 components are described in detail in Ref. 18. These included tests on every major component.

Reliability and Maintenance

The prevention of catastrophic failures and the reduction to acceptable levels of critical and less severe failures is a goal of reliability and quality assurance for large wind turbines. An exhaustive failure modes and effects analysis covering over 750 failure modes was conducted for MOD-2 to quantify the mean time between failures.[9] Redundancy and fail safe design was used to preclude catastrophic damage. An availability analysis was performed to verify that the MOD-2 availability goal of .96 is realistic. This can only be achieved if equipment failures are easily repaired and the system is quickly restored to operational status. The MOD-2 availability analysis revealed that approximately 12 system failures would occur per year with an average time to repair of 21 h per failure resulting in 250 h of downtime per year for unscheduled maintenance.

Maintenance concepts for a 25-unit wind farm of MOD-2 machines were developed during preliminary design and published in 1979.[12] Owing to the lack of operating experience, these concepts have not yet been verified. An estimate of maintenance requirements for an early commercial wind farm, consisting of megawatt-scale machines, is summarized in Table 2.

Maintenance will consist of preventative and unscheduled maintenance. An estimate of the labor requirements is summarized as follows:

	Early Commercial Operation	Mature Technology
Scheduled	100	73
Unscheduled	630	270
Administrative tasks	75	38
Total manhours	805	381

Table 2 Annual Maintenance Labor Estimates* (25-Unit Cluster, Repair and Inspect Time, Manhours per Year per Unit)

- Labor   (25-unit cluster)
  - Two-man crew
    - Two-shift coverage
    - Six days per week (20% overtime)
  - One-man, in-plant support
  - Major maintenance support crew*
    (Five men, 20 trips per year, 12 days per trip)
  - Travel*
    (Six men, 20 trips per year)

- Services
  - Use of outside services for shop repairs, special tasks
  - Heavy equipment rental

- Spares
  - 100% spares availability
  - Major items stored at facility

- Maintenance Equipment
  - Portable tools and fixtures
  - $350,000 per farm (1983 dollars)

Table 3   Estimated Annual Maintenance Requirements for a Wind Farm

---

*The estimate for mature technology is based on Ref. 12, while the early commercial operation is based on Table 3.

ECONOMICS

Development Costs
The total cost associated with engineering, fabrication, and prototype testing for a large-scale wind turbine varies considerably depending on the size of the turbine, degree of technical risk, and contractor. The range would be from about $1 to $2 million for an intermediate size turbine (30- to 40-m diameter) to about $15 to $25 million for a large size turbine (100- to 125-m diameter). These prices include procurement of a single prototype for testing and do not include any contingency for major modifications.

Series Production Costs
The key to an economic wind turbine is low unit cost. Second unit costs on the order of $3 million per megawatt installed do not result in attractive economics for commerical operation. For large wind turbines, significant price reductions result from mass production in which savings can be obtained by reducing the amount of labor required and by obtaining discounts from vendors which supply key components such as gears, generator, and blades. These price reductions begin to become noticable for an initial order of 25 to 50 machines, although significant reductions occur at about the one-hundredth unit in a high volume production run from a dedicated production facility which is capable of producing 10 to 20 units per month.

The capital investment required to establish dedicated production facilities for the Boeing MOD-2 is estimated to cost about $47 million (1982). In addition, approximately $5 million (1982) would be required for tooling.[19]

A detailed cost analysis has been made for one-hundredth production unit costs for MOD-2 assuming the following:

- 25-Unit wind farm cluster;
- one machine per month installation;
- generally flat sites with few natural obstacles;
- land cost not included;
- soil easily prepared for foundation;
- transportation distance 1,000 mi;
- no escalation or interest during construction.

The results escalated from the mid-1977 dollar estimates are summarized in Table 4.[2]

---

*The major maintenance crew would initially utilize personnel employed at the manufacturing facility. As the industry matures, this would be provided by local service personnel, and travel would be eliminated.

Turnkey Account	Cost
Site preparation	$ 240,000
Transportation	43,000
Erection & checkout	202,000
Rotor assembly	485,000
Drive train	559,000
Nacelle subassembly	271,000
Tower subassembly	400,000
Initial spares & maintenance equipment	52,000
Nonrecurring (amortization of plant and equip.)	52,000
Total initial cost	2,304,00
10% profit	231,000
Total Turnkey Cost	$2,535,000

Table 4. MOD-2 One-Hundredth Production Unit Costs (Mid-1982 Dollars)

It should be emphasized that these costs are for a mature technology in mass production. For the first 25 to 50 units in a limited quantity production run, the actual costs are likely to be two to two and a half times these estimates. For example, the estimated unit costs for 40 MOD-2 wind turbines for the Bureau of Reclamation's proposed 100-mega watt Medicine Bow Wind Project are $6.25 million per turbine (mid-1981 dollars).[20]

Site-Related Costs

In addition to these manufacturer's turnkey costs, site-related costs must be added. These include land, access road construction, electrical interconnection, buildings, and remote control equipment. These costs are highly site-dependent and are largely determined by terrain and remoteness. A range for MOD-2 site-related costs (1982 dollars) is $250,000[19] to $675,000.[20] Both of these estimates are for flat terrain with no obstacles.

Depending on the length of time required to complete the construction of the project and the construction cost, interest during construction could be a significant cost item. This interest cost can be estimated by assuming that the project requires two years to construct and that half the total cost must be financed for one year at 15% interest. A range of costs for a wind farm is summarized in Table 5.

Table 5 illustrates a range of possible costs depending on both the production volume of the turbine and the site-related costs. For both extremes, the site-related and interest costs represent about 16% of the total project cost.

Operation and Maintenance Costs

These costs include labor, travel, services, consumables, and spare parts replacement. Since there is very little operating history with large turbines, these costs are difficult to estimate. For mature technology, these costs are estimated to be less than 1% of the total capital costs. For the MOD-2, Boeing has estimated that these annual costs would be about 0.9% times the turnkey unit cost.[12] For an early commercial wind farm, these annual costs are likely to be 2% to 3% times the capital cost. For example, the labor for early commercial units is estimated in Table 2 to be more than double the labor for a mature technology. A major unknown is the frequency of unscheduled maintenance which results in increased costs for both labor and spare parts. A comparative figure of merit for annual maintenance costs for large commercial jet transport aircraft is 3% to 4% times initial capital cost; however, large wind turbines are considerably less complex and do not have to cope with the high temperature environment of the turbo-jet engine. If the jet engine maintenance costs are eliminated, the remaining annual maintenance cost for these aircraft is 1.5% to 2% times capital cost. This is a likely target for large wind turbines.

Cost of Electricity

The cost of electricity for a utility-owned wind farm is given by the following formula:

$$COE = \frac{FCR \cdot IC + \overline{AOM}}{AEP}$$

where   FCR = Levelized fixed charge rate which includes return on capital (profit), taxes, and in-

	High Volume Production Low Site Costs	Low Volume Production Moderate Site Costs
Turnkey turbine cost	$2,535,000	$6,250,000
Site-related cost	250,000	675,000
Interest during construction	225,000	520,000
Total unit cost	$3,010,000	$7,445,000

Table 5. Unit Cost for a MOD-2 Wind Farm

surance. FCR = 18% for an investor-owned utility.

IC = Initial capital cost.

$\overline{AOM}$ = Levelized annual operations and maintenance costs. For 30-year life and 8% annual inflation rate, the levelized costs are 2.5 times initial annual O & M costs, AOM.

AEP = Annual energy production.

A wide range of annual maintenance costs and initial capital costs are possible. This equation can be used to calculate cost of electricity given values for IC, AOM, and AEP. A summary of calculations made for assumed values of these parameters is given in Tables 6 and 7.

Assume the following values for AEP and AOM for Table 6:

AEP = 8 million kWh per year. This would be typical for a site with 7 m/sec annual average wind speed, and includes allowances for aerodynamic cluster losses and line losses.

AOM = 2% · IC.

The equation can also be used to study the sensitivity of cost of electricity to annual energy production.

Assume for Table 7:

IC = $7.5 million
AOM = 2% · IC.

Both of these examples illustrate the wide variations in economics given a range of capital cost or energy capture. Either high capital costs, low energy capture, or high annual O & M could prevent these systems from being economic. As the cost of energy increases, the breakeven time approaches the useful lifetime of the machine.

IC	COE	Time to Breakeven*
(Millions $)	(¢/kWh)	(Years)
3	8.6	5
4	11.5	9
5	14.4	12
6	17.3	14
7.5	21.5	17

Table 6. Cost of Electricity and Breakeven Time for a MOD-2

---

*Assumes initial fuel costs are equal to 6¢/kWh and escalate at the rate of inflation, 8%.

AEP	COE	Time to Breakeven*
(Millions kWh/yr)	(¢/kWh)	(Years)
10	17.3	14
8	21.6	17
6	28.8	21
4	43.2	26

Table 7. Cost of Electricity and Breakeven Time for a MOD-2

CURRENT STATUS

Three generations of large-scale wind technology are currently in various stages of development in North America, primarily under the sponsorship of the U.S. Department of Energy and the Canadian National Research Council, and a number of privately funded large turbine tests are currently under way. The status of the technology has passed from concept development to performance verification, prior to early commercial operation. Table 8 summarizes the major developments to date.

A great deal of experience has been gained in the DOE/NASA MOD-0A program. During three and one-half years and 28,000 hours of operation, these four machines have proven beyond any doubt that large wind turbines can operate reliably and fail-safe on an unattended basis, that the units can be maintained, and that there are no special problems with grid compatibility or public acceptance.[21] Based on this experience, the Westinghouse Electric Corporation plans to produce a commercial machine rated at 500 kilowatts. A 30-MW wind farm is planned for the San Gorgonio Pass in Southern California.

Currently, two companies are testing MW-scale machines that appear to have commercial viability. The key characteristics of these turbines are summarized in Table 9.

The Boeing MOD-2 has already experienced more than one year of testing, while the WTS-4 is just beginning its test experience. Boeing is gaining more confidence with its system, having successfully solved a problem with the hydraulic pitch control system that damaged a generator on one of its units. Barring any unforseen problems, the MOD-2 could be in limited quantity commercial production by late 1984. One of its first projects may be a planned 100-MW wind farm near Fairfield, Calif., with a wind farm development company. Site permitting is already underway, and the first units could be sited as early as 1984 with project completion by late 1985.

Model and Manufacturer	Size and Rating	Status
First generation MOD-0A Westinghouse	38 m dia. 200 kW downwind	Four units have accumulated over 28,000 h of valuable operating experience. Westinghouse plans to develop and market 500 kW units.
MOD-1 General Electric	61 m dia. 2 MW downwind	Experimental unit to test MW-scale design concepts and evaluate environmental impacts.
Second generation MOD-2 Boeing Engineering	91 m dia. 2.5 MW upwind	One unit in Sweden and one unit in U.S. beginning test. Commercial projects likely by 1984/85.
WTS-4 Hamilton Standard	78 m dia. 4 MW downwind	Three MW units in Sweden and four MW units in U.S. beginning test. Commercial projects likely by 1984/85.
Third generation MOD-5A General Electric	122 m dia. 6.2 MW upwind	Preliminary design complete. Prototype test to begin in 1984.
MOD-5B Boeing Engineering	128 m dia. 7.2 MW upwind	Preliminary design complete. Prototype test to begin in 1984.
AEOLUS To be determined	64 m dia. · 96 m ht. 3 MW	Preliminary design complete. First operation planned in late 1984.

Table 8.   Status of Large-Scale Wind Turbines in North America

A "third" generation of large wind turbines is presently under development that would result in a 25% improvement in cost of electricity compared to MOD-2. If full funding is provided by Congress, the MOD-5 program will result in the development and testing of two prototypes with rotor diameter in excess of 120 m. Preliminary design has already been completed, and the prototypes should be ready for testing starting in 1984.[8,16]

NATIONAL PROGRAMS

The United States has allocated to date $200 million in research and development on wind turbines. The program has been managed by the Department of Energy, with NASA Lewis Research Center acting as program manager for large wind turbines.

In 1980 Congress approved a Wind Energy Systems Act to accelerate the pace of wind energy commercialization and lead to the siting of 800

Feature	Boeing MOD-2	Hamilton Standard WTS-4
Hub height	61 m (200 ft)	80 m (262 ft)
Rotor diameter	91.4 m (300 ft)	78.3 m (256 ft)
Rated power	2.5 MW	4 MW
Rated wind speed*	12.3 m/s (27.5 mph)	15.1 m/s (33.9 mph)
Cut-in wind speed*	6.3 m/s (14 mph)	6.9 m/s (15.4 mph)
Cut-out wind speed*	20.1 m/s (45 mph)	27 m/s (60 mph)
Rotor placement	Upwind	Downwind
Rotor tip speed	84 m/s (188 mph)	123 m/s (275 mph)
Transmission type	Planetary	Planetary
Generator type	Synchronous	Synchronous
Tower type	Steel shell	Steel shell
Rotor material	Welded steel	Filament-wound fiber glass
Total weight	263,000 kg (580,000 lb)	268,000 kg (591,000 lb)
Estimated Cost**	$6.5 million	$6.8 million

Table 9.   Comparison of Early Commercial Megawatt-Scale Wind Turbines

* Hub Height
** Ref. 20

MW of systems by the end of 1988. The strategy involves government procurement and cost-shared projects. The current administration has slowed the pace of the Federal Wind Program, reduced the budget to a maintenance level, and emphasized research and development on a very modest scale. Other legislation which provides for a 15% business energy tax credit and a five-year depreciation for equipment has done a great deal to encourage private sector investments in wind projects.

The appropriate federal role for commercialization of large wind turbines was the subject of a workshop sponsored in the fall of 1980 by the National Science Foundation in which a panel of experts concluded that continued involvement of the federal government is needed in order to insure that wind energy reaches its potential. This involvement should include a mix of technology development, federal power purchases, information dissemination, and energy production subsidies to reward projects which actually produce energy.[22]

The activities managed by NASA will continue. This will include

- continued use of the MOD-0 test machine to evaluate new design concepts;
- continued testing of MOD-2 and WTS-4;
- cost-shared development with industry of MOD-5;
- continued technology development support.

RESEARCH AND DEVELOPMENT TOPICS

NASA will continue technology development activities in the following areas:

- aerodynamics
  better performing airfoils, aileron tip controls;
- structural dynamics
  better understanding of teetered rotor;
- materials
  wood rotor blades;
- generators
  variable speed generator development and testing.

The major problem in wind energy is cost-effectiveness which can be solved by improvements in energy capture, reliability, and cost. Energy capture improvement can be achieved partly by technological advances in machines and partly by improvements in selecting eco-nomic sites. This latter advance is often overlooked. Development of cost-effective wind measurement equipment is an important ingredient. Improvements in reliability can be made as more experience is gained with the existing machines, and modifications are made. This can be accomplished by rigorous testing, thorough documentation, and widespread dissemination of results. Standardized testing procedures and certification programs will help to remove some of the uncertainties about various machines being sold commercially. In the absence of a strong Federal Wind Program, this requires a more aggressive industry program than is currently in place under the auspices of the American Wind Energy Association.

Improvements in cost will come chiefly as units are produced in quantity, and as competition begins to develop. In addition, low-cost materials development and weight reduction in next generation designs will help to reduce the cost per kilogram as well as the total number of kilograms per wind machine.

REFERENCES

1. B. F. Habron. et al. 1980. Wind-turbine power improvement with modern airfoil sections and multiple-speed generators. AIAA/SERI Wind Energy Conference. April 9-11, 1980. Boulder, Co. AIAA Paper No. 0633, New York.
2. L. Viterna and D. Janetzke. 1981. Theoretical and experimental power from large, horizontal-axis wind turbines. *Proceedings of the Fifth Biennial Wind Energy Conference and Workshop*. (WWV). Washington, D.C.; October 5-7, 1981. SERI/CP-635-1340, CONNF-811043, Vol. 265.
3. P. B. S. Lissaman. et al. 1982 (June). Numeric modeling sensitivity analysis of the performance of wind turbine arrays. Department of Energy/Pacific Northwest Laboratory Contractor Report, D. E. 82027570. PNL-4183.
4. T. Sullivan. 1981. A review of resonance response in large, horizontal axis wind turbines. *Wind turbine dynamics*. Proceedings of a Workshop Sponsored by Department of Energy and NASA. Cleveland, Ohio; Feb. 24-26, 1981. DOE Publication CONF 810226, 237.
5. R. W. Thresher. 1981. Structural dynamic analysis of wind turbine systems. *Proc. of the 5th Biennial Wind Conf. and Workshop*. Vol. 3, 61.
6. B. M. Brooks. 1981. MOD-0 wind turbine dynamics test correlations. *Wind turbine dynamics*, 287.
7. J. C. Glasgow and R. D. Corrigan. 1982 (June). MOD-0 passive yaw test results. PIR No. 196. Cleveland, Ohio. NASA Lewis Research Wind Energy Project Office.
8. R. R. Douglas. 1981. Conceptual design of the 7 megawatt MOD-5B wind turbine generator. *Proc. of the 5th Biennial Wind Conf. and Workshop*. Vol. 1, 169.
9. W. E. B. Mason and B. G. Jones. 1981. Reliability and quality assurance on the MOD-2 wind system.

Proc. of the 5th Biennial Wind Conf. 2nd Workshop. Vol. 2, 151.

10. R. L. Simon. 1981 (July). Potential errors in using one anemometer to characterize the wind power over an entire rotor disc. *Proc. of Large Horizontal-Axis Wind Turbine Conference.* Sponsored by DOE and NASA. Cleveland, Ohio.

11. W. E. Holley and R. W. Thresher. Response of wind turbines to atmospheric turbulence. *Proc. of the 5th Biennial Wind Conf. and Workshop.* Vol. 2, 281.

12. MOD-2 wind turbine system concept and preliminary design report. Vol. 2. Boeing Engineering and Construction Company, contractor report. DOE/NASA 0002-80/2. July, 1979.

13. G. S. Doman. 1979. System configuration improvement. *Large Wind Turbine Design Characteristics and R & D Requirements, Workshop.* NASA Lewis Research Center; April 24-26, 1979. DOE Publication CONF-7904111, 385.

14. J. R. Faddoul. 1981. An overview of large horizontal axis wind turbine blades. *Proc. of the 5th Biennial Wind Conf. and Workshop.* Vol. 3, 113.

15. Large, horizontal axis wind turbine projects. NTIS # SERI/SP-732-730. Nov. 1981.

16. R. S. Barton and W. C. Lucas. 1981. Conceptual design of the 6 MW MOD-5A wind turbine genera-

tor. *Proc. of the 5th Biennial Wind Conf. and Workshop.* Vol. 1, 157.

17. S. H. Kratz and R. C. Metzger. 1981. High torque drive systems for large wind turbines. *Proc. of the 5th Biennial Wind Conf. and Workshop.* Vol. 3, 89.

18. J. S. Andrews and J. M. Baskin. 1981. Development tests for the 2.5 megawatt MOD-2 wind turbine generator. *Proc. of the 5th Biennial Wind Conf. and Workshop.* Vol. 2, 611.

19. A. Plaks, et al. 1981. (Jan.) Wind turbine system for high wind regions in California. Boeing Engineering and Construction Company contractor report. California Energy Commission Publication # P500-81-005.

20. L. L. Nelson. 1981. Medicine bow wind project. *Proc. of the 5th Biennial Wind Conf. and Workshop.* Vol. 2, 557.

21. A. G. Birchenough, et al. 1981. Operating experience with the 200 kW MOD-0A wind turbine generators. *Proc. of the 5th Biennial Wind Conf. and Workshop.* Vol. 1, 167.

22. J. I. Lerner. 1982 (July). Assessment of large scale wind system technology and prospects for commercial application. *Workshop on the Federal Role in the Commercialization of Large Scale Windmill Technology.* Sponsored by National Science Foundation. Washington, D.C.; Sept. 25-26, 1980. (Available from NTIS).

H. SELZER: LARGE WIND TURBINE SYSTEMS SEEN FROM THE EUROPEAN VIEWPOINT

BASICS OF WIND ENERGY CONVERTERS

Windmills have been used for more than 4,000 years, but only in this century has the knowledge of aerodynamics been able to provide the tools necessary to gain a high performance.

In 1926, Betz[1] published an analysis of the maximum power to be extracted out of an airstream:

$$P_{max} = 0,593 \; \frac{\delta V_0^3}{2} \; A_R$$

where P = power   in [W] = [Nm/s]
                              = [kgm²/s³]

  δ = ambient air
        density   in kg/m³

  V_0 = undisturbed   in $\frac{m}{s}$
        wind velocity

  A_R = rotor swept
        area   in m²

The "Betz coefficient" is limited to $C_{Pmax} = 0,593$ as any higher loss of energy would cause stagnation of the air flow. The optimum ratio of

air speed in front versus behind the rotor disc is two-thirds.

Later, Glauert[2] showed that the airstream behind the rotor has a rotational component inversely dependent upon the speed ratio of the rotor tip with respect to the airspeed. This influence induced the modern tendency to develop "fast runners" with high tip-speed ratios instead of the low tip-speed ratio used in old designs.

The next source for reducing the efficiency is the solidity of the rotor disc, for example, the number N of rotor blades. As all air molecules have to be slowed down, a low solidity will not succeed in reducing the total air speed optimally to two-thirds or will cause additional losses due to tip vortices.

A fourth influence stems from the aerodynamic lift-to-drag ratio of the profile C_L/C_D. As shown in Fig. 1, the aerodynamic quality has more importance for high than for low tip-speed ratios.

Profiles have been developed offering lift-to-drag ratios up to 200, as represented in Fig. 2. But they need an excellent surface quality in order to provide laminar flow. A rough surface will increase the drag, thus causing a reduction by 50% as exemplarily shown for the profile FX 77-W-153 used in the GROWIAN 3 MW WTS.

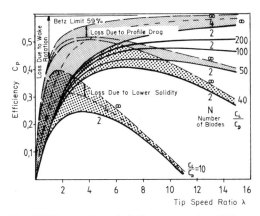

Fig. 1 Main aerodynamic influences on the efficiency of wind turbines.

In practice, the ideal shape of the rotor blade would cause high production costs as the dimension of the profiles in the inner section will increase due to the lower speed ratio. Therefore, linearized modifications are preferred, ranging from a simple linearization of the trailing edge through a linear trailing edge plus chord thickness distribution, to even a rectangular blade contour of the same chord length from root to tip.

There are three main reasons for this procedure:

- to simplify the production;
- to safe weight;
- to reduce air forces.

Computer programs exist in the aircraft and helicopter research and development centers to minimize the losses connected by such modifications. The open question is: What is optimal?

In a unit for energy production, the final measure will be the cost of the annual energy delivered. Losses in efficiency will only produce an effect below rated wind speed; this equals roughly 50% of the annual energy production. Hence, 5% losses due to the rough blade surface equal a decrease of annual energy by 2.5%. As two rotor blades may cost approximately 20% of the total WTS costs in a line production, the provision of an aerodynamically smooth blade surface might be worth 12.5% of increased production cost. This argument is often used the other way round, as the blades will collect some dirt during operation, the production quality can be adapted to the roughness level encountered during operation. Another example deals with the influence of blade linearization: the contour simplification of the blades for the AEOLUS 2MW WTS resulted in a loss of annual energy of 3.6%, outweighed by the savings in production.

Subjects for improvement include profiles with increased flap-bending stability, that is, great profile thickness, and data for thick profiles and high angle of attack.

DYNAMIC AND AEROLASTIC PROBLEMS

A high torque machine, like a WEC with 3 independent axis of rotation, is sensible to dynamic coupling of the various components, as Fig. 3 shows. The torsion and bending of the tower results in gyroscopic moments, Coriolis forces additionally being brought about by the bending of the tower. The blade deformation leads to changes in the aerodynamic loads, and this leads to changes in the torque of the main shaft and in the power generation.

Natural Vibrations

The analysis of all these deformations and motions starts with the calculation of the natural vibrations of the most important components in the absence of external forces. Normally, structural damping is included by empirical assump-

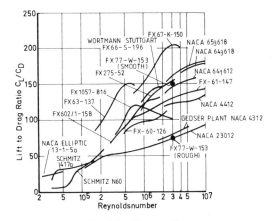

Fig. 2 Profile data.[60]

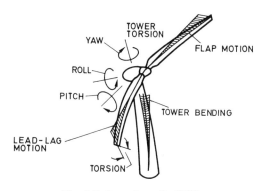

Fig. 3 Deformation of a WTS.

tions only. The mathematical treatment is to be found in the literature cited.

An example of the frequency spacing of these coupled components is represented in Fig. 4 for the 2 MW-WEK AEOLUS. At normal operation, the first tower bending was placed between the second and third mode of the rotor revolution, while the lowest blade bendings are found to be higher than the fourth respectively sixth rotor mode. The slight increase of the flap bending curve depends on the centrifugal forces.

The excitation forces may stem from several loads.

periodic loads as:

- rotor rotation (for example, centrifugal load);
- the air flow oblique to the rotor plane;
- wind shear;
- gravitational forces;

- out-of-balance condition if one blade is partially broken or iced;
- separation of the Karman whirls at the tower;
- short circuit of the generator;

impulse loads as:

- individual gusts;
- disturbances during operation;
- hard maneuvers (for example, braking);
- aerodynamic loads (for example, tower shadow);

random loads as:

- found in parked position.

The importance of the dynamic analysis stems from the magnification of loads if the frequency ratio approaches the resonance frequency. For example, the rotor rotation at 0.42 Hz would create an excitation force for the earlier design of the AEOLUS tower (first bending 0.6 Hz) with a magnification of 1.04 at frequency ratio 2 and a magnification of 50 if the rotation were increased to 0.60 Hz (frequency ratio 1.42).

Great efforts have been made during the last five years to better calculate models of the complete systems. Recently, König[3] succeeded in the development of a computer program coupling the loads, deformations, natural frequencies, generalized masses, damping (material and air), and joints and bearings of all major components, including five degrees of freedom into one model. If the test operation proves the validity of this program, a great improvement in the design of large WECs would have been achieved.

Aeroelasticity deals with interaction of elastic, aerodynamic, and inertial forces, of all constructions subject to air flow. The theoretical treatment is to follow the possible movements of the structure by:

- a linear system with one degree of freedom;
- a linear system with several degrees of freedom (SO) in a modal matrix;
- a system with nonlinear influences:
  large angles of movement
  gyroscopic moments in the Eulerian equations
  centrifugal accelerations
  Coriolis accelerations.

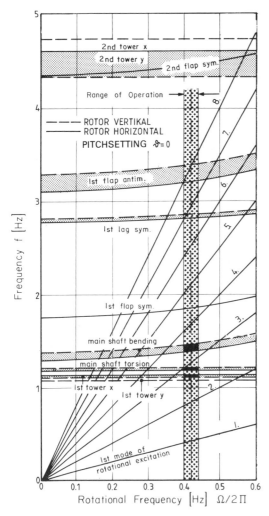

Fig. 4 Coupled frequency spacing of AEOLUS 2 MW-WTS.

Special attention must be given to the aerodynamic loads, as they might act as a driving force for undamped resonances because a torsional load to the blade might lead to a higher angle of attack causing higher lift, higher torsion, and so on. The fatal result could be the "flutter" instability which caused the rupture of the Tacoma Narrow Bridge on Nov. 7, 1940.

In Fig. 5, the aeroelastic load triangle is illustrated. The mathematical treatment cannot be described here in detail, and should be followed in the literature. The knowledge has been developed so far that tendencies for flutter problems can be detected already in the design phase.

Subjects for improvement include simplification of the programs to reduce computational time.

LOADS

The calculation of the loads is one of the most important tasks during the development of a WEC. Due to the modern engineering know-how in aircraft, and especially helicopter design, suitable computer programs are available for the determination of these forces and can be found in the relevant textbooks.

The essential problems are hidden in the assumptions for the various load cases (see Fig. 6), which have to include the external effects as well as the peculiarities of the special design:

- aerodynamics
- gravity
  structure mass
  ice
- inertial forces
  gyroscopic forces

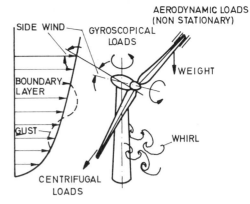

Fig. 6 Loads on a WTS.

Coriolis forces
control forces
- other
electric faults
bird collision
earthquakes.

The best collection of load cases has been published by the Swedish Board for Energy Source Development (NE)[4] and most of the following data stem from this source.

Load Cases During Powered Operation

Case 1 Normal operation;
periodic loads due to the wind duration profile and the design life time, including centrifugal and gravity loads, tower drag, cyclic actuation (if provided), and so on.
Case 2 Operation under turbulent wind conditions;
superimposed stochastic loads.
Case 3 Sharp gradient wind shear;
site dependent.
Case 4 Starting in normal operation, recommended number: 1,000 per year.
Case 5 Stopping in normal operation, recommended number: 1,000 per year.

Malfunctions during operation

Case 6 Blade angle fault.
Case 7 Wind turbine overspeeding;
recommended: 1.25 times normal rpm at 10 times per year.
Case 8 Emergency braking;
conditions: highest operational wind-speed and turbine overspeed.

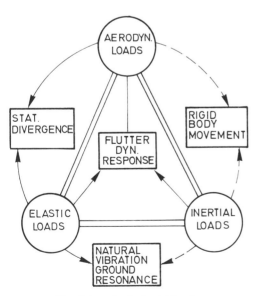

Fig. 5 Aeroelastic load-triangle.

Case 9 Loads due to electrical faults;
sudden cutoff of electric load;
short circuit.

Case 10 Ice loads during rotation;
ice formation up to 5% of blade weight,
sudden loss of ice deposits, balanced/
unbalanced cases.

Case 11 Loss of one blade;
might be reduced to partial loss of blade
structure.

Case 12 Bird collision;
should not cause damage to the primary
structure;
bird weight 4 kg.

Loads in Parked Position

Case 13 Extreme gale winds;
feathered blades parked horizontally
vertically/rotating freely; transversal
wind direction on WTS locked in yaw.

Case 14 Ice loads on parked WTS.

Case 15 Earthquakes;
site dependent; if possible, the WTS
should be stopped.

Additional load cases might result in special designs, vertical axis rotors (for example, vertical gusts), and WECs based offshore.

The great difficulty, or skill, for the load calculation of large WECs is to stay within realistic boundaries and not to include safety margins which are too high. Otherwise, loads of other cases might be influenced in an uncontrolled way, thus leading to difficulties in the resulting fatigue behavior. Typical mistakes also stem from the fact that forced movements of components (for example, teetering or hinged blades with automatic pitch compensation, or some operational conditions) have not been analysed which might be possible due to the control subsystem.

Subjects for improvement include investigation of gust load spectra and load optimization by control strategies.

DIMENSIONING

Once the shape of the blade and the external forces have been determined, the stress levels have to be calculated and converted into the adequate materials. For a quick survey, beam models are still suitable. It has been found that in most cases, the NASTRAN computer program can be used with sufficient accuracy. There are only rare cases which need to be calculated by Finite Element Methods.

Despite the availability of such tools, the design depends on assumptions still under discussion. Most components of a WEC have to be designed for fatigue loading. There are different practices in various countries for the final stress level of the Wöhler-curve after 10^6, 10^7, or 10^8 load cycles. For steel, 10^7 rather than 10^6 should be used. Corrosion will worsen the problem. For composite material, the fatigue values have to be still measured in most cases.

The applicable codes differ from country to country. Careful consideration must be given to satisfy the various norms for buildings, welds, bolts, and so on. The applicability of some details of these codes must be discussed in the future. As the codes have been developed for conventional structures and averaged quality level, they might conflict with the results of modern methods, like crack propagation calculation. Normally, the sensitive parts of WECs are designed and manufactured at such a high quality level and computational effort, that the right and appropriate compromise between a conservative and an aircraft industry-like approach, must still be found.

All stress levels exceeding the final stress level of the Wöhler curve will damage the material and have to be investigated by fatigue calculation. Although already published in 1945, the Miner rule[5] of cumulative damage is still the best tool for correlating the actual loading conditions with the fatigue life data. The total damage is gained by the sum of all ratios of the number of applied cycles to the number of cycles to failure of all relevant loads. Life of the material is exhausted if the sum equals one.

Figure 7 visualizes the relative damages within 30 years of the relevant load cases for the design life of the AEOLUS rotor blade.

While this Miner rule is a standardized and accepted procedure, a more sophisticated method might be applied, too: crack propagation calculation. Like FEM, the stress concentration at a surface or an internal flaw is followed until it has grown to total damage. The propagation of a crack is not influenced by stresses below the so-called threshold value and will be delayed by high stress pulses as they seem to smooth the sharp edges of such a crack. The application of the fracture mechanics to WECs needs more investigation as most of the published data refer to cases below the range of loads of rotor blades.

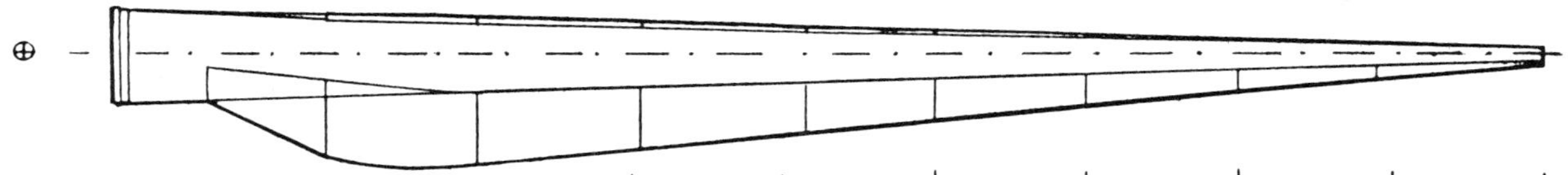

	0.1	0.2	0.3	0.4	0.5	0.6	0.7	0.8	0.9	CYCLES
START	0	0	0.0076	0	0	0	0	0	0	2.3×10^6
NORMAL OPER.	0	0	0	0	0	0	0	0	0	10^7
HIGH WINDSHEAR	0	0	0.0876	0.0931	0.0768	0.0942	0.0688	0.1145	0.1293	5.0×10^5
BRAKING soft	0.0048	0.0338	0.0866	0.0928	0.0750	0.0992	0.1004	0.0939	0.1119	2.7×10^5
BRAKING hard	0.0008	0.0070	0.0204	0.0244	0.0217	0.0296	0.0365	0.0272	0.0335	2.0×10^4
GUSTS, long	0.0995	0.2470	0.7979	0.7068	0.4143	0.4763	0.3050	0.4679	0.5249	
GUSTS, short	0	0	0	0	0.0001	0.0001	0	0.0002	0.0002	
	0.1043	0.2878	1.0	0.9171	0.5878	0.6994	0.5106	0.7037	0.7000	

Fig. 7 Relative damaging of the various stations within 30 years due to fatigue loads (Miner-Palmgreen method).

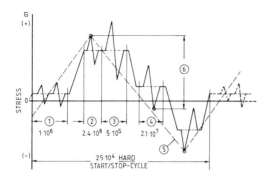

Fig. 8 Cycle of load collective for AEOLUS 2 MW-WTS.

The following legend applies to Fig. 8:

1. BLADE FEATHERED   90° BLADE →
                     45° BLADE →
2. NORMAL OPERATION 12,55 m/s
3. HIGH WIND SHEAR   12,55 m/s
4. NORMAL OPERATION 21,1 m/s
5. START /ROTATION / STOP-CYCLE
6. GUSTLOAD e.g. V = 17±4 m/s

With respect to inspection intervals, the fracture mechanics provides a very good basis for the determination of the allowable crack sizes and the applicable method of inspection.

In practice, the various loads are put together to form a load collective as a basis for the fracture mechanical and fatigue calculations. One representative load collective is shown in Fig. 8.

Subjects for improvement include the understanding of the Wöhler diagram after 10^7 load cycles and experimental and theoretical investigations for the crack propagation mechanism.

MATERIALS

A variety of materials can be chosen for the structure. In most cases, concrete or steel will serve for the tower with a slight tendency of favoring steel in the series production. But expected production and maintenance costs will decide the business.

For the blades, aluminium, steel, glass reinforced plastic, carbon reinforced plastic, and wood have been tested already, although with different intensity. The fatigue problems should favor a fiber material, but the lack of experience with the resulting thicknesses of composite material gave at least one reason that only one (WTS3) of the four types of large WECs used GRP. The other three (Mod. 2, AEOLUS, GROWIAN) selected steel for the main load carrying structure. Additional reasons stem from easiness of production and inspection, as well as

the mass of data collected through the centuries of application.

The rediscovery of wood caused a very high attraction, this time in laminated versions. Several experimental studies showed very promising results, thus, more field tests can be expected for the future.

WIND STRUCTURE

Heating, cooling, and rotation of the earth drive the movement of the atmosphere. Variations in the topography influence these streams of energy balancing.

For wind energy utilization, the turbulent boundary layer close to the surface is of greatest interest mainly depending on the "geostrophic" wind, that is, the undisturbed atmosphere above approximately 1,000 m over flat terrain in the Northern and Southern Hemisphere. The pressure differences drive the air towards the low pressure, but the initial momentum on Earth causes a deflection due to the Coriolis forces until the wind direction is parallel to the isobars. This situation can quite often be noticed on weather maps.

As the conventional wind mapping using the many data derived from measurements at 10 m height (see Fig. 9) yielded great difficulties in applying to the needs of large WECs, the Danish Wind Program funded an investigation to predict the wind profiles by starting from the geostrophic winds. This very successful approach cannot yet be used in mountainous areas, but might be applied to most of the coastal areas of interest.

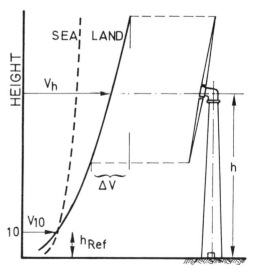

Fig. 9 Typical wind profiles.

The annual distribution of wind speeds, v, is mathematically expressed by the Weibull probability function:

$$f(v) = \frac{C}{A} \cdot \left(\frac{v}{A}\right)^{C-1} \cdot \exp\left\{-\left(\frac{v}{A}\right)^{C}\right\}$$

with C the shape parameter as
   C = 2 leads to the Rayleigh distribution,
   C = 3,5 approximates the Gaussion distribution
and A the scaling parameter.

Assuming a logarithmic scale for the height in a height versus windspeed plot, the extrapolation of the speed line to v = 0 results in the so-called roughness length. Four different roughness classes have been investigated.

The variation of those Weibull parameters A and C in accordance to the height was calculated for all four roughness classes (Fig. 10), and enables now a quick and rather precise determination of wind speed distribution everywhere in Denmark. A visual aid eases the classification (Fig. 11).

   As long as such a procedure is missing in other countries, the conventional approach has to be made, starting with the distribution at 10 m height and calculating the relevant wind speed at the hub height following a simple approximation like

$$\frac{v_{10}}{v_h} = \left(\frac{10}{h}\right)^{\alpha}$$

with α = 0.3 for rough terrain
and α = 0.12 for smooth surface

or more closely to the Weibull function, if sufficient data of the site are available.

   An open question still exists for the type of measurements applicable for WEC technology. Mean wind speeds are reported with averaging time intervals of a year, a month, a day, an hour,

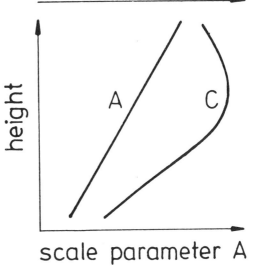

Fig. 10 Weibull parameter variation with height.

10 minutes, 1 second, and so on. The annual mean wind speed might be a good general classification of a site, but the actual performance of a rotor has to be based on the cumulative wind distribution during the year (see Fig. 12).

   In the Gedser test program (24 m ϕ) a good correlation between momentary performance and averaged wind speed was found in using the 10 min mean data. Yet, the reference basis needs more investigation, and more data on this subject are to be expected from the test and evaluation phases of the Nibe, Mod-2, WTS 3, and AEOLUS turbines.

   For the load assumptions, the gustiness of the wind has great importance. Precise measurements have been performed at several sites and correlated with power spectral forms given by Kaimal[6].

   Subjects for improvement include the extension of the Danish approach to other countries and mountain-regions, investigation of gustiness

Roughness Class	Terrain	Roughness Length	Relative Energy
0	Water areas	0–1 mm	10
1	Open country areas with very few bushes, trees, and buildings	1 cm	7
2	Farmland with scattered buildings and hedges with separation in excess of 1,000 m; strongly undulating terrain	5 cm	5
3	Built-up areas, forests, and farmland with many hedges	30 cm	3

Table 1: Types of terrain, roughness classes, and roughness lengths

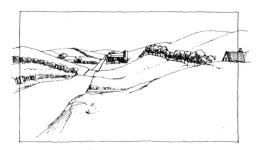

Fig. 11 Example of terrain corresponding to roughness class 2.[96]

of winds, and predictability of wind speeds for the next hour and the next day.

ENVIRONMENTAL IMPACTS

Mainly, two interferences have been observed: electromagnetic reflection and noise emission.

Large rotor blades with conducting surfaces may disturb electromagnetic signals. General observations are:

- The higher the frequency the more vulnerable is the transmission; the UHF channels of the TV signals are more sensitive than the VHF channels. AM/FM radio and TV audio are not effected.

- The interference severity decreases with increasing distance of the receiver from the WEC. In the worst cases, an influence was measured at 5 km (VHF) respectively 0.4 km (UHF) distance.

- Directional antennas, instead of omnidirectional ones, alleviate these effects.

Microwave communication link systems have also been investigated, and guidelines for negligible impact on the system performance have been developed.

Noise has quite often been cited since 1979 after the test of Mod-1 WEC at Boone, North Carolina.

There are two regions of interest: infrasound below 16 Hz, and audible sound between 16 Hz and 20 kHz.

The infrasound might originate from low speed interactions, like the rotor blades passing through the tower shadow in the case of Mod-1, while the audible sound preferably stems from the machinery and the air flow around the rotor blades. Stall, high tip speed, and flow around the tip are prime sources. Extensive measurements have been performed using the Mod-0 (38 m ϕ). The results are 75 dB sound pressure level for infrasound, and 64 dBA for audible sound at 30 m distance from tower, wind speed between 4.5 and 5.7 m/s.

At 250 m distance, the sound of the unit was indistinguishable from the background noise.

Sound pressure spectrum from the Mod-2 at Goodnoe Hills, Washington, showed almost the same results. At 137 m downwind, the pressure level fell continuously from 90 dB at 1 Hz to 40 dB at 100 Hz.

The infrasound annoyance of the Mod-1 was a specific problem and is not representative for

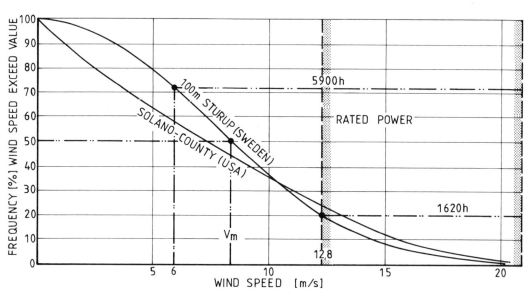

Fig. 12 Wind speed duration curve of Solano County (10 m height) and Sturup (100 m height).

wind turbines in general. Using the appropriate engineering tools, noise should not be a problem.

SAFETY

In principle, pilot plants should fail in order to learn where to place the boundary between safe lifetime and economy. Preferably, these failures should be detected in the workshop and, if on site, before they become catastrophic ones. Wind turbines from line production should be designed and built for such a low failure probability that the loss of one blade, or parts of the blades, is unlikely. If not, they cannot meet the economic target.

A real risk will come from ice particles thrown off the blade. Theoretical analysis of the trajectories of ice cubes and ice sheets have been performed resulting in safety distances of approximately 3 rotor diameters to the side and downward at a wind speed of 22 m/s (see Fig. 13).

Hit probability calculations have also been made, yet several years of practical experience will be needed for reliable assumptions as ice detectors will cause a stop to the turbines and the visitor density might be very small during such days.

NATIONAL AND INTERNATIONAL PROGRAMS

The last running unit of modern WECs, the Gedser windmill in Denmark, was stopped in 1967 and terminated a tradition of about 4,000 years.

The first oil crisis in 1973 caused a renaissance of this old technology, as the U.S. spent some $1.8 million for a new exploration of this potential energy source. Sweden followed, Denmark, West Germany, Canada, and the Netherlands set up programs, too, in order to investigate the technical and economic feasibility of large WECs, and, to a certain degree, of small WECs, too.

The American program took a leading role by the development and installation of 1 Mod–0 (1976), 4 Mod-0AS (1977–1978), 1 Mod–1 (1978), 5 Mod–25 (1980–1982), accompanied by the development of many small WECs and an extensive program of research and tests.

The Danish program started with a test program using the reactivated Gedser windmill and resulted in the development of two 630 kW-WECs, Nibe-A (40 m diameter), and Nibe-B (40

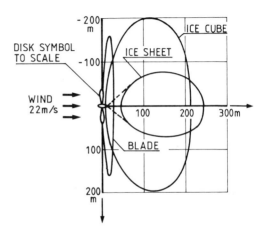

Fig. 13 Trajectories of ice debris.

m diameter), both commissioned in 1979/1980. A private initiative of a school enabled the installation of the Tvind-WEC (54 m diameter), in 1977, rated 1.3 MW.

West Germany began a program based on the advice of the wind pioneer Hütter, and headed for the development of the Voith WEC (54 m ϕ, 250 kW), GROWIAN I (100.4 m ϕ, 3MW), GROWIAN II, and (⅓ scale, 48 m ϕ, 350 kW). Test runs have been, or will be, available in 1982.

Sweden installed a small test machine in Kalkugnen and contracted the development of two large WECs, AEOLUS (75 m ϕ, 2 MW), and WTS 3 (78 m ϕ, 3 MW). Both units started test operation in 1982.

The Netherlands developed and tested a small Darrieus rotor, and a small horizontal axis rotor (25 m ϕ) as the basis for the development of large turbines.

Canada focuses on the development of vertical axis rotors. Several medium-sized turbines have been installed since 1977, and the development of a MW-sized unit has been initiated.

In the time being, other countries started activities in wind energy.

Greece is testing foreign-made WECs for the electricity supply of islands and has established a national program.

Italy installed two test fields on the isle of Sardinia and initiated a program for the development of large WECs.

Ireland is testing several small foreign units and started a first phase of a national program.

The United Kingdom ordered one small and one large WEC for the isle of Orkney and will possibly order a proven MW-unit for Wales in 1983.

West Germany behaves indifferently, as the large machines have not yet produced results to base a future program on.

Sweden moves steadily towards the goal of installing 100 units per year. There are plans to order several units of the prototypes in test and to increase the number continuously if the results are positive. The cooperation between a Swedish and German company (AEOLUS) respectively Swedish and American company (WTS3) offers a good basis for the international market, too.

Denmark already has installed more than 750 small WECs and set a goal to reach 15,000 machines until 1999.

On the other hand, there are plans to increase the size of the Nibe windmills to ÷ 70 m diameter and to get a capacity of 1,000 MW installed until 1999.

The Netherlands is already running the first phase of the second wind energy program to produce a wind energy power station of 10 MW, to design an offshore power station of 50 MW, and to establish a wind-hydro power station until 1984.

The final goal is the installation of 2,000 MW of large WECs, and the installation of 15,000 small WECs until 1999.

In the U.S., private enterprise's and some utilities' initiative replaced most of the needs for a federal program. But there is a great danger that this governmental withdrawal might be too early by one or two years to get the market developing by itself.

In addition to these national programs, bilateral and international programs take care of a further intensification to use this energy source. The main program to be mentioned is run by the Commission of the European Communities, with the first phase terminating early in 1983. This assessment study—accompanied by wind mapping, grid interface analysis, and data collection of existing WECs—will lead to the second phase of a broad European wind program in research, development, and demonstration in 1983.

ECONOMIC ASPECTS

Two criteria control all advances in solar energy: technical achievement and economy. While the technical progress opens an easy access for judgments, the economic aspects are embedded in a system not yet prepared to use the new units.

When the nuclear power plants started to penetrate the coal and oil fire power plant system, they were not cost effective, as they shifted the older stations from base load to peak load.

The same situation is true for the introduction of wind energy converters. On one hand, the optimal mix and operation of the various stations has to be learned. On the other hand, the utility companies receive various forms of subsidies in most countries and contribute to the environmental problems, thus distributing companies' expenses to the national, or even international counteractions.

A third consideration deals with the time of introduction. As it will take many years until the quantity of WECs will give a considerable contribution, the prices for WECs should be compared with those power stations in service then.

The future scenario might consist of:

- advanced coal technologies
  nonpolluting and high-efficient combustion
  conversion into liquid and gaseous fuels
- present nuclear reactors with completed recycling and waste disposal infrastructure
- integrated regional energy supply concepts
  photovoltaics
  hydrogen technology
  major share of decentralized energy supplies
  wind energy.

Hence, economic comparisons should rather be made between these energy lines than with the power stations existing now.

Nevertheless, cost figures have to be given. The published values follow two different lines:

- the annual electricity produced divided by the costs of the investment and operation;
- the relevant costs a utility would save in conventional production.

In Ref. 7, several detailed estimates were published and serve as a relative balanced collection.

Cost projections for the NIBE (Denmark) concept (1980 prices) are:

1. present NIBE concept
   (40 m ϕ) $2,200 $/kW
2. redesigned NIBE concept $1,870 $/kW
3. series production of (2) $1,500 $/kW
4. (3) with a 60 m rotor $1,200 $/kW
5. (3) with a 80 m rotor $1,120 $/kW

	1	2	3	4	5	6	7	8	9	10	11	12
	ROTOR DI-AMETER	ROTOR SWEPT AREA	HUB HEIGHT	RATED POWER	SPEC. RATED POWER	ANNUAL ENERGY PRODUCTION	SPEC. ANNUAL EN. PRO.	PRICE INCL. INSTALL.	PRICE PER SWEPT A	PRICE PER RATED P	(8) : (6)	UTILI-SATION FACTOR
	m	m²	m	kW m/s	W/m²	kWh/a	kWh/m²a	MioDM	DM/m²	DM/kW	DM/kWh	h
BONUS	15	177	18	55/16	310	116.000 87.000 46.000	655 490 260	0.112	650	2.035 ↑	0.97/20 a 1.29/20 a 2.43/20 a	2.110 1.580 836
GEDSER	24	450	24	200/15	444	300.000	660					1.500
NIBE A	40	1257	40	630/16	500	$1.12 \cdot 10^6$	887					1.770
NIBE B	40	1257	40	630/16	500	$1.44 \cdot 10^6$	1143	2.8	2255	4.500'	1.95/20 a	2.280
AEOLUS	75	4418	78	2000/12.8	450	$6.7 \cdot 10^6$	1516	12.0'	2716	6.000'	1.79/30 a	3.350
MOD-2	90	6362	60	2500/13	390	$10.0 \cdot 10^6$	1571	17.5	2750	7.000	1.75/30 a	4.000
GROWIAN	100.4	7917	100	3000/12	380	$10.8 \cdot 10^6$	1364	20.0''	2550	6.670	1.85/20 a	3.600

↑ from [144]
' from [145]
'' from [145] add. estimated installation

Fig. 14 Characteristic data of some wind energy converters.

The GROWIAN (West Germany, 100 m ϕ)
—data stand for

first unit	$3,100 $/kW
small series production (10 units)	$2,190 $/kW
large series production (100 units)	$1,700 $/kW

The Swedish WECs will cost (1981 prices):

10 units production	
WTS 3 (78 m ϕ)	$1,380 $/kW
AEOLUS (75 m ϕ)	$1,370 $/kW
100 unit production	
WTS 3	$1,200 $/kW
AEOLUS	$1,070 $/kW

The power companies quote additional costs for site preparation, connection, etc.:

Sydkraft	$200 $/kW
State Power Board	$330 $/kW

Unfortunately, the U.S. contribution did not cite equivalent figures for the Mod-2 unit. There is one value in a graph which might be interpreted as such:

Mod-2 (90 m ϕ)	$4,100 $/kW,

but this could only be the price of the first unit.

As the price of a WEC might be based on noncomparable specifications, a comparison was made between several types and several aspects of interest (Fig. 14).

As the costs per installed capacity depend on the somewhat arbitrary choice of the related power, a better comparison is enabled by the price per kilowatt hour, although there is still a strong site dependency included.

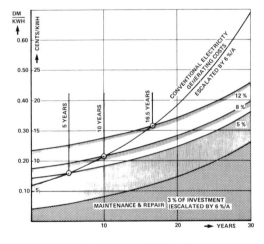

ENERGY PRODUCTION COSTS OF WTS-2000
ANNUAL ENERGY: 6000 MWH
INVESTMENT: 9 MILL. DM   ESCALATION RATE: 6 % PER YEAR

Fig. 15

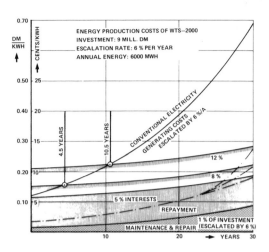

Fig. 16

The respective quotations are:

Nibe (40 m ϕ)	7.0 ¢/kW h
WTS 3 (78 m ϕ)	4.3 ¢/kW h
AEOLUS (75 m ϕ)	5.9 ¢/kW h
GROWIAN (100 m ϕ)	9.2 ¢/kW h

A great influence lies hidden in the operational costs for maintenance and repair, as Fig. 15 and 16 demonstrate. The capital costs (at interest rates 5%, 8%, 12%, respectively), together with the escalated maintenance and repair costs determine the specific wind cost per kW h. For comparison, the conventional production costs (exit of the power station) are also plotted, escalated by 6% per year. The great influence of the maintenance and repair costs should be noted. Hence, large WECs must be designed for rather low maintenance and repair, otherwise they will not pass the utility companies' selection criteria.

As the experience for the maintenance and repair costs is still missing, some indication might be gained by comparison with those of the wide body aircrafts, as tabulated in Fig. 17.

The other approach, the allowable investment expenditures, opens a wide range of controversial arguments. There is no doubt that WECs save fuel and, therefore, the fuel costs should be placed to the investor's credit, including the credit for the future escalation. Great differences exist in the actual selection of the fuel, as a power station mix consumes different types at different rates. A relatively neutral way of presentation was published in Ref. 8, a version of which is given in Figs. 18 and 19. The operational hours are calculated by dividing the annual energy production by the rated power, a value in

	DIRECT MAINTENANCE COSTS PER YEAR IN 10³ $			PRICE BREAKDOWN IN 10³ $			ANNUAL MAINTENANCE PER INITIAL INVESTMENT		
	BOEING 747-200	McDD DC 10-30	AIRBUS A300 B4	747-200	DC 10-30	A300 B4	747-200	DC 10-30	A300 B4
STRUCTURE	119.3	69.4	40.0	37.700	30.100	25.200	0.32%	0.23%	0.16%
EQUIPMENT	920.0	510.2	291.4	7.700	6.200	5.100	12.1 %	8.2 %	5.7 %
POWER PLANT	1.333.3	836.5	645.7	9.600	7.200	4.900	13.9 %	11.6 %	13.2 %
TOTAL	2.382.6	1.416.1	977.1	55.000	43.500	35.200	4.3 %	3.3 %	2.8 %

Fig. 17. Ratio of annual direct maintenance costs to price of wide body aircrafts from [146].

(SOURCE: AVIATION WEEK & SPACE TECHN., FEBR. 1980)

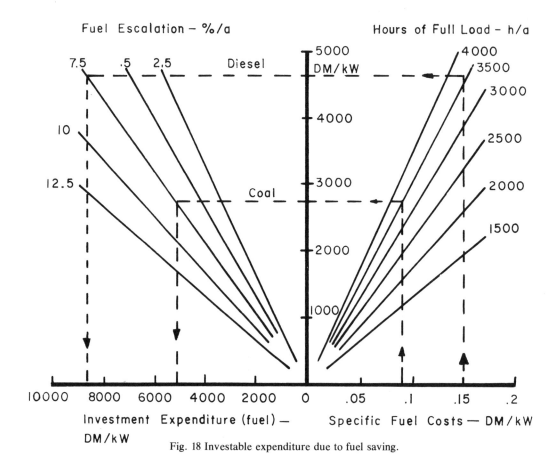

Fig. 18 Investable expenditure due to fuel saving.

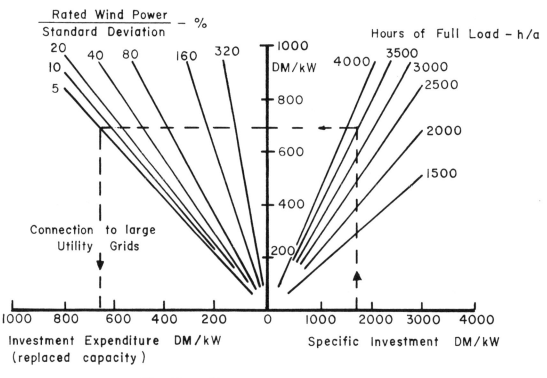

Fig. 19 Investable expenditure due to replaced capacity.

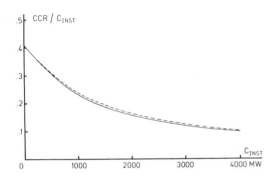

Fig. 20 Capacity credit CCR of wind plants at Den Helder (installed capacity C_{inst}) integrated with the power system of 1985. Calculations based on two reliability criteria:

loss-of-capacity probability during annual peak load, and

loss-of-load probability, LOLP (displaced conventional capacity: small (25 MW) units with 90% availability).

For the 1985 power system and load forecast (including uncertainty) a loss-of-capacity probability of 1% corresponds to an LOLP of 1.4 h/year.

the order of 3,500 to 4,000 hours. It has to be kept in mind that by the time of great installation rates, the energy prices will have escalated much more than the inflation rate.

Many arguments arise from the capacity effect of wind power stations, ranging from no credit at all to full credit. Of course, the station mix becomes more complicated as a new type of station with much lower availability (approximately 40%, instead of 70% to 90%) will be introduced.

But several calculations[9,10] showed that the same mathematical procedure as for a conventional mix could be applied, resulting in a capacity credit between 40% and 10%, depending on the penetration rate, as demonstrated in Fig. 20.

The other version of presentation (Figs. 17 and 18) offers a fast survey in the nomographic way.

In total, the contribution from the capacity credit is small in comparison to the fuel credit. Yet, the fact is important that already some utility companies look at wind power as a safe energy source with lowered availability.

REFERENCES

1. A. Betz. 1926. Wind-Energie und ihre Ausnutzung durch Windmühlen. In: *Naturwissenschaft und Technik*. Vol. 2. Göttingen: Vandenhoeck und Ruprecht.

2. H. Glauert. 1935. *Aerodynamic theory*. Vol. 6. Division L. Berlin: Julius Springer.

3. K. König. 1982. Vibrational and aeroelastic problems of large WECs and their computerized solution. Internal MBB-VFW Report. Bremen.

4. *Technical specification for design and installation of wind turbine systems in Sweden*. 1978. NE, Stockholm.

5. M.A. Miner. 1945. Cumulative damage in fatigue. *Trans. of ASME Journal of Applied Mechanics*.

6. U. Hütter. 1974 (July). Betriebserfahrungen mit der 100-kW-Windkraftanlage der Studiengesellschaft Windkraft. Brennstoff-Warme-Kraft, Band 16, Nr. 7, pp. 333–340.

7. *7th meeting of experts—costings for wind turbines*. 1982 (Mar.). IEA Impl. Agreement. Jül-Spez-147.

8. H. Nissen and W. Former. 1973. Wirtschaftlichkeit der Stromerzeugung aus Windenergie. Vol. 3. Energiewirtschaftliche Tagesfragen 29.

9. L. Jarass. 1981. Strom und Wind. Berlin: Springer-Verlag.

10. A.J. Janssen et al. 1981 (Oct.). Statistical methods for the assessment of wind power integration into the electricity supply system. ECN-101. 1979 (Aug.), Petten/Netherlands.

BIBLIOGRAPHY

GENERAL

Bonnefille. 1974 (Apr.). Les Réalisations d'Electricité de France Concernant l'Energie Eolienne Electricité de France F 40/74 Nr. 4.

F.R. Eldridge. 1975 (Oct.). *Wind machines*. The Mitre Corporation. (Also NSF-RA-N-75-051.)

E.W. Golding. 1956. *The generation of electricity by wind power*. New York: Philosophical Library.

H. Honnef. 1932. Windkraftwerke. Braunschweig: Friedrich Vieweg u. Sohn AG.

U. Hütter. 1947. Beitrag zur Schaffung von Gestaltungsgrundlagen für Windkraftwerke. Dissertation TH. Wien.

L. Jarass. *Wind energy*. New York: Springer Verlag.

J.P. Molly. 1978. *Windenergie in Theorie und Praxis*. Karlsruhe: Verlag C.F. Müller.

J. Juul. 1961. Design of wind power plants in Denmark new sources of energy. *Proceedings of The Conference, United Nations*. Rome; Aug. 21–31, 1961.

P.C. Putman. 1948. *Power from the wind*. New York: D. Van Nostrand Company.

D.M. Simmons. 1975. *Wind power*. Energy Technology Review No. 6. New Jersey: Noyes Data Corporation. 1975.

Wind energy report. 1979 (July). New York: Publishing Corporation.

DESIGN

J.S. Andrews and J.M. Baskin. 1981. Development tests for the 2.5 megawatt MOD-2 wind turbine generator. *WWV*. Vol. 2:611.

R.S. Barton and W.C. Lucas. 1981. Conceptual design of the 6 MW MOD-5A wind turbine generator. *WWV*. Vol. 1:157.

A.G. Birchenough et al. 1981. Operating experience with the 200 kW MOD-0A wind turbine generators. *WWV*. Vol. 1:107.

Jens Trampe Broch. 1968. Peak-distribution effects in random-load fatigue. From *Effects of environment and complex load history on fatigue life.* ASTM Special Technical Publication 462, pp. 104–126.

G.W. Brown and R. Ikegami. 1970. The fatigue of aluminum alloys subjected to random loading. SESA Spring Meeting. Huntsville, Ala.; May, 1970.

C.A. Cornell. 1969 (Dec.). A probability-based structural code. *ACI Journal.*

G.S. Doman. 1979. System configuration improvement. Large Wind Turbine Design Characteristics and R & D Requirements. Workshop held at NASA Lewis Research Center; April 24–26, 1979. DOE Publication CONF-7904111, p. 385.

R.R. Douglas. 1981. Conceptual design of the 7 megawatt MOD-5B wind turbine generator. *WWV.* 1981. Vol. 1:169.

H.A. El Maraghy and J.N. Siddall. 1978. Theoretical prediction and experimental determination of fatigue life distributions of SAE 1008 steel subjected to constant amplitude, block and narrow band random loading. Vol. 5 (No. 3). Trans Canadian Society of Mechanical Engineering. No. 78-CSME-9, EIC Accession No. 1769.

J.R. Faddoul. 1981. An overview of large, horizontal axis wind turbine blades. *WWV.* Vol. 3: 113.

R.W. Finger. 1980. Prediction model for fatigue crack growth in windmill structures. *Effect of load spectrum variables on fatigue crack initiation and propagation.* D.F. Bryan and J.M. Potter, eds. ASTM STP 714. American Society for Testing and Materials. p. 185–204.

J.W. Fisher et al. 1974. Fatigue strength of steel beams with welded stiffeners and attachments. Report No. 147. Highway Research Board, Washington, D.C.

J.W. Fisher et al. 1970. Effect of weldments on the fatigue of steel beams. Report No. 102. Highway Research Board, Washington, D.C.

J.T.D. Fritz. 1977 (Dec.). Accumulation of fatigue damage under random loading conditions. National Mechanical Engineering Research Institute. Reference No.: MES/4056, Pretoria, South Africa.

T.W. Graham and A.S. Tetelman. The use of crack size distribution and crack detection for determining the probability of fatigue failure. AIAA/ASME/SAE 15th Structures. Structural Dynamics and Materials Conference. Las Vegas, 1974. AIAA Paper No. 74-394.

D.S. Hoddinott. 1974. The effect of random loading near-term correlation on the fatigue behavior of a steel. *Engineering Fracture Mechanics* 6:163–164.

U. Hütter. 1977. Optimum wind energy conversion systems. *Ann. Rev. Fluid Mech.* 9:399–419. Vol. 9, S. 399–419.

U. Hütter. 1974. Optimum design concept for windelectric converters. Workshop on Advanced Wind Energy Systems. Stockholm; Aug. 29, 1974.

G.O. Johnston. 1980. Probabilistic fracture mechanics. *Symposium on Mechanical Reliability.* Edited by T.R. Moss. Guildford, England: IPC Science and Technology Press, p. 8–19.

M. Kawamoto; H. Ishikawa; N. Inoue; and Y. Yoshio. 1975. Fatigue test results and fatigue life estimation of hard steel and aluminum alloy under random loads. *Bulletin of the JSME* 18 (No. 122).

S.H. Kratz and R.C. Metzger. 1981. High torque drive systems for large wind turbines. *WWV.* Vol. 3: 89.

Large, horizontal axis wind turbine projects. 1981 (Nov.). NTIS #SERI/SP-732-730.

P. Lundsager. 1982 (Mar.). Experience with the Gedser windmill and small Danish windmills. Risö PL-ac 02.03.1982.

W.E.B. Mason and B.J. Jones. 1981. Reliability and quality assurance on the MOD-2 wind system. *WWV.* Vol. 2:151.

MOD-2 wind turbine system concept and preliminary design report. Vol. 2. 1979 (July). Detailed Report. Boeing Engineering and Construction Company, Contractor Report DOE/NASA 0002-80/2.

L.L. Nelson. 1981. Medicine bow wind project. *WWV.* Vol. 2:557.

J.C. Newman Jr. 1976 (June). Predicting failure of specimens with either surface cracks or corner cracks at holes. NACA TN D-8244.

P. Nielsen. 1981. *Measurements on the Nibe wind turbines.* DEFU, Report No. EEV 81-04.

F. Nilsson. 1977. A model for fracture mechanical estimation of the failure probability of reactor pressure vessels. *3rd Int. Conf. on Pressure Vessel Technology in Tokyo.* Part II—Materials and Fabrication. New York: American Society of Mech. Engineers, p. 593–601.

R. Olivier and W. Ritter. 1979. 1980. 1981. Catalogue of S-N-curves of welded joints in structural steels. Part 1: Butt Joints, Part 2: Transverse Stiffener, Part 3: Cruciform Joint. DVS Berichte 56/I-III. Düsseldorf: Deutscher Verlag für Schweisstechnik.

A. Plaks et al. 1981 (Jan.). Wind turbine system for high wind regions in California. Consultant report prepared by Boeing Engineering and Construction Company for California Energy Commission. CEC Publication #P500-81-005.

A. Raab. 1979. Combined effects of periodic and stochastic loads on the fatigue of wind turbine parts. Aeron. Research Inst. of Sweden. FFA, TN Au-1499, Stockholm.

J. Schijve. 1973. Effect of load sequences on crack propagation under random and program loading. *Engineering Fracture Mechanics* 5:267–280, Pergamon Press, 1973.

S.R. Swanson. 1968. Random load fatigue testing: a state of the art survey. *Materials Research & Standards* 8:10.

D.J. White and J. Lewszuk. 1970. (Nov.). Cumulative damage in fretting fatigue of pinned joints subjected to narrow band random loading. *The Aeronautical Quarterly.*

DYNAMICS AND AEROELASTICITY

B.M. Brooks. 1981. MOD-0 wind turbine dynamics test correlations. *Dynamics Workshop,* p. 287.

C.C. Chamis and T.L. Sullivan. 1976. Free vibrations of the ERDA-NASA 100 kW wind turbine. NASA TMX-71879.

A.W. Cherritt and J.A. Gaidelis. 1975 (June). 100-kW metal wind turbine blade basic data, loads and stress analysis. NASA CR-134956.

A.G. Davenport. 1966 (April). The treatment of wind loading on tall buildings. *Proceedings of a Symposium on Tall Buildings.* Southampton.

H.W. Försching. 1974. *Grundlagen der Aeroelastik.* Berlin, Heidelberg, and New York: Springer Verlag.

P. Friedmann. 1980. Aeroelastic stability and response analysis of large horizontal-axis wind turbines. *J. Industrial Aerodynamics* 5:373–401.

P. Friedmann. 1977a. Influence of modeling and blade parameters on the aeroelastic stability of a cantilevered rotor. *AIAA Journal* 15 (No. 2):149–158.

P. Friedmann. 1977b. Recent developments in rotary-wing aeroelasticity. *Journal of Aircraft* 14 (No. 11):1027–41.

P. Friedmann. 1976. Aeroelastic modeling of large wind turbines. *Journal of the American Helicopter Society.* 17–27.

P. Friedmann and C. Yuan. 1977 (July). Effect of modified aerodynamic strip theories on rotor blade aeroelastic stability. *AIAA Journal* 15 (No. 7): 932–40.

V. Giurgiutiu. 1977. Vibrations and dynamic stability of rotor blades. Ph.D. Thesis, Department of Aeronautics, Imperial College of Science and Technology, University of London.

J.C. Glasgow and R.D. Corrigan. 1982. MOD-0 passive yaw test results. PIR No. 196, NASA Lewis Research Center Wind Energy Project Office. Cleveland, Oh, June 30, 1982.

D.H. Hodges and E.H. Dowell. 1974 (Dec.). Nonlinear equations of motion for the elastic bending and torsion of twisted nonuniform rotor blades. NASA TN D-7818.

D.H. Hodges and R.A. Ormiston. 1977. Stability of hingeless rotor blades in hover with pitch-link flexibility. *AIAA Journal* 15 (No. 4):476–82.

J.A. Hoffman. 1977 (Feb.). Coupled dynamics analysis of wind energy systems. NASA CR-135152.

R.A. Johnston and S.J. Cessarino. 1976 (Jan.). Aeroelastic rotor instability analysis. USAMRDL-TR-75-40.

F. Kiessling. 1977. *Aeroelastische Probleme bei Windenergiekonvertern.* Tagungsbericht Energie vom Wind. Deutsche Gesellschaft für Sonnenenergie 4. Tagung, 7/8. Juni 1977, Bremen

S.B.R. Kottapalli; P. Friedmann; and A. Rosen. 1978. Aeroelastic stability and response of horizontal axis wind turbine blades. Presented at the Second International Symposium on Wind Energy Systems. Amsterdam, Netherlands; October 3–5, 1978.

B.S. Linscott; J. Glasgow; W.D. Anderson; and R.E. Donham. 1977. Experimental data and theoretical analysis of an operating 100 k W wind turbine. Presented at the Wind Energy Workshop. Washington, D.C.; Sept. 1977.

R.A. Ormiston. 1975. Dynamic response of wind turbine rotor systems. AHS Preprint S-993. Presented at the 31st Annual National Forum of the American Helicopter Society. Washington, D.C.; 1975.

R.A. Ormiston. 1973 (Dec.). Rotor dynamic consideration for large wind power generator systems. *Wind Energy Conversion Systems Workshop Proceedings.* National Science Foundation, NSF/RA/W-73-006.

R.A. Ormiston and D.H. Hodges. 1972. Linear flap-lag dynamics of hingeless helicopter rotor blades in hover. *Journal of the American Helicopter Society* 17 (No. 2):2–14.

C.P. Patrickson and P. Friedmann. 1976 (Dec.). A study of the coupled lateral and torsional response of tall buildings to wind loadings, University of California, Los Angeles, School of Engineering and Applied Science Report, UCLA-ENG-76126.

D.A. Spera. 1977. Comparison of computer codes for calculating dynamic loads in wind turbines. Presented at the Wind Energy Workshop. Washington, D.C.; Sept. 1977.

D.A. Spera and D.C. Janetzke. 1977 (July). Effects of rotor location, coning, and tilt on critical loads in large wind turbines. Prepared for *Wind Technology Journal.*

D.A. Spera; D.C. Janetzke; and T.R. Richards. 1977 (Aug.). Dynamic blade loading in the ERDA/NASA 100 kW and 200 kW wind turbines. NASA TMX-73711.

T. Sullivan. 1981. A review of resonance response in large, horizontal axis wind turbines. *Wind Turbine Dynamics, Proceedings of a Workshop.* Sponsored by Department of Energy and NASA. Cleveland,
Oh; Feb. 24–26, 1981. DOE Publication CONF 810226, p. 237.

R.W. Thresher. 1981. Structural dynamic analysis of wind turbine systems. *WWV.* Vol. 3:61.

A.J. Vollan. 1982. Aeroelastic stability and dynamic response for wind energy converters fourth int. symp. on wind energy systems. Stockholm; Sept. 1982.

A.J. Vollan. 1978. The aeroelastic behavior of large Darrieus-type wind energy converters derived from the behavior of 5,5 M rotor. *Second International Symposium on Wind Energy Systems.* 1978 Amsterdam; Oct. 3–6. P. C5-67-88.

ECONOMIC ASPECTS

T.S. Dillon et al. 1980 (Jan.). Stochastic optimization and modelling of large hydrothermal systems for long-term regulation.

L.L. Garver. 1966 (Aug.). Effective load carrying capability of generating units. *IEEE Trans. on Pow. Appar. & Syst.* Vol. PAS-85 (No. 8).

A guide to financial assistance for wind energy. CEC P500-81-014.

P.H. Jensen. 1982 (Feb.). En Vindmölles Privatökonomii. Risö-M-2335.

J.I. Lerner. 1982. Assessment of large scale wind system technology and prospects for commercial application. *Workshop on the Federal Role in the Commercialization of Large Scale Windmill Technology.* Sponsored by National Science Foundation. Washington, D.C.; Sept. 25–26, 1980. (Available from NTIS)

W.D. Marsh. 1979 (Jan.). Requirements assessment of wind power plants in electric utility systems. EPRI-ER-978.

B. Martin and M. Diesendorf. 1980. The capacity credit of wind power: a numerical model. *Proc. Third International Symposium on Wind Energy Systems.*

J.P. Molly. 1977. Nutzungs- und Speicherprobleme bei Windkraftanlagen. *Vortrag: Energie-politisches Forum der Landesregierung Baden-Württemberg und der Universität Stuttgart.* Stuttgart; May 9–12, 1977.

J.P. Molly. Möglichkeiten zur bedarfsorientierten Abgabe der Windenergie. *Tagungsbericht Energie vom Wind.* DGS-Tagung. Bremen; June 7–8, 1977.

A.P. Rockingham. 1979. A probabilistic simulation model for the calculation of the value of wind energy to electric utilities. *Proc. 1st BWEA Wind Energy Workshop.* April, 1979. London: Multi-Science Publ. Co. Ltd.

H. Selzer. 1981. Wind energy on its way to commercialization. *Int. Coll. on Wind Energy.* Brighton; Aug. 1981.

W.N. Sullivan. 1979 (Aug.). Economic analysis of Darrieus vertical axis wind turbine systems for the generation of utility grid electrical power. Vol. 2—the economic optimization model. SAND78-0962.

R.L. Sullivan. 1977. Power system planning. New York: McGraw-Hill.

M. Timm. 1978. Wirtschaftliche Windenergienutzung im Verbund mit herkömmlichen Kraftwerken. Statusreport Windenergie. Oct 23–24, 1978. Jülich

W. Weber. 1977 (June). Einflussfaktoren der Kostenanalyse von Windenergiekonvertern mit horizontaler Achse. *Tagungsbericht Energie vom Wind der DGS.* Bremen, June 7–8, 1977.

G.E. Whittle et al. 1980. A simulation model of an electricity generating system incorporating wind turbine plant. *Proc. Third International Symposium on Wind Energy Systems*. Copenhagen, Denmark; Aug. 1980. BHRA Fluid Engineering, Cranfield, Bedford, England, p. 545.

ENVIRONMENTAL EFFECTS

J.F. Balombin. 1980. An exploratory survey of noise levels associated with a 100 kW wind turbine. NASA TM-81486. Lewis Research Center. Cleveland, Oh.

H.D. Carden and W.H. Mayes. 1970. Measured vibration response characteristics of four residential structures excited by mechanical and acoustic loadings. NASA TN D-5776. Langley Research Center. Hampton, Va.

L.E. Ericsson and J.P. Reding. 1970. Unsteady airfoil stall review and extension. *Proc. AIAA Aerospace Sci. Meeting*. AIAA Paper No. 70-77. New York.

H. Fujita and L.S.G. Kovasznay. 1974. Unsteady lift and radiated sound from a wake cutting airfoil. *AIAA Journ*. 12:1216–1221.

W.E. Howell. 1978. *Environmental impact of large windpower farms*. Denver: U.S. Bureau of Reclamation.

G.F. Homicz and A.R. George. 1974. Broadband and discrete radiation from subsonic rotors. *Journal of Sound & Vibration* 32(2):151–177.

N.D. Kelley. 1981. Noise generation by large wind turbines. Wind Energy Technology Conference. Kansas City; March 1981.

R. Martinez; S.E. Widnall; and W.L. Harris. 1981. HAWT Noise. *Proceedings of 2nd DOE/NASA Wind Turbine Dynamics Workshop*. Feb., 1981.

S.E. Rogers et al. 1978. Wind energy conversion—environmental effects assessment. *Third Wind Energy Workshop* Vol. I. CONF-770921/1: 402–406. May 1978.

S.E. Rogers et al. 1976 (Aug.). *Evaluation of the potential environmental effects of wind energy system development*. Interim Final Report, ERDA/NSF/07378-75/1. Columbus: Battelle Columbus Laboratories. Available NTIS.

T.B.A. Senior et al. 1977 (Feb.). *TV and FM interference by windmills*. Final Report, C00-2846-76-1. Ann Arbor: Radiation Laboratory, Michigan University.

Solar program assessment: environmental factors, wind energy conversion. 1977 (Mar.). ERDA 77-47/6. Washington: Energy Research and Development Administration.

S.E. Wright. 1969. Sound radiation from a lifting rotor generated by asymmetric disk loading. *Journal of Sound and Vibration* 9(2):223–240.

LOADS

A.G. Davenport. 1964. Note on the distribution of the largest value of a random function with application to gust loading. *Proc. Inst. of Civil Engineers* 28:187–196.

A.G. Davenport. 1961. The application of statistical concepts to the wind loading of structures. *Proc. Inst. of Civil Engineers* 19:449–72.

K.R.V. Kaza; D.C. Janetzke; T.L. Sullivan. 1979. Evaluation of MOSTAS computer code for predicting dynamic loads in a two-bladed wind turbine. AIAA Paper 79-0733. *Proc. 20th Structures, Structural Dynamics and Materials Conference*. St. Louis, Mo.; April 1979, pp. 53–63.

A. Miller and R.L. Simon. 1980 (Sept.). Wind resource potential in California. San Jose State University, CEC Consultant Report. P500-80-052.

G. Törnkvist. 1980. Basic design recommendations for wind energy converters. Saab-Scania report FKL-V-80.9. Linköping, Sweden.

D.A. Spera. 1977 (Sept.). Comparison of computer codes for calculating dynamic loads in wind turbines. NASA TM-73773.

PERFORMANCE AND AERODYNAMICS

R.E. Akins. 1978. Performance evaluation of wind energy conversion systems using the method of bins—current status. SAND77-1375. Albuquerque, N.M.: Sandia Laboratories.

O. De Vries. 1979. Wind tunnel tests on a model of a two-bladed horizontal-axis wind turbine and evaluation of an aerodynamic performance calculation method. NLR TR 79071 L.

O. De Vries. 1978. The aerodynamic performance of a horizontal-axis wind turbine in a stationary parallel flow. NLR TR 78084 L.

B.F. Habron et al. 1980. Wind-turbine power improvement with modern airfoil sections and multiple-speed generators. AIAA/SERI Wind Energy Conference. Boulder, Colo.; April 9–11, 1980. AIAA Paper No. 0633.

P.B.S. Lissaman et al. 1982 (June). Numeric modeling sensitivity analysis of the performance of wind turbine arrays. Department of Energy/Pacific Northwest Laboratory Contractor Report. D. E. 82027570, PNL-4183.

L. Viterna and D. Janetzke. 1981. Theoretical and experimental power from large, horizontal—axis wind turbines. *Proceedings of the Fifth Biennial Wind Energy Conference and Workshop* (WWV). Vol. 2. Sponsored by Department of Energy. Washington, D.C.; October 5–7, 1981. SERI/CP-635-1340, CONNF-811043, p. 265.

R.E. Wilson and P.B.S. Lissaman. 1974 (July). Applied aerodynamics of wind power machines. Oregon State University.

PROGRAMS

The Danish wind power program. 1981. *Større elproducerende vindkraftanlaeg* (in Danish). Lyngby, Denmark: Polyteknisk Boghandel og Forlag.

M. Dubey; U. Coty; D. Bain; R. Donham; L. Vaughn; and R. Dickinson. 1980 (June). Impact of large wind energy systems in California. Solar Energy Conversion Systems, Inc., CEC Contractor Report. P500-80-031.

J. Lerner. 1978 (July). Wind-electric power: a renewable resource for California. CEC staff report. P500-78-025.

R. Windheim and R. Neumann. 1979. Wind energy R & D program of the Federal Republic of Germany and current wind energy projects. *4th Biennial Conference and Workshop ond Wind Energy Conversion Systems*. Washington; Oct. 29–31, 79.

SAFETY

S. Eggwertz. 1980. Study of WECs farm area and WECs safety limit requirements. *Minutes from the Expert Meeting JEA R&D WECS*. Annex I Sub Task A1. FFA Technical Note Hu-2218. Stockholm.

S. Eggwertz et al. 1981. Safety of wind energy conversion systems with horizontal axis. Aeron. Research Inst. of Sweden. FFA, TN HU-2229. Stockholm.

S. Eggwertz; I. Carlsson; A. Gustavsson; C. Lundemo; B. Mongomerie; and S-E. Thor. Safety of wind energy conversion systems (WECS), preliminary study. Aeronautical Research Institute of Sweden, Report HU-2126. Stockholm.

Implementing agreement for co-operation in the development of large scale wind energy conversion systems. 1981. *Fifth Meeting of Experts—Environmental and Safety Aspects.* Jül-Spez-100, Febr. 1981.

Implementing agreement. 1981. *Sixth Meeting of Experts—Reliability and Maintenance.* Jül-Spez-129, Sept. 1981.

WIND STRUCTURE

H.C. Chien; V.A. Sandborn; R.N. Meroney; and R.J.B. Bouwmeester. 1978 (Mar.). Preliminary measurements of flow over model, three-dimensional hills. Colorado State University, Research Memorandum 24.

W. Frost; B.H. Long; and R.E. Turner. 1978. Engineering handbook on the atmospheric environmental guidelines for use in wind turbine generator development. NASA Technical Paper 1359.

W. Frost and D.K. Nowak. 1976 (Sept.) Technology development for assessment of small-scale terrain effects on available wind energy. Monthly report for DOE Contract E(40-1)-5220.

W. Frost and A.M. Shahabi. 1977. A field study of wind over a simulated block building. NASA CR 2804.

W. Frost and C.F. Shieh. 1979. Guidelines for siting WECS relative to small-scale terrain features. FWG Associates, Inc., Tullahoma, Tenn. DOE Contract Report RLO/2443-78/1.

H. Gustavsson and M. Linde. 1979. The gust as a coherent structure in the turbulent boundary layer. FFA TN AU-1499 Part 5.

W.E. Holley and R.W. Thresher. Response of wind turbines to atmospheric turbulence. *WWV.* Vol. 2:281.

S.J. Kline; W.C. Reynolds; F.A. Schraub; and P.W. Runstadler. 1967. The structure of turbulent boundary layers. *J. Fluid. Mech.* 30:741.

R.N. Meroney; A.J. Bowen; D. Lindley; and J. Pearce. Wind characteristics over complex terrain: laboratory simulation and field measurements at Rakaia Gorge, New Zealand. Colorado State University, Colorado. Contract Report RLO/2438-77/2.

R.N. Meroney; V.A. Sandborn; R.J.B. Bouwmeester; H.C. Chien; and M. Rider. 1978. Sites for wind power installations: physical modeling of the influence of hills, ridges and complex terrain on wind speed and turbulence. Colorado State University, Colorado, Contract Report RLO/2438-77/3.

R.N. Meroney; V.A. Sandborn; R.J.B. Bouwmeester; and M.A. Rider. 1977. Sites for wind power installations: wind tunnel simulation of the influence of two-dimensional ridges on wind speed and turbulence. Colorado State University, Contract Report RLO/2438-77/1.

E.L. Petersen et al. 1981 (Jan.). Windatlas for Denmark. Risø-R-428.

D.C. Powell and J.R. Connell. 1980. Definition of gust model concepts and review of gust models. Battelle Pacific Northwest Laboratory, PNL-3138.

J.K. Raine and D.C. Stevenson. 1977. Wind protection by model fences in a simulated atmospheric boundary layer. *Journal of Industrial Aerodynamics* 2:159–180.

C.F. Shieh and W. Frost. 1979. Application of a numerical model to WECS siting relative to two-dimensional terrain features. Fifth International Conference on Wind Engineering. Fort Collins, Colo.; July 8–14, 1979.

C.F. Shieh; W. Frost; and J. Bitte. 1977. Neutrally stable atmospheric flow over a two-dimensional rectagular block. NASA CR 2926.

R.L. Simon. 1981. Potential errors in using one anemometer to characterize the wind power over an entire rotor disc. Large Horizontal-Axis Wind Turbine Conference. Sponsored by DOE and NASA. Cleveland, Oh; July, 1981.

A.S. Smedman. 1980. Turbulensförhållanden i Kalkugnen. Meteorologiska Inst. Uppsala Univ.

H.G.C. Woo; J.A. Peterka; J.E. Cermak. 1977. Wind-tunnel measurements in the wakes of structures. NASA CR 2806.

Advances in Solar Energy © 1983 American Solar Energy Society, Inc.

CONTROLS IN SOLAR ENERGY SYSTEMS

C. BYRON WINN

Colorado State University, Fort Collins, Colorado 80523

Abstract
The characteristics of bang-bang, proportional, integral, derivative, and PID controllers, and their applications to solar energy systems, are presented. Also included is a determination of the effects of temperature settings on cycling rates in systems using bang-bang controllers. A phase-plane representation is developed and an analytical representation for the number of cycles as a function of temperature settings, solar radiation, and physical parameters is presented.

Proportional and optimal control of mass flow rate in both low and high temperature applications is described. Examples are presented.

Finally, conventional, proportional, and optimal controllers for off-peak storage systems are described. This analysis includes the electric utility in the system models.

REVIEW OF CONTROL THEORY

The control of solar energy systems is quite similar to the control of any dynamic system that is subjected to disturbances. A great deal of work has been done in this area and it may be applied to controllers of solar energy systems. Before examining specific types of controllers used in solar energy systems, we shall first review some basic control theories. Consider first an example of a classical control problem as illustrated in Fig. 1.

The objective is to vary the energy input to the heater in order to control the enclosure temperature, T_E. Usually, the enclosure temperature is controlled in order to maintain some desired temperature, T_D. There are several ways in which the control could be effected.

The simplest form of control is "open loop" control. In this case, the heater is energized on some schedule regardless of the current enclosure temperature. Obviously, this type of control would not result in very high comfort levels. The comfort level is improved by using "closed loop" control, as illustrated in Fig. 2.

Bang-Bang Control

In this case, the thermostat serves as the first element of the controller and compares the actual room temperature, as sensed by the thermo-

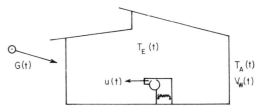

Fig. 1. Control of building temperature.

stat, with the desired room temperature, as set in the thermostat.

The final control element, the furnace, has only two states (full on or off) depending on the

Dr. Winn is a Professor of Mechanical Engineering and a member of the Solar Energy Applications Laboratory at Colorado State University. He has performed research and published articles in the areas of solar energy, orbital mechanics, satellite geodesy, remote sensing, hydrology and water resource systems, optimal control, and systems analysis. He has been actively involved in solar energy since 1973, served as General Chairman of the 1978 annual meeting of AS/ISES, has served on the Board of Directors of the Society and has chaired the Conference and Meetings Committee and the Nominations Committee. He is President of Solar Environmental Engineering Company of Fort Collins, Colorado, and has designed and installed a large number of solar systems and was the developer of the Solar Index.

Dr. Winn is also active in the American Society of Mechanical Engineering and serves as an Associate Editor of the ASME Journal of Solar Energy Engineering.

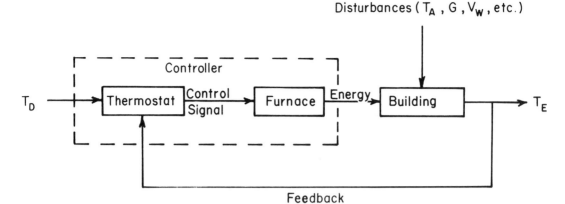

Fig. 2. Feedback control systems.

state (closed or open) of a switch (relay) operated by the thermostat. When the temperature at the thermostat falls below the desired temperature, the switch is closed, thereby causing the furnace to operate, and consequently increasing the room temperature, T_E. When T_E exceeds T_D, the switch is opened and the furnace is shut off.

This would also be an undesirable controller since it would lead to rapid cycling (on and off) of the furnace. That is, the controller would turn the furnace on whenever the enclosure temperature is sensed to be less than the desired temperature, no matter how small the difference. This could result in a rapid increase in T_E, depending on the furnace output and the ambient conditions, such that T_E exceeds T_D and the furnace is turned off. This type of controller is referred to as a bang-bang controller. It has only the two states, on or off. It is represented in Fig. 3. The building temperature response that would result from use of this controller is illustrated in Fig. 4.

The enclosure temperature is maintained at the desired value, but at the cost of excessive cycling of the furnace. The cycling rate may be reduced by introducing a deadband into the con-

troller. The effect of the deadband is that the controller will not turn the furnace on until the enclosure temperature drops below the bottom of the deadband; the controller will then keep the furnace on until T_E increases to beyond the top of the deadband. This is illustrated in Fig. 5. The control function (often referred to as the control law) is illustrated in Fig. 6, and may be expressed analytically as

$$u = \begin{cases} u_{MAX}, \; T_E - T_D \le -DB \\ 0, \; T_E - T_D \ge DB. \end{cases} \quad (1)$$

The bang-bang controller with deadband does not maintain as precise control of T_E as does the bang-bang controller without deadband, but it results in much less cycling. However, it also results in "overshoot"; that is, the enclosure temperature will exceed the top of the deadband and fall below the bottom of the deadband. Normally, this is of little consequence in building heating systems. However, with the advent of buildings having large thermal capacitances, the problem of overshoot becomes more significant. Cycling and overshoot are characteristics of bang-bang controllers, regardless of the system being controlled.

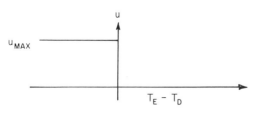

Fig. 3. Bang-bang control.

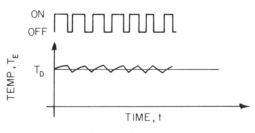

Fig. 4. Time variation of temperature (bang-bang controller).

210

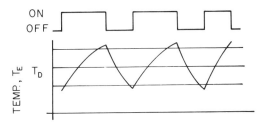

Fig. 5. Time variation of temperature (bang-bang controller with deadband).

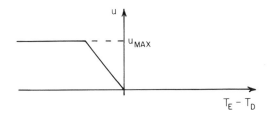

Fig. 7. Proportional controller (with saturation).

Proportional Control

The overshoot may be eliminated by using proportional control. In the case of proportional control, the output of the furnace is proportional to the differenece between the enclosure temperature and the desired temperature. Thus, if the room temperature is very low, the furnace will work hard to overcome the temperature error. However, as T_E approaches T_D, the output of the furnace will decrease, thereby preventing overshoot from occurring to the extent it occurs with a bang-bang controller. The control law for the proportional controller may be written as

$$u = \begin{cases} k_p(T_E - T_D), & T_E \leq T_D \\ 0, & T_E > T_D \end{cases} \quad (2)$$

and is illustrated in Fig. 7. The slope of the line, k_p, is referred to as the controller gain.

The principal disadvantage of the proportional controller, as illustrated, is that it would cause the furnace to operate continuously as long as T_E is less than T_D. This problem can be avoided by including an offset, as illustrated in Fig. 8. In this case, the controller will not turn the furnace on until T_E is less than T_D by the amount OS. The control is proportional to the magnitude of the temperature error, $|T_E - T_D|$, for $T_E - T_D < -OS$. It then remains constant at the value u_{min} until T_E exceeds T_D. There will still be some overshoot with this controller.

Integral Control

An additional type of control that may be used is integral control. An integral controller is designed to increase the controller output (the furnace output in our example) in proportion to the time integral of the error. That is, the longer the enclosure temperature remains below the desired temperature, the more the furnace output will be increased. This is represented analytically as

$$u = k_I \int_{t_o}^{t} [T_E(t) - T_D(t)]dt. \quad (3)$$

An integral controller does not result in a fast response, but does have a stabilizing effect on system response.

Derivative Control

If quicker responses are desired, then derivative control may be used. A derivative controller leads to a control output that is proportional to the rate of change of the error. That is

$$U = k_D \frac{d}{dt}(T_E - T_D). \quad (4)$$

Thus, the controller output increases as the time rate of change of the error increases. Derivative controllers lead to quick responses, but tend to be unstabilizing and are strongly affected by noisy signals.

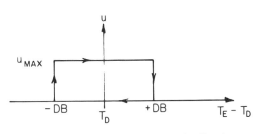

Fig. 6. Bang-bang control with deadband.

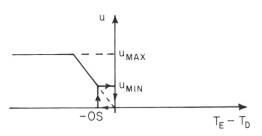

Fig. 8. Proportional controller (with saturation and offset).

PID Controllers

The above control types may be combined to result in a proportional-integral-differential controller, known as a PID controller. The control law is expressed as

$$u = k_p(T_E - T_D) + k_I \int_{t_o}^{t} (T_E - T_D)dt$$
$$+ k_D \frac{d}{dt}(T_E - T_D). \quad (5)$$

A PID controller can potentially reduce energy consumption in three ways. First, it reduces overshoot in heating the space (an improved method of thermostat "anticipation.") Second, if it is desired to implement a proportional actuator, the PID controller could increase the average steady state efficiency of a given size combustion heat exchanger by operating the burner at less than nominal rating most of the time. And third, the PID controller may increase transient efficiency by reducing on-off cycling, and by causing the heat exchanger to be at a temperature that is well below the nominal design temperature at the end of each "on" cycle.

Using the concept of a transfer function, which is defined as the ratio between the output and the input, both expressed in the frequency domain, the equation for the PID controller may be written as

$$\frac{U[s]}{E[s]} = K_p + \frac{k_I}{s} + k_D s. \quad (6)$$

The transfer function is shown in block diagram form in Fig. 9.

Design of Controllers

There are two basic approaches to the design of controllers. The first is referred to as trial and error, while the second is referred to as the analytical design approach. These are briefly described below.

Trial and Error Design

The trial and error method is illustrated in Fig. 10. The trial and error method is the method that is most often used in the design of control systems. This is further illustrated by Fig. 11, which depicts a unity feedback control system. In this figure, the output from the system, c(t), represents the controlled output. The desired response is represented by the input to the block diagram and denoted as r(t). The controller is a proportional feedback controller and is represented by the gain, K_p, acting on the error signal, e(t). The desired response could be interpreted as the enclosure temperature in our example. In the classical trial and error design procedure, one would select a set of performance specifications and then attempt to adjust the controller gain so that the performance specifications would be satisfied. A typical set of performance specifications relative to a step input to a second order system is illustrated in Fig. 12. For the building heating problem discussed previously, one would be primarily concerned with overshoot. The trial and error design process tends to be inefficient and expensive and, when one selects a set of performance specifications, one has no way of knowing whether or not those performance specifications can be satisfied. It is not at all uncommon for considerable amounts of money to be spent on computer simulations in attempting to select a set of gains for selected controllers in order to meet a set of performance specifications. It is often much more efficient to use an analytical design process in which the guesswork is eliminated and where one can determine directly the control strategy that will result in optimizing system performance.

Analytical Design

In applying the analytical design process, one selects a cost function to be optimized. Hence, it it often referred to as optimal control.

One of the issues raised by the integration of backup heaters with solar heating buildings is the interaction of controls. In particular, passive solar heated buildings present a number of problems. First, the time constants (T = RC) of rooms in such buildings are usually much longer than in conventional buildings because both the resistance to heat loss (R) and the thermal storage capacitance (C) are much greater. The thermal capacitance effect alone has a major significance: any overshoot of the set point by the backup heater controller represents a rela-

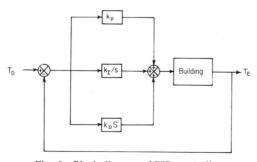

Fig. 9. Block diagram of PID controller.

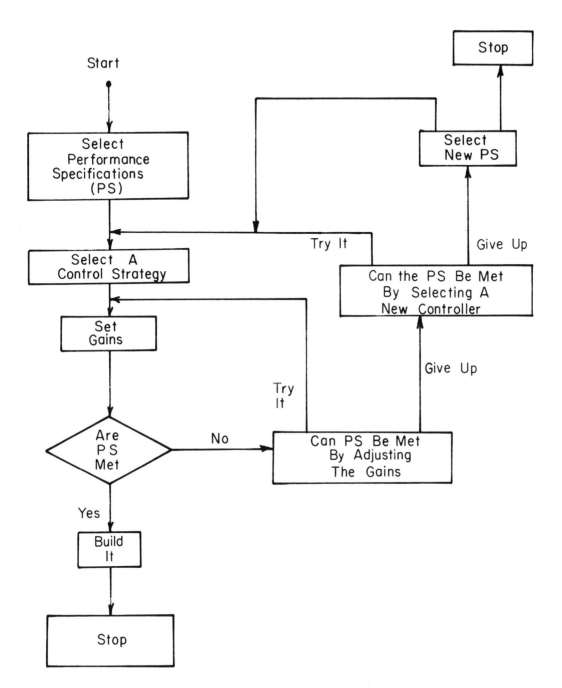

Fig. 10. Trial and error design procedure.

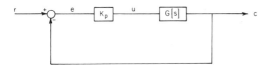

Fig. 11. Block diagram of a "classical" proportional controller.

tively large quantity of energy absorbed by thermal capacitance elements. This overshoot will then persist (decay very slowly) because of the large time constant of the heated space. It is therefore apparent that significant energy savings can be realized by reducing overshoot and by reducing thermostat deadbands. One means of accomplishing these goals is to replace the traditional deadband controller (thermostat) with a type of optimal controller known as an optimal linear regulator.

A linear regulator for this application could be designed to minimize the weighted sum of the root-mean-square deviation from the set point and the mean energy input:

$$J = \int_0^{t_f} (A(U(t))^2 + B(E(t))^2)\,dt \qquad (7)$$

where the deviation from the set point, E(t), is related to the controlled energy input, U(t) (and to other factors, such as weather and occupant behavior, which are both treated as noise) by a

set of differential equations that governs the dynamic behavior of the heated space and backup heating plant. The chief parameters of concern in these governing equations are the room parameters, R and C, defined above.

The ability of this type of controller to save energy has been demonstrated in previous work at Colorado State University [1,2,3,4] on active solar heating systems. However, it is expected that even larger savings will result from passive systems. Not only is the thermal capacitance, C, larger, but the resistance to heat loss, R, is often variable in passive solar heating systems that employ night insulation.

Adaptive Control

The variable R parameter makes it desirable to implement a controller that can identify changes in governing equation parameters and modify its control law accordingly. Such a device is known as an adaptive controller. The theoretical basis of adaptive controllers is well understood and considerable work has been done at CSU and elsewhere in applying identification and parameter estimation theory to solar heating applications. [5,6,7,8]

Because of the variable R parameter and generally longer time constant, RC, in passive solar heating systems, it is expected that the adaptive controller will provide even greater fuel

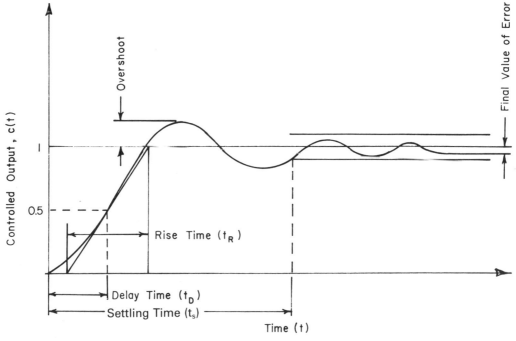

Fig. 12. Typical time domain performance specifications.

savings than the linear regulator did in active solar heating systems.

CONTROLLERS FOR COLLECTING SOLAR ENERGY

Low Temperature Applications *

Control Functions

The controller in a solar space heating and/or domestic hot water system is actually a very simple, but often misunderstood device. Its primary function is to control pump settings for the collection and distribution of solar energy. Secondary functions involve the control of pump and valve settings for freeze protection, high temperature protection, and for other special purposes.

The typical controller is composed of three component subsystems. These are the sensor system, the comparators, and the output devices. The purpose of the sensors is to measure temperatures at various points in the system, for example, in the thermal storage unit and at the collector outlet. The purpose of the comparator subsystem is to compare the differences between temperatures at various points and certain set point temperatures in order to determine whether or not the fluid mover should be turned on or valves should be opened or closed. The purpose of the output subsystem is to send the signals to the output devices (fluid movers and electronic valves) in order to control them. The controller subsystem is shown schematically below:

The most common controller used in the solar industry is the differential controller. More than 50 firms offer differential controllers for sale to the solar industry. While there is variation in design, all are functionally similar in that they have:

    (1) collector and storage sensors,
    (2) low voltage solid state control logic,
    (3) line voltage inputs, and
    (4) logically switched line voltage outputs.

---

* Much of the material in this section has been abstracted from Ref. 9.

The differential control is illustrated in Fig. 13. The collector and storage sensors measure temperatures. The temperature information is used in the controller to control the pump. The collector sensor may be located on the collector absorber plate or in or on the piping near the outlet from the collector. The storage probe may be located either in or on the storage tank (near the tank bottom), or in or on the piping near the tank outlet.

This differential control is popular because of its energy management capability and functional simplicity. The operation function of the control is obtained from the comparative thermostat logic as shown in Fig. 14. The operation is described as follows. It is not worthwhile to attempt to collect solar energy if the temperature of the fluid in the collector is not higher than that in storage. Therefore, ΔT_{OFF} is a preset temperature difference selected so that solar energy will be collected only when the difference between collector and storage temperature exceeds ΔT_{OFF}. If this were the only criterion used for controlling the pump, the controller would cause the pump to cycle on and off repeatedly. This is analogous to the cycling problem discussed above. This is not desirable due to wear on the pump and noise associated with startup and shutdown. Therefore, another preset temperature difference, ΔT_{ON}, is used in the controller to turn on the pumps. That is, when $T_{COLL} - T_{STOR}$ is greater than ΔT_{ON} the controller will turn on the pump and it will keep the pump running until $T_{COLL} - T_{STOR}$ is less than ΔT_{OFF}. This process is called hysteresis.

A typical daily system temperature history is illustrated in Fig. 15. The pump is initially off and the collector fluid is heated by solar radiation in the morning until $T_{COLL} - T_{STOR}$ exceeds ΔT_{ON}. This is indicated by point 1 in Fig. 15. When the pump is turned on it will cause cooler fluid to flow past the collector sensor causing a decrease in T_{COLL} as shown between points 1 and 2. If the temperature decrease is sufficiently great, so that at point 2, $T_{COLL} - T_{STOR}$ is less than ΔT_{OFF}, the controller will turn the pump off. The collector fluid temperature will then increase until $T_{COLL} - T_{STOR}$ is greater than ΔT_{ON}, at which time the controller will again turn on the pump. This is represented by point 3 on Fig. 15. There will again be a decrease in the collector fluid temperature and again the controller will turn off the pump if $T_{COLL} - T_{STOR} < \Delta T_{OFF}$ (point 4). This cycling will stop when there is sufficient

215

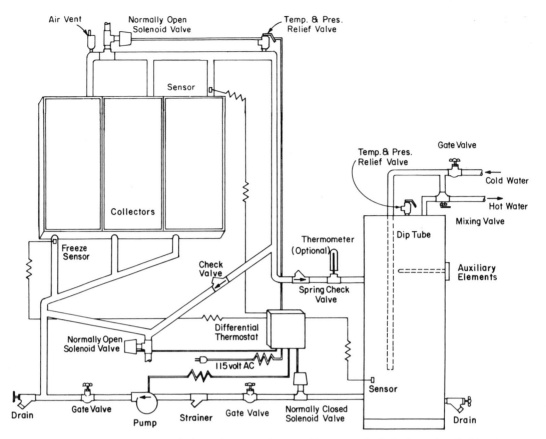

Fig. 13. Direct heating, pump circulation solar water heater with automatic drain-down (applicable also to a two-tank system.) From *Solar Heating and Cooling of Residential Buildings*, Colorado State University, 1980.

heating of the collector fluid so that, when the pump is turned on, $T_{COLL} - T_{STOR}$ remains greater than ΔT_{OFF}, as illustrated by point 5. The pump will remain on throughout the day until, late in the afternoon, $T_{COLL} - T_{STOR}$ becomes less than ΔT_{OFF}, as illustrated by point 6. If the subsequent heating of the collector fluid is not sufficient to raise the temperature enough, the pump will remain off; otherwise there could also be cycling in the afternoon.

In addition to controlling the fluid mover, the controller may also be used to control electronic valves that may be used for freeze protection or high temperature protection. The functions, descriptions, and rationale for the various choices of controller techniques and hardware follow.

Flow Control Techniques

The solar industry used two types of differential controller outputs. The first type, commonly called a bang-bang or on-off controller, provides make-or-break power switching to operate the collector pump either full on or off. The second type modulates the power supplied to the pump during the on time so as to vary the flow rate of the pump as a function of the collector to storage temperature difference. The flow characteristics of both types of operation are illustrated in Figs. 16 and 17.

The hysteresis function for a bang-bang (on-off) controller is illustrated in Figure 16. If the pump is off it will remain off until the temperature difference between the collector and the storage reaches the value ΔT_{ON}. At that point, the controller will turn the pump on and the pump will remain on until the temperature difference becomes less than ΔT_{OFF}, at which point the controller will turn the pump off.

The flow characteristics of a variable flow rate pump are illustrated in Fig. 17. The dashed line indicates the theoretical flow rate as a function of the temperature difference, ΔT. As indicated, the theoretical flow rate is a linear function of ΔT. In reality, however, the actual flow rate is a nonlinear function of ΔT, as illustrated by the solid line on the figure. This will result in less energy being collected than there would be at the slightly higher flow rate.

216

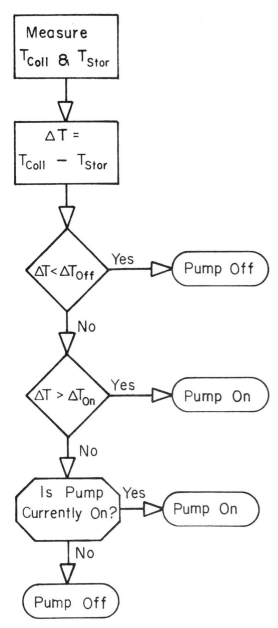

Fig. 14. Comparative thermostat logic of a differential controller.

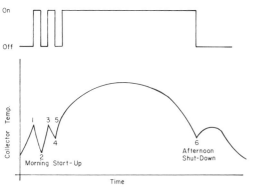

Fig. 15. Typical daily system temperature.

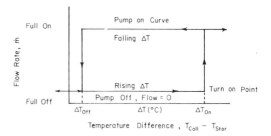

Fig. 16. On-off controller characteristics.

Bang-Bang Control

On-off operation of a fixed flow rate collector pump is the most widely used pump and control configuration. The make-or-break power switching implemented within the controller is familiar to the installer, electrician, and so on. On-off control can be used to switch any motor of appropriate voltage and current rating. Manufacturers use both electromechanical (relays) and solid-state outputs in these controllers.

The choice of the on and off temperature difference set points is critical if excessive pump cycling is to be avoided without significantly reducing the system performance. The coefficient of performance (COP) is a measure of how good the system is and is defined as the ratio between energy collected per unit time by the fluid in the collectors and external power delivered to the pump or fan motor. That is

$$COP = \frac{\dot{Q}_u \, (kw)}{P \, (kw)}$$

where $\dot{Q}$ represents the time rate at which energy is collected and P represents the power input to the fluid mover. Obviously, if COP $\leq$ 1, it is not worthwhile to have the fluid mover turned on. If ΔT_{ON} is set too high, the fluid mover will not be on as long as at a lower value for ΔT_{ON} and there will be less energy collected, thereby reducing the value for COP. If ΔT_{ON} is set too low there will be excessive pump cycling.

The preferred on-off temperature set points to be used in bang-bang control depend on many

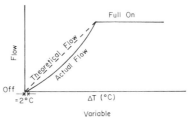

Fig. 17. Variable flow controller characteristics.

characteristics that affect the system. These include the collector performance, the heat exchanger features, the piping thermal capacitance, the system's flow rate, the system loss characteristics, probe placement, and probe accuracy and precision. Collector probe placement is particularly critical in its effect upon the system's transient thermal performance. Analytic results, generally confirmed by field experience, show that for efficient operation the collector probe should be near the collector outlet at approximately 90% of the length along the flow path. If the sensor is located too near the exit, it turns the fluid mover on later, and if it is located too near the entrance, excessive cycling will be introduced. An exception to this general rule is that a more downstream location may be preferable for a system located in a climate that experiences large amounts of radiation to the night sky.

The deadband (hysteresis) between the off settings and the on temperature settings minimizes pump "hunting," that is, frequent turn on and off, during morning start-up and afternoon shut-down or intermittent cloud conditions. The off setting must be sufficiently high to accommodate normal sensor inaccuracies, but low enough to otherwise allow collection of as much solar energy as possible, while assuring that collection will be stopped when parasitic power for the pumps exceeds the value of the solar energy gain; that is, if the COP is less than or equal to 1, collection should be stopped. It has been determined from the experience of the controller industry that if an on-to-off temperature ratio of from 4/1 up to 6/1 is maintained, with the ΔT_{OFF} temperature ranging from 1° to 2°C, satisfactory operation will result in most types of flat plate liquid transport systems. It is usually found that the higher ratio of ΔT_{ON} to ΔT_{OFF} is applied when ΔT_{OFF} is at its lower range of values. That is, if ΔT_{OFF} is approximately 1°C, then ΔT_{ON} would be approximately 6°C. Similarly, the lower value of the ratio of on-to-off temperature differentials is applied when ΔT_{OFF} is near its upper range of values. Thus, the hysteresis in liquid systems is normally in the range of 6° to 8°C, with some systems as high as 11°C.

Air transport systems normally require ΔT_{OFF} temperatures of 5°C and as high as 11°C, with an 8° to 14°C hysteresis. This slightly wider range for hysteresis is required to compensate for the effects of higher pressure drops in air systems, the relatively low specific heat of air,

and the relatively low efficiency of blowers (as compared to pumps). This is apparent from the following equation. From the relation presented earlier for COP, it may be shown that for COP = 1

$$\Delta T_{OFF} \propto \frac{\Delta p}{\eta C_p}$$

where Δp is the pressure drop in the collector loop, η is the efficiency of the fluid mover (fan or pump), and C_p is the specific heat of the transport fluid. Hence, for higher pressure drops and lower specific heats and lower efficiencies, the value for ΔT_{OFF} will increase. Also, because of the low specific heat of air, an on temperature twice the off temperature is usually adequate to satisfactorily reduce cycling. The actual set points are usually higher than the values given by the above equation in order to account for inaccuracies in the sensors.

Solar designers and installers should note that while pump cycling often arouses concern in homeowners, it causes very little (if any) equipment degradation if limited to a small number (approximately six cycles or fewer) during a 10–15 minute startup or shutdown period. Cycle rates of this magnitude have little impact on motor winding temperatures or on the pump's bearing life or on the controller output switches which are generally rated for more than 2 million closures. Conversely, pump cycling indicates a longer collection duty cycle and, hence, more stored energy. Nonetheless, should the designer or installer wish to limit the cycling in an on-off controlled installation, the system operation is generally best served if it is done by lowering the flow rate. The use of controllers, with manually adjusted or timing delays, is recommended only for experimental installations because of the very high field failure rate of components required for this type of application. Controllers with time delays, additional circuit logic, or variable flow rate outputs should be considered in installations where cycling is chronic.

Effects of Temperature Settings on Cycling Rates
The problem of selecting appropriate values for ΔT_{ON} and ΔT_{OFF} for a bang-bang controller to be used in the collection process in a solar energy system has been studied extensively by numerous investigators.[10,11,12,13,14] As discussed above, it is well known that if ΔT_{ON} is set too high, then the amount of useful energy collected will be reduced, and if ΔT_{ON} is set too low, or ΔT_{OFF} is

set too high, then excessive cycling of the fluid mover will occur. It is clear that it is not worthwhile to operate the fluid mover in order to collect energy if the rate at which energy is collected is less than or equal to the power required to collect the solar energy. This provides a useful criterion for determining the appropriate value for ΔT_{OFF}. The value for ΔT_{ON} may then be determined by setting the ratio of ΔT_{ON} to ΔT_{OFF} to a value based on the radiation and number of cycles acceptable. The development of these relationships for the temperature settings is described in the following paragraphs.

The relationship for ΔT_{OFF} is well known and has been presented in Refs. 9, 14, and 15. It is

$$\Delta T_{OFF} \geq k P_m / \dot{m} C_P \quad (8)$$

where P_m is the power delivered to the motor of the fluid mover (W), and k is a constant coefficient determined by the value of the energy being displaced. For electric resistance heating, one would normally choose k to be unity. Also, $\dot{m}$ represents the mass flow rate and C_P the specific heat of the fluid. The values for ΔT_{OFF} as determined from Eq. (8) are typically lower than those used in practice. The values used in practice are higher to account for sensor inaccuracies and drift, and are typically on the order of 1° to 2° for liquid based systems.

It has also been shown [9,14,16] that if

$$\Delta T_{OFF} \leq (F_R U_L A_c / \dot{m} C_p) \Delta T_{ON} \quad (9)$$

then cycling will not occur. Equation (9), however, is not necessary and sufficient to prevent cycling. It is very conservative and leads to extremely high values for ΔT_{ON}. Previous studies [10,14] relating to this problem have been based on numerical simulations and have provided useful results. An analytical development is presented in the following pages.

Analytical Formulation

More realistic values for ΔT_{ON} than those given by Eq. (9) may be obtained from the following analysis.

When the fluid mover is on, the rate of collection of energy may be represented by

$$\dot{Q}_u = \dot{m} C_p (T_o - T_i) = F_R A_c [G_T(\tau\alpha) \\ - U_L(T_i - T_A)] \quad (10)$$

where

$$\dot{m} = \text{mass flow rate, kg} \cdot \text{s}^{-1}$$

C_p = specific heat of refrigerant, kJ/kg · °C

T_o = collector outlet temperature, °C

T_i = collector inlet temperature, °C

F_r = heat removal factor

A_c = collector area, m²

G_T = instantaneous radiation incident on the collector, W/m²

$(\tau\alpha)$ = transmittance-absorptance product

U_L = collector loss coefficient, W/m² · °C

T_A = ambient temperature, °C.

Let

$$T_o = T_i + \Delta T_{ON} \quad (11)$$

and

$$T_i = T_s \quad (12)$$

where T_s represents the storage temperature (°C); then Eq. (10) may be written as

$$\dot{m} C_p \Delta T_{ON} = F_R A_c [G_T(\tau\alpha) \\ - U_L(T_s - T_A)]. \quad (13)$$

Solving for $T_s - T_A$ results in

$$T_s - T_A = F_R(\tau\alpha) G_T / F_R U_L \\ - (\dot{m} C_p / A_c) \Delta T_{On} / F_R U_L. \quad (14)$$

Phase Plane Representation

Equation (14) represents a family of straight lines in a "phase space" of $T_s - T_A$ and G_T, as illustrated in Fig. 18. The slope of each line is given by the ratio between $F_R(\tau\alpha)$ and $F_R U_L$. Now consider just two members of this family of lines: the first for $\Delta T_{ON} = \Delta T_{OFF}$ and the second for $\Delta T_{ON} = (\dot{m} C_p / A_c) \Delta T_{OFF} / F_R U_L$. These two lines define a cycling region, as illustrated in Fig. 19. To the left of the $\Delta T_{ON} = \Delta T_{OFF}$ line, the fluid

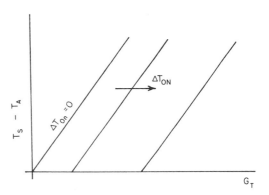

Fig. 18. Phase plane representation.

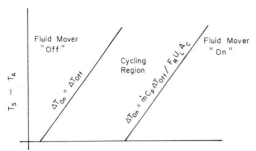

Fig. 19. Cycling region.

mover remains off, whereas to the right of the line for ΔT_{ON}, as determined from Eq. (9), the fluid mover will remain on. The cycling rate is constant along the straight lines within the cycling region. The cycling rate may be determined as follows.

Cycling Rate

The variation in fluid temperature along the length of the collector may be approximated by a linear relationship. The maximum value for radiation, $G_{T_{NC}}$, for which cycling will not occur is that value for which the outlet temperature, T_o, is just equal to $T_S + \Delta T_{OFF}$. This determines the slope of the T_o versus x line. An increase in G_T will increase the slope. To determine $G_{T_{NC}}$, consider a steady flow problem for which an energy balance equation results in

$$G_{T_{NC}} = [\dot{m}C_p \, \Delta T_{OFF} + F_R U_L A_c \\ (T_S - T_A)]/F_R(\tau\alpha) A_c. \quad (15)$$

The time to empty the collector of the fluid initially in the collector at the time the fluid mover is turned on is given by

$$T_E - m/\dot{m}. \quad (16)$$

The collector outlet temperature at the time t_E is given by

$$T_o = T_S + [F_R(\tau\alpha)G_T - F_R U_L \\ (T_S - T_A)](A_c/\dot{m}C_p). \quad (17)$$

The time required to raise the collector fluid outlet temperature from T_o to $T_S + \Delta T_{ON}$ may be determined from an energy balance equation for the "no flow" condition; the result is

$$(mC_p/A_c)(T_S + \Delta T_{ON} - T_o)/t_H \\ = A_c[G_T(\tau\alpha) - U_L(T_S - T_A)]. \quad (18)$$

where t_H represents the time to heat the fluid. Substituting for T_o results in

$$t_H = \left(\frac{m}{\dot{m}}\right) \left[\frac{(\dot{m}C_p/A_c) \Delta T_{ON}}{G_T(\tau\alpha) - U_L(T_S - T_A)} - F_R \right]. \quad (19)$$

The total cycle time, T, is given by

$$T = \left(\frac{m}{\dot{m}}\right)$$

$$\left[\frac{(\dot{m}C_p/A_c) \Delta T_{ON}}{G_T(\tau\alpha) - U_L(T_S - T_A)} - F_R + 1 \right]. \quad (20)$$

This expression is valid for $G_{T_{min}} < G_T < G_{T_{NC}}$, where

$$G_{T_{min}} = U_L(T_S - T_A)/(\tau\alpha) \quad (21)$$

and

$$G_{T_{NC}} = [\dot{m}C_p \, \Delta T_{OFF} + F_R U_L A_c \\ (T_S - T_A)]/F_R(\tau\alpha) A_c. \quad (22)$$

The cyling rate, ω, is given by

$$\omega = 1/T. \quad (23)$$

Number of Cycles

The number of cycles that will occur while the value for the radiation remains within the cycling region may be obtained by integrating the equation for the cycling rate with respect to time. That is,

$$N = \int_{t_o}^{t_f} \omega(t) \, dt \quad (24)$$

where N represents the number of cycles, and t_o and t_f represent the times corresponding to $G_{T_{min}}$ and $G_{T_{NC}}$, respectively. In order to perform this integration analytically, it is necessary ot have an analytical representation for the radiation, G_T, as a function of time. Schiller et al.[14] performed numerical studies using the following model:

$$G_T(t) = G_{T_M} \sin\left(\frac{\pi}{12} t\right). \quad (25)$$

220

This results in

$$N = \left(\frac{m}{\dot{m}}\right) \int_{t_o}^{t_f} \left[\frac{- U_L(T_S - T_A) + G_{T_M}(\tau\alpha) \sin(\pi t/12)}{(\dot{m}C_p/A_c) \Delta T_{ON} - (1 - F_R)U_L(T_S - T_A) + (1 - F_R)(\tau\alpha)G_{T_M} \sin(\pi t/12)} \right] dt \quad (26)$$

which may be integrated in closed form to give

$$N = INT \left\{ \frac{(\dot{m}C_p/A_c)}{(mC_p/A_c)} \frac{12}{\pi} \left\{ \frac{1}{(1 - F_R)} (x_f - x_o) - \frac{2(\dot{m}C_p/A_c)\Delta T_{ON}}{(1 - F_R)\sqrt{B}} \right. \right.$$

$$\left. \left. \tan^{-1}\left\{ \frac{C \tan(x_f/2) + E}{\sqrt{B}} \right\} - \tan^{-1}\left\{ \frac{C \tan(x_o/2) + E}{\sqrt{B}} \right\} \right\} \right\} \quad (27)$$

where

$$x_f = \sin^{-1}\left\{ \frac{1}{G_{T_M}} \left\{ \frac{(\dot{m}C_p/A_c)\Delta T_{OFF}}{F_R(\tau\alpha)} + \frac{F_R U_L}{F_R(\tau\alpha)} (T_S - T_A) \right\} \right\} \quad (28)$$

$$x_o = \sin^{-1}\left\{ \frac{1}{G_{T_M}} \frac{F_R U_L}{F_R(\tau\alpha)} (T_S - T_A) \right\} \quad (29)$$

$$B = \left\{ \frac{\dot{m}C_p}{A_c} \Delta T_{ON} - (1 - F_R)U_L(T_S - T_A) \right\}^2 - \left\{ (1 - F_R)G_{T_M}(\tau\alpha) \right\}^2 \quad (30)$$

$$C = \frac{\dot{m}C_p}{A_c} \Delta T_{ON} - (1 - F_R)U_L(T_S - T_A) \quad (31)$$

$$E = (1 - F_R)G_{T_M}(\tau\alpha). \quad (32)$$

Note that rather than using $\dot{m}/m$ in Eq. (27), the ratio has been expressed in terms of fluid capacitance and collector capacitance per unit collector area. If the term (mC_p/A_c) represents fluid plus collector mass and specific heat, then the equation for t_E is not exact. However, it will be shown that the error involved in this approximation is small for representative collectors.

Comparisons with Earlier Results
An earlier study[14] was performed in which a numerical simulation program was used in order to determine the number of cycles that would occur as a function of ΔT_{ON} for various values of G_{T_M}, and for given values of ΔT_{OFF} and collector parameters. The parameter values used were as follows:

High Gain, High Flow, Clear Day

$(\tau\alpha) = 0.84$
$F_R = 0.937$
$(mC_p/A_c) = 14.3 \text{ kJ/m}^2 \cdot °C$
$(\dot{m}C_p/A_c) = 511 \text{ kJ/h} \cdot m^2 \cdot °C$
$\Delta T_{OFF} = 1.7°C$
$G_{T_M} = 946 \text{ W/m}^2$
$T_S - T_A = 25°C$
$U_L = 3.97 \text{ W/m}^2 \cdot °C$
$\Delta T_{ON} = 5°C$

High Gain, Low Flow, Clear Day

$(\tau\alpha) = 0.84$
$F_R = 0.93$
$(mC_p/A_c) = 14.3 \text{ kJ/m}^2 \cdot °C$
$(\dot{m}C_p/A_c) = 306 \text{ kJ/h} \cdot m^2 \cdot °C$
$\Delta T_{OFF} = 1.7°C$
$G_{T_M} = 946 \text{ W/m}^2$
$T_S - T_A = 25°C$
$U_L = 3.97 \text{ W/m}^2 \cdot °C$
$\Delta T_{ON} = 5°C$

Low Gain, High Flow, Clear Day

$(\tau\alpha) = 0.84$
$F_R = 0.937$
$(mC_p/A_c) = 14.3 \text{ kJ/m}^2 \cdot °C$
$(\dot{m}C_p/A_c) = 511 \text{ kJ/h} \cdot m^2 \cdot °C$
$\Delta T_{OFF} = 1.7°C$
$G_{T_M} = 473 \text{ W/m}^2$
$T_S - T_A = 36.1°C$
$U_L = 3.97 \text{ W/m}^2 \cdot °C$
$\Delta T_{ON} = 5°C$

Low Gain, Low Flow, Clear Day

$$(\tau\alpha) = 0.84$$
$$F_R = 0.93$$
$$(mC_p/A_c) = 14.3 \text{ kJ/m}^2 \cdot {}°C$$
$$(\dot{m}C_p/A_c) = 306 \text{ kJ/h} \cdot \text{m}^2 \cdot {}°C$$
$$\Delta T_{OFF} = 1.7°C$$
$$G_{T_M} = 473 \text{ W/m}^2$$
$$T_S - T_A = 36.1°C$$
$$U_L = 3.97 \text{ W/m}^2 \cdot {}°C$$
$$\Delta T_{ON} = 5°C$$

The cycles reported for each of the above cases were 10, 2, 61, and 10, respectively. The analytical solution presented here in Eq. (27) leads to 8, 2, 37, and 7 cycles for each time the value for the incident radiation passes through the cycling region for each of the above cases. This agreement between the analytical and numerical results is sufficiently satisfactory that the analytical representation may be used for purposes of analysis. For example, for a given collector one may develop graphs of the number of cycles as a function of ΔT_{ON}, as illustrated in Fig. 20. Also, the analytical solution enables one to conduct sensitivity studies without having to resort to (possibly) lengthy numerical solutions using a computer. This is illustrated in the next section. Note that for the first case, cycling ceases at a value for ΔT_{ON} of approximately 14°C, whereas, use of Eq. (9) would result in a value of 38°C.

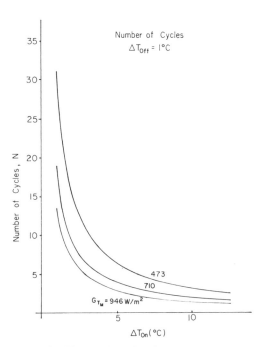

Fig. 20. Number of cycles versus ΔT_{ON}.

Sensitivity Analysis

The equation for the number of cycles is of the form

$$N = N\left\{ \frac{\dot{m}C_p}{A_c}, \frac{mC_p}{A_c}, F_R, \tau\alpha, U_L, \Delta T_{ON}, \right.$$
$$\left. \Delta T_{OFF}, T_S - T_A, G_{T_M} \right\}. \quad (33)$$

The sensitivity of N to any of the parameters may be determined analytically by partial differentiation with respect to that parameter. For multiple parameter variations, one has

$$\Delta N = \sum_{j=1}^{n} \left(\frac{\partial N}{\partial x_j} \right) \partial x_j \quad (34)$$

where ∂x_j represents the variation in parameter x_j. Since uncertainties exist in some of the parameter values, in particular ΔT_{OFF} and the collector capacitance term, there will be a corresponding uncertainty in N. The most likely value for the uncertainty in N is given by

$$N = \left\{ \sum_{j=1}^{n} \left(\frac{\partial N}{\partial x_j} \delta x_j \right)^2 \right\}^{1/2}. \quad (35)$$

The sensitivity of N to ΔT_{OFF} is given by

$$\partial N/\partial \Delta T_{OFF} = \frac{(\dot{m}C_p/A_c)(12/\pi)}{(1 - F_R)\sqrt{1 - A^2}\ G_{T_M} F_R(\tau\alpha)}$$
$$\left\{ 1 - \frac{(\dot{m}C_p/A_c)\ \Delta T_{ON}\ C}{[B + (C \tan D + E)^2] \cos^2 D} \right\} \quad (36)$$

where

$$A = \frac{1}{G_{T_M}} \left\{ \frac{(\dot{m}C_p/A_c)\ \Delta T_{OFF}}{F_R(\tau\alpha)} \right.$$
$$\left. + \frac{F_R U_L}{F_R(\tau\alpha)}(T_S - T_A) \right\} \quad (37)$$

$$D = 0.5 \sin^{-1}(A) \quad (38)$$

and B, C, and E are given by Eq. (30), (31), and (32).

The sensitivity to collector capacitance is simply

$$\frac{\partial N}{\partial(mC_p/A_c)} = -N/(mC_p/A_c). \quad (39)$$

A method for estimating the collector capacitance is presented in Ref. 17.

The sensitivity to ΔT_{OFF} for the case considered previously is illustrated in Fig. 21. It is apparent that the number of cycles is strongly sensitive to ΔT_{OFF} when ΔT_{ON} is relatively near

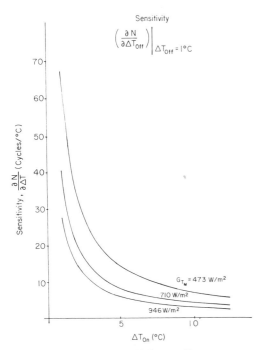

Sensitivity

$$\left(\frac{\partial N}{\partial \Delta T_{Off}}\right)\Big|_{\Delta T_{Off}=1°C}$$

$G_{T_M} = 473$ W/m^2

710 W/m^2

946 W/m^2

ΔT_{On} (°C)

Fig. 21. Sensitivity to ΔT_{ON}.

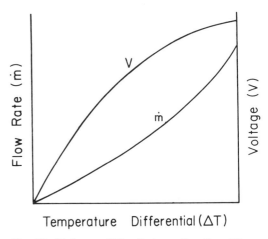

Fig. 22. Voltage and Flow Rate as a Function of Temperature Difference.

ΔT_{OFF}. Hence, the location of the collector outlet sensor is very important in determining the number of cycles.

Proportional Control

The variable flow controller (often marketed as a "proportional" controller) operates by controlling pump speed. It increases power to the pump as a function of collector to storage temperature difference, thereby increasing collector fluid flow rate. The major operational advantage produced by the variable output controller comes from the elimination of the typically large ΔT_{ON}, so that a longer collection duty cycle is obtained in intermittent cloudy weather without cycling. This advantage is achieved with low speed pump operation and without pump cycling that would occur in an on-off system with a low turn-on temperature and small deadband (hysteresis).

Variable flow is achieved by interrupting, on a cycle-to-cycle basis, the electrical power to the pump to generate the flow versus ΔT characteristics of Fig. 22 and shown by a dotted line on Fig. 17. The theoretical flow shown on Fig. 17 is not generally achieved because of the third power relationship of current to flow, nonlinear I^2R losses in the pump motor and highly variable system head characteristics. Instead, the variation in flow rate with ΔT is similar to that shown

in Fig. 22. The power interruption is accomplished by using a triac output that may be used either to chop the input power's individual half wave (see Fig. 23b) or to modulate the number of wave forms being delivered to the pump (Fig. 23a). The duty cycle of the interruption is a ramp function of collector-to-storage temperature differential, typically from 1° to 2°C low setting to 6° to 9°C high setting (varies with manufacturer) for the differential. This means that there will not be any flow until the temperature difference between collector and storage exceeds the lower value, say 1°C. At that point, the fluid will begin to flow at a low rate. As the temperature difference increases the flow rate will increase until, at the high limit, say 9°C, the fluid will be circulating at its maximum rate. There will be no further increase in the flow rate even though the temperature difference may increase. The low fluid flow rate that is initiated through the collector array as soon as the low turn-on temperature is reached, allows for a longer period of collection than would be realized with the use of a bang-bang controller with high ΔT_{ON}. Thermal shock to the collectors is also reduced because of the reduced flow rate; also, the requirement for a large on-off hysteresis is eliminated. Claims for increased energy collected, when compared to high offset temperature on-off systems with a low thermal capacitance collector (a lightweight collector) by as much as 6% to 8% in cloudy and overcast conditions, are made by some manufacturers. However, an on-off controller with a lower value for ΔT_{ON} would result in more energy collected but also more cycling.

The main advantage realized by the variable flow rate controller is in increasing the difference

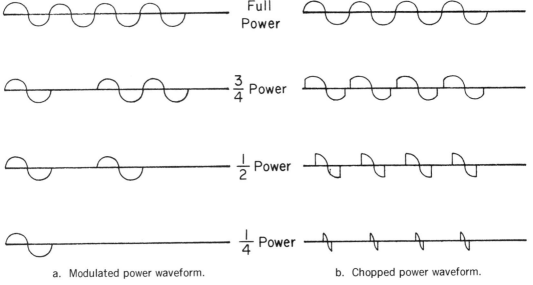

a. Modulated power waveform.

b. Chopped power waveform.

Fig. 23. Output waveforms for variable flow rate controller.

between useful energy collected and parasitic losses incurred in collecting the energy, that is, in increasing the COP. This increase in COP results from the nonlinear relationship between parasitic losses and flow rate. The parasitic losses are defined as the power required to circulate the fluid. Power required to move a fluid is the product of the flow rate and the head loss. This may be expressed as

$$P_F = \dot{m} \, \Delta p \qquad (40)$$

where Δp is the pressure rise in the fluid mover and is equivalent to the head loss for a constant flow rate $\dot{m}$. The head loss depends on whether the flow is laminar or turbulent. If the flow is laminar (smooth pipes, no abrupt transitions, and so on) then Δp is proportional to $\dot{m}$ to the first power. That is,

$$\Delta p_{\text{LAMINAR}} = c_1 \, \dot{m}. \qquad (41)$$

If the flow is turbulent, then the head loss is proportional to $\dot{m}$ to the second power. That is,

$$\Delta p_{\text{TURBULENT}} = c_2 \, \dot{m}^2. \qquad (42)$$

Hence, the limiting cases for the power delivered to the fluid are

$$P_F = \begin{cases} c_1 \, \dot{m}^2, \text{ Laminar flow} \\ c_2 \, \dot{m}^3, \text{ Turbulent flow.} \end{cases} \qquad (43)$$

The flow in any solar DHW system will be a mixture of laminar and turbulent flows. Therefore, as the flow rate increases the parasitic

losses typically increase as a function of the flow rate to the x power, where x is between 2 and 3. This is illustrated by the concave upward curve in Fig. 24. Hence, if a bang-bang controller is used and the controller turns the fluid mover on, it will be at the maximum flow rate and hence the parasitic losses will also be maximized. On the other hand, the variable flow rate controller will result in the fluid being moved at a lower rate during periods of low solar radiation, and thereby experiencing lower parasitic losses. The rate of collection of useful energy varies with mass flow rate as illustrated by the concave downward curve in Fig. 24. The difference between the two curves varies with flow rate, $\dot{m}$. The best flow rate would be at the point where

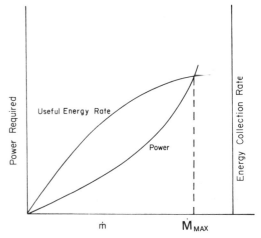

Fig. 24. Power required and rate of energy collected as functions of flow rate.

224

the separation between the two curves is the greatest. If one is only interested in maximizing energy collected without regard for parasitic losses, then the bang-bang controller should be used.

The above discussion is not altered if one considers the power delivered to the fluid. The efficiency of the motor may be defined as

$$\eta = \frac{\text{Power Delivered to the Fluid, } P_F}{\text{Power Delivered to the Motor, } P_M}. \quad (44)$$

Hence

$$P_M = \frac{P_F}{\eta}. \quad (45)$$

The efficiency, η, also varies with flow rate, but in a manner such that it continues to be a concave upward function of flow rate.

Since the pump is actually being turned on and off by the triac switch many times each second, producing a lower V_{rms} which may have a residual DC component, only pumps with series wound motors (permanent capacitor, shaded pole, and so on) are safe for this type of operation. The effects of either the low voltage chopped wave, with the resulting rapid on-off operation, or the full wave modulated outputs should be reviewed with the system's pump supplier. Other drawbacks of the variable controller include the lower reliability of triac output devices when field wired, and the particularly large electromagnetic interference (EMI) generated by the chopped wave modulated controller. This interference can cause problems with reception of radio and television sets. Also, the full wave modulated output is usually not recommended for air transport systems because of a high occurrence of pulsed duct noise. Installers have also noted that both modulation techniques yield low starting torques in high static pressure operation. Conversely, both aid the installer in filling or priming the system.

The actual performance gains realized by the variable flow rate controller may be less than the theoretical gains because the residual DC component may preclude the reduction of parasitic power even when the pump is running slowly. The DC component will result unless the solid-state switch chops or modulates the input power waveform precisely at the correct moment. If a DC component is present, this will increase the power delivered to the motor, and will therefore increase the parasitic power.

These criticisms and the industry's disagreement on efficiency superiority of on-off versus variable flow notwithstanding, many users feel that the noncycling characteristic and the space age technology image are highly desirable in today's marketplace.

Optimal Control

The rate at which energy is collected by a solar heating system can be increased by increasing the flow rate through the collectors. Increasing this flow rate, however, increases the power required to drive the fluid mover. This was illustrated in Fig. 24. The rate of energy collected is a concave downward function of mass flow rate, whereas the parasitic power is a concave upward function of mass flow rate. These two curves are shown intersecting at a flow rate of $\dot{M}$, which represents an upper bound for $\dot{m}$. Most systems will operate at a maximum flow rate, $\dot{m}_{max}$, less than $\dot{M}$. The flow rate in a bang-bang controller will be either zero or $\dot{m}_{max}$ (neglecting viscous effects). In either event, this would not normally maximize the difference between $\dot{Q}_u$ and P. Ideally, one would choose the flow rate, $\dot{m}_{opt}$, that maximizes the difference between the two curves shown in Fig. 24. This is not normally at $\dot{m}_{max}$.

The difficulty in determining $\dot{m}_{opt}$ stems from the fact that $\dot{Q}_u$ is a function of solar and weather conditions and is a dynamically changing variable. Therefore, $\dot{m}_{opt}$ also changes with respect to time. The methods of optimal control of dynamic systems should be applied to determine the value for $\dot{m}_{opt}$ at each point in time.

The pumping costs are small in relation to the energy collected in a well-designed solar heating system, and it could be argued that it is not important to determine $\dot{m}_{opt}$. Not all systems, however, are well designed. Some, in fact, have actually increased consumer's utility bills. A controller which selects $\dot{m}_{opt}$ reduces the possibility of this occurring because it controls collector flow rate so as to maximize the difference between the solar power and the associated pumping power.

The problem of maximizing the difference between solar power and fluid moving power was first approached by Kovarik and Lesse.[18] Their approach resulted in a two point boundary value problem which was solved numerically. Their solution, however, could not be implemented in a practical controller because it was not a function of measurable states of the system. Winn and Hull presented an approximate analytical solution to the problem which was possible to implement.[1] They showed close agreement between their simulation results and those of Kovarik and Lesse. Winn and Winn [19] developed a controller that implemented the optimal strategy and tested it in Solar House II at Colorado State University.

A review of the optimal control problem for various combinations of objective functions and system models is presented below.

Optimization

There has been a considerable amount of discussion in the past concerning bang-bang control versus proportional control versus optimal control of solar heating and cooling systems and the effects of various factors such as collector capacitance on the control. An analytical derivation of the optimal controller is presented in this section. First, consider a general statement of the Pontryagin Maximum Principle that will be applied to each optimization problem.

Consider a dynamic system described by

$$\frac{dx}{dt} = g(x, u, t) \qquad (46)$$

$$x(t_o) = x^o \qquad (47)$$

where x is an n-dimensional state vector, u is an m-dimensional control vector, and t is the independent variable (time).

The state is to be transferred from its initial state, x^o, to some final state x^f, which may be either free or fixed. The objective is to effect the transfer in such a manner that some scalar function

$$J = \int_{t_o}^{t_f} L(x, u) dt \qquad (48)$$

is maximized. That is, determine the optimal control, $u_{opt}(t)$, such that J is a maximum. It may be shown that the optimal control must satisfy the following set of necessary conditions.

The optimal control must maximize the Hamiltonian, H, where H is defined as

$$H = \lambda^T g + L \qquad (49)$$

where λ^T is a vector of adjoint variables given by

$$\frac{d\lambda^T}{dt} = -\frac{\partial H}{\partial x}. \qquad (50)$$

If the final state is free, the differential equation for λ^T has the folllowing boundary condition (referred to as the transversality condition):

$$\lambda^T(t_f) = 0. \qquad (51)$$

Any control law that does not satisfy this set of necessary conditions cannot be optimal. Now consider applying these necessary conditions to a solar energy system.

The optimal mass flow rate in a solar energy collection system depends upon the statement of what is to be optimized (the objective function) and the model used to describe the performance of the system (the constraints). Each combination of objective function and constraints has an optimal, time-dependent, flow rate. To demonstrate this, consider the following examples of analyses using Pontryagin's Maximum Principle.

A. *Maximize energy collected while ignoring collector and storage dynamics and pumping power.*

Maximize

$$J = \int_o^{t_f} \dot{Q}_u(\dot{m}, t) dt \qquad (52)$$

where $\dot{m}$, the mass flow rate through the collector, is the control. The Hamiltonian is

$$H = \dot{Q}_u$$

where $\dot{Q}_u$ is given by the Hottel-Whillier equation

$$\dot{Q}_u = F_R A_c [H_T \tau \alpha - U_L (T_s - T_A)] \qquad (53)$$

and F_R, the heat removal factor, is

$$F_R = \frac{\dot{m} c_p}{U_L A_c} [1 - \exp(-F' U_L A_c / \dot{m} c_p)]. \qquad (54)$$

The Hottel-Whillier equation is an accepted model for accurately depicting collector performance. With these substitutions, the Hamiltonian is

$$H = A_c [H_T \tau \alpha - U_L (T_s - T_A)]$$
$$\frac{\dot{m} c_p}{U_L A_c} [1 - \exp(-F' U_L A_c / \dot{m} c_p)] \qquad (55)$$

which is a monotonically increasing function of $\dot{m}$.

Clearly, to maximize H, the flow rate must be

$$\dot{m}_{opt} = \begin{cases} \dot{m}_{max}, & \text{if } [H_T \tau \alpha - U_L(T_s - T_A)] > 0 \\ 0, & \text{if } [H_T \tau \alpha - U_L(T_s - T_A)] \leq 0 \end{cases} \qquad (56)$$

In other words, to maximize the energy collected with no constraints, the optimal control is bang-

bang and the switching condition is based on available net energy according to Eq. (56).

Next, the effects of including storage and collector thermal capacitance are considered.

B. *Maximize energy collected subject to storage dynamics while ignoring collector dynamics and pumping power.*

Maximize

$$J = \int_0^{t_f} \dot{Q}_u dt \qquad (57)$$

subject to

$$\frac{dT_s}{dt} = \frac{1}{C_s} \dot{Q}_u. \qquad (58)$$

The Hamiltonian is

$$H = \lambda \frac{\dot{Q}_u}{C_s} + \dot{Q}_u = \left(\frac{\lambda}{C_s} + 1\right) \dot{Q}_u \qquad (59)$$

where λ is an adjoint variable defined by

$$\frac{d\lambda}{dt} = -\frac{\partial H}{\partial T_s} = \left(\frac{\lambda}{C_s} + 1\right) F_R U_L A_c \qquad (60)$$

where

$$\lambda(t_f) = 0. \qquad (61)$$

For the final condition on λ to be satisfied, it is necessary that

$$0 < \frac{\lambda}{C_s} + 1 \leq 1. \qquad (62)$$

Proof:

i) Suppose that at some arbitrary time, t_1, $\lambda > 0$. Then $\dot{\lambda} > 0$, for all $t \geq t_1$, and the transversality condition will be violated. Hence, necessarily, $\lambda \leq 0$, for all t.

ii) Suppose that at some arbitrary time, t_1, $\lambda \leq -C_s$. Then $\dot{\lambda} \leq 0$, for all $t \geq t_1$, and again the transversality condition will be violated. Hence, necessarily, $\lambda > -C_s$, for all t.

iii) Combining conditions (i) and (ii) provides that

$$-C_s < \lambda \leq 0. \qquad (63)$$

Dividing by C_s and adding 1 gives Eq. (62). Therefore, with the substitutions for $\dot{Q}_u$ and F_R as before, to maximize H, the flow rate must again be

$$\dot{m}_{opt} = \begin{cases} \dot{m}_{max}, & \text{if } [H_T \tau \alpha - U_L(T_s - T_A)] > 0 \\ 0, & \text{if } [H_T \tau \alpha - U_L(T_s - T_A)] \leq 0. \end{cases} \qquad (64)$$

It is clear that the addition of storage dynamics to the model does not affect the form of the optimal control. The optimal control remains bang-bang and the condition for switching between the two control levels (0 or $\dot{m}_{max}$) is the same as in the previous case.

C. *Maximize energy collected subject to storage and collector dynamics while ignoring pumping power.*

Maximize

$$J = \int_0^{t_f} \dot{Q}_u dt \qquad (65)$$

subject to

$$\frac{dT_s}{dt} = \frac{1}{C_s} \dot{Q}_u \qquad (66)$$

$$\frac{dT_c}{dt} = \frac{1}{C_c} [\dot{Q}_u - \dot{m} c_p(T_c - T_s)]. \qquad (67)$$

The Hamiltonian is

$$H = \lambda_1 \frac{\dot{Q}_u}{C_s} + \lambda_2 \left[\frac{\dot{Q}_u}{C_c} - \frac{\dot{m} c_p}{C_c}(T_c - T_s)\right] + \dot{Q}_u \qquad (68)$$

and

$$\frac{d\lambda_1}{dt} = \left(\frac{\lambda_1}{C_s} + \frac{\lambda_2}{C_c} + 1\right) F_R U_L A_c - \frac{\lambda_2}{C_c} \dot{m} c_p \qquad (69)$$

$$\frac{d\lambda_2}{dt} = \frac{\lambda_2}{C_c} \dot{m} c_p \qquad (70)$$

where

$$\lambda_1(t_f) = \lambda_2(t_f) = 0. \qquad (71)$$

Integrating Eq. (70) provides

$$\lambda_2(t) = \lambda_2(t_f) \exp\left[\int_t^{t_f} \dot{m}(\tau) \frac{c_p}{C_c} d\tau\right]. \qquad (72)$$

Since the exponential function is nonzero and $\lambda_2(t_f) = 0$, necessarily $\lambda_2(t) \equiv 0$.

Therefore, the problem reduces to one which is identical to B and the optimal flow rate is

$$\dot{m}_{opt} = \begin{cases} \dot{m}_{max}, & \text{if } [H_T \tau \alpha - U_L(T_s - T_A)] > 0 \\ 0, & \text{if } [H_T \tau \alpha - U_L(T_s - T_A)] \leq 0. \end{cases} \qquad (73)$$

Therefore, the addition of collector dynamics does not affect the form of the optimal control. The optimal controller is still a bang-bang controller; this changes only when parasitic power is included in the objective function.

227

D. *Maximize the difference between solar power collected and parasitic losses while including storage and collector dynamics.*

Maximize

$$J = \int_0^{t_f} (C_1 \dot{Q}_u - P) \, dt \qquad (74)$$

subject to

$$\frac{dT_s}{dt} = \frac{1}{C_s} \dot{Q}_u \qquad (75)$$

$$\frac{dT_c}{dt} = \frac{1}{C_c} [\dot{Q}_u - \dot{m}c_p(T_c - T_s)]. \qquad (76)$$

The Hamiltonian is

$$H = \lambda_1 \frac{\dot{Q}_u}{C_s} + \lambda_2 \left[\frac{\dot{Q}_u}{C_c} - \frac{\dot{m}c_p}{C_c}(T_c - T_s) \right]$$
$$+ C_1 Q_u - C_2 m^\alpha \qquad (76)$$

and

$$\frac{d\lambda_1}{dt} = \left(\frac{\lambda_1}{C_s} + \frac{\lambda_2}{C_c} + C_1 \right) F_R U_L A_c - \frac{\lambda_2}{C_c} \dot{m}C_p \qquad (77)$$

$$\frac{d\lambda_2}{dt} = \frac{\lambda_2}{C_c} \dot{m}C_p \qquad (78)$$

where

$$\lambda_1(t_f) = \lambda_2(t_f) = 0. \qquad (79)$$

From Eq. (78) and (79)

$$\lambda_2 \equiv 0.$$

Because H does not continuously increase with $\dot{m}$, the optimal control is found by setting

$$\frac{\partial H}{\partial \dot{m}} = 0 \qquad (80)$$

and solving for $\dot{m}$. To allow an explicit solution for $\dot{m}$, the exponential function in F_R is expressed as a Taylor series and truncated after second order terms so that

$$F_R \approx F' - F'^2 U_L A_c / 2\dot{m}C_p. \qquad (81)$$

For the system in Solar House II, this approximation for F_R is accurate to within 0.6% at the highest flow rate and to within 6.5% at the lowest flow rate. The resulting equation for the optimal mass flow rate is

$$\dot{m}_{opt} = \left[\frac{C_1 f F'^2 U_L A_c}{2\alpha C_2 C_p} \right]^{1/(\alpha + 1)} \qquad (82)$$

where f, the available energy, is zero if $(T_c - T_s)$ is negative or, if $(T_c - T_s)$ is positive, determined from either

$$f = \dot{m}C_p(T_c - T_s)/F_R \qquad (83)$$

if the collector fluid mover is on, or

$$f = U_L A_c(T_c - T_s) \qquad (84)$$

if the fluid mover is off.[35] Note that the optimal control is not bang-bang. Additional details of this derivation are presented in Ref. 19.

The 6.5% error in F_R resulting from the use of Eq. (81) causes some error in the selection of $\dot{m}_{opt}$; however, this error is small. Because the exponent in Eq. (82) is small, the choice of $\dot{m}_{opt}$ is within 2% of the desired value. This results in an insignificant reduction in the objective function, J.

It has been suggested that the optimal control for a system which includes the effects of collector capacitance is a bang-bang control.[14] We have shown analytically that this is true so long as parasitic losses are ignored. The preceding analysis clearly shows that the optimal control is not bang-bang where parasitic losses are considered.

Parasitic Power

An accurate analysis of this problem requires an accurate relationship between flow rate and power required. Some authors suggest a linear relationship between flow rate and power required.[14] The fan laws suggest a cubic relationship between air flow rate and fan power.[20] A theoretical analysis of each system will yield the proper relationship for each system.

The power required to move a fluid is proportional to the product of the mass flow rate and the pressure increase across the pump or fan. The pressure increase can be determined by analyzing the rest of the system. For an open system, such as with a trickle collector, the head against which the pump moves the liquid is predominantly the result of the increase in elevation from the storage tank below the collector array to the end of the pipe at the top of the array. This head loss (pressure drop) is independent of flow rate. For a closed system, the elevation change throughout the system does not affect the head required of the fan or pump.

Except during start up when the fluid is being accelerated, the pressure drop in a closed system is primarily due to viscous effects. If the flow is laminar in a particular section of the system, the pressure drop is linear with flow rate. If the flow is turbulent, the pressure drop is proportional to flow rate raised to some power. For flows just barely turbulent, the exponent is slightly greater than one; for flows with Reynolds numbers above 10^6, the exponent is two.[21] In a system with flow in some sections at high Reynolds numbers and some at low Reynolds numbers, the pressure drop is proportional to the flow rate raised to some power between one and two. With power delivered to the flow proportional to the product of flow rate and pressure change, the power is proportional to the flow rate raised to some power between two and three. This analysis of flow in a closed system applies to any fluid, either liquid or gas.

In the system in CSU's Solar House II, the Reynolds number varies widely from place to place. The Reynolds number, based on data taken in March 1980, is about 80,000 in the ducts to and from the collectors, about 3,000 in the collectors, and about 200 in the rock box. The head developed by the fan at the different flow rates is shown in Table 1. A least-squares fit of these data shows the increase in pressure to be proportional to the flow rate raised to 1.4; therefore, the power delivered to the air is proportional to the flow rate raised to 2.4.

Comparisons of Control Strategies
Analytical and experimental results obtained at Colorado State University indicate that, in a climate similar to that of Fort Collins, Colo., the optimal controller as derived above (case D) will result in an annual performance improvement on the order of 5% relative to a bang-bang controller with fixed ΔT_{ON} and ΔT_{OFF}. The effect of the optimal controller is more pronounced in regions having lesser amounts of solar radiation. This is illustrated by the results shown in Table 2, in which performance comparisons between bang-bang controllers and optimal controllers are presented for varying daily amounts of solar radiation.[35]

ΔQ_u represents the percentage change in useful energy collected by the system using the

Flow rate (kg/s)	Fan ΔP (mmH$_2$0)
.156	5
.263	11
.378	19
.510	33

Table 1 Solar House II flow rate—fan pressure relationship

H KJ/m^2 – day	T_A °C	AIR		LIQUID	
		ΔQ_u %	ΔJ %	ΔQ_u %	ΔJ %
14292	−7	~1	5	5	6
14190	+2	~1	4	4	5
9751	−5	~1	9	22	26
6482	−7	~1	23	74	82

Table 2 Comparison of Control Strategies for Varying Weather Data

optimal controller compared with that collected using the bang-bang controller. Similarly, ΔJ represents the change in the objective function defined in case D above. It is clear that the performance improvement of the optimal controller is more pronounced at lower levels of solar radiation.

SUMMARY
It has been shown that the optimal control strategy for the collection of solar energy is a bang-bang controller with a switching condition given by Eq. (56). However, if one considers parasitic losses and maximizes the difference between the rate of solar energy collected and the parasitic power, then the optimal controller is not bang-bang. Instead, it must follow the control law given by Eq. (82). A relationship between the number of cycles and the solar radiation has also been presented for a bang-bang controller for given collector parameters.

High Temperature Applications
The principal high temperature applications involve solar thermal power systems and solar industrial process heat systems. In both cases, the control problems typically involve controlling flow rates, pressures, temperatures, or the position of a receiver. The problem of positioning a heliostat, for example, is a classical linear servomechanism problem for which there is extensive literature available, primarily from the aerospace industry. The primary goal in this area as it relates to solar thermal power systems is cost reduction. Current efforts are underway at SERI in this regard. Problems of controlling flow rates, pressures or temperatures, arc classical problems in process control. There is also extensive literature available in this area, primarily from the chemical industry. Some papers in this area that relate specifically to solar applications are Ref. 22, 23, 24, 25, and 26. In addition, SERI personnel have compiled information for a handbook[27] which includes material on controls

for IPH systems. Some design guidelines presented in the handbook follow.

Design Guidelines
Flow control: Proportional plus integral (reset) controllers are used almost exclusively. The process is typically very fast and noisy, and the flow measurement is usually nonlinear (square). The controller has low gain and fast reset. Linear valves for differential pressure measurement and equal percentage valves for linear measurement are generally used.

Pressure control: For a liquid, the process is fast and noisy with most of the lags in the control system, and the measurement is nonlinear (square). Linear valves and proportional plus reset controllers with low gain and fast reset are used. For a gas, proportional controllers with high gain are sufficient and the valve characteristics are not critical. For vapor pressure control, equal percentage valves and proportional-integral-derivative controllers are employed. The process is slow compared to other pressure processes.

Temperature Control: Almost all the temperature control problems in solar applications are heat transfer problems, and are characterized by long time constants and slow reaction rates. Distance-velocity lag (also known as dead time) is common. The measurement lag can pose a serious problem, especially if the thermal system is protected with a well. The measurement time constant depends on the mass and surface area of the bulb (or the well), the fluid being measured, and its velocity past the bulb.

Processes dominated by one large capacity, for example, storage tanks in air heating systems, can be controlled with on-off controllers.

Proportional-plus-reset control is used in smaller capacity systems where load changes are large and where distance-velocity and measurement lags are important. Most shell and tube heat exchangers fall into this category. Derivative control becomes helpful provided the distance-velocity lag is not the dominant secondary dynamic element, for example, as with a batch reactor.

Controlling temperature by mixing hot and cold streams is more nearly a blending problem than a heat exchange problem. Good mixing and fast temperature measurement are the keys to simplifying the control job. Proportional-plus-integral controllers should be used.

In general, the following guidelines can be used in selecting controllers. Use proportional control where

- the cycling action, due to on-off control, is undesirable;
- set point changes are small or infrequent; and
- the steady-state deviation between the set point and the process variable, that is, the offset, can be tolerated.

Use integral control where

- the offset must be reduced or eliminated; and
- the set point changes are frequent.

Do not add integral control when

- startup overshoot must be eliminated; and
- the process can be controlled with high-gain proportional control.

Add derivative action to proportional control when

- the distance-velocity lag (for example, dead time in the pipes) is smaller than either of at least two linear lags (for example, storage tanks) in the process loop.

Do not use derivative control if

- the distance-velocity lag is significant;
- the process is noisy.

A Representative Problem
Control of the outlet temperature of a molten salt solar thermal central receiver is very important for satisfactory operation. A central receiver is simply a heat exchanger, usually mounted on a tower, which is used to absorb solar energy. The solar radiation is reflected onto the central receiver by heliostats, which are automatically controlled tracking reflectors. The outlet temperature from the central receiver must be controlled to within about 5°C, even though the solar energy input can vary significantly due to clouds, dust, and so on. DeRocher et al.[26] have examined this problem and have presented results for various control strategies, in particular for a PID type of controller. The problem is to vary the flow rate in the event of a disturbance (change in solar energy input) in order to maintain the outlet temperature within 5°C of its set point ($\sim$567°C). The control gains for the PID controller were selected through a trial and error design process, as previously discussed, and were

$$k_P = 1 \text{ kg/s} \cdot {}^\circ\text{C}$$
$$k_I = 0.01 \text{ kg/s}^2 \cdot {}^\circ\text{C}$$
$$k_D = 10 \text{ kg/}{}^\circ\text{C}.$$

Results from simulation studies of the system using the PID controller are shown in Fig. 25.

It is clear from Fig. 25 that the PID type of controller causes the system to respond satisfactorily to step disturbances. The outlet temperature remains within 2.5°C of the desired value on the high side and within 1.4°C on the low side. The overshoot and settling time for flow are relatively small. However, the temperature response is somewhat oscillatory. Improved performance could probably be obtained by either optimizing the gains for the PID controller or by using the analytical design process to determine the optimal control law, which may be different from PID control.

CONTROLLERS FOR DISTRIBUTING ENERGY (ACTIVE SPACE AND WATER HEATING SYSTEMS)

Bang-Bang Controllers

The simplest form of controller to be used for space and water heating systems is a bang-bang controller with deadband, as discussed above. A two-stage thermostat is used for these systems. The first stage is provided in order to first use the available internal energy of the thermal storage unit to satisfy the heating requirements. If there is insufficient energy available in storage, then the second stage of the thermostat will cause the auxiliary heater to operate. This controller will result in overshoot, as previously discussed, but in addition to the overshoot problem, it has a much more serious disadvantage in the case of electric auxiliary. This is illustrated in Fig. 26, which is based on an analysis of several systems in the service territory of Florida Power and Light.[28] It is apparent from the curves on Fig. 26 that the solar systems have peak demands that are in phase with the peaks in the utility system-wide load curve. Also, since the solar systems provide energy during the middle portions of the daylight periods, the load factors for the solar systems are considerably lower than for conventional systems, resulting in higher unit costs for the utility. This need not be the case however, and by using improved control strategies, the problem may be avoided and the total costs may be reduced.

The same problem exists for space and water heating systems as illustrated in Fig. 27.[29] The building electric load peaks are almost

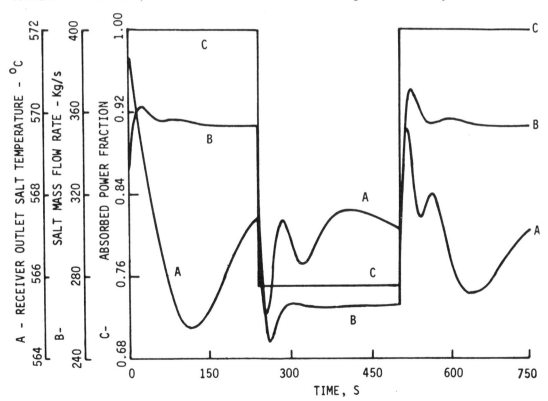

Fig. 25. System response to step function disturbances in absorbed flux (from Ref. 26).

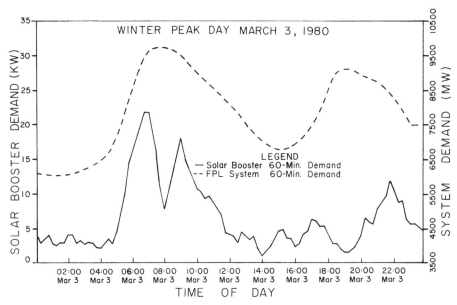

Fig. 26. Solar booster performance for 14 customers.

exactly in phase with the utility systemwide load. Also, as in the previous case, the solar system provides a significant percentage of the energy requirements during the day when the utility systemwide load is decreasing. Obviously, the cost to the utility to supply auxiliary

energy for these systems is higher than if the solar systems were not used.

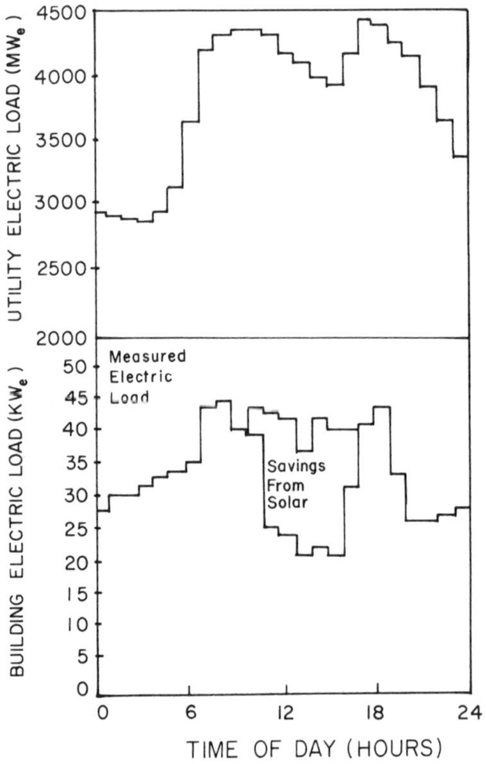

The utility's cost of supply is composed of energy- and capacity-related costs. Energy-related costs are composed mainly of the costs of fuel, taking into account the efficiencies of the generating units needed to satisfy the system-wide load. Energy costs generally increase with systemwide load, but are essentially independent of the power draw at the residence. Capacity-related costs are composed of the various costs associated with owning and maintaining generating equipment. The capacity related costs are charged whenever the building has an electric power draw coincident with the utility's on peak period. A methodology that may be used for computing the cost of supply is illustrated in Fig. 28.

The bang-bang control strategy for discharge from solar storage is to use the energy as required to satisfy the load until the temperature in storage reaches some limiting temperature. The problem with this strategy is that, if the available energy in storage is depleted before the end of the utility high demand period, a very large coincident demand may result. This is illustrated in Fig. 29, which shows the HVAC electrical demand for a residence in Albuquerque, New Mexico, on the day for which the utility experiences its maximum systemwide load for the heating season. The available internal energy from storage has been depleted by 5 p.m. and there is a resulting electrical demand of ap-

Fig. 27. Load profile for a delicatessen.

232

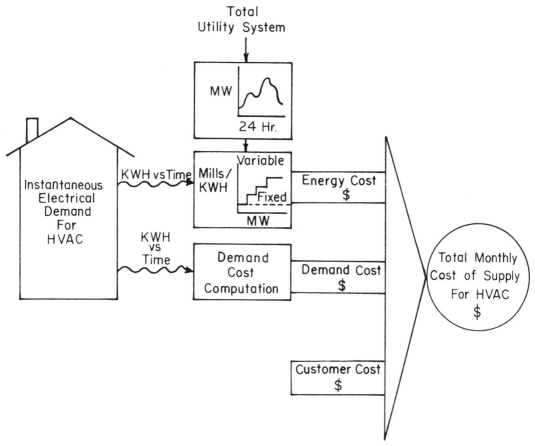

Fig. 28.   Methodology for computing the cost of supply (abstracted from Ref. 30).

proximately 10 kW at the residence at the time. The cost of supply for energy provided by the utility is shown in Fig. 30. The residential demand, shown in Fig. 29, is quite large at the same time that the cost of supply, shown in Fig. 30, is high. Use of the solar system with conventional

control has not led to much of a decrease in the generation capabilities required of the utility.

The bang-bang control strategy results in a very low electrical demand during the middle of the day. Comparison of the HVAC electrical demand with the utility fuel cost of supply shows

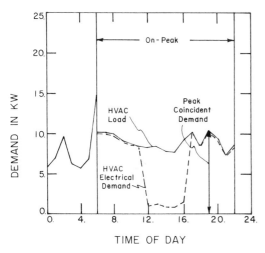

Fig. 29. Conventional strategy for solar storage on January 14.

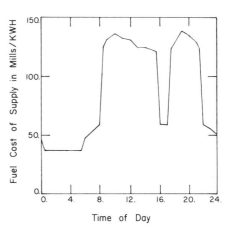

FUEL COST OF SUPPLY ON UTILITY PEAK DAY

Fig. 30.   Fuel cost of supply.

that the solar contribution occurs at a low utility demand time of day. In fact, the characteristics of the heating system under bang-bang control tend to accentuate the difference between the peaks and valleys on the utility load curve. An improved control strategy can significantly reduce the coincident demand and consequently reduce the maximum power required of the utility.

Georgia Institute of Technology, in work for the U.S. Department of Energy, performed a numerical study of the management of electric backup demand for solar heating and cooling in applications to buildings. It was concluded, on the basis of numerical simulations, that the application of load management controls results in financial benefits to the customer and to the utility, and it was recommended that a complete system study be conducted to determine the optimal mix of auxiliary energy sources.[31]

The Franklin Institute also contracted with the U.S. Department of Energy to examine control strategies for electrical peak shaving. It was concluded that reductions in peak loads imposed on electric utilities from the electrical backup energy requirements of active solar heating or cooling systems can be achieved with the use of the thermal storage device inherent in such systems and through the application of appropriate control strategies.[32] The Electric Power Research Institute (EPRI) contracted Colorado State University to develop appropriate control strategies.[33] Some results from that study are presented in the following sections.

Proportional Controllers
As indicated above, the bang-bang controller may lead to high coincident demands since the available energy in storage may be depleted before the end of the utility on-peak period. Therefore, it seems intuitively obvious that a controller that proportions out the available energy from storage so that storage is not depleted until the end of the on-peak period would result in a lower overall cost.

There are two variations to this proportional discharge strategy. In variation one, if energy is available it is discharged from storage during the off-peak period. In variation two, no energy is discharged during the off-peak period. Variation two is designed to keep more energy available for on-peak use. Simulations have been performed comparing these two variations of the proportional control.[33] The use of variation one

resulted in an annual coincident demand cost of supply of $1,809 compared to $1,704 for variation two for a residence in Albuquerque. This decrease in coincident demand cost results from the fact that variation two causes more energy to be available for on-peak use. The fuel cost of supply was $648 with 9,808 kWh of solar energy collected when variation one was used. For variation two, the fuel cost of supply was $637 with 9,238 kWh of solar energy collected. By having more energy available for on-peak use with variation two compared to variation one, the average storage temperature is higher, and, therefore, less solar energy is collected. However, the fuel cost of supply does not necessarily decrease as more solar energy is collected. The cost of fuel depends on the overall utility demand at the time of use as shown earlier in Fig. 30. Because the use of variation two causes the collected solar energy to be saved for use only during the on-peak period, it replaces only expensive energy. For that reason, even though less solar energy is collected and more electrical energy is required, the fuel cost of supply is lower for variation two.

The result of using the proportional controller in place of the bang-bang controller is shown in Fig. 31.

Optimal Controllers (Space and Water Heating)
To determine the best control strategy for the discharge of solar storage, the methods of dynamic optimization may be employed. The optimization problem is formulated to determine the on-peak power draw to minimize

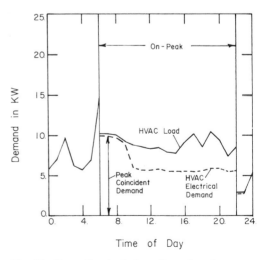

Fig. 31. Proportional strategy for solar storage on January 14.

$$J = \int_{t_0}^{t_f} [\dot{Q}_{\text{ON-PEAK}}(t)]^2 dt \qquad (85)$$

subject to the dynamic equations of the enclosure and the storage,

$$C_E dT_E/dt = \dot{Q}_{\text{ON-PEAK}} + \dot{Q}_{ST} - \dot{Q}_{\text{LOAD}} \qquad (86)$$

$$C_S dT_S/dt = \dot{Q}_{\text{SOL}} - \dot{Q}_{ST} - \dot{Q}_{\text{LOSS}} \qquad (87)$$

where:

t_0 = time at the beginning of the on-peak period

t_f = time at the end of the on-peak period

C_E = thermal capacitance of the storage, kJ/°C (Btu/°F)

T_E = enclosure temperature, °C(°F)

C_S = thermal capacitance of the storage, kJ/°C (Btu/°F)

T_S = storage temperature, °C(F)

$\dot{Q}_{\text{ON-PEAK}}$ = on-peak resistance heat rate, W (Btu/h)

$\dot{Q}_{ST}$ = rate of energy removal from storage W (Btu/h)

$\dot{Q}_{\text{SOL}}$ = rate of supply of solar energy to storage, W (Btu/h)

$\dot{Q}_{\text{LOAD}}$ = heating load, W (Btu/h)

$\dot{Q}_{\text{LOSS}}$ = rate of energy loss from storage, W (Btu/h).

Minimizing J will minimize the product of coincident demand ($\dot{Q}_{\text{ON-PEAK}}$) and on-peak energy consumption ($\dot{Q}_{\text{ON-PEAK}}dt$) while satisfying the dynamic equations which describe the thermal performance of the residence and the storage. The dynamic equations used in this problem formulation assume a well-mixed storage and a uniform (but not constant) enclosure temperature.

Several reasonable assumptions make this a very straightforward problem to solve. First, in a well-designed system, the storage losses are small and safely can be ignored in formulating a control strategy. Second, the enclosure temperature is held nearly constant during the on-peak period by the thermostat in the enclosure. Certainly, when diversity is considered, the utility sees the enclosure temperature as constant, and, therefore, the time derivative of enclosure temperature is effectively zero. Third, the boundary conditions on the storage temperature can be specified. For heating, the storage temperature at t_0 is the maximum storage temperature, $T_{S_{\text{MAX}}}$, which is determined by reasonable design practice. The temperature at t_f is the temperature, $T_{S_{\text{MIN}}}$, below which the pump will be deactivated. With these assumptions, the problem may be solved using Pontryagin's Maximum Principle as described earlier.

Let the control, u, be defined as

$$u = \dot{Q}_{\text{ON-PEAK}}. \qquad (88)$$

Minimize

$$J = \int_{t_0}^{t_f} u^2(t) dt \qquad (89)$$

subject to

$$C_S \frac{dT_S}{dt} = \dot{Q}_{\text{SOL}} - \dot{Q}_{\text{LOAD}} + u. \qquad (90)$$

Let

$$x_1 = T_S \qquad (91)$$

$$\dot{x}_2 = u^2, \ x_2(0) = 0. \qquad (92)$$

Then we want to maximize the scalar function

$$\phi = -x_2(t_f) \qquad (93)$$

as this is equivalent to minimizing J. The Hamiltonian is constructed as

$$H = \frac{\lambda_1}{C_S}(\dot{Q}_{\text{SOL}} - \dot{Q}_{\text{LOAD}} + u) + \lambda_2 u^2 \qquad (94)$$

where the adjoint variables are defined by

$$\dot{\lambda}_1 = -\frac{\partial H}{\partial x_1} = 0 \qquad (95)$$

$$\dot{\lambda}_2 = -\frac{\partial H}{\partial x_2} = 0. \qquad (96)$$

The optimal control is found by

$$\frac{\partial H}{\partial u} = 0 = \frac{\lambda_1}{C_S} + 2\lambda_2 u \qquad (97)$$

or

$$u_{\text{OPT}} = \frac{-\lambda_1}{2\lambda_2 C_S}. \qquad (98)$$

To find λ_1 and λ_2 use

$$\lambda^T(t_f) = \frac{\partial \phi}{\partial x_f} - \mu^T \frac{\partial \psi}{\partial x_f} \qquad (99)$$

where ψ represents terminal conditions on the states; then

$$\frac{\partial \phi}{\partial x_f} = (0 \ -1) \qquad (100)$$

$$\frac{\partial \psi}{\partial x_f} = (1 \ \ 0). \qquad (101)$$

235

Therefore,

$$\lambda_1 = -\mu \tag{102}$$

$$\lambda_2 = -1 \tag{103}$$

and

$$u_{OPT} = \frac{-\mu}{2C_S}. \tag{104}$$

To evaluate μ, use the constraint equation,

$$C_S \dot{x}_1 = \dot{Q}_{SOL} - \dot{Q}_{LOAD} - \frac{1}{2C_S}\mu. \tag{105}$$

Integrating,

$$C_S(T_{S_{MIN}} - T_S) = \int_t^{t_f} \left[\dot{Q}_{SOL} \right.$$
$$\left. - \dot{Q}_{LOAD} - \frac{1}{2C_S}\mu \right] dt. \tag{106}$$

Solving for μ and substituting gives

$$u_{OPT} = \frac{1}{t_f - t} \left\{ \int_t^{t_f} [\dot{Q}_{LOAD} - \dot{Q}_{SOL}] dt \right.$$
$$\left. - C_S(T_S - T_{S_{MIN}}) \right\} \tag{107}$$

or

$$\dot{Q}^*_{ON-PEAK} = \overline{\dot{Q}}_{LOAD} - \overline{\dot{Q}}_{SOL}$$
$$- \frac{C_S(T_S - T_{S_{MIN}})}{(t_f - t)} \tag{108}$$

where the overbar indicates an average over the entire on-peak period and * represents the optimal value. The optimal rate of energy delivery from storage is

$$\dot{Q}^*_{ST} = \dot{Q}_{LOAD} - \overline{\dot{Q}}_{LOAD} + \overline{\dot{Q}}_{SOL}$$
$$+ \frac{C_S(T_S - T_{S_{MIN}})}{(t_f - t)}. \tag{109}$$

Implementation of this strategy requires a knowledge of the current heating load, the total heating load during the on-peak period, and the total amount of solar energy delivered to storage during the on-peak period. For a simple off-peak storage heating system, $\overline{\dot{Q}}_{SOL}$ is zero.

A proportional discharge strategy results if one assumes that the heating load is constant throughout the day and no solar energy will be collected. In that case,

$$\dot{Q}^*_{ST} = \frac{C_S(T_S - T_{S_{MIN}})}{(t_f - t)}. \tag{110}$$

This is the same control law that was obtained earlier by intuition.

The load profile that results from use of the optimal controller is shown in Fig. 32.

Implementation of the optimal discharge strategy for solar storage is not easily achieved. In addition to needing an estimate of the heating load for the entire on-peak period, the optimal discharge of solar storage requires advanced knowledge of the amount of solar energy to be collected. This points up the need for accurate short term weather predictions. The Prototype Regional Observing and Forecasting Service (PROFS) program may prove to be useful in this regard.[33]

Cost Comparisons
Cost of supply comparisons for the bang-bang, proportional and optimal controllers are shown in Table 3. These costs were developed for the Public Service Company of New Mexico for the 1990 time period. The baseline heating system was considered to be an electric central heating system.

Optimal Control of DHW Systems
It was noted in the above discussion that, in order to implement the optimal control strategy for space heating systems, it is necessary to predict the future load for the building. In the case of domestic hot water systems, this is much more easily realized. The daily hot water load profile for DHW systems is relatively uniform and is not influenced by the weather conditions to any appreciable extent. A typical daily hot water load profile is shown in Fig. 33. We ob-

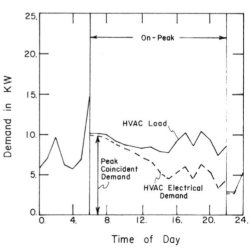

Fig. 32. Optimal strategy for solar storage on January 14.

	Baseline Heating System	Solar Heating System		
		Conventional Strategy	Proportional Strategy	Optimal Strategy
Annual peak HVAC coincident Demand	13.32 kW	13.07 kW	10.29 kW	9.59 kW
HVAC coincident Demand annual Cost of supply in 1990 dollars and % of baseline	2551 100%	2227 87%	2020 79%	1905 75%
HVAC Fuel annual Cost of supply in 1990 dollars and % of baseline	1396 100%	705 51%	730 52%	691 49%
HVAC Total annual Cost of supply in 1990 dollars and % of baseline	3946 100%	2932 74%	2749 70%	2596 66%

Table 3 Solar storage discharge strategy comparisons

serve that there is a morning peak and an evening peak. One might assume that, in the case of a solar plus off-peak storage system, the best strategy would be to provide the energy for the morning peak from the use of the off-peak power and collect the energy necessary for the evening peak from solar. The question that arises is, what happens if there is no solar energy available on a given day? Perhaps a better control strategy then would be to use enough off-peak power to increase the internal energy available in the hot water storage tank to a level to satisfy the entire daily load.

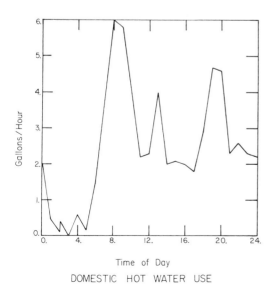

Time of Day

DOMESTIC HOT WATER USE

Fig. 33. A typical daily hot water profile.

Rather than use a trial and error type of approach, again we consider the analytical approach.

System Model (1 Tank System)

$$C_S \frac{dT_S}{dt} = \dot{Q}_{\text{ON-PEAK}} + \dot{Q}_{\text{SOLAR}} - \dot{Q}_{\text{HW}} - \dot{Q}_{\text{LOSS}}. \quad (111)$$

We want to minimize

$$J = \int_{t_o}^{t_f} [\dot{Q}_{\text{ON-PEAK}}(t)]^2 dt \quad (112)$$

subject to the above constraint.

The terminal constraint on T_S is

$$T_{S_f} = T_{S_{\text{MIN}}}.$$

Let $u = \dot{Q}_{\text{ON-PEAK}}$;

then

$$J = \int_{t_o}^{t_f} u^2(t) dt \quad (113)$$

and

$$C_S \dot{T}_S = u + \dot{Q}_{\text{SOL}} - \dot{Q}_{\text{HW}} - \dot{Q}_{\text{LOSS}} \quad (114)$$

where

$$\dot{Q}_{\text{LOSS}} = UA(T_S - T_E) \quad (115)$$

$$\dot{Q}_{\text{SOL}} = F_R A_C [H_T \tau \alpha - U_L(T_S - T_A)] \quad (116)$$

$$\dot{Q}_{\text{HW}} = F(t) \quad \text{(known)}. \quad (117)$$

Simplify: Assume $\dot{Q}_{\text{SOL}} = 0$, $\dot{Q}_{\text{LOSS}} = 0$.

Then we have

$$C_S\dot{T}_S = u - \dot{Q}_{HW}(t) \qquad (118)$$

Let

$$\dot{x}_1 = \dot{T}_S$$

$$\dot{x}_2 = u^2, \ x_2(0) = 0$$

and

$$\phi = -x_2(t_f).$$

Then

$$H = \lambda_1 \frac{1}{C_S}[u - \dot{Q}_{HW}] + \lambda_2 u^2 \qquad (119)$$

where

$$\dot{\lambda}_1 = -\frac{\partial H}{\partial x_1} = 0$$

$$\qquad\qquad\qquad (120)$$

$$\dot{\lambda}_2 \equiv -\frac{\partial H}{\partial x_2} = 0$$

and

$$\lambda^T(t_f) = \frac{\partial \phi}{\partial x_f} - \mu^T \frac{\partial \psi}{\partial x_f}$$

$$= [0 \quad -1] - \mu[1 \quad 0] \quad (121)$$

$$= (-\mu - 1).$$

Control equation:

$$\frac{\partial H}{\partial u} = 0.$$

Therefore,

$$\frac{\lambda_1}{C_S} + 2\lambda_2 u = 0 \qquad (122)$$

$$u_{opt} = -\frac{\lambda_1}{2\lambda_2 C_S} = -\frac{\mu}{2C_S}. \qquad (123)$$

Determine μ from the constraint equation:

$$\dot{x}_1 = \frac{1}{C_S}\left(\frac{\mu}{2C_S} - \dot{Q}_{HW}\right). \qquad (124)$$

Integrate

$$x_1(t_f) - x_1(t) = \frac{1}{C_S}\int_t^{t_f}\left(\frac{\mu}{2C_S} - \dot{Q}_{HW}\right)dt \quad (125)$$

Substituting

$$C_S(T_{S_{MIN}} - T_S(t)) = \frac{\mu}{2C_S}(t_f - t)$$

$$+ \int_t^{t_f} - \dot{Q}_{HW}(t)\,dt \quad (126)$$

$$\frac{\mu}{2C_S} = C_S\frac{C_S(T_{S_{MIN}} - T_S)}{t_f - t} + \int_t^{t_f}\dot{Q}_{HW}(t)\,dt \quad (127)$$

Hence,

$$u_{opt} = \dot{Q}^*_{ON\text{-}PEAK} = \frac{1}{t_f - t}\left[\int_t^{t_f}\dot{Q}_{HW}(t)\,dt\right.$$

$$\left. - C_S(T_S(t) - T_{S_{MIN}})\right]. \quad (128)$$

As developed here, u_{opt} is the amount of charge to be put into the hot water tank during the on-peak period. If we include solar inputs and tank losses, or both, we have a modification to the above control law, but that is easily determined. Since $\dot{Q}_{HW}$ is known, the optimal control is a known function of time. This can be easily implemented using electronic devices.

RECOMMENDED AREAS FOR RESEARCH AND DEVELOPMENT

Weather Prediction in Control

Weather forecasts are useful in controlling off-peak storage systems and stratified storage SHAC systems. Linear programming models have been proposed to do this.[34] Two such models are developed below.

Consider an off-peak storage system operating under a time-of-day rate structure, c_i, $i = 0, 1, \ldots, 23$ where c_i is in \$/kWh. Each hour the controller updates its 24-hour auxiliary load forecast.

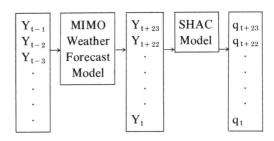

The controller next solves the linear programming model:

$$\text{minimize} \quad f(x) = \sum_{i=0}^{23} c_j x_i$$

$$\text{subject to} \quad S_0 + \sum_{i=0}^{k} x_i - q_j \geq 0$$

$$k = 0, 1, \ldots 23$$

$$\text{and} \quad S_0 + \sum_{i=0}^{k} x_i - q_i \leq S_{max}$$

$$\text{and} \quad x_k < x_{max}$$

$$x_k = q_k < \dot{S}_{max}$$

where

S_o = initial storage charge
S_{max} = maximum permissible storage charge
$j = t + 1$; mode 24 is the hour of day corresponding the ith lead forecast.

The optimal control policy for the current hour is the amount of energy to add to storage, $x_o - q_o$.

Useful variants of the foregoing linear programming model exist. The structure of the model makes it possible to optimize deviations in room temperature instead of requiring a fixed room temperature schedule. This is done by defining the cost, p_j, per degree-hour of temperature deviation for each hour of the day (thus reflecting the fact that the penalty for temperature deviations is less severe during low occupancy periods). The linear programming model then incorporates 24 additional variables, r_i, proportional to the deviation ($r_i UA$) from nominal auxiliary demand, q_i, $i = 0, 1 \ldots, 23$, and the model becomes:

$$\text{minimize} \quad f(c) = \sum_{i=0}^{23} (c_j x_i + p_j r_i)$$

subject to

$$S_o + \sum_{i=0}^{23} x_i - (q_i - r_i UA) \geq 0$$

$$S_o + \sum_{i=0}^{k} x_i - (q_i - r_i UA) \leq S_{max}$$
$$k = 0, 1, \ldots, 24$$

$$r_k UA \leq q_k$$

$$r_k \geq 0$$

where S_o, S_{max} and j retain their former definitions.

The same model may be used during the cooling season if the sense of both q and r are reversed (that is, positive q is a cooling demand and positive r is a deviation of room temperature above the nominal value).

Controllers in Passive Systems

Mention has already been made of the effects that the high resistance and thermal capacitance of passive systems have on overshoot. The determination of optimal controllers for auxiliary heaters in passive systems is an area that requires further development. This also includes the determination of the appropriate distribution of auxiliary heaters in passive systems. The interactions between convective and radiative heat transfer in passive systems are important in their effects on thermal comfort; consequently, the control of air flow becomes more significant in passive systems than in active systems.

Additional areas for controller research and development in passive systems include optimal control of air-to-air heat exchangers, movable insulation, shading, and off-peak thermal storage units for heating or cooling.

REFERENCES

1. C. B. Winn and D. E. Hull. 1979. Optimal controllers of the second kind. *Journal of Solar Energy* 23.
2. C. B. Winn and D. E. Hull. 1979. Optimal control of active solar systems. *Proceedings of the First Workshop on the Control of Solar Energy Systems for Heating and Cooling.* Newark, Del.: American Section of the International Solar Energy Society, Inc. (ASÈS), Univ. of Delaware.
3. C. B. Winn and R. C. Winn. 1982 (June). Optimal control of passive solar buildings with load managed storage. Progress in Solar Energy, Vol. 5.
4. C. B. Winn and D. E. Hull. 1979. (Eds.) Singular optimal solar controller. *Proceedings of the 1978 IEEE Conference on Decision and Control*, San Diego, Calif.; Feb. 1979.
5. D. V. Pryor; P. J. Burns; and C. B. Winn. 1980. Parameter estimation in passive solar structures. *Proceedings of the Fifth National Passive Solar Conference.* Amherst, MA (1980).
6. D. V. Pryor and C. B. Winn. 1982. A sequential filter for parameter estimation in a passive solar system. *Solar Energy* 28:65.
7. D. V. Pryor; P. J. Burns; and C. B. Winn. 1980 (Aug.). Parameter estimation in solar structures by the method of least squares. *Proceedings of the Joint Automatic Controls Conference.* San Francisco, Calif.; Aug., 1980.
8. T. M. McDonald; D. R. Farris; and J. L. Melsa. 1979. Energy conservation through adaptive optimal control for a solar heated and cooled building. *Proceedings of the First Workshop on the Control of Solar Energy Systems for Heating and Cooling.* Newark, Del.: ASES. University of Delaware.
9. T. B. Kent; C. B. Winn; and W. G. Huston. 1980. Controllers for solar domestic hot water systems. Prepared for the Solar Energy Research Institute. Contract No. AH-9-8189-1. SERI/TR-9819-1A.
10. P. R. Herczfeld et. al. 1978. Study of pump cycling in the control of solar heating and cooling systems. *Proceedings of the First Workshop on the Control of Solar Energy Systems for Heating and Cooling.* Newark, Del.: ASES, Univ. of Delaware.
11. R. Lewis, Jr. and J. B. Carr. 1978. Comparative study of on/off and proportionally controlled systems. *Proceedings of the First Workshop on the Control of Solar Energy Systems for Heating and Cooling.*
12. T. B. Kent; M. J. McGavin; and L. Lantz. 1978. Development of a novel controller. *Proceedings of the First Workshop on the Control of Solar Energy Systems for Heating and Cooling.*
13. R. C. Winn and C. B. Winn. 1981. Optimal control of mass flow rates in flat plate solar collectors. *Journal of Solar Energy Engineering* 103 (No. 2).

14. S. R. Schiller; M. L. Warren; and D. M. Auslander. 1980 (Nov.). Comparison of proportional and on/off solar collector loop control strategies using a dynamic collector model. *Journal of Solar Energy Engineering* 102 (No. 4).

15. J. Alcone and R. Herman. 1981. Simplified methodology for choosing controller set-points. *Solar Engineering—1981: Proceedings of the ASME Solar Energy Division Third Annual Conference on SSEA and Operational Results*. Reno, Nevada.

16. J. A. Duffie and W. A. Beckman. 1980. *Solar engineering of thermal processes*. New York: John Wiley and Sons, Inc.

17. B. J. Huang and H. J. Lu. 1982. Performance test of solar collector with intermittent output. *Journal of Solar Energy* 28 (No. 5).

18. M. Kovarik and P. F. Lesse. 1976. Optimal control of flow in low temperature solar heat collectors. *Solar Energy* 18.

19. R. C. Winn and C. B. Winn. 1981. Optimal control of mass flow rates in flat plate solar collectors. *Journal of Solar Energy Engineering* 103.

20. *ASHRAE Guide and Data Book, Equipment*. 1974. New York: American Society of Heating, Refrigerating and Air Conditioning Engineers.

21. R. M. Olson. 1961. *Essentials of engineering fluid mechanics*. Scranton: International Textbook.

22. H. J. Gerwin. Field experience with solar concentrating collector control systems. *Proceedings of the 1980 Joint Automatic Control Conference*.

23. J. L. Logan. A control scheme for solar industrial process steam used for potato frying. *Proceedings of the 1980 Joint Automatic Control Conference*.

24. J. D. Wright. 1980. Analytical modeling of line focus solar collectors. *Proceedings of the 1980 Joint Automatic Control Conference*.

25. R. Schindwolf. 1980. Fluid temperature control for parabolic trough solar collectors. *Proceedings of 1980 Joint Automatic Control Conference*.

26. W. L. de Rocher, Jr.; D. K. Melchior; and R. K. McMordie. 1982. Control simulations of a molten salt solar thermal central receiver. *Solar Engineering—1982*.

27. C. F. Kutscher et. al. 1981. Design approaches for solar industrial process heat systems. SERI/TR-253-1356.

28. J. J. Whalin and D. F. Paxson. 1981. Residential solar water heaters, energy use study. Miami, Flor.: Florida Power and Light Co.

29. Correspondence from Richard Merriam. 1981. Cambridge, Mass.: Arthur D. Little Co.

30. R. Merriam et. al. 1978. EPRI methodology for preferred solar systems (EMPSS) computer program documentation. Report ER-771 prepared by Arthur D. Little, Inc. for EPRI, Palo Alto, Calif.

31. A. S. Debs. 1978. Management of electric back-up demand for solar heating and cooling applications. DOE/CS/31595–11.

32. H. G. Lorsch. 1980. Novel control systems for solar assisted systems that reduce electric utility peak loads. Report No. F-C4827. Philadelphia, Pa.: Franklin Research Center.

33. C. B. Winn et. al. 1982. Preferred systems controls for optimizing the performance of solar heating and cooling and heat and cool storage installations. EPRI Project 1670–1.

34. C. B. Winn and N. Duong. 1978. An optimal control strategy for peak load reduction applied to solar structures. *Proceedings of the First Workshop on the Control of Solar Energy Systems for Heating and Cooling*.

35. C. B. Winn and D. E. Hull. 1978. Optimal control studies of solar heating systems. DOE Report No. COO-451901.

Advances in Solar Energy. © 1983 American Solar Energy Society, Inc.

PASSIVE AND HYBRID COOLING RESEARCH

JOHN I. YELLOTT, P. E.

Professor Emeritus, College of Architecture, Arizona State University, Tempe, Arizona 85287

Abstract

The term "passive cooling" generally denotes the dissipation of heat from buildings by the natural processes of radiation, convection, and evaporation which do not require the expenditure of any nonrenewable energy. In many cases, evaporation and convection can be significantly enhanced by the use of motor-driven fans or pumps, which consume small amounts of electrical energy and the word hybrid has been adopted to characterize such processes.

Much of the early work in passive cooling, notably that of Harold Hay in Arizona and California, was empirical in nature, carried out at the investigator's expense to prove the effectiveness of systems which were the result of intuition rather than analysis. With the entrance of the Department of Energy into the arena of solar space conditioning, funds became available to finance research into the basic nature of the passive cooling processes. Among the most important of those studies has been the work at Trinity University, under the direction of Dr. Eugene Clark, in the field of radiative cooling under both clear and cloudy skies. The Trinity team has shown, by validated computer simulations, that nocturnal heat rejection from roof tops can successfully dissipate solar heat gains for residences in virtually the entire United States.

The warming effect of the high nocturnal atmospheric temperatures which prevail in the southern part of the United States has been studied by Trinity and by Martin and Berdahl from the Lawrence Berkeley Laboratory of the University of California. The term "convective intrusion" has been added to the vocabulary of passive cooling and means of combatting this phenomenon has been reported. Givoni has shown, by work done in Israel, that the use of thin polyester films which are transparent to infrared radiation can be helpful until the film temperature falls to the local dewpoint. At this condition, the film becomes covered with moisture which effectively blocks the escape of radiant heat.

Much of the work done in recent years has been analytical and quantitative in nature, relying upon extensive and expensive instrumentation. Combined with the earlier empirical work carried out before the DOE came into existence, a firm foundation has been laid for the use of passive cooling processes in the residences which will be built in the post-DOE era.

INTRODUCTION

Passive cooling may be defined as the removal of heat from an indoor environment by utilizing the natural processes of rejecting thermal energy to the ambient atmosphere by convection, evaporation, and radiation or to the adjacent earth by conduction and convection. Hybrid cooling utilizes those same heat sinks, with the assistance of pumps or fans to circulate a heat transfer fluid, such as air or water, between the cooled space and the heat sink. The exterior surfaces of all buildings begin to be cooled by radiation to the sky as soon as the sun goes down and this cooling continues until the sun rises again. Much of the recent research in passive cooling deals with the quantitative evaluation of this phenomenon. The usefulness of convection-suppressing screens has also been investigated, with results which are reported here.

John I. Yellott is Professor Emeritus, College of Architecture, Arizona State University, and President of the Yellott Solar Energy Laboratory, Phoenix, Arizona. He is recipient of the American Society of Mechanical Engineers' Gold Medal for Education and the 1980 ASHRAE Distinguished Fellows Award. His publications include chapters on solar energy utilization in McGraw-Hill's Encyclopedia of Science *and* Marks' Mechanical Engineers Handbook, *also in the* Encyclopaedia Britannica *and the 1974 ASHRAE* Handbook of Applications, *as well as numerous research articles. Professor Yellott has served as advisor to the Department of Energy Passive Systems Committee and the Arizona State Solar Energy Research Commission, and as past-director and vice-president of the International Solar Energy Society.*

The most potent of the passive and hybrid cooling processes involves the evaporation of water into an air stream for the purpose of cooling the water, as in a cooling tower, the air (as in an evaporative cooler), or a surface supporting the evaporating water, as in the "Skytherm" roof. A recent development in this area is the application of corrosion-resistant plastic plate-type heat exchangers to the old concept of the indirect evaporative cooler. Convective cooling involves the introduction of relatively cool outdoor air into a warmed space. It is effective in those fortunate areas where warm days are succeeded by cool nights, but it is ineffective where daytime temperatures are excessively high and the diurnal variation does not result in nighttime temperatures which are low enough to produce cooling. Enhancement of the relatively feeble air flows caused by small temperature differences is a goal which continues to be sought.

In the more humid regions of the world, removal of moisture from the ambient air is just as important as lowering its temperature, but this is much more difficult to accomplish. The conventional method of dehumidification involves reduction of the air temperature below the dew point, thus bringing about moisture removal by condensation. This requires some form of active cooling because the passive processes are generally unable to reach the dew point. Desiccation can be used to remove moisture, but only at the cost of increasing the dry bulb temperature to the point where cooling is still required.

The moderate temperatures which always prevail in subsurface earth have been used for millennia by the people who dwell in caves and other subterranean spaces. More recently, efforts have been made with varying degrees of success to cool excessively warm air by causing it to flow through tubes or other conduits buried at varying depths below the surface of the earth. Problems have arisen from the fact that the heat which is transferred to the earth causes its temperature to rise and thus the cooling capability of the unaltered earth is diminished, often to the point that the system becomes ineffective.

Research in passive cooling continues with enthusiasm on the part of researchers who live in overly warm or overly humid regions because they are well aware of the discomfort which results from excessive heat and humidity. Little of this work has commercial sponsorship and much of the progress which has been reported in ASHRAE and ISES literature is due to Depart-

ment of Energy (DOE) support which is likely to diminish in coming years.

Radiative Cooling

Historical Summary

Cooling by exchange of radiant energy between terrestrial services and the sky above them was first studied quantitatively more than 80 years ago by Anders Ängstrom, the son of the Swedish astronomer for whom the unit of wavelength was named. The younger Ängstrom invented the electrical comparison pyrheliometer and he used the same principles to develop a longwave radiometer by which he could determine accurately the rate of heat loss from an electrically warmed surface to the sky. He found that the primary variable in this process was the amount of water vapor in the atmosphere,[1] which, fortunately, can be related accurately to the dew point temperature near the earth's surface.[2]

Nocturnal radiation, as this phenomenon is generally, but incorrectly, called, was studied subsequently by many meteorologists, including G. V. Parmalee[3] who initiated the ASHRAE research program in this area. One of the most valuable and readily accessible of the pre-DOE reports on this subject is the ISES paper by R. W. Bliss[4] which resulted from his study of the heat rejection capability of an unglazed tube-in-sheet roof through which warm water was circulated.[5] This concept was pioneered 20 years earlier in Tokyo by Yanigamachi[6] who employed a water-to-air heat pump to cool his residence during the day, storing the absorbed heat in a very large underground tank of water. This water was subsequently circulated during the night through an unglazed roof-mounted collector which dissipated the day's collection of heat to the sky.

Figure 1 (adapted from Bliss) gives a good idea of the order of magnitude of the radiation heat loss to a clear sky from a ground-level, horizontal surface, emittance e = 1.0, at the same temperature as the ambient atmosphere. A black surface exposed to ambient air at 70° F (*21° C*), with a dew point temperature of 30° F (*-1.1° C*), at 30% relative humidity, will lose heat to a clear sky at the rate of about 27 Btu/ft² (85 W/m²). This is about 10% of the rate at which solar radiation impinges on that same surface at noon on a clear summer day.

The most recent and comprehensive work on radiant cooling is that undertaken for the

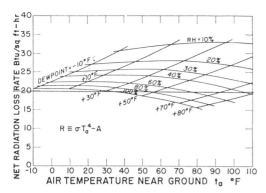

Fig. 1. Values calculated by Bliss[4] for radiative heat loss from a blackbody at local air temperatures to a clear sky with varying moisture content in the atmosphere.

Department of Energy by Gene Clark, Fred Loxsom, and their colleagues at Trinity University in San Antonio, Texas, and by Marlo Martin and Paul Berdahl at the Lawrence Berkeley Laboratory of the University of California. Thanks to them, we now have a good understanding of the ability of both dry and wet surfaces to be cooled by radiation and evaporation under both clear and cloudy skies. The work at Trinity began with the study of heat loss from relatively small exposed surfaces and it has now progressed to the point where two full-scale buildings with flat roofs have been instrumented to enable accurate measurements of heat flow rates to be made under the high humidity which prevails in San Antonio during most of the year.

The first comprehensive demonstration of the ability of a flat roof to provide adequate cooling for a building located in the extremely hot Arizona desert was provided by Harold Hay's prototype Skytherm building which was tested extensively in Phoenix by Mr. Hay and the writer[7, 8] during the summers of 1966 and 1967. Subsequently, the first full-scale Skytherm residence was built by Mr. Hay at Atascadero, California, and tested by P. W. Niles and Kenneth Haggard of California State Polytechnic University at San Luis Obispo. These two buildings demonstrated conclusively that the combination of enclosed roof ponds and horizontally movable insulation could provide both winter heating and summer cooling in areas where ice and snow are rarely experienced.

Clear Sky Cooling

Turning now to cooling by radiant energy exchange alone, with no evaporation, the basic equation for the effective temperature of a clear sky, acting as a blackbody radiator, is:

$$T_{sky} = e_{at}^{0.25} \cdot T_{at} \qquad (1)$$

where

T_{sky} = absolute radiative temperature of the sky, R or K;

T_{at} = absolute temperature of the ambient atmosphere, R or K;

e_{at} = emittance of the atmosphere.

Table 1 shows the Bliss relationship between the emittance of the atmospheric and its dew point temperature, with clear sky conditions.

For dew point temperatures between $-20°$ F ($-28.8°$ C) and $60°$ F ($15.6°$ C), this is virtually a linear relation, following the equation:

$$e_{at} = 0.73 + 0.0022 \, t_{dp}$$
$$e_{at} = \text{atmospheric dry bulb temp. F}$$

As an example of the use of this relationship, when the ambient air temperature is $80°$ F ($26.7°$ C), the dew point temperature is $50°$ F ($10°$ C), the relative humidity at sea level is about 35%, and the emittance of a clear sky is close to 0.84, the effective temperature of the sky, acting as a backbody radiator, is:

$$T_{sky} = 0.84^{0.25} \cdot (80 + 459.6) = 0.957 \cdot 539$$
$$= 516.6 \, R;$$
$$t_{sky} = 57° \, F \, (13.9° \, C).$$

It will be seen later in this section that the Trinity University values for e_{at} versus T_{dp} do not agree precisely with the Bliss equation but for the high dew point temperatures encountered in summer, the difference is within the limits of accuracy for this technology.

The rate of heat loss by radiation from a

t_{dp}	°F	−20.0	−10.0	0.0	+10.0	+20.0	+30.0	+40.0	+50.0	+60.0
	°C	−28.8	−23.3	−17.8	−12.2	−6.7	−1.1	+4.4	+10.0	+15.6
e_{at}		0.68	0.71	0.73	0.76	0.77	0.79	0.82	0.84	0.86

Table 1: Effective emittance of a clear sky as a function of ground-level dew point temperature (from Bliss[4])

flat horizontal roof to a clear sky is found from the Stefan-Boltzmann equation:

$$Q_{rad} = A \cdot 0.1713 \cdot e_r \cdot (T_r^4 - T_{sky}^4) \cdot 10^{-8} \text{ Btu/h ft}^2 \quad (2)$$

where A = roof area, ft^2;
e_r = emittance of the roof for long-wave radiation;
T_r = roof surface temperature, R.

Since the roof area is infinitesimally small compared with the dome of sky above it, the simple equation shown above is adequate. It is somewhat unwieldy, however, and so it is generally written as:

$$Q_{rad} = A \cdot 0.1713 \cdot e_r \cdot \left[\left(\frac{T_r}{100} \right)^4 - \left(\frac{T_{sky}}{100} \right)^4 \right]. \quad (3)$$

For all ordinary roofing materials which are exposed to the weather, the emittance, e_r, is very high, in the range of 0.90 to 0.95.

The Stefan-Boltzmann constant, which is $0.1713 \cdot 10^{-8}$ in U.S. units, is $5.67 \cdot 10^{-8}$ in SI units.

When the temperatures of the roof and the sky are not far apart, a simplified version of Eq. (3) may be used:

$$Q_{rad} = A \, 0.00686 \cdot \left(\frac{T_m}{100} \right)^3 \cdot e_r \cdot (T_r - T_{sky}) \quad (4)$$

where

$$T_m = \frac{T_r + T_{sky}}{2}.$$

Over the limited range of temperatures which are actually encountered in radiative cooling, Clark and Berdahl[10] have shown that a still simpler version of Eq. (4) may be used with confidence:

$$Q_{rad} = e_r \cdot 4 \cdot 0.1713 \cdot T_o^3 \cdot (T_r - T_{sky}) \cdot 10^{-8};$$
T_o = reference temperature, 527 R or 67° F (19.4° C). $\quad (5)$

If the emittance of the radiating surface may be assumed to be unity, which is generally within

5% of being a correct assumption, Eq. (5) may be still further simplified to give:

$$Q_{rad} = h_{rad} \cdot (T_r - T_{sky}) \text{ Btu/h} \cdot \text{ft}^2, \text{ or } (W/m^2); \quad (6)$$
$$h_{rad} = 4 \cdot (\text{Stefan-Boltzmann constant}) \cdot T_o^3. \quad (7)$$

Using the reference temperature cited above, this becomes:

$$h_{rad} = 1.0029 \text{ Btu/h ft}^2 \text{ °F } (5.705 \, W/m^2 \cdot K). \quad (8)$$

Accepting the simplifications embodied in Eq. (7), the rate of heat loss from a blackbody to the sky, in U.S. units, is numerically equal to the sky temperature depression, $T_r - T_{sky}$ (expressed in degrees F). For the example given above, with an atmospheric temperature of 80° F (26.7° C) and a sky temperature of 57° F (13.9° C), a blackbody at the air temperature would lose $1.00 \cdot (80 - 57) = 23$ Btu/h ft^2.

Figure 2 shows Clark and Berdahl's correlation of the calculated emittance values found by Bliss with values determined by measurements of sky radiation made by Clark and Allen at San Antonio during the period from October 1976 through September 1977 using a Fritchen net radiometer, and similar measurements made by Berdahl and Fromberg, using Eppley pyrgeometers, at Gaithersburg, Maryland, St. Louis, Missouri, and Tucson, Arizona. Clark and Berdahl[10] concluded that the values of water vapor emittance used by Bliss[4] were too high,

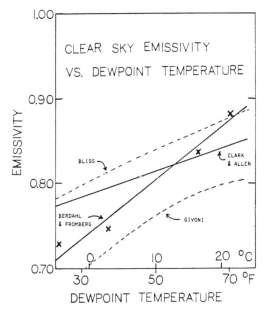

Fig. 2. Clear sky emittance versus dew point temperature, as given by four authorities (adapted with permission[10]).

thus making his sky emittances from 2 to 5% too high. Givoni's[13] values came from measurements made in Israel prior to 1937 and they are now recognized as being too low. Using the Berdahl and Fromberg correlation, the rate of heat rejection from a blackbody at atmospheric temperature to a clear sky, with dew point temperatures ranging from 80° F (very high) to 40° F (26.7° to 4.4° C), is shown in Fig. 3.

For surfaces which are at temperatures other than that of the local atmosphere and those with emittances differing widely from 1.00, Eq. (2) should be used. Uncovered collectors which face the sky can become potent radiators at night if they are used to cool fluids at temperatures significantly above the local air temperature. Using the conditions of an earlier example (air at 80° F dry bulb and 50° F dew point the sky temperature was found to be 57° F (13.9° C). If an unglazed collector is supplied with water at 100° F (37.8° C), and its collector surface emittance is 0.93 for longwave radiation, the rate of that loss would be:

$$Q_{rad} = A \cdot 0.1713 \cdot 0.93 \cdot (5.60^4 - 5.17^4)$$
$$= A \times 42.86 \text{ Btu/h.}$$

Radiative Cooling Under Cloudy Skies

When the sky is partly covered with clouds, its effective radiative temperature rises until, with complete coverage by relatively low clouds, the sky temperature becomes virtually the same as the air temperature and no radiative cooling can take place. The researchers at Trinity University[10] have correlated their measured rates of radiative heat loss, made during 1,433 hours of cloudy conditions, with the San Antonio National Weather Service observations of the simultaneous degree of cloud cover, expressed in tenths of the sky area. The cloud correction factor, C_S, given by Fig. 4, shows the results of their work, while Eq. (9) gives the correlation between C_S and the cloud cover number, n. The cloud correction factor, C_S, is defined as the ratio of the atmospheric radiation with n tenths opaque sky cover to the clear sky atmospheric radiation under the same temperature conditions. An equation which fits this data reasonably well is:

$$C_S = 1.0 + 0.022\,n - 0.0035\,n^2 + 0.00028\,n^3. \quad (9)$$

If the sky is completely overcast, it radiates towards the earth about 16% more energy than a clear sky would emit at the same dry bulb and dew point temperatures. Equation (9) was developed from data obtained when the cloud cover was at a low altitude. Higher level clouds are colder and so they have less effect than comparable low level clouds. High thin cirrus clouds have little, if any, effect.

Estimation of average night meteorological conditions presents many of the same problems as those encountered when average values of solar irradiation are sought. The National Weather Service has data for long periods of time on atmospheric temperatures (dry bulb, wet

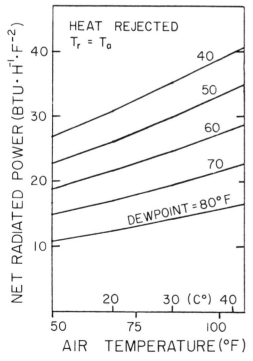

Fig. 3. Net radiative cooling power of a black surface (e = 1.0) at the local air temperature, radiating to a clear sky with varying dew point temperature.

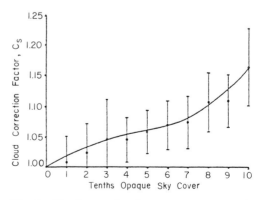

Fig. 4. Cloud correction factors versus opaque sky cover, expressed in tenths.[10]

bulb, and dew point), as well as cloud cover, wind speed, and direction. Very little correlation of this information has been done, although Clark and Berdahl[10] have compiled temperature data for the month of July for the years 1953 through 1964 for Albuquerque, San Antonio, and Miami. Figure 5 shows their temperature profiles for the high altitude, low humidity city of Albuquerque, New Mexico. The sky temperatures follow the dry bulb temperatures throughout the day, as one would expect from Eq. (1) and the relatively small variation throughout a 24-hour period of the dew point and wet bulb temperatures. The sky temperature is usually about 20° F (*11.1° C*) below the dry bulb temperature at Albuquerque; for the more humid cities of San Antonio and Miami, the depression of the sky temperature is usually about 15° to 20° F (*8° to 11° C*).

Convection's Influence on Radiative Cooling

The effect of radiative cooling is to reduce the temperature of exposed surfaces below the local air temperature and so convective heating sets to work to establish an equilibrium temperature at which the loss by radiation is just equal to the gain by convection. The convection coefficient, h_C, depends primarily upon wind speed, but it is also affected to an indeterminate degree by turbulence imparted by obstructions, such as parapets and other architectural features, which protrude above the roof line.

Values of the surface coefficient for relatively smooth surfaces were measured more than 50 years ago at the University of Minnesota using small sheets of metal in a wind tunnel. The radiative component of the heat transfer was about 0.9 Btu/h ft² ° F and so the following equation can be derived from data published in the *ASHRAE Handbook of Fundamentals*[11]:

$$h_{Con} = 1.0 + 0.28 \, V \; \text{Btu/h ft}^{2\circ}\,\text{F} \qquad (10)$$

where V = wind speed, mph.

The coefficient of the velocity term in Eq. (10) is often rounded to 0.3. In SI units, Eq. (10) becomes:

$$h_{Con} = 5.7 + 3.8 \cdot V^* \; \text{W/m}^2 \cdot {}^\circ\text{C} \qquad (10a)$$

where V^* = wind speed, m/s.

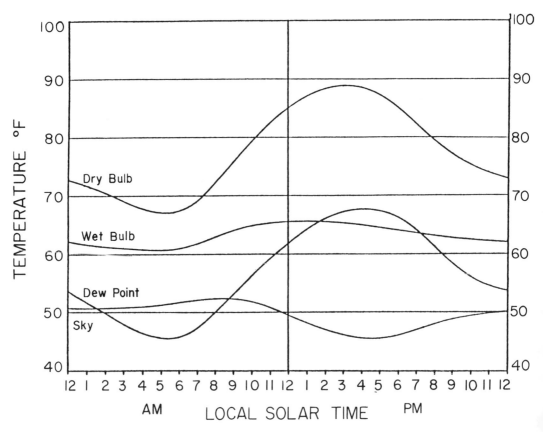

Fig. 5. Atmospheric temperature variation for average July days at Albuquerque, New Mexico, for the period from 1953 through 1964.

The term stagnation temperature has been used by Clark and Berdahl to denote the temperature attained by a well-insulated roof (no upward or downward heat flow from the underside of the roof) when the heat loss by radiation to the sky just equals the heat gain from the air. Using their simplifying equations for the radiative loss and their modification of Eq. (10), they predict that the stagnation temperature for a sky temperature depression of 20° F and a wind speed of 7.5 mph should be about 5° F, which agrees with the writer's measurements of the early morning surface temperature for the glazing on a large solar collector.

In aerodynamic terminology, stagnation temperature denotes the temperature attained by an air stream which is suddenly stopped by an obstruction or by a surface which moves through it at high speed. For our purposes, we will adhere to the Clark-Berdahl definition.

Net Cooling Rates for Exposed Dry Surfaces

Prediction of the net cooling, which can be accomplished by horizontal surfaces exposed both to heat loss by radiation to the sky and heat gain or loss by convection, can be done by applying the following equation:

$$Q_{net} = A \cdot 0.1713 \cdot e_r \cdot$$
$$\left[\left(\frac{T_r}{100} \right)^4 - \left(\frac{T_{sky}}{100} \right)^4 \right] - h_c \cdot (t_{at} - t_r). \qquad (11)$$

When cloud cover is present, the downward radiation from the atmosphere is increased by the Cloud Correction Factor given by Eq. (9), thus reducing the radiative cooling portion of Eq. (11).

Clark and Berdahl[10] have generated maps of the U.S. similar to Fig. 6 showing the net cooling effect, expressed in Btu/ft^2 per night, that may be expected during average July weather. Their maps take into account all of the meteorological factors which govern outward radiation and inward convection. If the radiating surface is at 66° F (19° C), little, if any, cooling can be expected in Arizona, the southern half of New Mexico, and most of Texas, as well as the Gulf states. The northeastern and Great Lakes states may expect as much as 350 Btu/ft^2 per night and even higher rates, up to 750 Btu (3,571 to 7,652 kJ/m^2) should be experienced in the high Rocky Mountain and Pacific Coastal regions.

If the dry surface temperature rises to 74° F, (21°C), at least 100 to 200 Btu/ft^2 per night (1,000 to 2,000 kJ/m^2 per night) will be experienced in all of the southern tier of states, except for the very hottest parts of western Arizona where daytime temperatures rise to 117° F (47° C). The rate will rise as high as 1,000 Btu/ft^2 per night (10,200 kJ/m^2) in the higher Mountain and the Pacific Coast states.

If the dry surface temperature is permitted to rise to 76° F (24° C), the distribution of cooling capability will be approximately as shown in Fig. 6. The hottest parts of the southern tier of states will still have about 100 Btu/ft^2 per night (1,000 kJ/m^2) of cooling capability, except for parts of Arizona's western deserts where the extremely high dry bulb temperatures will overcome the radiative cooling tendency. Most of the rest of the country should experience from 400 to 1,000 Btu/ft^2 per night (4,080 to 10,200 kJ/m^2), with a maximum net cooling effect of 1,500 Btu/ft^2 (15,307 kJ/m^2) along the coasts of Oregon and Washington.

A major problem which remains to be solved before the application of these cooling rates can become widespread is the nature of the coupling which can exist between the roof surface and the rooms sheltered by, but generally insulated from, that roof. The Skytherm principle, invented and patented by Harold Hay of Los Angeles, remains the only method which has thus far demonstrated its feasibility, but it has geographical limitations imposed by the necessity for freedom from snow and ice. For northerly climates, Skytherm North can accomplish much of the heating, but the enclosure of the roof-mounted Thermoponds removes their cooling capability. The coupling problem will be addressed later in this report.

The Spectral Basis for Radiative Cooling

The ability of terrestrial surfaces the lose heat by radiation to the sky is based on the fact that there is a relatively large "window" through which radiant energy in the wavelength band between 8 and 13 μm is transmitted quite freely when the water vapor content of the atmosphere is low. Fortunately, the energy radiated by sun-warmed surfaces occurs principally in this band and it is this outward radiation which enables the earth to achieve its relatively comfortable (compared to the other planets) temperature balance. Berdahl and Martin[12] have published charts which show how the extent of openness of the atmospheric window varies with the zenith angle of the portion of the sky dome from which the radiation

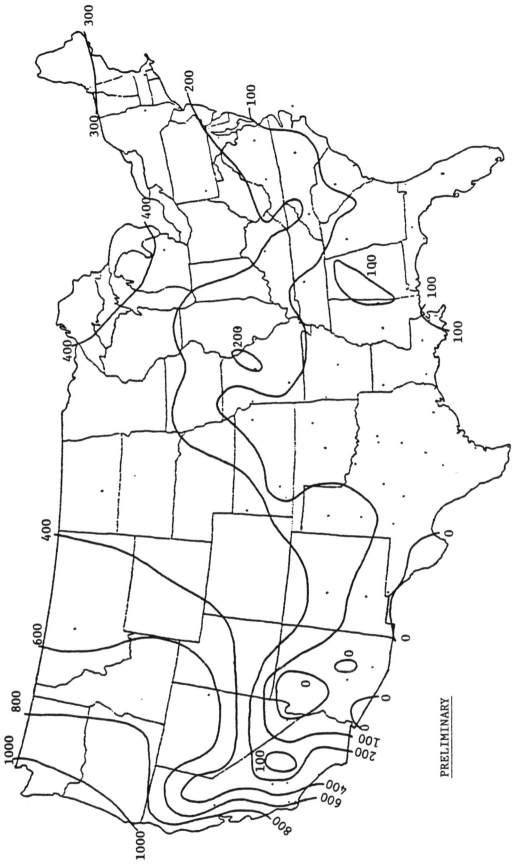

Fig. 6. July nocturnal cooling rates for dry surfaces at 70° F (21° C) for the continental U.S.[10]

PRELIMINARY

248

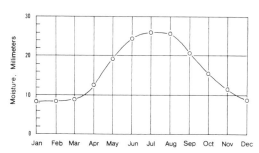

Fig. 7. Average amounts of precipitable moisture across the U.S. throughout the year.

comes and with the amount of precipitable moisture at any given time. The window opening, to use a simplistic analogy, is widest when the zenith angle is $0°$, which means that a horizontal surface enjoys the most favorable situation for radiant cooling. The back radiation from the sky increases as the zenith angle rises and it reaches a maximum for a surface which is vertical, and thus, its field of vision is primarily horizontal.

The openness is also dependent upon the amount of moisture in the atmosphere, which is usually expressed in terms of the amount of precipitable water vapor in the atmosphere above a given location. This is expressed in terms of depth of the water column which would result if all of the moisture in a column of air of unit area were condensed and collected in a graduated cylinder (see Fig. 7). There is excellent correlation between the height of this imaginary water column and the dew point of the air at the earth's surface. Reitan[2] has established the following empirical equation:

$$Ln\ W = 0.081 + 0.034 \cdot t_{dp} \qquad (12)$$

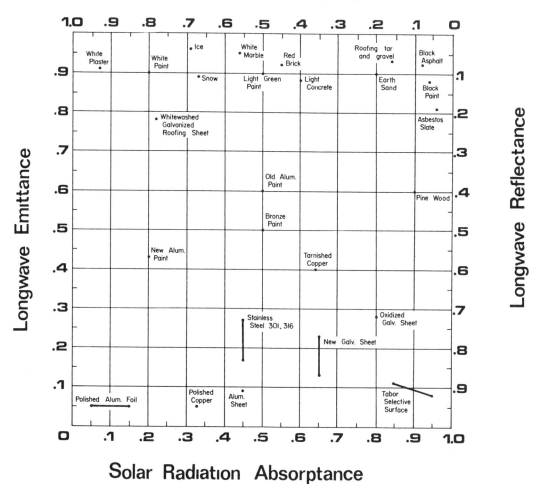

Fig. 8. Shortwave absorptance and reflectance of building materials versus longwave absorptance and reflectance.

249

where W = amount of moisture in the air, expressed in cm of precipitable moisture;

t_{dp} = dew point temperature, $°F$.

Berdahl and Martin[12] have correlated the precipitable moisture with the atmospheric radiance and they have found that doubling the amount of water vapor results in approximately doubling the sky radiation at 10 μm wavelength. Bliss[4] came to approximately the same conclusion and his charts of the variation of the spectral intensity of water vapor irradiation are particularly enlightening. The absorptance of the water vapor is particularly marked for wavelengths below 8 μm. while carbon dioxide affects the longer wavelengths so vigorously that the window closes almost completely beyond 13 μm.

Methods of Reducing the Convective Intrusion

Clark and Berdahl have coined the term "convective intrusion" to denote the warming effects of convection as radiation lowers the temperature of an exposed roof. Much attention has been given to the effectiveness of windscreens which would be transparent to the 8 to 13 μm radiation emitted by an exposed surface and would still be able to minimize the effects of convection. There are a number of plastic films which possess high transmittance (60 to 80%) in this part of the infrared spectrum, including polyethylene and polypropylene. Baruch Givoni and his colleagues at the Ben-Gurion University of the Negev[13] have studied the practical application of convection screens and they have noticed an interesting and significant phenomenon.

The anticipated improvement in cooling performance was indeed obtained at the beginning of night-long tests and the polyethylene-protected radiator attained temperatures which were considerably lower than those reached by the bare surfaces. However, as the surfaces cooled below the prevailing dew point temperature, moisture began to condense on them and the cooling rate immediately diminished. Givoni explains this reduction by noting that the emittance of polyethylene for longwave radiation is quite low, but when it is covered by even a thin layer of water, the emittance immediately rises into the 90% range which characterizes an open water surface. He concludes that the "net effect of the windscreen may be so small as to be insignificant or even counterproductive." He also points out that the use of the windscreen is even less beneficial when the radiator is at work, trying to dissipate warmth which has been accumulated during the day by the building beneath the radiator. Only when the radiator is at a temperature more than $4°F$ ($2°C$) below or above the ambient air is the cooling rate from a screened surface higher than that from one which is exposed to the sky.

A very ancient method of reducing convection heat gain is the erection of a substantial windbreak to shield the radiating surface from the prevailing nocturnal wind. This technique was used effectively by the Persians who were able to produce ice on summer nights by radiative cooling of shallow ponds of water.

Selective Cooling Surfaces

For most surfaces, emittance and absorptance vary with wavelength; only highly polished metals or black nonmetallic surfaces have radiative properties which are virtually independent of wavelength. For the most effective cooling, a surface should be highly reflective to shortwave (0.3 to 3.0 μm) solar radiation and highly absorptive, and hence highly emissive, for longwave radiation, particularly in the band between 8 and 13 μm. White paint is a good example of this, since its absorptance for solar radiation can be as low as 0.20 while its emittance for longwave radiation is about 0.90 to 0.96. Figure 8 shows emittances and absorptances for a number of typical opaque building materials. Polished aluminum foil has the lowest absorptance and emittance at all wavelengths, while black asphalt, as its name suggests, is thermally "black" at all wavelengths.

The coolest building material is white plaster, which has a solar absorptance as low as 0.08 and a longwave emittance above 0.90. Desert dwellers long ago recognized the qualitative effectiveness of white plastered structures, while oxidized, galvanized sheet roofs abound in the tropics where high atmospheric moisture diminishes the intensity of both sunshine and nocturnal radiation. Whitewash is often applied to these roofs as summer approaches, to reduce their solar absorptance and increase their longwave emittance, but heavy rains soon remove the whitewash and its beneficial effects.

Whiter white paint has been sought by a number of investigators and the use of titanium dioxide as the pigment can reduce solar absorptance to as low as 0.10. The whiteness may not be maintained, however, in an industrial environment. The ideal selective surface would have low absorptance and emittance in the region from 0.3 to 8 μm, high emittance in the 8 to 13 μm range, and low emittance from 13 to 30 μm. Such a surface is still being sought but, as Givoni points out in his definitive treatment of radiative cooling,[13] as soon as the cooled surface reaches the dew point, moisture begins to condense and the high longwave emittance of water dominates the situation. Condensed moisture and dust can make a bad combination, since the coating which they produce generally does not blow away with the wind.

Summary of Radiative Cooling
A flat roof will generally absorb a large amount of solar radiation during the daylight hours and this will raise its temperature so that longwave exchange with the sky can dissipate the absorbed solar energy in the late afternoon and at night. If the roof is coated with a white substance which remains white in spite of weathering, dust, and so on, the solar absorptance is materially reduced while the nocturnal cooling is not impaired. As soon as the sun sets, cooling begins to be felt, particularly on evenings when the dew point temperature is low and so the sky temperature is also well below the ambient air temperature.

The beneficial effect of this nocturnal radiation can be offset by incoming convective heat transfer which begins as soon as the roof temperature drops below the ambient air. The convection heat transfer coefficient is a linear function of wind speed, following a relation like:

$$h_{Con} = 1.0 + 0.3 \cdot \text{wind speed, mph}$$
$$\text{Btu/h ft}^2 \text{°F}. \tag{13}$$

The wind speed coefficient, 0.30, is appropriate for surfaces which are slightly rough and it is somewhat lower for completely smooth surfaces such as glass. In many parts of the country, winds speeds are relatively low at night, which is an asset for nocturnal cooling. On cold clear nights in the Southwest, freezing of fruit trees is frequently prevented, or at least minimized, by the use of gigantic fans which stir up the air and thus enhance the convective warming and diminish the net cooling effect.

Windscreens which are relatively transparent to radiation in the 8 to 13 μm wavelength band can enhance radiative cooling until the temperature of the screen reaches the dew point temperature of the ambient air. When this occurs, moisture begins to condense on the screen

and its absorptance and emittance immediately rise to the point where the beneficial effect of the screen vanishes.

Selective surfaces, which emit strongly in the 8 to 10 μm region and reflect all other radiation, are still being sought, but they will also encounter the condensation barrier and Givoni[13] questions whether they will be cost effective in actual use. A roof painted white, which reflects sunshine effectively, is likely to give just as good results as a far more expensive selectively surfaced roof.

EVAPORATIVE COOLING

Psychrometric Background Information
The most potent natural cooling process is evaporation, which occurs to some extent whenever the vapor pressure of water in the form of droplets or a wetted surface is higher than the partial pressure of the water vapor in the atmosphere. It has been pointed out earlier that the amount of precipitable moisture in the atmosphere is a function of the dew point temperature and so this meteorological property takes on added importance. Fortunately, it is reported on a regular basis by most weather stations and the *Climatic Atlas of the U.S.*[14] gives maps which show the dew point temperatures that prevail across the U.S. for each month of the year.

The relation between the dew point temperature, in degrees F and C, and the vapor pressure in inches and millimeters of mercury is given in Table 2.

The latent heat is designated by H_{fg} in both ASHRAE and ASME terminology. The following equation relates the dew point temperature

Temperature at the Dew Point		Vapor Pressure in. (p$_v$) mm		Latent Heat Btu kJ	
°F	°C	Hg	Hg	lb	kg
30	− 1.1	0.165	4.216	1219	2835
35	+ 1.7	0.204	5.18	1074	2498
40	4.4	0.248	6.29	1071	2491
45	7.2	0.300	7.62	1068	2484
50	10.0	0.363	9.22	1066	2480
55	12.8	0.436	11.1	1063	2473
60	15.6	0.522	13.3	1060	2466
65	18.3	0.622	15.8	1057	2459
70	21.1	0.739	18.8	1054	2452
75	23.9	0.875	22.2	1052	2447
80	26.7	1.032	26.2	1049	2440

Table 2: Variation with dew point temperature of vapor pressure in in. and mm hg and latent heat in Btu/lb and kJ/kg

(in °F) and the vapor pressure in in. Hg; for temperatures from 32° to 150° F (see Ref. 11, p. 5.4):

$$T_{dp} = 79.047 + 30.579 \, (\ln p_v)$$
$$+ 1.889 \, (\ln p_v)^2. \quad (14)$$

The rate of evaporation from a wetted surface depends upon the air velocity and the difference between the vapor pressures of the moisture and the air adjacent to the moist surface. An empirical equation, given below, relating these quantities with the cooling rate was derived by Willis H. Carrier from extensive tests made on cooling ponds:

$$Q_{Evap} = A \cdot 0.093 \cdot h_{fg} \cdot (1.0 + 0.38 \cdot mph) \cdot \Delta p_v$$
$$\text{Btu/h} \quad (15)$$
$$\Delta p_v = p_v \text{ of moisture} - p_v \text{ of atmosphere, in. Hg.}$$

This equation contains all of the necessary variables (the latent heat, the wind speed, and the vapor pressure difference) but there is some difference of opinion as to the values of two constants, which are obviously dependent upon the degree of turbulence which is experienced by the air as it flows over the wetted surface.

A more analytical approach[15] makes use of the Lewis number:

$$Le = h_{Con}/h_{Dif} \cdot C_{pa} \quad (16)$$

where h_{Con} = convection heat transfer coefficient; $\text{Btu/h ft}^{2\circ}$ F

h_{Dif} = convection mass transfer (diffusion) coefficient; lb/h ft^2

C_{pa} = specific heat of air, Btu/lb/F.

Loxsom and Kelly[15] use the following equation for the evaporative heat loss from a wetted surface at temperature T_{sur} (degree Rankine), to air at dry bulb temperature T_{air}:

$$Q_{Evap} = \left(\frac{h_{fg}}{Le \, Cp_a}\right) \cdot h_{Con} \cdot (W_s - W_a) \cdot A; \quad (17)$$

W_{sat} = humidity ratio for saturated air at existing air temperature;

W_a = humidity ratio for the actual ambient air, lb vapor/lb dry air;

A = area, ft^2.

These authors then combine the convection and the evaporation heat losses in the following equation:

$$Q_{Evap} + Q_{Con} =$$
$$\left[3 \cdot h_{Con} \cdot \frac{T_{DB} + T_{WB}}{64.7} \right.$$
$$\left. -1.0 \cdot (T_{DB} - T_{WB}) \right] \cdot A. \quad (18)$$

For the following conditions: $t_{db} = 77.°$ F, $T_{db} = 537°$ R; $t_{wb} = 70.8°$ F, $T_{wb} = 530.8°$ R; $t_{surf} = 75°$ F $= 535°$ R, 7.5 mph wind speed, Eq. (18) gives

$$Q_{Evap}/\text{Area} = 52 \text{ Btu/h ft}^2.$$

The Carrier equation, 15, would give for these conditions:

$$Q_{Evap}/\text{Area} = 0.093 \cdot 1052 \cdot (1.0 + 0.38 \cdot 7.5) \cdot$$
$$(0.876 - 0.691)$$
$$= 69.7 \text{ Btu/h ft}^2.$$

To make these two agree, the Carrier coefficient, 0.093, would have to be reduced to 0.069. One of the reasons for the difference may lie in the effective length of the path across the wetted surface, which Loxsom and Kelly took to be 20 ft, while it was closer to 100 ft for the carrier ponds. A definitive equation remains to be found.

The Loxsom and Kelly approximate equation (see Ref. 15, Fig. 2) is apparently linear for the combined evaporative and convective heat loss in terms of the wind speed, expressed in mph:

$$(q_{Evap} + q_{Con}) = 16 + 4.8$$
$$\cdot \text{ wind speed, mph Btu/h ft}^2. \quad (19)$$

The exact equation is also linear in the wind speed range between 7.5 mph (summer) and 15 mph (winter).

$$(q_{Evap} + q_{Con}) = 10 + 4.7 \cdot \text{mph Btu/h ft}^2. \quad (20)$$

The exact relation departs from linearity for wind speeds below 4 mph so that it can pass through zero when the wind speed falls to zero. At this point, natural convection and diffusion would take over and there would probably be a small heat loss.

Loxsom and Kelly (See Ref. 15, Fig. 6) have combined their approximate equation with the radiation equation for heat loss and, using meteorological data from 77 cities for average July weather, they have produced the map shown here as Fig. 9. This gives the typical nocturnal net cooling for a wet surface at 70° F;

252

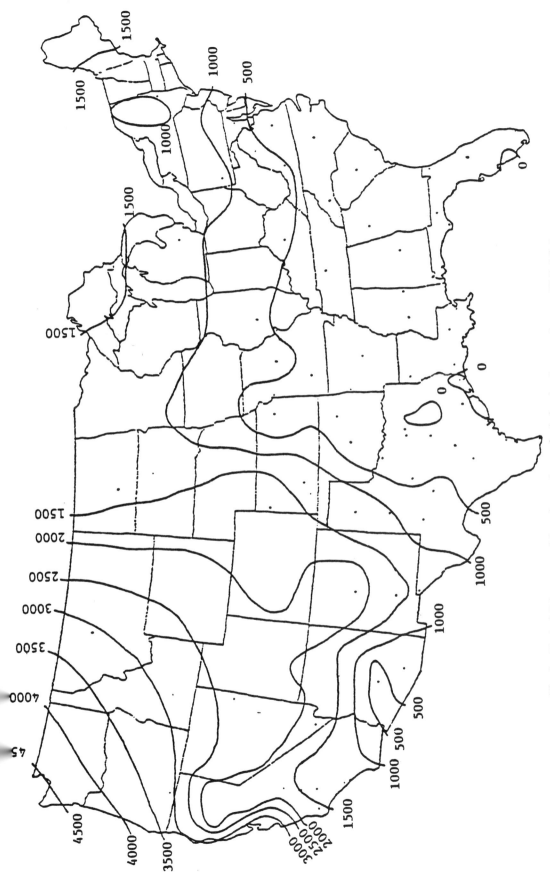

Fig. 9. Typical July nocturnal heat losses for perfectly insulated wet surfaces at 70° F (21° C).

253

comparing this with Fig. 6, it is seen that the effect of the evaporation is great almost everywhere in the U.S. Only in the warmest and most humid parts of Texas and Florida does the net cooling, expressed in Btu/ft² day, drop to zero. Throughout most of the U.S., the net daily cooling exceeds 1,000 Btu/ft². Figure 9 includes the effect of the convective heat flux which is beneficial in the northern parts of the U.S. where summer air temperatures at night are low enough to provide some nocturnal cooling. Convection is regarded as an intrusion in the cooling process in the southern half of the U.S. where nighttime temperatures do not fall into the cooling range. The beneficial effects of convection will be discussed later.

The major problem which must now be confronted is the matter of coupling the passively cooled roof to the space which needs cooling and to the occupants who are our real concern.

Methods of Coupling a Cool Roof to a Warm Space
The beneficial effects of wetting the roof in hot climates has long been known and practiced. Before the advent of glass fiber insulation, flat roofs were frequently flooded in summer to minimize solar heat gains, and early editions of the *ASHRAE Guide and Data Book* (the predecessor of today's *Handbook of Fundamentals*) contained tables to provide "equivalent temperature differentials for calculating heat gains through sunlit and shaded roofs." For light roofs, completely exposed to the sun, the temperature differential across the roof was assumed to reach 62° F on a clear August 1. The same roof with 1 in. of water standing on it would have only a 22° F differential at 2:00 p.m., and the use of a

water spray would reduce that to 18° F. No mention was made of nocturnal cooling.[16]

Intermittent water sprays, operated as frequently as needed to keep the roof damp, were shown by the writer[17] to be more effective than ponds in reducing roof temperature for both horizontal and sloping roofs. During the exceptionally hot Phoenix summer of 1965, it was found that a sprayed roof could be kept within 12° F of the dry bulb temperature of the ambient air at solar noon. By midafternoon, the roof surface temperature would fall below the dry bulb temperature, and by sunset the roof would approach the wet bulb temperature. Unfortunately, these tests were not run throughout the night and so there is no information as to the beneficial effects of radiation plus evaporation from a sprayed roof during Arizona's extremely hot summer nights.

David Hansen, working at the Solar Test Yard of Arizona State University, found that the exposure to the sky of a surface moistened by a very fine water spray would result in temperatures which approched the dew point of the ambient air.[18]

As Loxsom and Kelly[15] point out, the Hay Skytherm concept represented a "thermally superior application of passive cooling" (p. 86). This system combines enclosed Thermoponds in contact with a steel ceiling roof with horizontally movable insulation capable of protecting the roof from the heat of the summer sun and the cold of a winter night. The Thermoponds are relatively small and shallow, and they are essentially in thermal contact with the space beneath them because the steel roof deck has virtually no resistance to heat transfer. In spring and fall, all

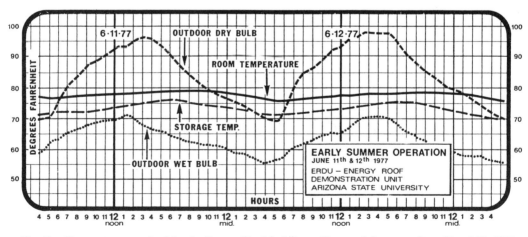

Fig. 10.   Temperatures attained by the Energy Roof building at Tempe, Arizona, on June 11 and 12, 1977.

of the required cooling in the hot desert climate of the Southwest can be provided by convective and radiative cooling. Evaporation can be added during the hottest weeks of the summer by simply flooding the roof so that about 0.5 in. (12 mm) of water covers the Thermoponds. Extensive testing of the small prototype building in Phoenix and the full-scale residence at Atascadero demonstrated the ability of the Skytherm system to provide cooling in summer and heating in winter with minimal requirements for standby cooling or heating.

The work at Trinity University by Gene Clark, Fred Loxsom, and their colleagues, cited above, has shown that the *sensible* cooling requirements of most residences in the U.S. can be met by roof ponds. The latent cooling load cannot be met by any system which cannot reduce temperatures below the wet bulb temperature and this is the fundamental limitation of the Skytherm system. Dehumidification by passive means is still an unsolved problem, and so supplemental cooling must be provided if completely comfortable conditions are to be attained in the most humid parts of the U.S.

The Energy Roof, invented by A. L. Pittinger and W. R. White and reported by them and the writer,[19] used a different method to achieve good coupling between a heat dissipating roof and the room beneath it. Once again, a moderate amount of water was stored in waterproof ponds supported by a metal ceiling, but the insulation in this case floated on the water and cooling was accomplished by using a small pump to circulate water from the pond and allowing it to flow across the floating insulation at night. Cooling was accomplished by radiation and convection during the spring and fall, and evaporation could be added in midsummer by spraying water over the thin plastic film which was in contact with the upper surface of the flowing water. Figure 11 shows the temperatures which were attained by the stored water and the room beneath the steel roof deck on June 11 and 12, 1977, when the outdoor air temperature rose to 96° and 98° F (35.5° and 36.7° C). The room temperature remained below 80° F).

Later in the summer (August 29 and 30, 1977) when the dry bulb rose to 107° and 105° F (41.7° and 40.4° C) and the wet bulb temperature reached 79° F (26.1° C), the room remained some 22° F below the maximum outdoor air temperature. The room temperature was nearly constant at 83° to 85° F throughout those two days, despite the very high outdoor air temperature. Supplementary cooling would be needed to attain complete comfort during the peak of the summer season.

Other methods have been proposed for coupling a passively cooled roof surface with storage and the cooled space. Harold Hay has proposed and patented the idea of connecting the bottom of one of his Thermoponds to a waterwall containing a vertical partition which divides the water into an inner downcomer and an outer riser. The sun shining on the outer surface of the riser warms the water and a thermosyphon action takes place which brings cool water downward to remove heat from the indoor wall surface. The rising warmed water is carried up to the top of the Thermopond where the heat can be readily dissipated to the sky at night.

Karen Crowther[20] has studied the operation of the Cool Pool, which consists of a shaded

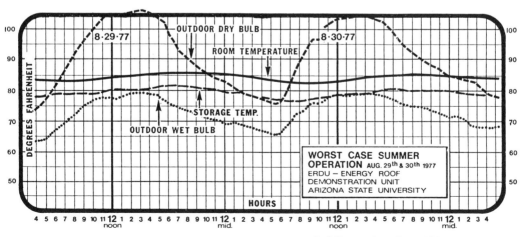

Fig. 11. Temperatures attained by the Energy Roof building on Aug. 29 and 30, 1977.

evaporatively cooled roof pond from the bottom of which cool water flows downward through a central tube in columns located within the room beneath the pool. The water in the outer portion of each column is warmed and it rises by thermosyphon action back to the upper portion of the roof pond. A computer program was prepared to predict the performance of this type of cooling system and good agreement was obtained between measured and predicted temperatures. The water velocity was measured by timing the motion of colored dye which was injected into a clear tube attached to the riser. The flow was always laminar since the measured velocity ranged from 0.04 to 0.07 ft/sec. Significant cooling of the air in the space was reported.

Givoni[13] has studied a number of different means of coupling roof cooling devices with the occupied space beneath them. He reports that the movable insulation, which is the key to the success of the Skytherm system, can be replaced by hinged panels which are positioned vertically during the night to expose the roof mass. If the panels are covered with a highly reflective coating, such as polished aluminum foil, the roof can "see" effectively the entire vault of the sky. Thus, the shading effect of the panels, due to their thickness, can be overcome. He considers such rotatable panels would be suitable for use on large buildings.

Givoni has also discussed the performance of a radiative air cooler, with outdoor air being cooled as it flows under a corrugated steel radiator. The cooled air is then encouraged to fall by gravity into the space below the radiator, where it picks up heat before it is discharged through an open window. A wind scoop, similar to those used in ancient Persia and today in Iran and Pakistan, aids in withdrawing the warmed air. No performance data were given in Givoni's report (see Ref. 13, p. 58). He points out that a fan could be used to induce air flow in regions where there is no persistent night wind. If the radiator is painted with a dark color and a fan is provided to ensure air movement, the system could be used for winter heating since the roof could absorb solar radiation during winter days. Its ability to radiate at night would not be impaired and if the radiator has a low thermal mass, it will cool off very rapidly as soon as the sun goes down.

Other workers have considered the use of open roof ponds which would be dry during the daylight hours and provided with water at night

from a storage tank mounted at ground level. A second circulating system would be used to circulate the cooled water through fan coils or radiant panels within the enclosed space. No test results have been reported for such systems and it is likely that the power required by the pumps and fans would not be negligible.

In Japan, Y. Saito[21] has tested an ingenious combination of an open-flow, glazed, corrugated copper collector and two blowers. One blower causes air to flow over the corrugated copper surface, below the single glazing, thus producing evaporative cooling which, during the daylight hours, virtually eliminates solar heat gain through the roof. The second blower forces a second stream of air to flow under the corrugated plate where it is cooled before it enters the enclosed space.

At night the evaporative cooling of the water flowing downward through the corrugations is used to abstract heat from the second air stream which flows in contact with the lower side of the copper. The cover glass is cooled by radiation to the sky and this minimizes the convective intrusion mentioned earlier. The air entering the room can be cooled during the hours of darkness by 1.8° to 7.2° F (1° to 4° C). During winter days, the water flowing down through the corrugated copper is heated by absorbed solar radiation in much the same manner as is accomplished by the Thomason Solaris system which is now widely used in the U.S.

There is apparently still plenty of opportunity for an ingenious designer to propose the ideal method of coupling a roof-mounted radiator/evaporator to the space which needs to be cooled. The field of desiccant dehumidification is still one which offers plenty of room for fruitful development. A point which is often overlooked when passive cooling processes are discussed is the desirability of filtering the outdoor air to remove the dust which it would otherwise bring into the enclosed space. In the hot, dry areas where passive cooling is needed, the air is generally dust laden and the pressure drop encountered in ordinary filters would effectively prevent thermally induced ventilation.

HYBRID EVAPORATIVE COOLING

Direct Evaporative Cooling

The most widely used evaporative device is the direct air cooler, similar to the unit shown in Fig. 12. Here, outdoor air is drawn through

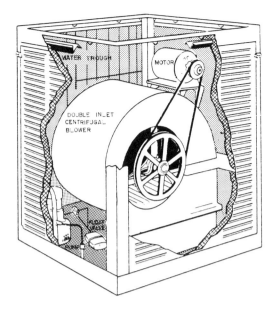

Fig. 12.   Direct-effect evaporative cooler.

wetted pads by a blower (giving the unit its *hybrid* character) which delivers the cooled and humidified air to the space where it is used. This type of cooler is so widely used and its characteristics are so well known that only a brief mention is needed here. The pads were originally made of shredded wood (excelsior) and these usually required replacement every spring because of the inevitable deterioration of the wood under the combined influences of the moisture and the dissolved salts which were left behind as the water evaporated. The container was made of galvanized steel which was subject to rust, the fan belt deteriorated after a few years of service, and the life of the units was generally about 10 to 15 years.

Recently, improvements have been made in constructing the conventional evaporative cooler by using stainless steel or high strength fiber glass reinforced polymers for the casing, thus minimizing or eliminating the corrosion which steel experiences. The pads in the newer units are also made of a hydrophilic plastic material with a far longer lifetime than the shredded wood used in the older coolers. Recirculating pumps which draw water from the sump at the bottom of the unit are useful in reducing water consumption. Water distribution is now accomplished in smaller coolers by using belts which pick up water from the sump and raise it into the air stream where the evaporative cooling takes place.

Most modern coolers are designed to produce temperature reductions of about 80% of the wet bulb depression. This term is used to designate the difference between the dry bulb and the wet bulb temperatures of the inlet air, t_{dbi} and t_{wbi}. The dry bulb temperature of the outlet air, t_{dbo}, is then found from:

$$t_{dbo} = t_{dbi} - 0.80 \cdot (t_{dbi} - t_{wbi}). \qquad (21)$$

The wet bulb temperature is the limiting factor in determining the cooling capability of any evaporative device which does not also possess the possibility of radiating heat to the sky. Evaporation is a process in which the wet bulb temperature remains constant as the dry bulb temperature falls. The objective of most cooling processes is to reduce the temperature of uncomfortably hot outdoor air until it comes within the ASHRAE comfort zone shown in Fig. 13. The limiting conditions of that relatively small zone are temperatures between 72° and 78° F (*22° and 26° C*) and vapor pressures of 0.20 and 0.55 in. Hg (*5 and 14 mm*). If the outdoor wet bulb temperature is below 67° F, a direct evaporative process can bring such air into the comfort zone. At higher outdoor wet bulb temperatures, direct evaporation cannot do so and some other process must be used to attain a comfortable condition. Note that the midsummer condition is sure to produce discomfort, while comfort can be attained during the spring and fall seasons in the Southwest.

Indirect Evaporative Cooling

In past years, there have been many attempts to produce a satisfactory indirect evaporative air cooler by using a cooling tower to obtain relatively cool water with an air-to-water heat exchanger to reduce the dry bulb temperature of the outdoor air without increasing its humidity ratio. The term humidity ratio, HR, denotes the ratio of weight of water vapor to the weight of the dry air in an air-moisture mixture. It may be expressed as lb/lb (kg/kg) but the numbers become inconveniently small and so it is usually expressed in the U.S. in grains/lb, where 7,000 grains = 1 lb. In SI units, HR is often expressed in g/kg.

None of the earlier attempts to accomplish indirect evaporative cooling led to success because the thin sheet steel or aluminum heat exchangers had very short lives due to the rapid

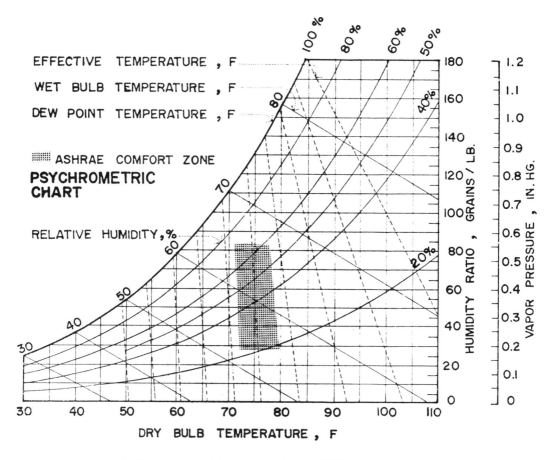

Fig. 13. Psychrometric chart showing the ASHRAE comfort zone.

rate of corrosion and deposition of salts. Tube-type heat exchangers could not provide enough surface area to ensure a high effectiveness and they also encountered other problems. In seeking to overcome the heat transfer problem, engineers at the Commonwealth Scientific and Industrial Research Organization (CSIRO) in Melbourne, Australia, developed a plate-type heat exchanger which used dimpled sheets of a hydrophilic polymer which was not subject to corrosion. Early tests of these devices were so encouraging that a commercial development was undertaken leading to the type of unit shown in Fig. 14.

In this unit, now known as the Dricon, hot dry outdoor air is drawn in through filters by a supply fan which delivers the air to the dry passages of the heat exchanger. The spacing of the plates is about ⅛ in. (3 mm) and the dimples in the plastic sheets provide both turbulence and spacing to maintain the relatively narrow air passages. The psychrometric processes involved are shown in Fig. 15, with the numeral 1 denoting the hot, dry outdoor air, and 2 showing the con-

dition of the air as it leaves the heat exchanger and enters the cooled space. In the space, the air picks up both heat and moisture to reach condition 3, at which it is exhausted from the space by the second blower in the Dricon unit. As the spent air moves upward through the alternate passages in the heat exchanger, it encounters the downwardly flowing water sprays and the fine droplets evaporate within the rising air stream and on the plate surfaces. The exhaust and the plates are thus evaporatively cooled to condition 4, and the heat exchange process warms the exhaust air to condition 5, at which it is discharged back to the atmosphere.

As long as the humidity ratio of the outdoor air is below about 80 gr/lb (11.4 g/kg) and the dew point temperature is below 65° F (18° C), the indirect cooling process can produce air which is within the ASHRAE comfort zone. If the vapor pressure exceeds 0.6 in. Hg (15 mm), dehumidification becomes necessary and this can be accomplished by a relatively small compression chiller with the sensible cooling being done by the Dricon unit.

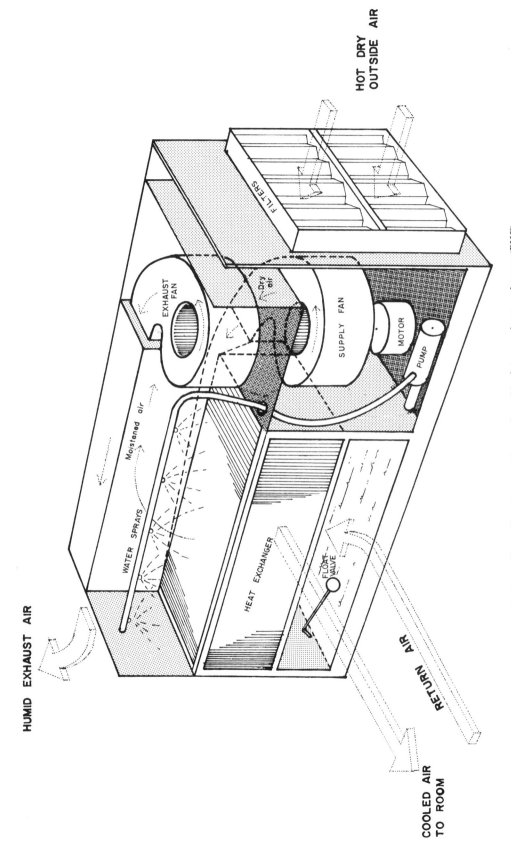

HOT DRY OUTSIDE AIR

FILTERS

EXHAUST FAN

Dry air

SUPPLY FAN

MOTOR

PUMP

Moistened air

WATER SPRAYS

HEAT EXCHANGER

FLOAT VALVE

HUMID EXHAUST AIR

RETURN AIR

COOLED AIR TO ROOM

Fig. 14. Australian indirect evaporative cooler with plate-type heat exchanger (PHE).

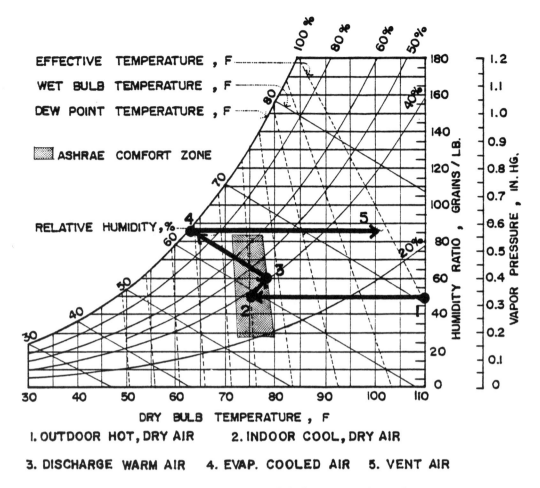

EFFECTIVE TEMPERATURE , F

WET BULB TEMPERATURE , F

DEW POINT TEMPERATURE , F

ASHRAE COMFORT ZONE

RELATIVE HUMIDITY , %

DRY BULB TEMPERATURE , F

1. OUTDOOR HOT, DRY AIR   2. INDOOR COOL, DRY AIR

3. DISCHARGE WARM AIR   4. EVAP. COOLED AIR   5. VENT AIR

Fig. 15. Psychrometric processes in indirect evaporative cooling.

Since the indirect evaporative cooling process requires two relatively large air streams, it must be quite large and so it is generally mounted on the ground outside of the cooled space. In winter, it can be used as a heat recovery unit with the water sprays turned off, to preheat the incoming outdoor air by transfer through the dry plates. Additional heat can be provided by a conventional means located within the warm air duct. The first large-scale use of the original CSIRO units was in schools where the amount of supply air must be relatively large. The PHE unit, shown in Fig. 14, is now being used in telephone exchanges and other commercial applications where air cooling is essential to satisfactory operation as well as in single and multifamily residences. The first Dricon unit in the U.S. was extensively tested in Phoenix, Arizona, with satisfactory results during the summers of 1981 and 1982.

CONVECTIVE COOLING

When the temperature and humidity ratio of the outdoor air are within the ASHRAE comfort zone, passive cooling may be accomplished by the simple expedient of opening windows and allowing a prevailing breeze to bring in fresh air. If there is no breeze at all, a condition which is often encountered during hot summer nights, some means must be found to encourage the cool outdoor air to enter and the warm indoor air to depart. H. J. Sobin,[22] of the University of Arizona, Tucson, has reported the results of extensive studies of window designs and their effect on passive ventilative cooling. The objective of his study was to convert the kinetic energy of natural wind into useful indoor air movement. He found that windows with a large horizontal dimension are more effective than those with a high vertical extent, and that, as might be expected, cross ventilation is highly desirable. In-

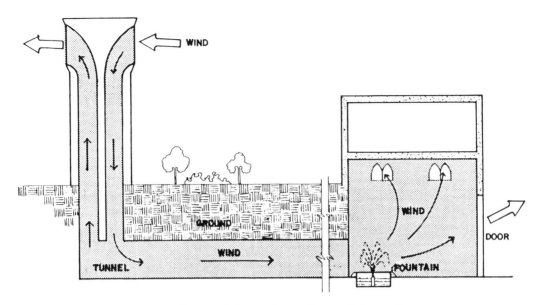

Fig. 16.   Persian wind tower for passive ventilative cooling.

let and outlet areas should be approximately equal, and indoor air velocities can be produced which are 25% or better of the prevailing outdoor wind speed.

Bahadori[23] has made a careful study of the operation of the wind towers, Fig. 16, which have been used for centuries in the Middle East and the Orient. These towers can operate in several different modes, depending upon wind direction, to draw or force air into the buildings to which they are attached. By using a small model and a wind tower, he has produced pressure coefficients which allow ventilative air quantities and velocities to be estimated with good accuracy. In many cases the wind-tower-induced air flow is combined with the evaporative cooling effect of a fountain to provide cooling.

It is well known that, within certain limits, comfort is enhanced by increasing the velocity of air flow and that a moderate excess of dry bulb temperature above the rather arbitrary limits imposed by the ASHRAE comfort zone can be offset by increased air velocity. If the outdoor air is significantly warmer than the desired indoor temperature, open windows are sources of discomfort and one of the other means of cooling must be sought.

Ventilative cooling on sunny summer days can be enhanced by the use of solar chimneys which are heated by the sun, and can thus produce the difference in air density which is required to produce air flow. Unfortunately, they require sunshine to make them function, and this is obviously absent at night when cooling is most urgently needed in residences. A hybrid approach to the provision of air velocity in an enclosed space is the use of a small electric fan which can produce a sensation of comfort which more than offsets the small amount of energy which the fan's motor uses.

DESICCANT DEHUMIDIFICATION
In hot, humid climates, removal of moisture is necessary if comfortable indoor conditions are to be attained. The conventional way to dehumidify air is to chill the air below its dew point temperature, thus causing the moisture to begin to condense on the cooling surface. Conventional air conditioning units have drip pans beneath their cooling coils for the purpose of catching and then disposing of the condensate. Dehumidification can also be accomplished by spraying cold water into the air stream and as long as the vapor pressure of the droplets is lower than that of the air stream, condensation will occur on the drops and the air will emerge both lower in humidity and cleaner than it was when it entered the air washer. This type of system does not qualify as either passive or hybrid because it requires compression or absorption refrigeration to produce the chilled water, pumps to circulate the water, and fans to move the air.

Moisture can be removed from flowing air by allowing it to come into contact with a dry desiccant such as silica gel, which can, by the process of adsorption, remove much of the water vapor from the air stream. The problem here is twofold. First, the air motion must be established by a fan since pressure drops in desiccant beds are usually too great to be overcome by the moderate chimney action which may be available in a passively cooled building. Secondly, the desiccant must be regenerated to remove the moisture which it adsorbs and this usually involves heating the desiccant. The process of moisture removal is exothermic, which means that the air stream follows a path of constant wet bulb temperature, but the path goes from warm to hot on the psychrometric chart. Thus, before the air can be used, the air must be cooled after dehumidification has taken place.

An indirect evaporative cooler can be used to do this cooling since hot, dry air can be brought into the comfort zone by the constant vapor pressure cooling which is accomplished in the heat exchanger of the cooler. Regenerating and cooling the spent desiccant offers a greater challenge since granular solids are far more difficult to handle than are air or water. Solar air heaters have been used successfully for daytime operation of desiccant dehumidifiers[24] using at least two beds or containers for the granular silica gel. The system becomes "active," in the sense of the word used here, since a fan is needed to move the air through the collector array and then through one or the other of the desiccant beds. Four dampers also must be moved simultaneously.

Liquid desiccants have been used commercially for many years to produce very low humidity ratios in air required for industrial processes, and a similar process can be used for comfort dehumidification. In some cases, the desiccant liquid has been freed of its load of water by solar heating on an open surface. This is still in the experimental stage and many problems must be solved, not the least of which is removal of the dust which is accumulated by an unglazed, open-flow solar heater.

CONCLUSIONS

Passive cooling is considerably more difficult than passive heating because the available cooling processes are far less potent than the sunshine which produces heating. Convection is by far the oldest of the passive cooling processes because the use of cool night air to remove the day's heat and to provide comfortable indoor conditions is as old as windows. For convective cooling to be successful, there must be a happy combination of some outdoor wind movement and an air temperature which is low enough to provide cooling, but neither too cold nor too dusty. If there is no wind at all, then fans must be used to bring the air into the space and to exhaust the warm indoor air. If the outdoor air is too hot, as it often is at night during the height of summer, then bringing such air indoors will only compound the problem and the windows should remain shut.

Radiation to the sky is an effective way of removing heat from the exterior surfaces of a building and particularly from the flat roofs which are used in many overheated parts of the world. Quantitative information on nocturnal radiation is now available for both clear and cloudy nights, thanks to the research team at Trinity University.

Evaporation from water droplets or from a moistened surface is the most potent of the passive cooling processes and the use of sprayed or ponded roofs can keep roof temperatures close to the dry bulb temperature on even the hottest and most sunlit days in midsummer. At night, these expedients can cool the roof well below the dry bulb temperature of the air and so "convective intrusion," to use the Trinity terminology, will work against the cooling process as heat flows from the air by convection to the cooled surface.

One passive system which uses all three cooling processes, as they may be available, is Harold Hay's Skytherm concept, in which roof-mounted and enclosed Thermoponds, with horizontally movable insulation, are in thermal contact with the ceiling of the space beneath them. By moving the insulating panels to expose the Thermoponds at night, they can be cooled by convection and radiation. When additional cooling is needed, the effect of evaporation can be added by spraying or flooding the ponds. This system, covered at this point by Mr. Hay's patents, has been proven to be effective for both winter heating and summer cooling in the snow-free Southwest. Other versions of the Skytherm concept can do the heating required in the cold northern parts of the U.S.

The problem of coupling the cool roof with the warm interior of a building, solved so readily by Skytherm, still awaits other solutions when the roof-top water storage is not feasible. Givoni has proposed air systems which would use fans to draw air under a cooled radiating surface and deliver it to the space. Thermosyphon systems have been tested for bringing cooled water down from an exposed pond to pick up heat from an enclosed space and then to rise again back to the pond. A small pump would probably do that job effectively and use so little energy that it would almost qualify as passive.

Where nature cooperates by providing cool air at night, open windows are the time-honored answer to nocturnal cooling and if the wind does not cooperate, small fans can do the job adequately. Wind scoops and wind-powered ventilators are also effective means of introducing cool air. When the air is too warm for comfort, as it is during much of the summer in the arid and overheated Southwest, recourse must be had to radiation and evaporation.

REFERENCES

1. A. Ängstrom. 1915. A study of the radiation of the atmosphere. *Smithsonian Misc. Collection*. 65 (3).
2. C. H. Reitan. 1963. (Dec.) Surface dewpoint and moisture aloft. *J. App. Met.* 2 (6): 776–778.
3. G. V. Parmalee and W. W. Aubele. 1952. Radiant energy transmission of the atmosphere. *ASHRAE Trans.* 58: 85.
4. R. W. Bliss. 1961. Atmospheric radiation near the surface of the earth. *Solar Energy*. 5 (3): 103.
5. R. W. Bliss and Mary D. Bliss. 1961. Performance of an experimental system using solar energy for heating and night radiation for cooling. Proceedings U. N. Conf. on New Sources of Energy, Rome. U.N. Pub. No. E/Conf. 35/6. Vol. 5, Solar Energy II, New York, 1964.
6. M. Yanigamachi. 1961. Report on two years of experimental living in the Yanigamachi solar house. *Ibid.*, pp. 233–247.
7. H. R. Hay and J. I. Yellott. 1969. Natural air conditioning with roof ponds and movable insulation. *ASHRAE Trans.* 75 (1): 1965.
8. J. I. Yellott and H. R. Hay, Thermal analysis of a building with natural air conditioning, *ibid.*, 175.
9. P. W. Niles, K. Haggard; and H. R. Hay. 1976. Nocturnal cooling and solar heating with water ponds and movable insulation. *ASHRAE Trans.* 82: 793.
10. Eugene Clark and Paul Berdahl. 1980. Radiative cooling: resource and applications. *Passive Cooling Handbook: Proceedings of the Passive Cooling Workshop.* Amherst, Mass.; Oct. 20, 1980. Edited by Harry Miller. Berkeley, Calif.: Center for Energy Efficient Design; pp. 177–212.
11. Handbook of Fundamentals, 1981 ed., *ASHRAE*, Atlanta, p. 23.3.
12. P. Berdahl and M. Martin. 1979. Spectral measurements of infrared sky radiance. *Proceedings of the Third National Passive Solar Conf.*, AS/ISES, Newark, Del., p. 443.
13. B. Givoni. Cooling buildings by longwave radiation. *Ann. Res.* Report, No. 2, part 3, Institute for Desert Research, Ben-Gurion University of the Negev, March 1982. (To be published in the *Passive Solar Journal*.)
14. Climatic atlas of the U.S. 1969 ed. U.S. Government Printing Office for the Environmental Science Service Administration, U.S. Department of Commerce.
15. F. M. Loxsom and B. Kelley. Evaporative cooling; roofs as dissipators. *Passive Cooling Handbook*, 1980, p. 81.
16. *Guide and Data Book*, 1963 ed., ASHRAE, Atlanta, p. 473.
17. J. I. Yellott. 1966. Roof cooling with intermittent water sprays. *ASHRAE Trans.* 72: 486.
18. D. Hansen and J. I. Yellott. 1978. A study of natural cooling processes. *Proceedings of the Second National Passive Solar Conf., AS/ISES*, Philadelphia, 2, 1978, p. 653.
19. A. L. Pittinger; W. R. White and J. I. Yellott. 1978. The energy roof, a new approach to solar heating and cooling. Ibid., p. 773.
20. Karen Crowther. 1979. Cooling from an evaporating thermosyphon roof pond. *Proceedings of the Fourth National Passive Solar Conference*, AS/ISES, Newark, Del., 1979, p. 499.
21. Y. Saito. 1981. A simple air conditioning system using a double roof. Paper No. B2:75, 1981 ISES Conference, Brighton, U.K.
22. H. J. Sobin. 1981. Window design for passive ventilative cooling. *Passive Cooling:* AS/ISES, Newark, Del.
23. M. N. Bahadori. Pressure coefficients for evaluating flow patterns in wind towers. p. 206.
24. C. E. Francis. A dual-bed desiccant dehumidifier for use with passive hybrid cooling systems. p. 292.

PASSIVE SOLAR HEATING RESEARCH

J. DOUGLAS BALCOMB
Los Alamos National Laboratory, Los Alamos, New Mexico 87545

Abstract

Key elements of research into the passive solar heating of buildings are described and examples are given to illustrate how research in the field has been approached. The major emphasis of the research has been on devising mathematical models to characterize heat flow within buildings, on the validation of these models by comparison with test results, and on the subsequent use of the models to investigate both the influence of various design parameters and the weather on system performance. Results from both test modules and monitored buildings are given. Simulation analysis and the development of simplified methods are described.

INTRODUCTION

This chapter presents a brief discussion of the key elements that constitute research into passive solar heating. It is certainly not a comprehensive review of all research that has been done; this would require a much more extended treatment. Instead, it is intended to present examples to illustrate how research in the field has been approached. The chapter should be considered as an introduction; those wanting more detail are referred to the literature.

Passive solar heating research has been as diverse as the participants; these are people who come from a variety of technical backgrounds and have brought a variety of analytical and experimental techniques to bear on various problems. We are in a dynamic and evolutionary period, and more work remains before it would be fair to say that the field is mature. Perhaps in time common approaches will be accepted, but for the moment, this growth and diversity is both challenging and invigorating.

Passive solar buildings are structures that use environmental energies to meet some or all of the need for heating or cooling. The environmental energy of major interest for heating is solar radiation incident on the building; for cooling, it is heat flow to the air surrounding the building, perhaps in combination with wind ef-

fects, long wave radiation from the building to the environment, and earth contact.

Artificial heating of the building by means of a furnace or boiler, and artificial cooling by means of an air conditioner, are referred to as auxiliary energy. Although both conservation and passive solar strategies normally have the effect of reducing the auxiliary energy requirements, we distinguish between conservation strategies, which tend to reduce auxiliary heating by reducing the total energy requirement, and passive solar or natural cooling strategies that reduce the auxiliary energy by promoting energy transfer through the external surfaces of the building. Although the distinction is not very precise, conservation consists of a "demand side" strategy, while passive solar or natural cooling consists of a "supply side" strategy. As we shall see, these strategies work well together

J. Douglas Balcomb is scientific advisor to the Solar Energy Group at the Los Alamos National Laboratory. He has played a leadership role in passive solar heating research, both at Los Alamos and nationally; is a principal author of DOE Passive Solar Design Handbook *Vols. II and III; and has lectured on passive solar in 20 countries. Dr. Balcomb is a past Chairman of the American Solar Energy Society (ASES) and currently serves on the Board of Directors of both ASES and the International Solar Energy Society.*

to achieve building designs that minimize auxiliary energy requirements for a fixed total added building cost.

In the 10 years since the terminology was coined, there has been considerable discussion over definition of terms, particularly the words "passive" and "natural," to describe the approaches being taken. Passive is generally meant to include approaches in which the energy flows are by natural means. By definition, this precludes the use of a fan or a pump to augment energy flow. Energy flow is by radiation, convection, or conduction. Passive solar normally refers to the use of solar radiation for its heating or daylighting value. Natural cooling generally refers to other passive, nonsolar approaches. Mixed designs combine two or more passive solar strategies (such as direct-gain and Trombe wall) into one building. Hybrid designs combine passive and active strategies in one building.

The emphasis in this chapter will be on the supply side strategies, and particularly on passive solar heating. The analysis of the building energy load is often simplified, perhaps oversimplified, to avoid distraction from an accurate evaluation of the effectiveness of various passive solar heating strategies. This is not to underestimate the importance of correctly analyzing heat flow through walls by conduction and convection, air leakage through cracks in the external fabric of the building (infiltration), conduction through the ground surrounding the building, or any other conservation strategies. We assume that the researcher or designer who needs a more precise analysis of loads will undertake the necessary, accurate calculations. One can still use the methods described in this chapter to perform an assessment of the passive solar heating aspects by simply replacing the simplified loads with the more accurate ones. In this way, the procedures can be extended, for example, to earth-coupled buildings. However, when used sensitively and sensibly, the simpler load models are often surprisingly accurate, especially for smaller buildings where the cost of a more complex analysis is not justified.

HISTORICAL PERSPECTIVE

Although passive solar applications in buildings have been extensively practiced by several different civilizations throughout human history, modern interest in passive solar heating research is quite recent, starting in the early 1970s. The number of passive solar installations has grown incredibly from perhaps a half dozen prior to 1970 to well over 100,000 currently; most are located in the United States. We are also seeing major interest developing in many other countries throughout the world, notably in France. It is certainly conceivable that within 10 years, most new construction within the United States will use passive solar techniques for heating and that this trend will become worldwide, especially in the more temperate climates.

It is interesting to note that much of this development has taken place without the aid of a strong research program. Although passive systems developed in parallel with active systems, a large research budget stimulated active system development, whereas minimal funding was devoted to passive system research. The reasons for this dichotomy are many and we will not belabor them here, but its existence is a matter of record. Even when government funding did materialize, it was directed largely toward commercialization activities rather than research.

This historical difference has strongly colored the development of passive solar compared with active solar systems. Passive solar heating has developed as a grass roots technology in which designers transfer concepts into practice, not only without the benefit of sophisticated experiments or detailed numerical analysis, but often without even the most rudimentary engineering calculations. In this regard, passive solar heating development has been more characteristic of the building industry as a whole than of the engineering industry. This process has not necessarily been detrimental to the orderly development of passive solar techniques. Much of the evolution has focused on intangible issues that are not particularly conducive to a research-oriented analysis. These issues include concerns about aesthetics, marketability, and the integration of passive solar construction techniques into the infrastructure of the conventional building industry. Passive solar buildings are quite forgiving of design variations and flaws; most such buildings have worked quite well, requiring a fraction of the heating requirements of conventional buildings, and their owners have been quite satisfied.

So where does research fit into the passive solar heating picture? Is it unnecessary and irrelevant? Probably no more or less so than in any other technology. Doubled performance can be expected from improved materials and techniques. With the aid of analysis tools, designers will be able to complete a project with the certainty that they have both selected the most appropriate solution and possess full knowledge and confidence of the performance that can be expected.

Examples of early research on passive solar heating are rare and isolated. Researchers such as Dietz[1] at MIT and Neubauer[2] at the University of California, Davis, conducted a few experiments, but the results of these were neither widely known nor appreciated. A patent was taken out by E. M. Morse[3] in 1882 for a thermosiphoning air panel, but there was no quantitative prediction or evaluation of performance.

This is not to say that there had not been significant efforts put to the thermal evaluation of buildings. Quite the contrary; major efforts in many countries were devoted to evaluation of building materials and the analysis of heat flow through building elements. This work laid an important foundation for passive solar research today. However, it cannot be considered passive solar research in itself because of several characteristics: (a) solar gains were usually regarded as a nuisance, (b) heat storage in building elements was treated peripherally, if at all, (c) the emphasis of the investigation was to provide equipment sizing information, and, most importantly, (d) the results were rarely used to guide building design toward better use of natural energy flows.

Passive solar heating research really began in the 1970s in direct response to passive solar designs that had been built. The three early buildings of greatest significance are the Wallasey School[4] by Morgan in England, the Trombe house[5] in France, and the Atascadero house[6] by Hay in California. Each was analyzed and evaluated by a team not associated with their construction. M. G. Davies and A. D. M. Davies analyzed the Wallasey School from a thermal and a sociological standpoint, respectively.[7] J. F. Robert, M. Cabanet, and B. Sesolis[8] evaluated, but did not analyze, the Trombe house (Robert lives in the Trombe house). Prior to its construction, Yellott and Hay had built and evaluated a test structure based on the Skytherm principle.[9] Following the construction of the Hay house in Atascadero, California, Niles and Haggard[10] analyzed and evaluated its performance. Significantly, both Davies and Niles used thermal network simulation analyses to investigate the dynamic behavior of heat flow in the structures. However, even in these cases, the work was confined to an investigation of the particular building in its particular climate, and no attempt was made to generalize the results.

The turning point for passive solar research was the first passive solar conference in Albuquerque in May 1976.[11] Most of the early research work was reviewed, and most of the individuals who have subsequently contributed to the research effort were present. The work at Los Alamos had been under way about one year at that time but already showed the potential for broad applicability of passive solar concepts throughout the United States. Since that time, national passive solar conferences, sponsored by the American Solar Energy Society, have served as the focal point for the presentation of research results.

RESEARCH DIRECTIONS

The main focus of research in passive solar heating has been on the performance evaluation of buildings. Knowledge gained through the understanding of the behavior of existing buildings can be used both to predict the performance of future buildings and to devise strategies to make them more effective. Thus, a major emphasis of the work has been threefold: (1) devising mathematical models that characterize heat flow and, thus, thermal behavior, (2) the validation of these models by comparison with test results, and (3) the subsequent use of the models to investigate the influence of both various design parameters and weather characteristics on performance.

The explanation that follows illustrates how analytical modeling work has become the cornerstone of the research effort. This relationship is indicated clearly in the schematic diagram of Fig. 1, which shows the key elements of the research program and the relationship between those elements. The logical progression of activity flows from left to right in this schematic, beginning with experimental results obtained in test modules, special experiments, or monitored buildings. Based on these results and known physical principles, analytical models are developed and validated. Using weather and solar data from a particular locality, the analytical models can then be used to predict performance in a variety of climates for a variety of proposed designs. The models can also be used for sensitivity analyses, to develop simplified prediction methods, and to explore the relationship between passive solar and conservation strategies.

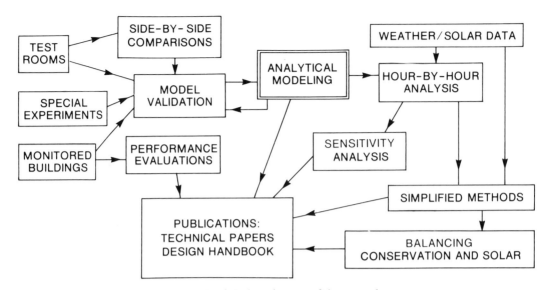

Fig. 1. Schematic of the key elements of the research program.

Results are published both in technical papers and as user-oriented manuals.

In the remainder of this chapter, we will explore these various elements in more detail, giving examples of each, and end with an assessment of major research directions that remain for the future.

THE PHYSICS OF PASSIVE SOLAR HEATING

We will start with a brief overview of the analysis approach that has been most fruitful. In a fundamental sense, the analysis of energy flow by natural means is a very complex and formidable task. Fortunately, the most complex of these problems can be bypassed without loss of significant accuracy by the use of simplifying assumptions. The three most important assumptions follow.

- Radiation flow can generally be regarded as either short-wave or long-wave. The short-wave solar radiation spectrum from the sun, as altered by its transmittance through the atmosphere, is quite complex; however, building analysis does not usually require a detailed spectral analysis. The most important thing to know accurately is the glazing transmittance, and this is usually determined as a function of the angle of incidence based on measured data. For the simple case of glass, a classic calculation using the Fresnel equations and measured absorption characteristics is quite adequate. By far the knottiest problem has been the prediction of the solar radiation,

as modified by the atmosphere. This problem is normally treated by separating direct or beam radiation and diffuse radiation scattered within the atmosphere. A variety of atmospheric models has been used with varying success.

- Once the short wave solar radiation enters the building through the glazing, very little is scattered back out, normally less than 5%. The effect of energy transport within the building by long-wave thermal radiation is so dominant that the energy moves quickly to all of the enclosing surfaces of a direct-gain space. Although some analysts have developed detailed radiation flow models and view factors for thermal radiation flow between surfaces, it is seldom worth the effort. Surfaces directly in the sun may be temporarily warmer than others, but within a few hours this energy will be found quite uniformly distributed around the space.[12] In special cases of extreme geometries or large aspect ratios, some detailed analysis may be warranted; however, even in these cases, simplified models based on Wye-delta transformations are very useful.[13]

- When a building is separated into a series of different spaces connected by doorways, energy flow by convection through the doorways becomes a significant and important mechanism, tending to aid the distribution of energy throughout the building. The energy flow rates obtained by convection are significant enough (with modest temperature differences between spaces) that they are often adequate for heating north rooms. Engineering correlations have proved to be a reasonably accurate method of predicting these important convective effects.[14]

Research work has been directed mainly at developing an understanding of the physics of energy flow in passive solar elements. When this is accomplished, the physics can be converted into algorithms that will reside in computer codes designed to analyze buildings.

By far the most widespread analytical approach has been mathematical simulation of the energy flow in the building using finite difference techniques. The building can be regarded as a simple thermal network with significant heat-storing masses within the building connected by thermal conductances. The situation is analogous to an electrical network with resistive and capacitive elements only. The result is a set of simple first-order differential equations that must be solved simultaneously. Some of the energy transport mechanisms result in nonlinear terms; most of the radiation flow terms, however, can be reasonably well characterized using linearized equations because the total temperature range, on an absolute temperature scale, is relatively small. This is not true, however, of the convective effects.

These equations are normally solved using either implicit or explicit finite difference techniques. The computation marches through time, simulating the energy flows in the building and predicting temperatures. Auxiliary energy is supplied, as required, to maintain desired comfort conditions.

The simulation is normally driven using hourly temperature and solar radiation data for the site of interest. Calculations are typically done first for selected periods of both extreme and average weather conditions in different seasons to determine the general behavior and comfort characteristics, and then for an entire year to predict overall auxiliary energy requirements.

In larger buildings, simulation of the performance of heating, ventilating, and air conditioning equipment, and internal heat sources can be very important. The more complex computer building codes usually accomplish these simulations with a library of routines representing different equipment packages.

TEST MODULES

Test modules range from small elements, known as test boxes, to larger structures (that have a

Fig. 2. Los Alamos test rooms.

small floor area but are full height), known as test rooms or test cells, to yet larger structures dedicated to passive testing, known as test buildings. A comprehensive review of test modules in the United States is given by Moore.[15] He identifies 77 test rooms that have been built by 14 research groups and 31 test buildings built by 8 research groups; of course, there well may be others. In addition, there are many test modules in other countries such as France, Argentina, Italy, Israel, Peru, and China. The best known test modules are those at the Los Alamos National Laboratory in New Mexico. Figure 2 shows a photograph of the test rooms as they were set up in early 1981, and Fig. 3 shows drawings of a typical test room. Figure 4 shows test boxes set up at Los Alamos, and Fig. 5 shows the test buildings at the University of California at San Diego.

Los Alamos Test Room Results

Side-by-Side Comparisons
If test rooms are built to have the same net load

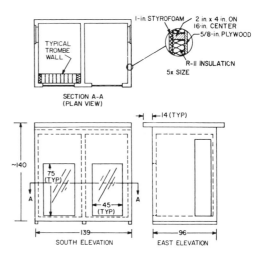

Fig. 3. Schematic and section drawings of the Los Alamos test rooms.

Fig. 4. Test boxes at Los Alamos.

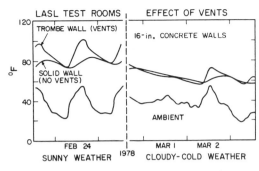

Fig. 6. Globe temperatures measured in two free running Trombe wall test rooms, both with and without thermocirculation.

coefficient (exclusive of the solar aperture), the same collection area, and are operated in the same way, their results can be compared directly. Figure 6 shows room globe temperatures measured inside two adjacent Trombe wall rooms, one with thermocirculation vents, and the other without vents.[16] The effect of thermocirculation in overheating the room is directly apparent. One room experiences $10°$ F temperature swings and the other $27°$ F swings. These rooms were operated in a free running mode, that is, with no auxiliary heat.

A second example, in Fig. 7, shows the performance of several rooms each thermostatically controlled to a minimum $75°$ F room globe temperature using electric heat. Each is double glazed. The Trombe wall is 16 in. thick, unvented, and has a black chrome selective surface ($\alpha = 0.94$, $\epsilon = 0.07$) glued to the exterior surface. The direct-gain room has the same mass as the Trombe wall room distributed to have an exposed surface area of 3.2 times the glazing area and a thickness of 5.62 in. The waterwall is 12 in. thick. The phase-change (PCM) wall is composed of 4.75-in. diameter steel cans filled with

sodium sulphate decahydrate contained within a solid matrix made by Boardman Energy Systems. The 27 tubes each have an advertised latent heat of 2,000 Btu occurring at $89°$ F. Both the waterwall and phase-change wall have black chrome exterior surfaces and are sealed to prevent air thermocirculation between the glass/wall space and the room.

Figure 7 shows room globe temperatures and auxiliary energy use in each room for January 25, 1982, one of several clear sunny days in a sequence. The outside temperature is about $40°$ F. The direct-gain room performed badly. It had high daytime temperatures and needed much

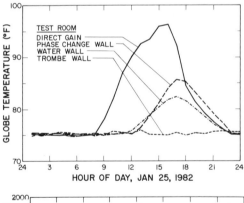

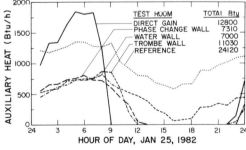

Fig. 7. Temperatures and powers measured during one sunny day in five controlled test rooms. The net load coefficient of each room is 624 Btu/$°$F day. The average outside temperature is approximately $40°$ F.

Fig. 5. Test buildings at the University of California at San Diego.

auxiliary heat at night. Although the Trombe wall needed continuous auxiliary heat, the total auxiliary used during the day was less than for the direct-gain room. The water wall room and phase-change room are closely matched and need much less auxiliary than the others. Auxiliary heat used by a reference room is also shown. This room is identical to the other rooms, except that the entire south wall is well insulated and there are no significant solar gains.

Performance Comparisons

As expected, the results of test rooms in the United States and around the world have been qualitatively quite similar. Data that have been presented by McFarland are given as an example of typical results that are obtained,[17,18,19]

Fourteen Los Alamos test rooms have been operated for several years; this section presents summary results for the winters of 1980-81 and 1981-82. Extensive data have been taken and computer analyzed to determine performance parameters such as efficiency, solar savings fraction, and comfort index. Configurations include direct gain, unvented Trombe walls, waterwalls, phase-change walls, and two sunspace geometries. Strategies for reducing heat loss include selective surfaces, two brands of superglazing windows, a heat pipe system, and convection-suppression baffles. Significant differences in both backup heat and comfort are observed among the various rooms.

The rooms are built in seven side-by-side pairs with an insulating wall between rooms as shown in Fig. 3. The general construction uses 2×4-in. wood frame walls insulated with fiberglass and a 1-in. sheet of expanded polystyrene insulation on the inside. Thus, the rooms themselves deliberately have very low thermal mass; the predominant mass is composed of the added passive solar element such as a Trombe wall, a waterwall, or internal concrete blocks used in direct gain.

With the exception of one free-running Trombe wall, each room is maintained at or above a minimum setpoint temperature of 65° or 75°F, using calibrated light bulbs as a heat source operated on a thermostatic-type control. To minimize the uncertainty associated with unknown infiltration, each room is pressurized using a small fan that introduces a calibrated three air-changes-per-hour continuously into the space.

Data scans are made 180 times per hour.

Hourly averages are computed, recorded on tape, and then transferred to disk computer storage for analysis. Approximately 179 data channels are recorded including thermocouples, pyranometers, auxiliary heat, and weather information.

To achieve comparability, most of the test rooms have the same net load coefficient of 26 Btu/h °F (excluding the south aperture) and the same net projected area of 23.44 ft^2 (45 in. $\times$ 75 in., net). The ratio of these numbers (times 24) is the load/collector ratio (LCR) of 26.6 Btu/ft^2 °F. This is a reasonable LCR value for a cold climate like Los Alamos.

The configurations of the test rooms are shown in Fig. 8. (Rooms 9 and 10 were used for special experiments aimed at the retrofit of existing U.S. Navy buildings and are not included in this comparison.[20,21]) In 1980-81, Room 12 had no auxiliary heat or forced ventilation (free-running) and Rooms 3 and 4 and Rooms 7 and 8 were configured as double sunspaces and, thus, constitute single experiments. Detailed descriptions of the room configurations for 1980-81 are given in Refs. 17 and 19. Note that the glazing area used to calculate LCR is the projected area, that is, the same as the area measured in an elevation view of the building and not the actual collection area. This is consistent with the convention established at Los Alamos for sunspace analysis.[22]

Three main performance indices are calculated based on the observed hourly data averages. The first is the "useful" efficiency, η_u, defined as

$$\eta_u = S/(I_v A_p),$$

where

$$S = \Sigma[L - Aux] + M\Delta T,$$
$$I_v = \Sigma[\text{vertical solar radiation}],$$
$$A_p = \text{projected area},$$

where

$$L = NLC (TI - TA) + CLC (TG - TGN),$$

where

NLC = net load coefficient, measured in reference Room No. 13,

  TI = inside globe temperature or 65°F, whichever is smaller,

  TA = ambient temperature,

CLC = cell-to-cell load coefficient (3.5 Btu/ °F h),

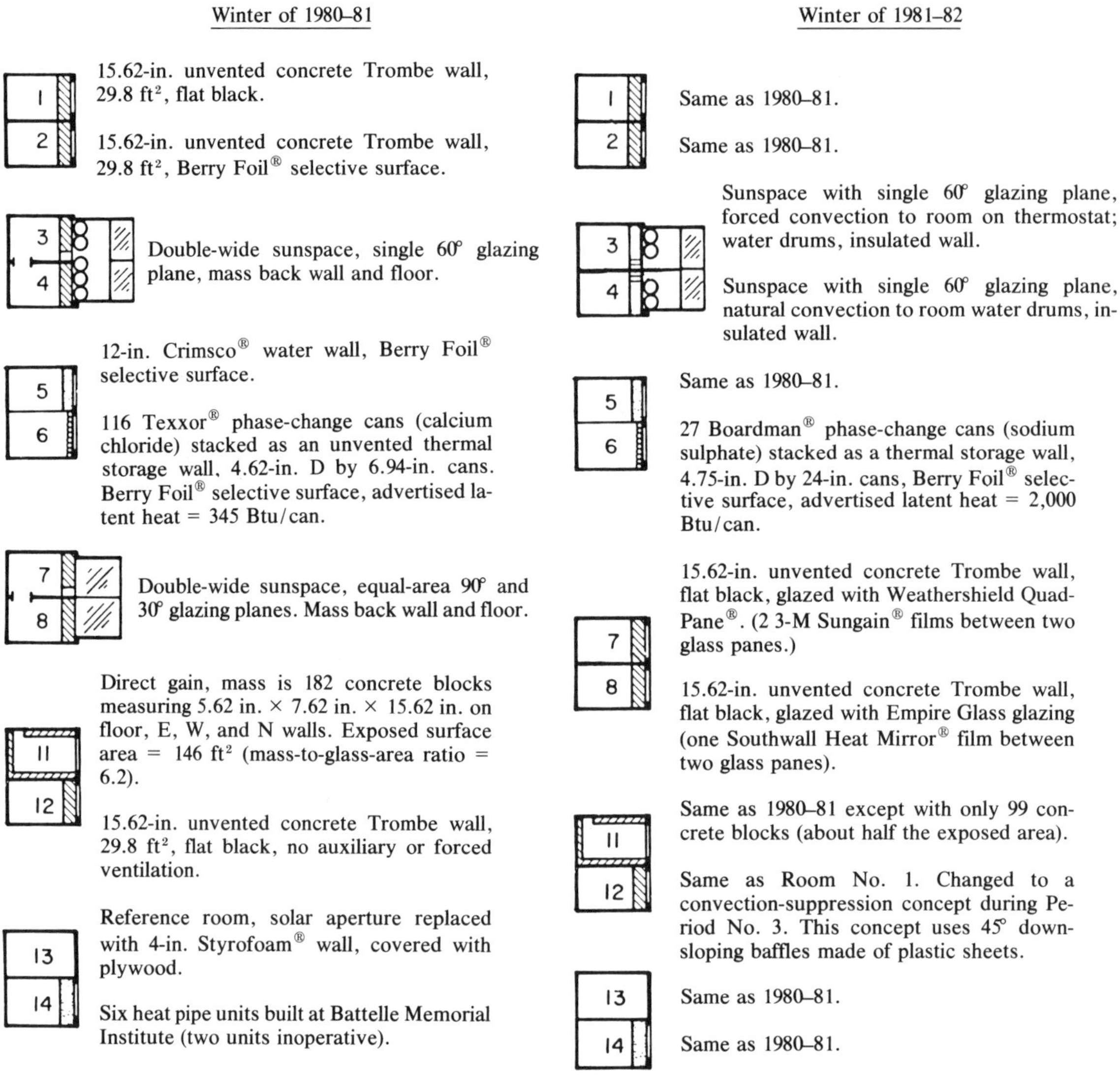

1 15.62-in. unvented concrete Trombe wall, 29.8 ft², flat black.

2 15.62-in. unvented concrete Trombe wall, 29.8 ft², Berry Foil® selective surface.

3 4 Double-wide sunspace, single 60° glazing plane, mass back wall and floor.

5 12-in. Crimsco® water wall, Berry Foil® selective surface.

6 116 Texxor® phase-change cans (calcium chloride) stacked as an unvented thermal storage wall, 4.62-in. D by 6.94-in. cans. Berry Foil® selective surface, advertised latent heat = 345 Btu/can.

7 8 Double-wide sunspace, equal-area 90° and 30° glazing planes. Mass back wall and floor.

11 Direct gain, mass is 182 concrete blocks measuring 5.62 in. × 7.62 in. × 15.62 in. on floor, E, W, and N walls. Exposed surface area = 146 ft² (mass-to-glass-area ratio = 6.2).

12 15.62-in. unvented concrete Trombe wall, 29.8 ft², flat black, no auxiliary or forced ventilation.

13 Reference room, solar aperture replaced with 4-in. Styrofoam® wall, covered with plywood.

14 Six heat pipe units built at Battelle Memorial Institute (two units inoperative).

1 Same as 1980–81.

2 Same as 1980–81.

3 Sunspace with single 60° glazing plane, forced convection to room on thermostat; water drums, insulated wall.

4 Sunspace with single 60° glazing plane, natural convection to room water drums, insulated wall.

5 Same as 1980–81.

6 27 Boardman® phase-change cans (sodium sulphate) stacked as a thermal storage wall, 4.75-in. D by 24-in. cans, Berry Foil® selective surface, advertised latent heat = 2,000 Btu/can.

7 15.62-in. unvented concrete Trombe wall, flat black, glazed with Weathershield Quad-Pane®. (2 3-M Sungain® films between two glass panes.)

8 15.62-in. unvented concrete Trombe wall, flat black, glazed with Empire Glass glazing (one Southwall Heat Mirror® film between two glass panes).

11 Same as 1980–81 except with only 99 concrete blocks (about half the exposed area).

12 Same as Room No. 1. Changed to a convection-suppression concept during Period No. 3. This concept uses 45° down-sloping baffles made of plastic sheets.

13 Same as 1980–81.

14 Same as 1980–81.

Fig. 8. Test room configurations.

TG = inside globe temperature,

TGN = inside globe temperature of adjacent room,

Aux = auxiliary heat,

M = heat storage mass x heat capacity,

ΔT = change in average mass temperature over the time period, and

Σ represents an hourly summation over the time period.

The useful efficiency does not count as useful any heat that results in raising the globe temperature above the thermostat setpoint of 65°F; that is, credit is not given for overheating the room. The useful efficiency can be significantly less than the total efficiency (computed with TG instead of TI) for test rooms that are prone to overheating, such as the direct-gain room. This can alter the rank ordering of room performance.

The second performance measure is the useful solar fraction, F_u, defined as

$$F_u = 1 - \Sigma[Aux]/\Sigma[L].$$

This fraction is as close as one can come experimentally to the conventional solar savings fraction used in performance prediction.

The third performance measure is the discomfort index, DI, described by Carroll.[23]

$$DI = \Sigma[E^2 W]/\Sigma[W],$$

where

W = weighting factor
  = 1,   7 a.m. — 11 p.m.
  = 0.5,   11 p.m. — 7 a.m.,
E = temperature error
  = 0.93 TG + 0.04 TA + 1.1 − PT
PT = preferred temperature,
   = 0.91 TB − 0.09 TA − DN,

where

DN = 0,   7 a.m. to 11 p.m.
     2,   11 p.m. to 7 a.m., and
TB = base temperature = 72.5°F.

Units of discomfort index are (°F)². As a measure of discomfort, a zero value indicates perfect comfort, and a doubling of DI indicates a condition of roughly twice the personal discomfort.

Environmental data and test room configurations for the two years are given in Table 1 and the results are given in Table 2.

The following conclusions have been drawn from the comparisons.

- Reasonable useful efficiencies in the range of 20 to 40% were obtained in all of the test rooms.
- Very significant performance and comfort variations were evident between various test rooms.
- Good solar fractions (in the range of 40 to 90%+) were achieved in the test rooms.
- A significant performance increase was obtained with the use of a selective surface. A direct comparison was made only for Trombe walls, but earlier data (1979-80) indicated a similar effect for waterwalls. A 40% increase in useful efficiency was observed during 1981-82 through the use of a selective surface on a double-glazed Trombe wall. During 1980-81, problems with both foil adhesion and foil qual-

ity had been noted. New foil was installed before 1981-82, and data taken thereafter are considered to be more representative of the performance enhancement that can be realized. The enhancement was more pronounced during colder weather, as expected.

- Trombe walls (without vents) had better comfort characteristics than the other systems tested.
- The waterwall room (No. 5) had consistently excellent performance. This is because of the combination of high mass (63 Btu/°F ft$_g^2$) and selective surface.
- The 1980-81 (Texxor®) PCM wall had reasonably good performance, but overheated badly. Apparently, the advertised phase-change potential (1,725 Btu/ft$_g^2$) was not being used effectively. Leaks and corrosion were noted in many of the cans.
- The 1981-82 (Boardman Energy Systems®) PCM wall had the best performance of all the test rooms, in addition to reasonable comfort characteristics. This behavior is thought to be associated with a high latent heat (2,330 Btu/ft$_g^2$) and immobility of the melted salt in the can.
- The Quad-Pane® glazing worked well in conjunction with a Trombe wall. However, the Heat Mirror® glazing did not show significant improvement over ordinary double glazing. The advertised U-value of both glazings is very low (~0.25 Btu/h °F ft²). The difference in performance is thought to be caused by the higher transmission of the Quad-Pane® glazing. We also note that the Quad-Pane® application could benefit significantly from the use of a selective surface on the Trombe wall because its low U-value is based on convection suppression (three cavities). However, the Heat Mirror® application would probably not benefit from a selective surface on the

1980-81	Ave Temp (°F)	Min Temp (°F)	Max Temp (°F)	Vertical Insolation (Btu/ft²)	Horizontal Insolation (Btu/ft²)	Wind Velocity (mph)
1. Dec. 20—Jan. 2	38.3	28.0	51.1	1895	938	3.4
2. Jan. 6—Jan. 19	31.4	22.9	42.8	1464	864	3.5
3. Feb. 14—Feb. 27	41.7	27.4	54.6	1610	1355	4.6
4. Mar. 15—Mar. 28	39.5	28.4	50.4	1305	1708	6.7
1981-82						
1. Jan. 12—Feb. 15	28.3	19.3	37.4	1550	979	3.8
2. Feb. 16—Mar. 22	38.7	29.4	49.4	1301	1364	4.7
3. Mar. 2—Mar. 22	38.7	30.0	49.7	1233	1420	5.3

Table 1A   Average Daily Weather Data for Selected Periods   *continued*

Table 1 (cont.)

Room	NGL[a]	NI[b]	Other
			Period 1: December 20, 1980—January 2, 1981
1	2	no	Unvented Trombe wall, flat black.
2	2	no	Unvented Trombe wall, selective absorber.
3/4	2	no	Sunspace with masonry wall, sunspace doors always closed.
5	2	no	Waterwall, selective absorber.
6	2	no	Phase-change cans.
7/8	2	no	Sunspace with opaque end walls, sunspace door always closed.
11	2	no	Direct gain unpainted.
14	—	—	Not operational.
			Period 2: January 6–19, 1981
1	2	yes	Same as Period 1.
2	2	no	Same as Period 1.
3/4	2	yes	Same as Period 1, except sunspace door opened daily.
5	2	no	Same as Period 1.
6	2	no	Same as Period 1.
7/8	2	no	Same as Period 1, except sunspace door opened daily.
11	2	yes	Same as Period 1.
14	—	—	Not operational.
			Period 3: February 14–27, 1981
1	1	yes	Same as Period 1.
2	1	no	Same as Period 1.
3/4	2	yes	Sunspace with insulated wall, five water drums for storage. Sunspace door opened daily until February 22, open at all times after February 22.
5	1	no	Same as Period 1.
6	2	no	Same as Period 1.
7/8	2	no	Sunspace with glazed end walls. Sunspace door opened daily.
11	2	yes	Direct gain painted dark brown.
14	2	no	Heat-pipe collector/water storage.
			Period 4: March 15–28, 1981
1	1	no	Same as Period 1, except with reflector.
2	1	no	Same as Period 1, except with reflector.
3/4	2	no	Same as Period 3, except with 10 water drums.
5	1	no	Same as Period 1, except with reflector.
6	2	no	Same as Period 1.
7/8	2	no	Same as Period 3 except sunspace door open at all times.
11	2	no	Same as Period 3.
14	2	no	Same as Period 3.

[a] NGL = number of glazing layers
[b] NI = night insulation

Table 1B   Cell Configurations During Selected Time Periods, 1980-81.   *continued*

Trombe wall because its low U-value is based on the low emittance properties of the enclosed film. Thus, the combination of a low-convection, high-transmission glazing, selective surface, and waterwall, Trombe wall, or PCM wall could be expected to show exceptional performance.

- The heat pipe room showed excellent performance despite having only three of the four exposed units operational. We believe that the failure of the units was caused by freezing of the small amount of water used in the heat pipes; this problem could be alleviated by using a different working fluid, for example, Freon®. Performance may benefit from the use of a selective surface.
- Although the direct-gain room results are somewhat ambiguous, the following conclusions are fairly clear.

  a. Performance with night insulation, a 6:1 mass-to-glass area ratio, and light-colored surfaces was quite good, and comfort was marginal (Periods 1-2, 1980-81).

  b. Performance without night insulation, a 3:1 mass-to-glass area ratio, and dark-colored surfaces is among the lowest of all rooms, and discomfort is extreme (1981-82).

  The use of night insulation, we believe, is the major determinant in performance and the higher mass-to-glass area ratio is the major determinant in comfort.

- The performance of the sunspace rooms in 1980-81 was ordinary, although overheating was a problem. Presumably, venting the sun-

Table 1   (cont.)

Room	NGL	Other
		Period 1: January 12—February 15, 1982
1	2	Unvented TW, flat black.
2	2	Unvented TW, selective surface.
3		Not operational.
4		Not operational.
5	1	Water wall, selective surface, single glazed.
6	2	Phase change tubes, selective surface.
7	4	Trombe wall, Quad Pane®.
8	3	Trombe wall, Heat Mirror®.
11	2	Direct gain, 3:1, dark brown.
12a	2	Same as No. 1.
14	2	Heat pipe collector/water storage.
		Period 2: February 16—March 22, 1982
1	2	Same as Period 1.
2	2	Same as Period 1.
3	2	Sunspace with water drums, forced convection to room on thermostat.
4	2	Sunspace with water drums, natural convection to room.
5	1	Same as Period 1.
6	2	Same as Period 1.
7	4	Same as Period 1.
8	3	Same as Period 1.
11	2	Same as Period 1.
12		Not operational.
14	2	Same as Period 1.
		Period 3: March 3—March 22, 1982 (overlaps Period 2)
1–11, 14		Same as Period 1.
12b	2	Convection suppression scheme.

Table 1C   Configurations During Selected Time Periods, 1981-82

RESULTS FOR 1980-81

Room	Useful Efficiency, %				Useful Solar Fraction, %				Discomfort Index (°F)^2			
	P1*	P2	P3	P4	P1	P2	P3	P4	P1	P2	P3	P4
1	30	39	31	27	72	69	90	51	16	23	10	13
2	33	34	29	38	79	63	86	71	13	26	10	10
3/4	26	36	28	40	58	66	79	72	23/13	55/29	175/75	23/10
5	36	41	31	45	85	74	93	82	23	16	36	13
6	32	45	33	36	83	78	88	68	116	49	91	6
7/8	30	30	28	39	70	60	83	70	19/32	65/55	97/45	61/23
11	36	39	36	30	82	77	98	55	32	32	45	6
12	(25)	(31)	(22)	(29)	(100)	(100)	(100)	(100)	(39)	(68)	(26)	(52)
14	—	—	31	36	—	—	86	67	—	—	19	6

RESULTS FOR 1981-82

Room	P1	P2	P3	P1	P2	P3	P1	P2	P3
1	21	24	23	26	48	42	1	8	11
2	33	31	31	40	57	54	1	4	5
3	—	30	32	—	54	53	—	16	19
4	—	30	30	—	51	50	—	21	16
5	38	32	32	48	62	59	3	10	5
6	40	34	34	51	67	63	5	9	5
7	35	32	33	41	56	55	2	3	3
8	23	26	25	28	51	45	1	3	2
11	21	22	21	25	39	36	49	50	36
12	25	—	25	32	—	47	2	—	11
14	37	27	26	45	48	45	2	5	4

*P1 refers to Period 1, etc.

Table 2 Los Alamos test room results.

275

space in fall and spring, plants in the sunspace, and occupant control of sunspace-building convective openings would all help to mitigate overheating. No major difference is noted between the two configurations tested. Two expected effects are clear in the results:

(a) added sunspace mass increases comfort (Period 4, 1980-81), and

(b) night insulation increases performance. (Periods 2-3, 1980-81).

Many different variations were tested, and we rely primarily on the validation of simulation models and the use of these models to sort out the many differential effects.

- In 1981-82, sunspace Rooms 3 and 4 were separated. The only difference between them was control of air flow from sunspace to room. Air flow was fan-forced on a room temperature thermostat set at 75°F in Room 3 and through vents by natural convection in Room 4 (both vents have backdraft dampers). This operation improved performance somewhat and improved comfort greatly in Room 3 compared with Room 4.

- Although data exist for only a very limited period quite late in the year, the convection suppression scheme tested in Room 12b does not seem to have improved performance (Period 3, 1981-82).

Test Boxes

Test boxes are much cheaper and simpler to build than test rooms and serve a very useful purpose for certain types of testing.[24] If one is concerned with a situation in which the flow of heat is essentially one dimensional through the wall, and height effects are not of particular importance, a test box may give satisfactory results. Boxes may also be quite useful for the comparative testing of various materials or surface treatments.

Test boxes are generally built of expanded polystyrene. If the side walls are made quite thick and the back wall less thick, heat loss from the test box will be primarily through the back wall. This configuration helps in maintaining a more one-dimensional heat flow.

The most critical parameter in a test box or test room configuration is the load/collector ratio (LCR), which was defined earlier. A test room or a test box will respond to the weather in much the same way as a building having the same LCR.

Whereas it is usually not very important in a larger structure, the two-dimensional nature of the heat flow through the insulation of a test box must be correctly considered in calculating the net load coefficient. We can achieve a convenient approximation by calculating the area, A, in the U × A formula (where U is the conductance through the wall) as the logarithmic average of the inside area, A1, and the outside area, A2:

$$A = \frac{A1 - A2}{\ln (A1/A2)}.$$

Moore has used test boxes at Los Alamos to study the surface adhesion of various selective surface foils to concrete Trombe walls.[25] The test box Trombe walls were built of solid cast blocks of 7.5-in.-thick concrete. Thermocouples were carefully placed on the outside surface by milling small slots in the surface of the block and grouting in the thermocouple with a fine plaster. Various methods of surface adhesion were tried, side by side, to determine if a significant resistance to heat flow occurred between the selective surface foil and the surface of the block. It is much faster to build six test boxes for this purpose than six test rooms.

Configurations tested are as follows:

1. Flat black paint on concrete.
2. Manufacturer's recommended selective surface installation procedure (prior to 7/1981), consisting of surface treatment, a rubber-based cement, and the selective surface foil. (The foil has a cement on the back covered by a peel-off paper layer.)
3. Surface treatment plus a thin layer of rubber cement plus a 1/16-in. layer of neoprene plus the selective surface foil.
4. Surface covered with a smooth finished aluminum-loaded epoxy cement plus the selective surface foil.
5. Surface covered with a smooth finished commercial grout plus a rubber-based adhesive plus the selective surface foil.
6. Surface covered with a smooth finished aluminum-loaded epoxy plus an experimental aluminum foil adhered directly to the wet epoxy. The foil has a partially selective paint on the exterior surface.

Test results are shown in Fig. 9 for four of the test boxes. Figure 9a shows outside surface temperatures measured with a thermocouple adhered to the back of the selective surface foil and temperatures inside the test boxes for one sunny day in December. A more direct measure of the

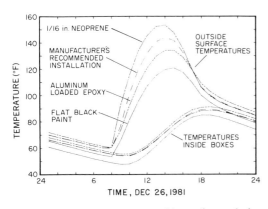

Fig. 9a. Temperatures measured in test boxes during a sunny December day. The outside surface temperatures are measurements made with a small thermocouple fastened to the back of the selective surface foil, except for the flat black paint case, where the thermocouple is grouted into the concrete block surface.

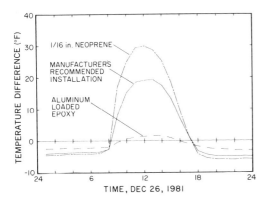

Fig. 9b. Delta T across the selective surface adhesive bond. Shown plotted are the differences between measurements made by thermocouples fastened on the back of the selective surface foil and grouted into the concrete block surface.

surface contact resistance is the temperature difference between the foil and the concrete surface as shown in Fig. 9b. The figures show a significant contact resistance between the foil and the concrete for both the manufacturer's recommended procedure and the 1/16-in. neoprene case but only a small contact resistance for the aluminum-loaded epoxy case. Note, however, that the temperature difference reverses at night, partially offsetting the daytime disadvantage of the contact resistance.

Because the test boxes are free-running (no auxiliary heat), a direct comparison of performance can be obtained by observing the difference ($\Delta \overline{T}$) between the average temperature in the test boxes ($\overline{T}$) and the average outside temperature. These average temperatures are given in the following table for the two days of

December 26 and 27, 1981 (the average outside temperature is 39.8°F).

Box	$\overline{T}$	$\Delta \overline{T}$	Ratio
Flat black	69.4	29.6	1.00
SS—manufacturer's recommended installation	75.3	35.5	1.20
SS—1/16-in. neoprene	75.3	35.5	1.20
SS—aluminum-loaded epoxy	76.9	37.1	1.25
SS—grout	76.0	36.2	1.22
Selective paint—aluminum-loaded epoxy	71.9	32.1	1.08

If each of the test boxes is constructed identically, the heat transferred through the back of the Trombe wall will be directly proportional to the difference between the average temperatures, $\Delta \overline{T}$. The last column in the table gives the ratio of $\Delta \overline{T}$ to the $\Delta \overline{T}$ for the flat black case. This shows that most of the performance advantage of the selective surface is being realized with the manufacturer's recommended installation and that performance is not very sensitive to contact resistance. Nonetheless, the manufacturer has since modified the recommended installation procedure to achieve a lower contact resistance.

SPECIAL EXPERIMENTS
As examples of special experiments, three different cases will be described; the first leads into the second, but the third is on an unrelated topic.

Similitude Experiments
Because many passive solar buildings gain solar heat on the south side of the building only, a concern arises about the distribution of that heat to rooms without direct solar gain. Many designs rely on convection through doorways to provide heat for isolated north rooms, especially small rooms such as a bathroom. One way of studying the effectiveness of such convection is through a similitude experiment. Because it is less expensive and simpler to work under laboratory conditions, we studied the convection in a small-scale model. However, to obtain comparable results, it was necessary to maintain a similitude in certain dimensionless parameters, namely, the Grashof number, Gr, and Prandtl number, Pr. If this is done, the streamlines in the experiment will scale with those in the larger building, and the Nusselt number, Nu, will be the same.[26] That this is so can be shown from the steady state

Navier-Stokes equations that govern the convection within the building and the experiment.

A reasonably convenient way to implement this is to use a Freon gas within the experiment. The fact that Freon is much heavier than air means that the experiment can be scaled to one-fifth size. Weber and Kearney conducted experiments of this type using a laboratory mockup of two rooms with a separating door.[27,28] Heat was convected from a hot plate on the far side of one room through the doorway to a cold plate on the far side of the opposite room. By measuring heat transfer rates and the corresponding temperature differences, correlations were determined.

Figure 10 shows some results from this experiment. The different points on the curve represent two different heights of doorway. The data were found to correspond reasonably well to a standard correlation often used to describe convective flow in other geometries as follows:

$$Nu/Pr = 0.3 \, Gr^{1/2}.$$

Spot checks were made by Weber and Kearney to ascertain the reasonableness of this correlation for convection by air in actual buildings. If the properties of air are put in, the equation takes the form

$$Q = 4.6 \, w[h(\Delta T)]^{3/2}$$

where

Q = heat flow from room to room through the doorway, Btu/h,

h = the height of the doorway, ft,

ΔT = difference between the average temperature in the two rooms measured at mid-door height, °F, and

w = door width, ft.

One would expect the leading coefficient in this correlation would be somewhat sensitive to geometry, and further work has been undertaken to determine how pronounced the effect is.

Heating of an Actual North Room

How accurate is the above correlation in predicting actual temperatures in a north room heated only by convection through the doorway? To answer this question, we analyzed data taken in a bathroom located against the north wall of the Balcomb solar residence.[29] A simple mathematical simulation model of the room was devised which accounted for heat flow through the doorway using the above equation, heat flow from the room to the outside based on the room load coefficient, and storage of heat in both the light-weight elements of the room and the heavier plaster walls of the room. The model was driven with temperatures measured in the adjacent living room and outside the house. A comparison was then made between the predicted and measured room temperatures. This comparison is shown in Fig. 11, indicating excellent agreement over a 19-day period. A few anomalies are observed when the door may have been closed. Overall, however, the agreement is quite good with a root-mean-square error of 0.5°F over this time period. This must be compared with the 1.6°F average temperature difference between the living room and bathroom.

Solar Radiation Transmission Through Bare Tree Branches

Another special experiment that illustrates a completely different issue was conducted by Holzberlein.[30] He was concerned about the commonly accepted premise that deciduous trees

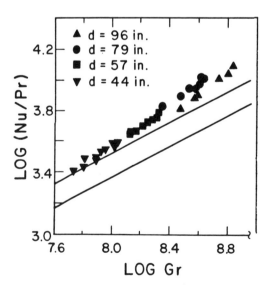

Fig. 10. Doorway correlation results from convection similitude tests.

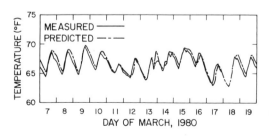

Fig. 11. Comparison of the measured and predicted temperature in the downstairs north bathroom of the Balcomb solar home.

could be placed on the south side of a passive solar building. The premise is that the trees provide an ideal shading mechanism, shielding the house from the summer sun and allowing penetration of winter sun through the tree after the leaves have dropped. Holzberlein's concern involved how much the bare branches would impede the solar transmission. The experiments were done by measuring the solar radiation from a large diffuse reflecting surface located in the shade of the tree in winter and comparing this with measurements made in the unshaded direct sun. The results are startling and cast doubt on the entire premise of south deciduous shade control. Figure 12 shows the transmission measured through various parts of two different trees that have bare winter branches. The penalty in solar radiation is 20 to 40%.

To see the full implications of these results, suppose that the oak tree of Fig. 12 were to be located 20 ft to the south side of a passive solar building. Figure 13 shows the result by superimposing a sun chart over the profile of the tree.[31] This chart shows the path of the sun across the

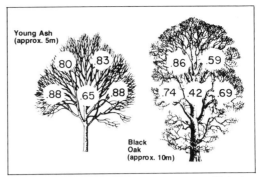

Courtesy of *Solar Age*.

Fig. 12. Measured transmittance of two deciduous trees with bare branches.

sky during different months showing the location at different times of day. One can readily see that the shading by the bare branches is most severe just at the time when the solar radiation is needed most. To make matters worse, one can also see that the tree is relatively ineffective for summer shading.

The strategy that evolved out of this study shows that deciduous trees can be used most

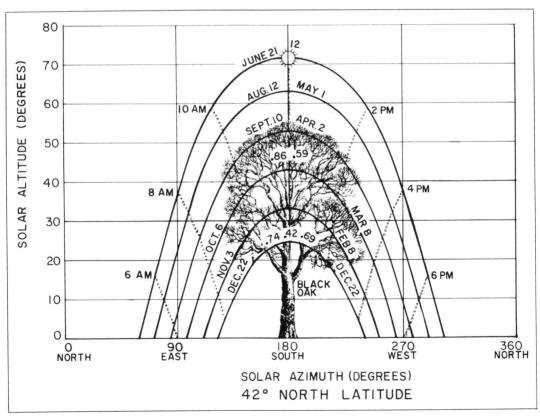

Courtesy of *Solar Age*.

Fig. 13. Superposition of the oak tree of Fig. 12 over a sun chart showing the relationship between tree shading and time.

effectively on the east and west sides of the building within an angle of 60 to 90°, measured from true south. Here the trees balance the sun's shift during the seasons, compensating for spring and fall, but do not impair performance during the coldest winter months. The south 120° arc should be free of shading.

MONITORED BUILDINGS

Buildings are monitored for two rather different reasons. The first and most straightforward reason is to determine the actual performance of the building in terms of heating required and comfort achieved. The second reason is to provide data for the further validation of mathematical models. Two examples of performance evaluation will be given to illustrate the principles involved.

Fowlkes Residence

Bozeman, Montana, is located at a latitude 45.7° N and at an elevation of 4,795 ft in a broad mountain valley. Heating degree days are approximately 8,300. The Fowlkes residence in Bozeman is a simple structure, in thermal terms, consisting of a large waterwall on the south side heating a well-insulated frame building. Although the building has two floors, it is basically a single-zone situation. A movable insulation curtain is dropped between the south double glazing and the waterwall at night.

The waterwall consists of 20 tubes, 15.75 in. in diameter, each containing 1,486 lb of water for a total heat storage capacity of 29,723 Btu/°F. It is estimated that this constitutes 87% of the total heat capacity of the building.

To determine an overall heat loss coefficient for the building, the inside-outside temperature difference was integrated over a long period and compared with the total input energies consisting of solar gains, incidental electric heat inputs, heat from the residents, and the wood stove. The measured heat loss coefficient of 437 Btu/°F h is in reasonable agreement with the estimates based on standard handbook procedures.

Solar energy was measured by positioning a transducer behind the solar glazing. Electric power input was monitored by measuring the current to the input mains with a clamp-on ammeter. Accuracy is estimated at 3 to 5%. The woodburning furnace output was determined from the temperature difference in the supply-and-return air ducts and the air flow rate. The accuracy of this measurement is estimated at about 10%.

Hourly data were recorded between November 28 and March 6. The heating degree days during this test period were 27% above average, creating a challenging test environment. The average house temperature was 66° F, with an average difference of 6.7° F between the upper and lower floors. The average daily temperature swing in the house was 11.5° F.

During the test period, electrical power dissipated in the house was 7.3 MMBtu. Thirteen fires were built in the auxiliary wood furnace, which added a total of 9.8 MMBtu to the house. Total heat requirement of the house was 49.3 MMBtu, based on the experimental heat loss factor discussed above. Based on these measurements, the sun provided 32.2 MMBtu, or 65% of the total heating energy during December, January, and February. Because the electrical power dissipation would occur anyway, it is more helpful to quote solar savings in terms of a fraction compared with the total energy minus the internal heat. This calculation yields a solar savings of 77%.

A number of other factors were evaluated during the course of the analysis. In one case, measurements of the insulating characteristics of the movable curtain revealed a much higher heat transfer coefficient than anticipated; this high value was probably caused by inadequate sealing around the curtain edges.

Daily and weekly data were also investigated to determine the general response characteristics of the building. The results are summarized in Refs. 32 and 33.

Balcomb Residence

The Balcomb residence is a more complex structure in which the solar heating is primarily by means of a two-story sunspace located on the south side of the building.[34] Not only are sunspace temperatures significantly different than those in the house, but there are room-to-room temperature variations within the house itself because of air flow isolation between walls and between rooms. The house is located in Santa Fe, New Mexico, at an elevation of 7,200 ft, latitude 36° N, with 6,000 heating degree days.

Although hourly data have been recorded for several years, the particular time period from November 1, 1978, through April 24, 1979, was selected for detailed analysis.[35] In the course of this investigation, it was found that several heat flows within the house could not be determined with sufficient accuracy. Additional channels were installed during a six-week period in spring 1980 to investigate these heat flows. These included heat flow through the doors separating the greenhouse from the house, heat lost by transpiration of water from plants within the sunspace, heat flow in the floor of the sunspace, and heat flow in a fan-forced rock bed under the house floor.

Based on subsystem models and correlations developed from the 1980 data, the information taken during the winter of 1978-79 was reanalyzed to produce overall energy balances.[36]

The overall heat loss coefficient for the house was determined using data taken during a 16-day period in December 1978. Total electrical energy was measured, and the integrated inside-outside temperature difference was determined for both the house and the greenhouse. Solar gains were determined by using data taken on a horizontal pyranometer, separating these data into direct and diffuse components, and then reassembling solar radiation incident on each of the six glazing planes of the house. Glass transmittance was calculated using the Fresnel relationships and a typical extinction coefficient. Ground reflection was taken at 0.3 without snow and 0.7 with snow cover.

The overall heat loss coefficient for the house was determined to be 445 Btu/$^\circ$F h and for the greenhouse, 224 Btu/$^\circ$F h. This is 3% lower than the overall calculated value. These same heat loss coefficients were then used to calculate house and greenhouse heat losses during the entire winter. Solar radiation was also calculated by the same process.

In addition, daily energy flows were calculated for the following elements.

- Heat flow through the adobe wall that separates the house from the greenhouse was calculated hourly during a dynamic method described in Ref. 37 based on temperature measurements made at the inner and outer wall surfaces at three locations in the upper and lower walls.
- Convection through the doorways that separate the house from the greenhouse was determined hourly using a correlation validated by Weber and Kearney. These values were calculated separately for the lower floor, the center bedroom upstairs, and the end bedrooms upstairs.

- Heat storage in the plaster walls, in the wood roof beams, and in the furniture of the house was estimated hourly based on air temperatures measured downstairs, in the center bedroom, and the end bedrooms. This analysis was done with a simple dynamic model that has been validated based on the detailed data taken in 1980.
- Heat required for the evaporation of water from plants and other sources of water within the house was estimated and found to correlate well with the average greenhouse temperature; the evaporation rate averages approximately 55 lb of water per day, corresponding to an energy requirement of 57,000 Btu/day.
- Heat transported by the fans from the greenhouse to the rock beds, heat flow up through the floor slabs covering the rock beds into the dining room and living room, and heat flow into the ground underneath the rock beds were calculated using a pair of coupled two-dimensional models. One model was used for the rock beds and another for the heat flow into the ground, around and through the perimeter insulation, and up through the north berm. These models have been partially validated by comparison with both the 1978-79 and the 1980 data.
- Heat generated by a small woodburning stove was estimated based on the hourly average flue gas temperatures using an empirical correlation determined during a controlled burn.
- Heat flow from the water heater to the house was estimated as 11,780 Btu/day plus 25% of the electrical energy into the water heater. (During most of the analysis period the solar water heater was shut down for modifications.) Heat from people in the house is estimated as 11,000 Btu/day.

Energy Balance

Monthly and annual energy flows are given in Table 3 and a few are summarized in Fig. 14. To estimate energy savings, the concept of a useful load was developed. The useful load is computed based on the degree hours computed between the measured house or greenhouse temperature and the outside temperature, but degree hours above an arbitrary fixed reference temperature are discarded. Thus, one does not count as useful load any energy required to keep the space above the fixed reference level, which was set to 70°F for the house and 45°F for the greenhouse.

KBtu for Balcomb House, 1978–79	NOV	DEC	JAN	FEB	MAR	APR	YEAR
Solar Gains	13790	15416	15718	19800	18265	13224	96217
House	2360	2490	3114	3941	3593	3059	18559
Greenhouse	11430	12926	12604	15859	14672	10165	77658
Heat Losses	16628	22482	22827	20043	18407	11996	112394
House	9589	13916	14625	11632	10307	6483	66555
Greenhouse	4697	6428	6533	5968	5364	3315	32307
Evaporation	1772	1564	1159	1810	2082	1676	10066
Greenhouse to Ground	570	574	510	633	654	522	3466
Useful Load	12652	18799	19987	15662	13548	8652	89309
House	9506	13914	14625	11589	10104	6300	66041
Greenhouse	3146	4885	5362	4073	3444	2352	23268
Vented Energy	1503	0	491	3675	3752	3995	13415
Auxiliaries	563	2379	3734	524	230	27	7457
Baseboard Electric	262	894	1193	222	119	27	2718
Stove	0	302	273	0	0	0	576
Fireplace	301	1183	2268	302	111	0	4163
Internal Gains	3779	4685	3865	3392	3663	2740	22133
DHW Retained	752	820	729	593	442	275	3614
People	324	334	334	302	334	259	1890
Other Electric	2703	3531	2802	2497	2887	2206	16629
Greenhouse to House	3444	4356	2596	5069	3237	3075	21785
Convection through Open Doorways	2519	3709	2397	3550	2504	2251	16932
Conduction through Doors	−177	−287	−347	−187	−187	−111	−1298
Adobe Wall	405	424	350	553	210	211	2155
Stairwell Wall	−112	−240	−414	−27	−83	−38	−911
Forced Convection to Rockbed	809	750	610	1180	793	762	4907
Rockbed	809	750	610	1180	793	762	4907
Upward through Floor	491	366	365	479	375	297	2375
Downward into Ground	311	330	271	492	389	363	2159
Other Rockbed Losses or Gains	7	54	−26	209	29	102	373
Heating Required	8873	14114	16122	12270	9885	5912	67176
Solar Savings	8310	11735	12388	11746	9655	5885	59719
SOLAR LOAD RATIO PREDICTION							
Auxiliary Heat Predicted	417	2320	3927	953	416	59	8092
Auxiliary Heat Observed	563	2379	3734	524	230	27	7457
AVERAGE TEMPERATURES (F)							
Dining Room	67.7	67.4	65.7	68.3	69.0	69.4	67.9
West Bedroom	68.5	65.7	63.0	70.1	71.7	70.2	68.2
Center Bedroom	65.1	61.7	59.6	67.2	69.0	70.0	65.4
Greenhouse	64.8	61.5	57.6	66.4	67.1	68.1	64.2
Outside Ambient	38.1	26.2	22.2	31.4	38.6	44.9	33.6
Rockbed	74.5	72.4	69.9	75.5	73.8	74.4	73.4
Floor Surface above Rockbed	69.2	68.6	66.7	70.0	70.2	70.5	69.2

Table 3 Balcomb House Results for the Period November 1, 1978—April 24, 1979

The heating requirement for the house is the useful load minus the internal energy generation by lights, people, water heater, and appliances. Solar savings is the heating requirement minus the auxiliary heat and totals 57 MMBtu for the year, or 89% of the heating requirement.

Thermal Comfort

Plots of the hours of occurrence in each one-degree temperature band are given in Fig. 15 for the dining room and greenhouse. The effect of mass wall buffering is very apparent in the dining room, which has a small daily temperature swing

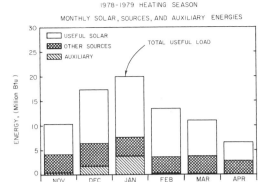

Fig. 14. Bar chart of monthly energies in the Balcomb solar home, 1978–79 heating season.

of only 5 to 6° F. By contrast, the uncontrolled greenhouse space has large temperature swings (30° F typical), clearly showing the two-zone nature of the house.

Conclusions

Results are summarized in Refs. 35, 36, 38, 39, and 40. Some conclusions about the performance of the house are as follows.

- The overall performance of the house has been very good. It has provided good comfort conditions in a cold climate with very small requirements for auxiliary heat. Operation is simple and reliable.
- The greenhouse is an efficient solar collector. Approximately 31% of the solar radiation transmitted into the greenhouse is subsequently transferred to the house. In addition, the greenhouse is adequately heated, maintaining conditions well above freezing, with-

out auxiliary heat. A critical design feature that leads to greenhouse effectiveness is the ability to thermally isolate the house from the greenhouse by closing doors.

- The predominant mode of heat transfer between the greenhouse and the house is by convection through doorways that are opened during the daytime. The fact that the greenhouse serves as a major traffic area is important to the effectiveness of this control mechanism. Typical convection through a doorway is 50,000 Btu on a sunny day for the upstairs bedrooms and 23,000 Btu for the downstairs; this difference is caused by the slightly colder room temperatures and higher greenhouse temperatures upstairs. The typical driving ΔT upstairs is 15° F. Much of this heat goes to satisfying daytime loads, but about 40% is stored in plaster walls, wood-beamed ceiling, and house furnishings.
- The primary utility of the massive adobe wall between the house and the greenhouse is for direct-gain storage in the greenhouse. Most of the heat absorbed by the wall is released back to the greenhouse at night and is essential to maintaining reasonable temperature conditions in the greenhouse. The amount of heat transmitted through the wall to the house is 1.9 MMBtu for the year. This effect is larger upstairs because of less shading of the wall, slightly lower room temperatures, and a thinner wall (10 in. versus 14 in. downstairs).
- Heat storage in the plaster walls, wood-beamed ceiling, and furnishings of the house is significant. Carry-over heat from one day to the next is observed on 89 of the 176 days of the analysis period, averaging 49,200 Btu per

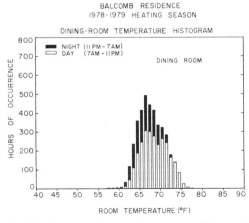

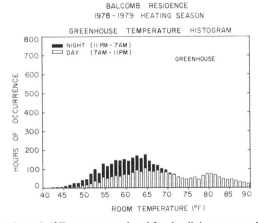

Fig. 15. Number of hours of occurrence of the temperature in each 1° F temperature band for the dining room and greenhouse in the Balcomb solar home during the 1978–79 heating season. The top of the bar is the total number of hours of occurrence; this is divided into two time periods, as indicated.

day. Diurnal heat storage (heat stored and released during the same day) occurs nearly every day and averages 89,800 Btu per day.

- Water evaporation in the greenhouse significantly improves the living quality by increasing the humidity into the 40 to 50% comfort range, but this is at the expense of about 57,000 Btu per day of energy.
- The rock bed definitely appears to have a positive effect on house heating, although less than originally estimated; nevertheless, it is considered especially important to the comfort characteristics of the house.
- Summer weather in Santa Fe is mild with large diurnal swings. Maximum house temperatures are 82°F upstairs and 78°F downstairs without air conditioning. Overheating that might be caused by the greenhouse is prevented by sun control, good ventilation, and night vent cooling of the large house mass. The greenhouse roof and second-floor balcony effectively shade the adobe wall as cross ventilation and stack ventilation remove excess heat.

Other Monitored Buildings

Many passive solar buildings have been monitored. Several have been evaluated in detail under the National Solar Data Network, where a standard procedure is prescribed to process and analyze data. Published results have generally shown very good performance.[41] In addition, a Class B monitoring program has been started that will emphasize the evaluation of passive solar buildings.[42]

A full discussion of the much-publicized double envelope concept is beyond the scope of this chapter; however, there has been considerable experimental evaluation. Although thermal performance has been good, this is largely attributed to the superinsulated character of the building rather than to the supposed convection and heat storage mechanisms. These evaluations reveal that natural convection is very weak and that heat storage under the building is minimal.[43, 44, 45] Various design solutions have been proposed, some involving the use of a fan, but there has been little follow-up investigation.

MATHEMATICAL MODELING

Simulation Analysis

The response of a building to any type of heat input is simulated by solving a set of differential equations that describe the heat flow from point to point within the building. One must first select a reasonably small set of elements within the building whose temperature will be calculated. Elements that can be expected to be about the same temperature can be lumped together into one element. It is of particular importance to include all of the important heat storing mass within the building in one or another of these mathematical elements. Massive portions expected to be at rather different temperatures should be characterized as different elements.

Having made this selection, the analyst then writes an ordinary differential equation describing the heat balance for each element. This heat balance includes heat flow to neighboring elements by radiation, conduction, or convection, solar energy inputs, and other heat inputs. This set of differential equations can then be solved as an initial value problem with several independent variables including solar gain, outside temperature, and thermostat setting. Auxiliary heat input is adjusted to maintain a desired temperature of one or more of the elements (the room air temperature is usually the controlled element).

The use of simulation can perhaps be best illustrated through the example of an actual residence outside of Denver, Colorado, which is a mixed Trombe wall and direct-gain design. The house is of slab-on-grade construction with massive interior partitions and exterior walls, insulated on the outside.

Figure 16 shows a schematic of the simulation model of the house. Each point represents an element whose temperature is simulated by a differential equation. The number beside the point represents the heat capacity in Btu/°F. The resistor connections between points represent thermal conductance paths; the numbers along side indicate heat flow in Btu/h°F.

Because all of the solar glazing faces south and is vertical, a single calculation can be made each hour for the solar radiation transmitted through 1 ft² of glazing. The source numbers on the diagram (shown in the circles) give the number of square feet of solar glazing associated with solar gains into each of the appropriate elements. Direct gain is distributed among the various sunlit elements in the building; all of the Trombe wall solar gain is absorbed on the outside surface element.

To describe temperatures at various depths within a massive wall, the wall is mathematically sliced into sections. Because the surface ele-

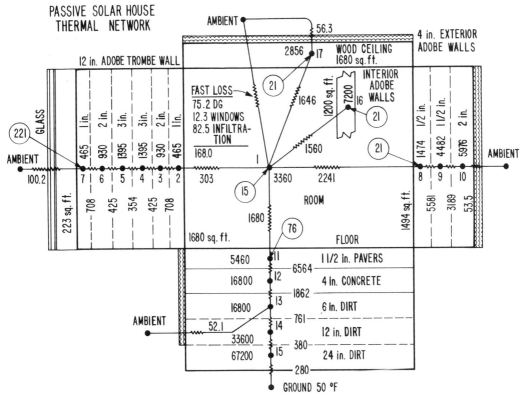

Fig. 16. Thermal network for the simulation example. On the left are the elements that describe the Trombe wall and its glazing, on the bottom are elements describing the floor and underlying earth, on the right are elements describing the exterior walls, on the top is the element describing the cciling. Interior walls are shown as Element 16.

ments can respond more quickly to external effects, these elements are made thinner than the interior elements. The floor is represented by several elements that simulate the behavior of the various layers; the earth layers are made progressively thicker at greater depths because the response will be increasingly sluggish. The earth is assumed to be at a constant temperature of 50°F at a depth of 42 in. This is probably not a very good assumption, but because floor heat flows are relatively small, extra earth detail is not deemed to be necessary.

The house has a fast loss coefficient that connects the room air (element No. 1) to the outside. This represents both heat flow back through glass to the outside and heat loss by infiltration. Other elements of the load coefficient are through massive walls, through the mass of the wood ceiling, and through the mass of the perimeter. The mass of the air in the house and all of the furniture are associated with the room air (element No. 1). Auxiliary heat is introduced into this same element.

Clearly, the selection of elements and the calculation of associated heat capacities and conduction coefficients is a matter of some engineering judgment; handbook values for most of these are readily available.

The differential equation that describes the energy balance of element No. 1 can be written from inspection of Fig. 16 and is given below as an example. The left side of the equation describes the rate of heat storage and the right side describes the rates of heat flow into the element.

$$3360 \, (dT_1/dt) = 15 \, S + Q_i + Q_a +$$
$$1646 \, (T_{17} - T_1) +$$
$$1560 \, (T_{16} - T_1) +$$
$$2241 \, (T_8 - T_1) +$$
$$1680 \, (T_{11} - T_1) +$$
$$303 \, (T_2 - T_1) +$$
$$168 \, (T_a - T_1),$$

where

T_1 = temperature of element No. 1,

T_n = temperature of element n,

T_a = ambient temperature,

S = solar radiation transmitted through 1 ft² of south glaxing,

Q_i = internal gains,

Q_a = auxiliary heat, and

d = indicates derivative.

The equations for the other 16 temperatures can be written from inspection of the network. One way to solve these differential equations is to write them as difference equations as follows:

$$3360 \, (T_1 - T_1')/\Delta t = 15 \, S + Q_i + Q_a +$$
$$1646 \, (T_{17} - T_1) +$$
$$1560 \, (T_{16} - T_1) +$$
$$2241 \, (T_8 - T_1) +$$
$$1680 \, (T_{11} - T_1) +$$
$$303 \, (T_2 - T_1) +$$
$$168 \, (T_a - T_1),$$

where

T_1' = temperature of element No. 1 at the previous time step, and

Δt = time step.

If the temperatures on the right side of the equation are all primed, that is, if they are all taken at the previous time step, the equations are said to be explicit. This is also called forward differencing because the differentials are based on conditions at the previous time. Although this is a convenient way to solve the problem, the solution is numerically unstable unless the time step is quite small (typically 10 minutes, depending on the heat capacities and thermal conductances). If the quantities on the right side of the equation are all taken to be at the end of the time step, the solution is implicit because none of the quantities is known and the entire set of 17 differential equations must be solved together. This is also called backward differencing. The advantage here is that the solution is unconditionally stable. Although energy is conserved by the form of the equations, the transients are not necessarily accurately represented if the time step is too long because the derivative changes significantly during the time step.

A commonly used approach sets the variables on the right side of the equation to the average of their previous and present values. This is called central differencing, and the equations are still implicit. Although the formulation is more complex, the solution is numerically hardly more difficult than backward differencing. The advantage is that the transient solution is about four times as accurate for the same time step. A one-hour time step can usually be used if one uses some care in the selection of elements and associated masses.

An implicit solution was used to solve the difference equations in the above example so that there are 17 equations in 17 unknowns. A further complication arises with the use of aux-

iliary heat. When the room temperature is within the range of 65 to 75° F, the auxiliary heat, Q_a, is zero and T_1 is a dependent variable. However, when the room temperature tends to stray outside the 75 to 85°F temperature band, Q_a is adjusted to maintain T_1 at the control value. Mathematically this means that Q_a becomes the dependent variable and T_1 becomes an independent variable. There are still 17 equations in 17 unknowns.

If the heat capacities and thermal conductances do not change with time, the equations are linear and can be solved by linear algebraic techniques such as progressive substitution or matrix algebra. (The matrix approach is convenient because the main seventeenth-order matrix needs to be inverted only once; thereafter the solution consists of matrix multiplications.) The inputs to the equations are ambient temperature and solar radiation; these are read from a weather tape. S is calculated by the conventional approach of separating horizontal solar radiation into direct and diffuse components. These components are then used to estimate transmitted solar radiation ascribable to direct, diffuse, and reflected components based on the angle of incidence (for direct) and solid angle (for diffuse and reflected).

Typical Meteorological Year (TMY) data were obtained from tapes supplied by the United States Weather Service for the purpose of building energy performance calculations. They are based on actual historical data measured at the site, but different month-long data periods are selected (based on obtaining the correct long-term averages); these months are spliced together to form a Typical Meteorological Year. Data from Colorado Springs were used because it is at nearly the same elevation as the building site in our example and is nearby.

The speed of calculation depends largely on the type of computer used. An annual calculation may take 30 seconds on a large mainframe computer, 4 or 5 minutes on a minicomputer, and 2 hours or longer on a microcomputer. This problem was calculated on a Hewlett-Packard 9845® desktop microcomputer, which codes in BASIC language. Computation was facilitated with a matrix Read Only Memory (ROM) unit in the computer.

The simulations were run for 200 days of the heating season extending from November 1 through May 19. Figure 17 shows characteristic results for a 20-day period during January that

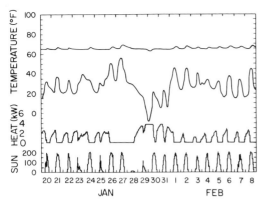

Fig. 17. Simulation results for a 20-day time period. Room temperature, exterior temperature, backup heat, and solar radiation are shown.

included the coldest weather of the year. This figure is very illuminating in itself because it shows the basic response characteristics of the building. Note that during sunny weather when no auxiliary heating is needed, the room temperature tends to have a dual characteristic response. There is a daily swing of about 7° F; this is superimposed on a much more gradual trend that follows the average outside temperature and average solar radiation. Note also that the backup heating is not at all uniform and is required primarily at night during the hours from midnight to sunrise.

These two characteristics are very typical of most passive solar buildings. The daily response is called a diurnal response and is driven primarily by the fast solar heating effect in the house, in this case the direct gain. The magnitude of the diurnal temperature swing is determined by the direct gain and the amount of diurnal heat capacity, which is the effective heat capacity of the house for a 24-hour sine-wave response.

The slow response of the house is determined by the loss coefficient of the house and the solar gain averaged over the "time constant" of the house. This time constant, τ, is the characteristic response time of the building to changes in outside temperature. It is given by the equation

$$\tau = \frac{\text{total heat capacity}}{\text{total loss coefficient}}.$$

In the case of this example building, the total heat capacity is 53,200 Btu /° F and the total loss coefficient is 8750 Btu /° F day, which can be ascertained from the thermal network. Thus the time constant is 53,200 / 8750 = 6 days. Note that in computing the total heat capacity, the heat capacity of the dirt under the house was omitted because it is very weakly coupled to the house. Had it been included, the time constant would be 19 days indicating that there will be a very slow drift with this characteristic time.

Simulation is a very powerful tool because it allows the analyst to estimate how the building will respond to a variety of different normal or extreme weather conditions before it is built. One can also see the influence of changing design parameters on the building performance. A further example of this computing power is obtained by simply rerunning the same simulation, but without any backup heat. This result is shown in Fig. 18. In this case, the building inside temperature drops to its yearly low value of 51° F on the night of January 30 at a time when the outside temperature is − 9° F. This exercise gives confidence that, with passive solar gains alone, the building is never likely to freeze inside. Other "what-if" scenarios can be played out on the computer to test the response of the building to a variety of conditions.

When the simulation was run for an entire year with no backup heat, the results shown in Fig. 19a were obtained. This shows the number of hours of occurrence in each 1° F temperature band of both the room temperature and the outside temperature. One can see that the effect of solar gains and good insulation is to transform outdoor temperatures into indoor temperatures. Figure 19b shows Trombe wall interior temperatures during a clear day showing how heat diffuses through the wall with a resulting decrease in peak temperature and delay in the time of the peak at points progressively further from the wall surface.

As a matter of course, one can also estimate

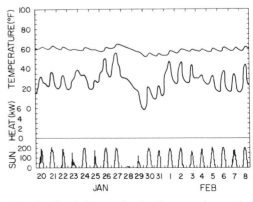

Fig. 18. Simulation results for the same time period as Fig. 17 except without backup heat.

287

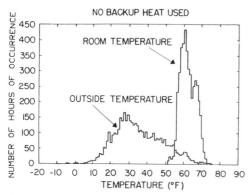

Fig. 19a. Number of hours of occurrence in each 1°F temperature band for both the inside temperature and exterior temperature for the entire 200 winter days simulated.

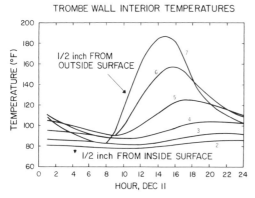

Fig. 19b. Calculated temperatures at various locations in the Trombe wall during a sunny winter day. Numbers refer to the node numbers in Fig. 16.

the monthly and annual auxiliary heat requirements of the house. For this case, the answer is 7.4 MMBtu for a 6,473 degree day heating season. This is less than 10% of the typical heating requirements of a modern, moderately insulated house of this size in this climate. In the course of the simulation one can also ascertain that a backup heating capacity of 12,970 Btu/h (3.8 kW) would be adequate to maintain a minimum temperature of 63.1°F if the thermostat is set at 65°F (see Fig. 17). This is also much less than the backup capacity required for a contemporary house.

Advantages and Disadvantages of Simulation
Simulation is a powerful tool because so much information is revealed about the thermal behavior of the building. The information that can be gleaned is limited primarily by the time and patience of the analyst. As a research tool, it is unequaled, and research has been its primary application.

As a design tool, simulation has very serious drawbacks. It requires a computer and people trained in its operation—not a normal part of a design office. Although software on the front end of the computer code can greatly simplify the process, describing the building to the computer can be a very time consuming and laborious process. This generic problem, common to all thermal analysis, but a particularly difficult issue with respect to simulation, has been addressed by Arumi,[46] Palmiter,[47] and Niles[48] in their codes, DEROB, SUNCODE, and CALPAS3.

Another difficulty with simulation, ironically, is the sheer mass of output information. Of course, the user can limit the output if desired, but it seems like analytical overkill to perform billions of calculations to produce millions of results only to peruse one number, perhaps the annual auxiliary heat, before repeating the calculation.

Validation
The primary purpose of validation is to assure that the algorithms contained in the simulation models are in reasonable conformance with the physics of what is occurring in the building. A straightforward method is to compare simulation results point by point with observable, measured values in a structure.

Although some validation has been done against data taken in actual occupied buildings, characterizing all the inputs in such a situation is so difficult and expensive that test room data have generally been used for validation. In time, enough data will be available from the Class A building monitoring program to allow a more comprehensive validation effort.

An example of test room validation has been given by McFarland.[49] Figures 20 and 21 show a sunspace test room and its thermal network. The

Fig. 20. Sunspace test room. Note that this is a double wide test room.

288

SUNSPACE THERMAL NETWORK

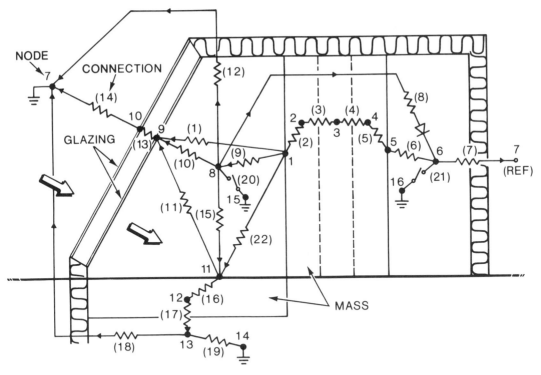

Fig. 21. Thermal network of the sunspace test room.

simulation was driven using both measured values of solar radiation incident on the glazing and ambient temperature. Figures 22 through 26 show comparisons of temperatures at various points inside the test rooms. Note that temperature extremes and variations in the test room are larger than those that one would tolerate in a passive building; the test rooms are intentionally somewhat overdriven to provide a somewhat more significant test for the analytical validations.

The graphs show comparison of predicted and measured temperatures at each point. Figure 27 shows a comparison of the predicted and measured auxiliary heat; no constants in the model were fudged to achieve the results shown. The results indicate good agreement between predictions and measurements. The integral error in auxiliary heat is approximately 4% of the total thermal requirements of the room. This was accepted as sufficient accuracy because other uncertainties in performance prediction, such as occupant effects and year-to-year variations in weather patterns, would result in significantly larger deviations.

Other Analysis Approaches

A standard approach to building energy analysis has been the use of weighting functions or transfer functions (these are not to be confused with Laplace transforms or Fourier analysis, to be discussed later). Because the heat flow through a wall is generally assumed to behave according to a set of linear equations, the response on one side can be characterized as a convolution of the inputs on the other side. The convolution is an integral, usually performed as a summation, of the input at each previous time interval multiplied by a weighting function, which is essentially the response of the wall to an impulse input. The technique is described in mathematical detail by Muncey.[50] Both the inputs and outputs can be either temperatures or heat fluxes.

The weighting function approach has been used in most large building analysis codes because it is particularly amenable to the handling of lightweight frame construction walls that may have considerable structural detail. The weighting functions can be precalculated, incorporating as much detail about the wall as desired, and

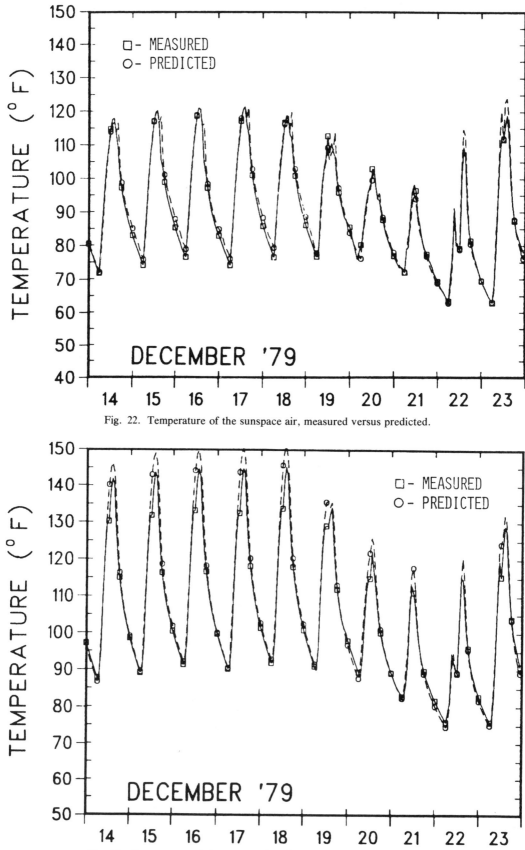

Fig. 22. Temperature of the sunspace air, measured versus predicted.

Fig. 23. Temperature of the outer wall surface, measured versus predicted.

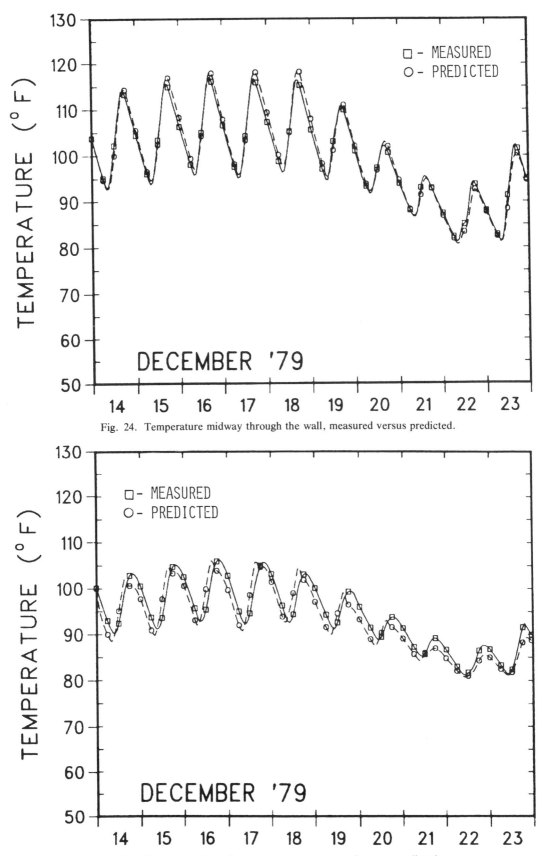

Fig. 24. Temperature midway through the wall, measured versus predicted.

Fig. 25. Inner wall surface temperature, measured versus predicted.

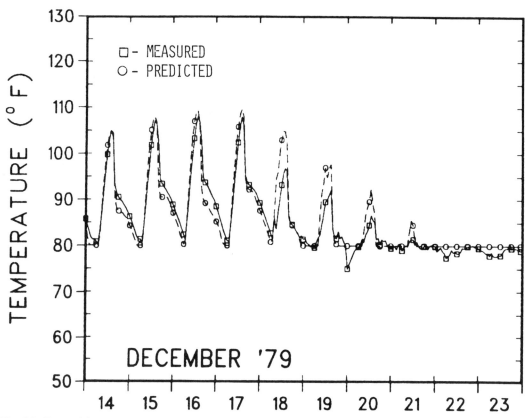

Fig. 26. Room globe temperature, measured versus predicted. Note that the thermostat was set at 80° and this controls the room temperature during the latter part of the period.

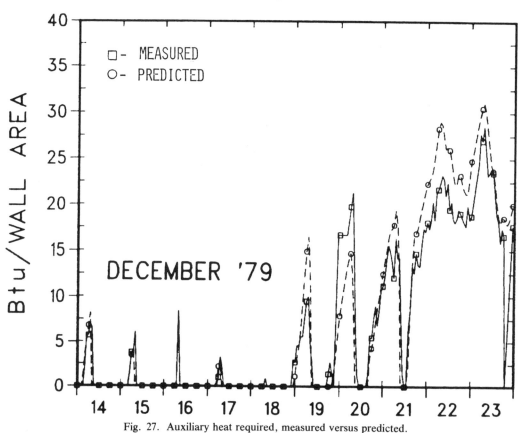

Fig. 27. Auxiliary heat required, measured versus predicted.

entered into a table to be used throughout the rest of the calculation. Generally there are only a few weighting functions because the time response characteristic of the wall is fairly short. For massive walls, this computational advantage is not so pronounced because the number of weighting functions becomes much larger; in terms of computational efficiency, thermal network analysis then becomes competitive.

Weighting function analysis is particularly convenient when the inside temperature is held constant; however, if it is allowed to float, account must be made with another set of weighting functions, increasing the computational complexity. By contrast, thermal network analysis is hardly more complex for a floating inside temperature than for one that is fixed.

Harmonic analysis, another computational technique that has been used by several analysts and researchers, provides yet another way of solving the same set of differential equations.[51, 52] The Fourier or Laplace transform of each equation is computed, and the equations are then solved algebraically in terms of the Laplace operator. Time solutions can be obtained by inverse transformation. Used to calculate the time response of a building, this approach has been primarily used by afficionados of the technique; it offers no particular computational advantage over network techniques.

The primary disadvantage of harmonic analysis techniques is their inappropriateness in cases where major nonlinearities exist. Although methods of dealing with nonlinearities have been developed, they are both awkward and complex, and this offsets any advantage of harmonic analysis. For smoothly varying nonlinearities such as radiation heat flow, the equations can easily be linearized around the operating point; the inaccuracy implied by this process is trivial for the size of temperature swings that would usually be experienced. An example of the type of nonlinearity that cannot be accommodated is the off-on character of a thermostat; therefore, harmonic analysis is not suitable for determining auxiliary heat requirements.

The primary advantage of the Fourier transform approach is in investigating the response of the system at a particular frequency. The frequency of greatest interest is the 24-hour cycle (the diurnal cycle) because it is such a dominant and important part of the weather input. By such techniques one can easily study the clear day temperature swings to be expected inside the building.[53, 54]

SYSTEMS ANALYSIS

In the context of this chapter, systems analysis means any of four types of analytical investigation: the first is the systematic study of how climate affects passive solar heating performance, the second is the study of how changes in design parameters affect system performance in a particular locality, the third is the development and use of correlation techniques as a simplified method of performance estimation, and the fourth is the development and use of a methodology for determining the optimal mix between conservation and solar strategies. Examples of the first three are given below.

The Effect of Weather on Performance

Once a simulation model has been developed for a particular building configuration, it can be run using hourly weather input data for any location where such data are available. Within the United States, there are 35 sites where weather data, including solar data, have been taken and compiled by the U.S. Weather Service. In addition, based on correlations developed from this primary data set, hourly solar data sets have been generated for all of some 240 sites where hourly temperature and other weather data have been taken (these are called "ersatz" data). This enormous data set is available from the United States Weather Service (National Oceanographic and Atmospheric Administration, Asheville, North Carolina) in a variety of computer-compatible media. It is easy to see that one is limited more by the time and cost of doing computer simulations than by available data.

Table 4 shows the results of an hour-by-hour computer simulation done by Gloria Lazarus using the PAssive SOlar Energy (PASOLE) computer code. The passive solar system is a semienclosed sunspace (meaning that the building extends around the ends of the sunspace so that they share three common walls). Sunspace glazing is double and is tilted at 50° to the horizontal. The common walls between the sunspace and the building are 12-in.-thick masonry. Auxiliary heat is supplied to maintain the building room temperature at a minimum of 70° F, and heat is vented from the room as necessary to prevent the room temperature from exceeding 80° F. Sunspace temperature is maintained between 45° and 90° F in the same way.

The building being simulated is a nominal 1,500 ft^2 single-family residence. Secondary mass within the thermal insulation envelope of the building is 9,000 Btu/°F, as might be asso-

City	Net Load Coefficient Btu/°F day	Sunspace Glass Area ft^2	Auxiliary Heat 10^6 Btu	Solar Savings Fraction %	Solar Savings Btu/ft^2
Albuquerque, New Mexico	6372	244	2.4	90	68,300
Apalachicola, Florida	9938	134	4.4	68	53,000
Bismarck, North Dakota	3695	129	17.4	23	31,100
Boston, Massachusetts	4688	146	13.0	30	29,800
Cape Hatteras, N. Carolina	7115	174	4.7	68	44,800
Caribou, Maine	3591	110	18.2	17	26,400
Charleston, S. Carolina	7964	157	5.0	68	52,000
Columbia, Missouri	4995	177	10.5	44	35,700
Dodge City, Kansas	5516	250	6.6	71	49,900
El Paso, Texas	7915	197	2.0	89	64,800
Ely, Nevada	4821	302	4.7	84	59,800
Fort Worth, Texas	7570	168	5.7	65	47,800
Fresno, California	7305	154	6.8	62	56,500
Great Falls, Montana	4163	161	13.7	35	35,300
Lake Charles, Louisiana	9231	127	6.4	53	43,400
Madison, Wisconsin	3992	134	14.9	26	30,300
Medford, Oregon	5251	139	11.5	42	45,300
Nashville, Tennessee	5795	149	7.9	50	40,500
New York, New York	5000	135	12.1	30	29,800
Omaha, Nebraska	4389	201	9.6	49	35,400
Phoenix, Arizona	9934	150	2.3	85	67,100
Santa Maria, California	8420	202	1.4	94	87,800
Seattle, Washington	5122	106	13.0	25	32,200
Washington, D.C.	5011	169	9.0	48	38,000

Table 4 Simulation Analysis Results for a Sunspace

ciated with normal wood-frame construction. Internal heat generation from people, lights, and appliances is 60,000 Btu/day, as might be expected from normal occupancy by three people. This is scheduled according to an hourly profile having two levels. A rate of 4,000 Btu/h is used between the hours of 5:00 p.m. and 11:00 p.m. and a rate of 2,000 Btu/h is used between the hours of 11:00 p.m. and 5:00 p.m.

The insulation and infiltration levels of the building and the size of the sunspace are varied from city to city to maintain the same cost effectiveness for both the conservation and solar measures. Thus, the insulation levels are greater (and the building net load coefficient is smaller) in colder climates, and the sunspace is larger in sunny climates.

Results are listed in Table 4 for 24 U.S. cities based on TMY hourly data. The following numbers are given:

- net load coefficient: the load coefficient of the building, excluding the sunspace and the common wall, Btu/°F day;
- sunspace glass area: actual net area of the sunspace glazing, ft^2;
- auxiliary heat: for both the sunspace and building, MMBtu/yr;
- solar savings fraction: defined by the equation

$$1 - \frac{\text{auxiliary heat}}{\text{net reference load}}$$

where the net reference load is the energy requirement of the building (excluding the sunspace) not met by internal heat generation;

- solar savings: the difference between the net reference load and the auxiliary heat, divided by the sunspace actual glass area, Btu/yr/ft^2.

Sensitivity Analysis

Another application of the hour-by-hour simulation is the study of the effect on performance of changes in one or another of the building design parameters. As an example of this, Fig. 28 shows the effect of changing Trombe wall thickness and density on annual performance in Los Alamos. For this calculation, the LCR was fixed at 16 Btu/°F day, internal gains were 0, and the thermostat was set at 65°F. The specific heat of the wall material was set at 0.2 Btu/°F lb, and the thermal conductivity, k, was varied according to the density, ρ, using the following equation:

$$k = 0.049e^{0.02\rho}, \text{ Btu/ft h°F}.$$

Simplified Methods

It is now generally accepted that computer simulation will give an accurate representation of the performance of passive solar buildings, a condi-

TROMBE WALL MODEL

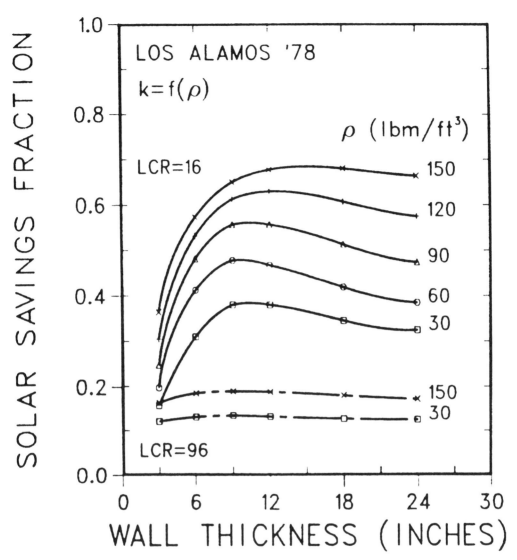

Fig. 28. Results of simulation analysis showing the sensitivity of solar savings fraction to variations in Trombe wall thickness and density in Los Alamos.

tion that makes simulation a desirable design tool if the designer has the equipment, the capability, and the inclination to take this approach. But even under the best of circumstances, it is costly and time consuming. Most designers ask for simpler techniques that are amenable to the use of hand calculators or desktop microcomputers on which estimates can be generated in a few minutes.

Correlation techniques that meet these requirements and give reasonable accuracy have emerged as practical procedures. These methods are particularly useful early in the design process when quick feedback is essential; they can be applied to either residential or commercial buildings. Both a monthly calculation—the solar load ratio (SLR) method—and an annual calculation—the load collector ratio (LCR) method— are described. The annual method uses tables precalculated by the SLR technique and is more appropriate to hand analysis, whereas the monthly method is more versatile and is more appropriate to programmable calculator or microcomputer-aided analysis.

Correlation Methods

A correlation technique is used to relate a desired result in terms of one or more correlating parameters (generally dimensionless). Success is much more likely if the chosen correlating parameters preserve some essence of the overall physics governing the energy balances. The F-chart technique, developed at the University of Wisconsin for active solar systems, is an example of a correlation technique that uses two correlating parameters.[55] Researchers at the Los Alamos National Laboratory independently developed the SLR method for active systems, which uses one correlating parameter, the solar load ratio. Since then, the SLR method has been applied extensively to passive solar systems, and the University of Wisconsin has developed the un-utilizability method for passive systems.[56]

These methods have two things in common. First, they use monthly weather data to predict monthly performance. A month has been found to be a particularly convenient time interval, being long enough that statistical variations tend to average out, and short enough that the basic weather statistics are stationary. Furthermore, only 8 to 12 calculations are required to predict annual performance. The prediction of monthly performance leads to relatively high errors (typically ±8%) but annual performance can be predicted with an error of only ±3%. This degree of accuracy is adequate for design purposes, as it is significantly less than the year-to-year variation that can be anticipated.

A second common feature of the methods is that the correlations are derived using data developed from hour-by-hour computer simulations. For the passive SLR correlations, the PASOLE[57] program was used. Thus, the correlation techniques are second generation analytical procedures intended to give reasonably good correspondence with simulation analyses; as such, their results are intrinsically no better than those obtained from simulations. The correlation techniques, however, require several thousand times fewer calculations to complete a yearly estimate and can be done using only a hand-held calculator.

The Solar Load Ratio Method Applied to Passive Solar Systems

The SLR method has been applied extensively to a variety of passive systems, each system requiring a different correlation. The method is the basis for the design techniques described in Refs. 16, 22, and 58, and the results are being widely used within the passive solar design community. The method leads to an estimate for passive solar heating only; it provides no insight into cooling issues.

The parent set of monthly performance data for the SLR correlations has been generated using the PASOLE hour-by-hour computer simulation code. The method uses a single correlating parameter (SLR) defined as follows:

$$SLR = \text{(solar energy absorbed)}/ \text{(building heating load)}. \qquad (1)$$

Different definitions of "solar energy absorbed" and "building heating load" have been used in the past; the precise definition of SLR used now is given below. The correlation time is one month so that each of the parameters in the above equation is for a one-month period. Both the numerator and denominator of SLR are in energy units, so that SLR itself is dimensionless. Physically, it relates the monthly net solar energy available to the building to the monthly net heating load that would be experienced by a comparable building without the passive solar element.

The parameter that is correlated to the SLR is the solar savings fraction (SSF) defined as follows:

$$SSF = 1 - \text{(auxiliary heat)}/ \text{(net reference load)}. \qquad (2)$$

In this equation the net reference load is equal to the degree-day load of the nonsolar elements of the building as follows:

$$\text{net reference load} = (NLC) \cdot (DD), \qquad (3)$$

where NLC is the net load coefficient. The NLC is computed leaving out the solar elements of the building. Nominal units are Btu/°F day or W h/°C day. The quantity DD is the degree days, computed for an appropriate base temperature, as will be discussed later. A building energy analysis based on the SLR correlations would then begin with a calculation of the monthly SSF values. The monthly auxiliary heats are then calculated by inverting Eq. (2):

$$\text{Auxiliary heat} = (NLC) \cdot (DD) \cdot (1 - SSF). \qquad (4)$$

The annual auxiliary heat is the sum of the monthly values.

By definition [Eq. (2)], SSF is the fraction of the degree-day load of the nonsolar portions of the building that is saved by the solar element. To the extent that the net reference load can be considered to be representative of the auxiliary heating requirement of a nonsolar building, the savings achieved by the solar elements would be the net reference load [Eq. (3)] minus the auxiliary heat [Eq. (4)]:

$$\text{Solar savings} = (NLC) \cdot (DD) \cdot (SSF). \quad (5)$$

Although very simple, this is a reasonable approximation because in many situations and climates, it is approximately true that energy flows through a conventional solar facing wall, with a normal complement of opaque walls and windows, will balance out to zero over the whole heating season. Thus, the term solar savings fraction, while having the precise definition given in Eq. (2), is also a reasonable indicator of the actual savings (as a fraction of the net reference load) that can be expected.

Alternatively, one can define a nonsolar comparison building in any desired way (although it is often difficult to obtain agreement on a single definition). Then the auxiliary for this comparison building can be calculated using the SLR method to estimate the direct gain solar effect. In this case, solar savings is the difference between the auxiliary heat of the nonsolar and solar buildings. This approach requires a second complete building calculation.

In any case, SSF is not a figure of merit in itself, but is more appropriately considered as a convenient correlation parameter useful as an intermediate number needed in the process of estimating the required auxiliary heat.

Reference Designs

The hour-by-hour simulations that are used as the basis for the SLR correlations are done with a detailed model of the building in which all the design parameters are specified. The only design parameter that remains variable is the load collector ratio (LCR), defined later.

The correlations allow the designer to estimate performance variations caused by changes in a certain limited group of design parameters: the thermostat setpoint, the internal heat generation rate, the glazing orientation and configuration, shading, and other modifiers of the solar radiation. There is no allowance, however, for estimating the effect of changes in the many other design parameters. Thus, the

correlations relate only to the reference design used in the simulations.

Correlations have been generated for 94 different configurations. Nine different direct gain correlations have been developed representing different numbers of glazings, different values of storage-surface-to-glazing-area ratios, and different wall thicknesses. Fifty-seven different thermal storage wall correlations have been developed representing Trombe wall and water wall, use or nonuse of night insulation, different numbers of glazings, use or nonuse of a selective surface, different Trombe wall thicknesses and thermal conductivities, different waterwall masses, and both vented and unvented Trombe walls. Twenty-eight different sunspace correlations have been developed representing five different configurations, glazed and unglazed end walls on the linear configurations, use or nonuse of night insulation, and masonry wall or water drum storage.

Example of an SLR Correlation

As an example of the correlation results, we give simulation results in Fig. 29 and the corresponding correlation accuracy in Fig. 30 for one reference design, the case of a semienclosed sunspace with 50° sloping glazing, masonry thermal storage between sunspace and house, and no night insulation.

Procedure

A solar load ratio correlation is generated by a statistical least-squares fit of an ensemble of monthly data points that represent the monthly performance of a particular passive solar system. These points are computed using a detailed hour-by-hour computer thermal network simulation model of the passive system. Each model has been validated based on comparisons with test room data. The PASOLE computer code and its derivative, SUNSPOT,[59] are the codes used. Annual simulations are done using typical meteorological year (TMY) data for ten or more cities representing a broad spectrum of U.S. climate types. Typically, annual simulations are performed for each city for five different values of LCR (load collector ratio, defined below). This gives a reasonably diverse ensemble of data points. The functional form of the SLR correlation allows for the selection of several different coefficients. The entire set of monthly simulation results is used to determine a set of coefficients that results in a minimum

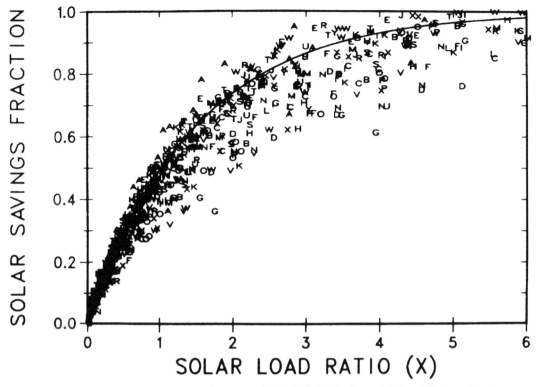

SEMI-ENCLOSED SUNSPACE — SSD1

Fig. 29. Solar load ratio correlation for passive system SSD1. Individual values of SSF and SLR resulting from the hourly simulations are shown for months having greater than 100 degree days. See Fig. 30 for city identification. The curve is the SLR correlation, chosen to yield a minimum standard deviation in annual SSF.

square error in the prediction of the annual solar savings expressed as a dimensionless solar savings fraction. The scatter in the monthly points (Fig. 1, for example) appears severe, but the greatest deviations are for the warmer months (larger SLRs) that have little effect on the annual results. The standard deviation of the error in prediction of the annual solar savings fraction, compared with the hour-by-hour simulations, is 2 to 4%.

Two SLR correlation forms were used, one for direct gain and other for all other systems. To avoid confusion, we have adopted a general, slightly more complicated correlation form that encompasses both. The monthly SSF is given by

$$SSF = 1 - (1 - F)K, \qquad (6)$$

where

$$K = 1 + G/LCR, \qquad (7)$$

and

$$F = \begin{array}{l} B - C \exp(-D \cdot X), X > R \\ A \cdot X, X < R. \end{array} \qquad (8)$$

Equation (8) requires the additional condition that the maximum value of F is 1. The quantity X is the generalized solar load ratio,

$$X = (S/DD - LCR_s \cdot H)/(LCR \cdot K). \qquad (9)$$

The quantity LCR is defined as follows:

$$LCR = \frac{\text{net load coefficient}}{\text{projected area}} = \frac{NLC}{A_p}, \qquad (10)$$

where net load coefficient (NLC) refers to the load coefficient of all of the building except the solar radiation aperture. The projected area (A_p) is the projected area of the net solar glazing on a vertical plane. This is the same as the area measured on an elevation drawing normal to the glazing azimuth.

The parameter LCR_s is the load collector ratio of the solar aperture. Nominal values of LCR_s are given in the correlation coefficient tables, however, LCR_s can be modified in practice to reflect changes in system design compared with the reference design. The dimensionless parameter H is determined in the correlation process.

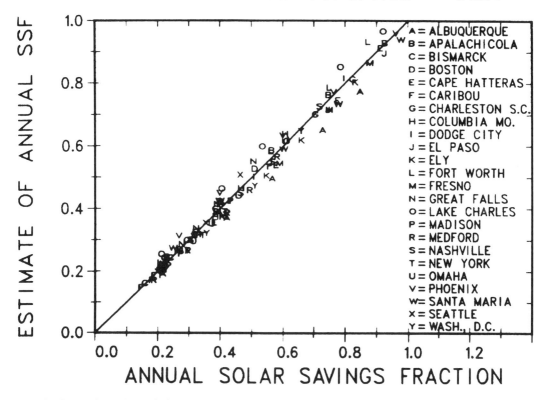

Fig. 30. Comparison of correlation vs simulation results for passive system SSD1. The X-axis is the simulation result, and the Y-axis is the result of the SLR monthly method using the SLR curve on Fig. 29. The root-mean-square error of these 120 points is 2.8%.

The parameter S is the solar radiation absorbed in the building per month per unit of projected area. It can be estimated using solar radiation correlations or by more complex trigonometric relations.

Values for the correlation coefficients A, B, C, D, G, R, and H for 94 passive system types are given in Refs. 58 and 60 together with system identifications and a description of the reference designs.

Performance Tables for Particular Locations (Annual Method)

Because the correlation curves are developed using weather data from a variety of different locations, these curves can be used in locations that have a climate type encompassed by the data for the original cities. However, an annual solar savings fraction calculation involves summing up the results of twelve monthly calculations. For a particular city, the resulting SSF depends only on LCR (the ratio of the net load coefficient to projected area), on the system type, and on the temperature base used in the calculation of the degree days. Thus, it is possible to make up tables for a particular city that relate SSF to LCR for the various systems, assuming one particular degree-day base temperature. These tables are much easier for hand analysis than are the SLR correlations; a simple four-function calculator is quite adequate, and one can complete an estimate in just a few minutes—quickly enough to be useful as a design aid. The disadvantage is that fewer parameters can be varied. The tables require that the orientation, degree-day base, and LCR_s all be specified and do not allow one to assess the implications of shading. Nonetheless, they are very useful and provide a very easy way to assess differences in performance for the various system options.

Annual SSF versus LCR tables have been made up for 209 different locations in the U.S. based on the SOLMET weather data and 10 cities in southern Canada. These are given in Ref. 58.

Mixed Systems

A simple methodology has been developed, using the SLR method, for dealing with designs that include more than one passive type. The technique treats the building net load coefficient as if it were divided into portions in the same ratio as the relative projected areas of the various passive system types. This amounts to the simple assumption that each of the system types serves a portion of the load with no exchange of heat across an imaginary boundary within the building. Normally one would expect that transfers which do take place would be beneficial and, therefore, the calculations based on this assumption might be somewhat conservative.

The procedure is as follows.

1. Calculate one LCR for the entire building using NLC for the entire building and the sum of all projected areas for A_p.
2. Calculate individual values of SSF for each passive type using the appropriate correlation or table and the LCR determined in (1) above. In this calculation use the appropriate separate values of S in Eq. (9) for each system.
3. Calculate the whole building SSF by computing an area-weighted average of all of the SSFs determined in (2) above.

Accounting for Internal Heat

Energy generated within the building by people, appliances, lights, and equipment partially offsets the need for auxiliary heat. A convenient method to account for this is to make the base temperature for calculating degree days equal to the balance point temperature. The procedure is as follows:

$$T_{base} = T_{balance\ point} = T_{set} - \frac{daily\ internal\ heat}{total\ load\ coefficient}. \quad (14)$$

Daily internal heat is the total internal energy deposited during one day in the building, and the total load coefficient is the net load coefficient, NLC, plus the steady state heat load coefficient of the solar aperture. The average thermostat setting is T_{set}.

To insure accuracy, it is important to include only that portion of the internal heat that stays in the building. The heating and ventilating system may discharge some of the internal heat directly; this discharged portion should not be counted.

In practice, the reduction in annual auxiliary that results from an increase in internal heat is often less than one-half the annual internal heat. It is normally much more efficient to use auxiliary heat just when needed, and to put internal heat into the building only as necessary for lighting and other vital functions.

Performance Variation Caused by Living Habits

All of the simulation analyses used to develop the correlations are based on a building that is used in a very specific and regular manner. The auxiliary heating thermostat is assumed to be set at a particular fixed level. A 10°F floating band is assumed. If the temperature in the house exceeds the thermostat setting by more than 10°F, it is assumed that the excess energy is vented so as to maintain the temperature less than or equal to the upper setting. This energy is not stored and is, therefore, lost.

The manner in which the house is operated greatly affects the energy consumption. The thermostat setting for auxiliary heat is by far the most important effect. Other operating characteristics of the house can also be important, such as (1) movable insulation, if provided, should be operated properly, (2) a building with much traffic in and out may experience unduly large infiltration caused by multiple door openings, (3) a building with doorways between the heated areas and a sunspace might be much more comfortable and use less auxiliary heat if some attention is paid to the appropriate opening and closing of these doors, and (4) the internal heat can have a major effect on the required auxiliary. Therefore, in interpreting the results from monitored buildings, or in stating the predicted performance of new buildings, one must be very careful to specify the operating conditions.

Design Analysis Versus Performance Prediction

As pointed out in the previous section, various aspects of the operation of a building have an important effect on the auxiliary heat requirement. Likewise, various building construction characteristics and the weather to which the building is exposed will have important effects on the building's auxiliary consumption. Because many of these elements are only approximately predictable, analysis tools like the SLR method cannot predict the performance of a particular building during a particular year. Instead, the analysis estimates the average annual performance using long-term normal weather statistics and certain building and operation assumptions. This is not a defect in the method, rather it is exactly what a design analysis tool should do.

Deciding Between Conservation and Passive Solar Options

A technique has been developed to determine the optimal mix between conservation and solar strategies.[61] To obtain an answer, one needs the cost characteristics of both the passive solar system and the energy conservation features. This information will generally be in the form of the cost per R per ft^2 for the wall and ceiling insulation, the cost per additional glazing for windows, the cost of reducing infiltration (including the cost of adding an air-to-air heat recovery unit if needed), and the cost per ft^2 for the passive solar collection aperture. Given this information, the method provides simple equations that can be used to trace the economic optimal-mix line for a particular locale.

This methodology was used to determine the appropriate sunspace glass area and building conservation levels for the example given earlier in Table 4.

LARGER BUILDING ISSUES

In large buildings the energy situation is often much more complex than in smaller buildings because internal energy generation by lights and equipment often provides much of the energy required for heating, and also because of the multi-zone character of many large buildings. The energy issues in the building may be dominated by cooling requirements instead of heating; moreover, the operation and control of the heating, ventilating, and air conditioning equipment may well be quite complex and significantly alter the overall energy-use pattern of the building. The following issues are of particular importance.

Daylighting

Use of natural light during the day can significantly reduce not only the requirements for artificial lighting, but somewhat reduce the cooling requirements because of the higher efficiency of daylighting. The analytical treatment of daylighting has dissuaded most practitioners from attempting to use detailed analysis as a method for design, because of the complex geometries that are typical of most buildings. However, a number of specialized techniques have been developed, and ongoing work will probably result in development of suitable analytical methods. Many of the methods that have been developed are suitable only for diffuse sky conditions.

The most widespread design technique is now, and probably will remain, the use of scale models. These are relatively easy to build, quite representative of full-scale conditions, and give both qualitative results (such as evaluation of light quality) and quantitative results.

Balancing

Thermal balancing of a building must be done both on a space-to-space basis and also on a time-of-day basis, considering patterns of internal heat generation and occupancy. Mismatches between the availability of natural energies and the building's energy needs result in the need for transporting heat from space to space and for providing auxiliary heating and cooling energy.

Analysis Methods

Because both building configuration and energy issues tend to be complex, analysis codes for large buildings must be extremely complex. The major computer codes that exist have evolved out of earlier codes and use weighting-function analysis techniques. They contain large libraries of materials properties, standard wall sections, and HVAC system characteristics. Great effort has gone into making the building description as architectural as possible, leaving the translation into mathematical terms to specific routines within the program.

Input to the computation consists of hourly measurements of weather and solar parameters for a location of interest. Output consists of building energy consumption, which is generally summarized according to a number of different user-specified options. The codes require large computers and a very sophisticated user.

SUMMARY

Research in passive solar heating is centered around the investigation of energy flow in buildings, emphasizing the collection, storage, and distribution of solar gains, and the extent to which these effects can reduce the need for auxiliary heat within a building. Far from being esoteric research, it has been highly motivated to produce results that are of direct relevance to building designers. Results of the research have found an eager audience among practitioners of passive solar heating, and the translation of those results into design tools has been both rapid and effective.

The basic physics of energy flow in buildings is well understood, and the behavior of the basic passive solar heating types is very predictable, given knowledge of the relevant solar and weather conditions and the factors that influence building load. It is doubtful whether further refinements and accuracy are warranted by the inherent uncertainties of weather and occupancy characteristics. These remarks apply to direct gain, thermal storage wall, and sunspace geometries.

For mixed systems and multizone situations, there has been less experimentation and analysis. Results from many buildings indicate, however, that there are not likely to be many surprises. Various passive solar strategies seem to work well together and complement one another effectively.

Exotic designs, with the exception of the double envelope concept, have not been carefully studied either analytically or experimentally. A variety of hybrid designs have been built in which air heated in one portion of the building (frequently a sunspace) is ducted either through a rock bed or through holes formed in the masonry structure of the building. These seem to have worked well, but there has been little generalized investigation.

To date passive solar design has consisted largely of reconfiguring traditional building elements to make more effective use of the building for the collection, storage, and distribution of solar heat function. There has been little research into improving those elements so that they might better serve these functions. Recent systems analysis at Los Alamos has indicated that net performance can be approximately doubled through the use of elements with improved performance characteristics.

Continued research can be expected to play an important role in the future development of passive solar heating. Three areas should be of particular importance: materials research, building evaluations, and performance analysis.

Materials research and development can play a significant role in developing new products or variations on old products that will significantly improve thermal performance and comfort. The area offering the greatest opportunity for improvement is in glazings and apertures. It

should be possible to improve transmittance, to greatly reduce heat loss, and to increase the controllability of the aperture. It may also be possible to more effectively use opaque sections of the building skin for solar collection.

Improvements in heat storage characteristics of the building will probably be realized largely through improved use of conventional materials within the building obtained through a better understanding of the role of building configurations on heat storage and distribution. Although phase-change materials remain an ever-popular field for research effort, they have had negligible impact on the use of passive solar heating to date.

The most important role of building evaluations is in the assessment of the effectiveness of techniques under field trial. There seems to be no lack of ideas being tried by inventors, and it is important that these ideas receive a fair and dispassionate evaluation. A case in point is the double envelope house that was mentioned earlier.

Performance prediction, based on sound physical principles, is essential if passive solar heating is to develop in an orderly fashion. Research results combined with practical experience will merge into design tools that are simple to use but comprehensive enough for widespread application. Intuition plays an important role in the evolution of new concepts, but it is only through the application of scientific techniques that research can sort out wishful thinking from sound and effective methods.

REFERENCES
1. A. G. H. Dietz and E. L. Czapek. 1976 (May). M.I.T. solar house 2 south-wall collection, storage, and heating. *Passive Solar Heating and Cooling Conference and Workshop Proceedings.* Albuquerque, N. M.; May 18–19, 1976. Los Alamos Scientific Laboratory report LA-6637-C, pp. 171–182.
2. A list of papers by Loren Neubauer, University of California at Davis, can be found in *The passive solar energy catalog.* 1978. Passive Solar Institute, 1625 Curtis, Berkeley, California.
3. Scientific American, May 13, 1882, carries a notice of the invention, by E. S. Morse, with the following description: "His invention consists of a surface of blackened slate under glass fixed to the sunny side or sides of a house, with vents in the walls so arranged that the cold air of a room is let out at the bottom of the slate, and forced in again at the top by the ascending heated column between the slate and the glass. The out-door air can be admitted, also, if desirable. The thing is so simple and apparently self-evident that one only wonders that it has not always been in use."

4. E. A. Morgan. 1966. Improvements in solar heated buildings. Patent Specification 1 022 411. Application date April 6, 1961. Specification published March 16, 1966.

5. F. Trombe. 1973. Le chauffage par rayonnement solaire. Techniques Francaises, Batiments, Travaux Publics, Urganisme no. 1, pp. 1–4.

6. H. Hay. 1976 (May). Atascadero residence. *Passive Solar Heating and Cooling Conference and Workshop Proceedings*, pp. 101–107.

7. J. E. Perry, Jr. 1976 (May). The Wallasey school. *Passive Solar Heating and Cooling Conference and Workshop Proceedings*, pp. 223–237.

8. F. Trombe; J. F. Robert; M. Cabanat; and B. Sesolis. 1976 (May). Some performance characteristics of the CNRS solar house collectors. *Passive Solar Heating and Cooling Conference and Workshop Proceedings*, pp. 201–222.

9. H. R. Hay and J. I. Yellott. 1970 (Jan.). A naturally air conditioned building. *Mechanical Engineering* 92(No. 1):19–23.

10. K. L. Haggard et al. 1975 (Jan.). *Research evaluation of a system of natural air conditioning*. California Polytechnic State University. HUD Contract No. H 2026R.

11. 1976 (May). *Passive Solar Heating and Cooling Conference and Workshop Proceedings*. Los Alamos Scientific Laboratory report LA-6637-C.

12. W. O. Wray; N. M. Schnurr; and J. E. Moore. 1980. Sensitivity of direct gain performance to detailed characteristics of the living space. *Proceedings of the Fifth Passive Conference*. 5.1. Edited by John Hayes and Rachel Snyder. Amherst, Mass.; Oct. 19–26, 1980. Newark, Del.: American Section of the International Solar Energy Society, Inc. ASES, University of Delaware, pp. 92–95. (LA-UR-80-2266.)

13. J. A. Carroll. 1980. An "MRT Method" of computing radiant energy exchange in rooms. *Proceedings of Systems Simulation and Economic Analysis*. San Diego, Calif.; Jan. 23–25, 1980. Solar Energy Research Institute report SERI/TP-351-431, pp. 343–348.

14. D. D. Weber and R. J. Kearney. 1980. Natural convective heat transfer through an aperture in passive solar heated buildings. *Proc. of the 5th Passive Conf.* 5.1, pp. 1037–1041. (LA-UR-80-2328.)

15. E. F. Moore and R. D. McFarland. 1982 (June). Passive solar test modules. Los Alamos National Laboratory report LA-9421-MS.

16. J. D. Balcomb et al. 1980 (Jan.). *Passive solar design handbook Vol. II*. U.S. Department of Energy report DOE/CS-0127/2, pp. 93–95.

17. J. C. Hyde. 1981 (Sept.). Passive test cell experiments during the winter of 1979-80. Los Alamos National Laboratory report LA-9048-MS.

18. R. D. McFarland and J. D. Balcomb. 1982. Los Alamos test room results. *Proceedings of the Seventh National Passive Solar Conference*. Knoxville, Tenn. Aug. 29–Sept. 1, 1982. Newark, Del.: ASES, University of Delaware. (LA-UR-82-1836.)

19. R. D. McFarland. 1982 (May). Passive test cell data for the solar laboratory winter 1980-81. Los Alamos National Laboratory report LA-9300-MS.

20. W. O. Wray and C. R. Miles. 1981. A passive solar retrofit study for the United States Navy. *Proceedings of the Sixth National Passive Solar Conference*. Edited by John Hayes and William Kolar. Portland, Ore.: Sept. 8–12, 1981. Newark, Del.: ASES, University of Delaware. (LA-UR-81-2200.)

21. W. O. Wray; C. R. Miles; and C. E. Kosiewicz. 1981 (Nov.). A passive solar retrofit study for the United States Navy. Los Alamos National Laboratory report LA-9071-MS.

22. J. D. Balcomb et al. 1982 (July). *Passive solar design handbook vol. III*. U.S. Department of Energy report DOE/CS-0127/3. p. 3.

23. J. A. Carroll. An index to quantify thermal comfort in homes. *Proc. of the 5th Passive Conf.*, 5.2, pp. 1210–1214.

24. D. P. Grimmer; J. D. Balcomb; and R. D. McFarland. 1977. The use of small passive solar test boxes to model the thermal performance of passively solar-heated building designs. *Proceedings of American Section of ISES, 1977 Annual Meeting*. 1.2. Orlando, Fla.; June 6–19, 1977. Newark, Del.: ASES, Univ. of Delaware, p. 15. (LA-UR-77-1323)

25. S. K. Reisfeld and D. A. Neeper. 1982. Solar energy research at Los Alamos April 1,—September 30, 1981. Los Alamos National Laboratory report LA-9473-PR, pp. 17–19.

26. W. O. Wray and D. D. Weber. 1979. LASL similarity studies: part I. Hot zone/cold zone: a quantitative study of natural heat distribution mechanisms in passive solar buildings. *Proceedings of the Fourth National Passive Solar Conference*. Edited by Gregory Franta. Kansas City, Mo.; Oct. 3–5, 1979. Newark, Del.: ASES, Univ. of Delaware, pp. 226–230.

27. D. D. Weber; W. O. Wray, and R. Kearney. 1979. LASL similarity studies: part II. Similitude modeling of interzone heat transfer by natural convection. *Proc. of the 4th National Passive Solar Conf.*, pp. 231–234. (LA-UR-79-2225.)

28. D. D. Weber. 1980. Similitude modeling of natural convection heat transfer through an aperture in passive solar heated buildings. Ph.D. diss., University of Idaho. (LA-8385-T.)

29. J. D. Balcomb. 1981. Heating remote rooms in passive solar buildings. *Proceedings of International Solar Energy Society Solar World Forum*. Brighton, England; Aug. 23–28, 1981. Oxford: Pergamon Press, 1835–1839. (LA-UR-81-2518.)

30. T. M. Holzberlein. Don't let the trees make a monkey of you. *Proc. of the 4th Passive Conf.*, pp. 416–419.

31. J. Kohler and D. Lewis. 1981 (Nov.). Passive principles: let the sun shine in. *Solar Age* 6:45–49.

32. C. W. Fowlkes. Measured performance of a passive solar residence in Bozeman, Montana. *Proceedings of the International Solar Energy Society Silver Jubilee Congress*. Atlanta, Ga.; May 1979. New York: Pergamon Press, 1600–1605.

33. C. W. Fowlkes. Experimental problems in measuring the thermal performance of passive solar systems. *Proc. of the 5th Passive Conf.* pp. 399–403.

34. W. D. Nichols. 1976. Unit 1, First Village. *Passive Solar Heating and Cooling Conference and Workshop Proceedings*. Los Alamos Scientific Laboratory report LA-6637-C, pp. 137–149.

35. J. D. Balcomb; J. C. Hedstrom; and S. W. Moore. 1979. Performance data evaluation of the Balcomb solar home (SI units). *Proceedings of the Second Annual Heating and Cooling Systems Operational Results Conference*. Colorado Springs, Colo. Nov. 27–30, 1979. (LA-UR-79-2659.)

36. J. D. Balcomb; J. C. Hedstrom; J. E. Perry, Jr. 1981. Performance summary of the Balcomb solar home. *Solar Rising 1981 Annual AS/ISES Meeting*. Philadelphia, Penn.; May 26–30, 1981. New-

ark, Del.: ASES, Univ. of Delaware, (LA-UR-81-1039.)

37. J. D. Balcomb and J. C. Hedstrom. 1980. Determining heat fluxes from temperature measurements made in massive walls. *Proc. of the 5th National Passive Solar Conf.* 5.1, pp. 136–140. (LA-UR-80-2231.)

38. Passive solar buildings. 1979 (July). Sandia Laboratories report SAND 79-0824, pp. 19–37.

39. J. D. Balcomb; J. C. Hedstrom; and J. E. Perry. 1980. Performance evaluation of the Balcomb solar house. *Colloque Solaire International*. Nice, France; Dec. 11–12, 1980. (LA-UR-80-3453.)

40. S. K. Reisfeld and D. A. Neeper. 1981 (Sept.). Solar energy research at Los Alamos October 1—December 31, 1980. Los Alamos National Laboratory report LA-8984-PR.

41. B. D. Howard and E. O. Pollock. 1982. Comparative report: performance of passive solar space heating systems. U.S. Department of Energy report Solar/0022-82/39.

42. D. Frey, J. Swisher, and M. Holtz. 1982. Class B performance monitoring of passive/hybrid solar buildings. ASME 1982 Solar Energy Conference. Albuquerque, N. M.; April 26–30, 1982. Solar Energy Research Institute Report SERI/TP-254-1492.

43. R. F. Jones; G. Dennehy; H. T. Ghaffari; and G. E. Munson. 1981 (May). Case study of the Mastin double-envelope house. Brookhaven National Laboratory report BNL 51460.

44. C. W. Fowlkes. 1982. Thermal performance of an envelope house in a cold climate. *Proceedings of the 1982 Annual Meeting of the American Solar Energy Society*. Houston, Tex.; June 1–5, 1982. Newark, Del.: ASES, Univ. of Delaware, pp. 691–696.

45. B. Chen; T-C. Wang; J. Maloney; and S. Chutin-taranond. 1982. The effect of deep earth heat migration to the performance of continuous thermal envelope structures. *Proc. of the 1982 Annual Meeting of the American Solar Energy Society*, pp. 697–702.

46. F. Arumi-Noe. 1980. Expansion of the simulation capabilities and user access to the computer program DEROB/PASOLE, final report. School of Architecture, University of Texas, Austin. April 25, 1980.

47. L. Palmiter and T. Wheeling. 1981. SUNCODE, a program user's manual. Seattle, Wash.: Ecotope Group. August 26, 1981.

48. CALPAS3 Manual. 1981. Berkeley, Calif.: Berkeley Solar Group.

49. S. K. Reisfeld and D. A. Neeper. 1980 (Nov.). Solar energy research at LASL: October 1, 1979—March 31, 1980. Los Alamos Scientific Laboratory report LA-8450-PR, pp. 20–23.

50. R. W. R. Muncey. 1979. *Heat transfer calculations for buildings*. London: Applied Science Publishers Ltd.

51. D. B. Goldstein; M. Lokmanhekim; and R. D. Clear. 1979 (Aug.) Design calculations for passive solar buildings by a programmable hand calculator. Lawrence Berkeley Laboratory report LBL-9371.

52. A. T. Kirkpatrick and C. B. Winn. 1982. A frequency response technique for the analysis of the enclosure temperature in passive solar buildings. *Proc. of the 1982 Annual Meeting of the American Solar Energy Society*, pp. 773–777.

53. J. D. Balcomb et al. 1980 (Jan.). *Passive solar design handbook vol. II.* U.S. Department of Energy report DOE/CS-0127/2, pp. 178–191 and G.1–G.6.

54. K. Raman. 1982. Analytic method for performance modelling of passive solar systems. *Proceedings of the 1982 Annual Meeting of the American Solar Energy Society*, pp. 679–684.

55. W. A. Beckman; S. A., Klein; and J. A. Duffie. 1977. *Solar heating design by the F-chart method*. New York: Wiley-Interscience.

56. W. A. Monsen; S. A. Klein; and W. A. Beckman. The un-utilizability design method for collector-storage walls. *Solar Rising 1981 Annual AS/ISES Meeting*, pp. 862–866.

57. R. D. McFarland. 1978 (Oct.). PASOLE: A general simulation program for passive solar energy. Los Alamos National Laboratory report LA-7433-MS.

58. J. D. Balcomb et al. *Passive solar design handbook volume III*. R. W. Jones, ed. Newark, Del.: ASES, Univ. of Delaware.

59. W. O. Wray. 1981 (June). Design and analysis of direct gain solar heated buildings. Los Alamos National Laboratory report LA-8885-MS.

60. J. D. Balcomb; R. W. Jones; R. D. McFarland; and W. O. Wray. 1982. Expanding the SLR method. *Passive Solar Journal* 1(No. 2):67–90.

61. J. D. Balcomb. Conservation and solar: working together. *Proc. of the 5th Passive Conf.*, pp. 44–48. (LA-UR-80-2330.)

TOPICAL INDEX

A

additives, 91
aeroelasticity, 190
air pollution, 15
alcohol dehydrogenase, 119
Alich, 62
Allen, 244
Analysis
  sensitivity, 294
  simulation, 265, 284
  systems, 293
Anderson, 62
Andersson, 41, 45, 46, 47
Angstrom, 242
anhydrosugars, 95, 97
Arseneau, 71, 84
Arthur, 75
Arumi, 288
ash, 91
Aspergillus oryzae, 118
Atascadero house, 267
atmospheric transmission, 4

B

Bacillus subtilis, 119, 128
back surface field, 138, 167
Bahadori, 260
Baker, 71, 86
balance point, 300
Balcomb, 278
Balcomb residence, 280
Bar-Gadda, 101
Barker, 76
Basch, 71, 74
Battelle Memorial Institute, 272
Battelle process, 153-154
Bedford, 95
Berdahl, 244, 245, 246, 247, 250
Berkowitz-Mattuck, 86
Berry Foil, 272
Biggs, 11
biomass
  characteristics of, 63
  comparison of, 66
  pyrolysis, 62
Bird, 13
Bliss, 242, 243, 244
Boardman Energy Systems, 270
Bradbury, 74
Bridgman growth, 147, 158
Broido, 71
Browning, 64

Bryce, 96
Buckius, 13
building temperature response, 210
Burwell, 62
Byrne, 79

C

Cabanet, 267
calibration, 15, 33, 52
California Solar Data Manual, 4
CALPAS3, 288
Candida tropicalis, 117
carbohydrates, 72
Carbon, 125
Cardwell, 82
Carrier, 252
Carroll, 272
cassette editing, 8
Cellulomonas fimi, 117, 128
cellulose, 64
  utilization, 128
central receiver, 230
Cerny, 92
char formation, 76
charring, 78
Chatterjee, 75
chemical route, 135
chemistry, 62
Chin, 96
circumsolar radiation, 10, 11
Circumsolar telescope, 10, 14
Clark, 125, 243, 244, 246, 247, 250, 255
Class B, 284
Cleveland, 92
cloning, 127
cloning vectors, 120
closed loop, 209
cloudy skies, 245
coefficient of performance, 217
Colstridium thermocellum, 117
concentrating photovoltaic, 4
conjugation, 120
Conrad, 75
conservation, 265
control
  adaptive, 214
  bang bang, 209
  deciduous shade, 279
  derivative, 211
  differential, 215
  integral, 211
  optimal, 225

proportional, 211
regulatory, 116
controller, PID, 212
convection, 246
convective intrusion, 250
Cool Pool, 255
cooling
    clear sky, 243
    convective, 260
    direct evaporative, 256
    evaporative, 251
    hybrid, 241
    hybrid evaporative, 256
    indirect evaporative, 257
    natural, 265
    passive, 241
    radiative, 242, 245, 246, 247, 251
    rate, 247
COP, 217
correction, shadow band, 8
costs
    electricity, 184
    maintenance, 184, 200
    operation, 184
    production, 183
    repair, 200
    site related, 184
    solar cell, 134, 168-169
    supply, 232
Coulson, 13
coupling, 254
crack propagation, 192
credit
    capacity, 203
    fuel, 203
Crimsco, 272
Crommelynck, 23
cross linking, 75
Crowther, 255
crucible material, 157
crystallinity, 74
crystallization technique, 135, 157, 165-166
cycling, 210
Czrochralski growth, 158, 166

D

data
    collection, 2
    handling, 4
    infrared, 13
    solar radiation, 1
    spectral, 13
data base, 5
data collection, 3
Dave, 13
Davies, 267
daylighting, 301
deadband, 210
Deglise, 72
dehumidification, 261
dehydration, 78
Demmitt, 91
Department of Energy, 4, 16
DEROB, 288

DeRocher, 230
desiccant, 261
design
    analytical, 212
    trial and error, 212
developing countries, 62
Diebold, 68
Dietz, 267
direct arc reactor, 155
direct beam, 6
disaccharides, 96, 98
discomfort index, 272
dislocations, 144-145
disproportionation, 82
Dollimore, 84, 101
Doman, 180
Doree, 64
DSET Laboratories, 5, 14
DTA data, 81

E

economics, 183, 198
edge-defined film fed growth, 162
edge-supported pulling, 156, 162
Empire Glass, 272
energy crisis, 62
energy flows, 1
Environmental Data Service, 2
environmental impact, 196
Environmental Research Laboratory, 5
epitaxial cells, 163-164
Eppley Laboratory, 2
Erwinia amylovora, 114
Escherichia adecarboxylata, 117
Escherichia coli, 116, 117
Estey, 28, 30
ethanol, 113, 114, 129
evapotranspiration, 1
Eventova, 86
extrachromosomal elements, 124

F

Fairbridge, 85
fatigue loading, 192
fermentation pathways, 116
Finkelstein and Schafer, 9
fission, 82
flocculent strains, 119
Flowers, 21, 45, 47
Fowlkes residence, 280
Fromberg, 244, 245
Fung, 74
Furneaux, 97

G

Gardiner, 79, 95
gaseous environment, 91
gasification, 63
Gavalas, 69
genetics
    development of, 119
    yeast, 122
Givoni, 245, 250, 251, 256

glucose, 64
Goldstein, 64
Golova, 73
grain boundaries, 144
grain size, 147, 151
Greenwood, 94, 96
grit structure, 147

H

Haggard, 243, 267
halogenide process, 153
Halpern, 71
Hamiltonian, 227
Hansen, 254
Hanson, 6
Hatfield, 13
Hay, 243, 262, 267
Healey, 13
heat pipe, 272
heat-exchanger method, 156, 158, 166
heat, internal, 300
heating
    passive solar, 265
    rate, 89
hemicelluloses, 64
hemlock process, 153-154
heterolytic bond cleavage, 79
Heyns, 96
high temperature phenomena, 85, 101
Hinnen, 125
Hinojosa, 75
Holt, 84, 101
Holzberlein, 278
homolytic bond cleavage, 78
Hon, 76
Hopkins, 69
horizontal ribbon growth, 159
horizontal shading disk, 33, 47
Hottel-Whillier equation, 226
Hounimer, 96
Howard, 69, 78
Hoyt, 6
Hubbert, 61
Hulstrom, 13
hybrid systems, 266
hysteresis, 216

I

impurities
    diffusion, 140
    segregation, 146, 156, 165
    solubility, 139-140, 167
Ingram, 119
Inman, 62
interface control crystallization, 163
International Energy Agency, 32
International Standards Organization, 35
inverted Stepanov technique, 163
IPC
    Fifth, 27
    Fourth, 26
irrigation, 1
irrigation scheduling, 9

J

Jahn, 64

K

Kala, 85
Kamorita, 75
Kato, 75
Katz, 92
Kearney, 278
Kelly, 252, 254
Kerr, 92
Kilzer, 74
kinetics, 74, 82, 89, 92, 97, 101
King, 13
Kipp & Zonen, 2
Kislitsyn, 79
Klier, 96
Klutcher, 13
Kluyveromyces lactis, 126
Kondratyev, 13
Koomanoff, 16
Kovarik, 226

L

Lactobacillus homohiochi, 129
Lai, 98
Lawrence Berkeley Laboratory, 10
Lazarus, 293
LeBaron, 8
Leckner, 13
Lede, 72
Lesse, 226
levoglucosan, 78, 81
Lewin, 71, 74
Li-Cor, 14
lignin, 64
Lincoln, 86
linear programming, 238
Lipska, 69
Lissaman, 176
Liu-Jordan, 13
living habits, 300
load cases, 191
load collector ratio, 298
Los Alamos National Laboratory, 269
low temperature phenomena, 72, 92
LOWTRAN 5, 13
Loxsom, 243, 252, 254, 255
Luner, 82

M

Madorsky, 71, 73, 79
Major, 72
management
    agricultural, 9
    load, 9
mannose, 64
manure, 66
Martin, 86, 250
McCarter, 71
McFarland, 271
measurement
    errors, 5

meteorological, 4
performance testing, 21
radiation, 4
resource assessment, 20
solar irradiance, 12
solar radiation, 19
spectral, 15
mechanisms, 74, 86, 92, 97, 101
Mehta, 76
metabolic pathways, 117
metallurgical route, 136, 165
methods
  calibration, 15
  continuous, 70
  correlation, 296
  discontinuous, 69
  experimental, 69
  hybrid, 72
  simplified, 265, 294
  solar load ratio, 296
Milne, 86
Minerleau-Puijalon, 127
model
  atmospheric transmission, 13
  mathematical, 4, 284
  regression, 5, 6
  solar cell, 137
moderate temperature phenomena, 94
Mok, 89
mold casting, 147, 159, 166
Molton, 91
monitored buildings, 265
monomers, 95, 97
monosaccharides, 96, 98
Moore, 269
Morgan, 267
Morris, 125
Morse, 267
Muncey, 289
Munroe, 14
Murphy, 100

N

National Aeronautics and Space Administration, 12
National Climatic Center, 2, 6, 14
National Digital Facsimile Network, 10
National Solar Data Network, 284
net load coefficient, 298
net reference load, 296
Neubauer, 267
Neurospora crassa, 126
New River, 27, 31, 54
Niles, 243, 267, 288
Nimbus 7, 11, 12
Noguchi, 86
NRIPs
  comparison of, 30
  operation of, 28

O

occulting tube, 47
Office of Technology Assessment, 62
open loop, 209

operation, 24
optimal mix, 301
orientation, 75
Orsi, 94
osmo-tolerance, 118
overshoot, 210

P

Pachysolen tannophilus, 117
Palmiter, 288
parasitic power, 228
Parder, 69
Parikh, 76
Parker, 82
Parmalee, 242
particle size, 90
PASOLE, 293
passivation, 148, 152
Patai, 71
Paucault, 74
PCM, 270
phase change, 270
phase plane, 219
Pictet, 77
Pimentel, 62
pitch control, 181
Pittinger, 255
plasma torch technique, 163
polymerization, 74
polysaccharides, 96, 100
Pontryagin Maximum Principle, 226
Poole, 62
precipitates, 142, 151, 159
Precision Spectral Pyranometer, 2
pressure, 89
products, 77, 78, 86, 92, 95, 101
Pseudomonas aeruginosa, 119
Puddington, 100
pulsed annealings, 167
purification techniques, 135, 156
pyranometer, 2, 15, 45, 52, 54, 56
  absolute cavity, 47
  calibration, 47
  comparisons, 32
pyranometry, 32, 45
  calibration standards, 40
  temperature response, 46
pyrheliometer, 2, 15, 26, 54, 55
pyrheliometer, calibration of, 40
pyrolysis, 62, 69, 70, 71, 72, 73, 92
pyrometallurgical process, 155

R

radiation measurement scales, 19
radiometry
  absolute cavity, 23, 24, 26, 40
  data, 1
  physical principles, 23
Randall, 6
rate laws, 83
recombination, 137, 139, 142-143, 147
recombination velocity, 144, 146

Reeves, 76
reference cells, 3
regulator, 214
Reitan, 249
Rensfelt, 78
research facilities, 4
resource estimates, 4
ribbon against drop, 164
ribbon from cylinder, 159
ribbon from powder, 159, 161
ribbon to ribbon, 145, 159, 166
Riches, 16
Robert, 267
Roberts, 82
Rodrig, 75
rotor blades
  design concepts, 181
  materials, 180
  number of, 181
roughness classes, 195
Russell, 124

S

Saccharomyces cerevisiae, 114
Saccharomyces uvarum, 118
safety, 197
Saito, 256
Salmonella typhimurium, 117, 120
Sandia National Laboratories, 8
Sarasen, 77
Sarcina ventriculi, 114
Sauret, 74
Schulten, 86
Schwanniomyces alluvius, 128
Schwanniomyces castelli, 128
Seaman, 28
segregation coefficients, 140, 155
sensitivity, 222
Sesolis, 267
sewage, 66
Shafizadeh, 71
Shepherd, 14
Shimazu, 74
Siemens process, 153, 156
silicon
  amorphous, 168
  metallurgical grade, 154, 156
  on ceramics, 163
  solar grade, 133, 156
Simon, 179
Sinclair, 124
Sklarov, 23
Skytherm, 242
Smith, 78
Smithsonian Radiation Biology Laboratory, 14
Sobin, 260
solar beam, 4
solar collection, 2
solar constant, 11
Solar Energy Research Institute, 4, 14
Solar Energy Meteorological Research and
  Training Site, 13
Solar Maximum Mission, 11

solar process heat systems, 229
Solar Radiation and Radiation Balance Data, 14
solar radiation, 19
  availability, 1
  data, 2
  data availability, 14
  data uses, 1
  diffuse, 4, 11
  direct, 11
  distribution, 4
  facility, 5
  forecasting, 9
  isotropic distribution, 13
  measurements, 1
  measuring network, 6
  monitoring network, 8
  nature of, 4
  spectral distribution, 3
  terrestrial, 4
  workshops, 4
solar savings fraction, 294
solar spectrum, 3
solar thermal power systems, 229
solar-cell market, 168
solar-cell technology, 167
SOLMET, 3, 6, 8
SOLTRAN 5, 13
Soltys, 86
Southwall Heat Mirror, 272
spectral basis, 247
Spencer, 13
Stamm, 91
Standard Year Irradiance, 6
standards, 31
  calibration, 22, 40
  International Standards Organization, 35
  measurement, 22
  pyrheliometry, 40
Stanek, 92
Stanwick, 84
starch utilization, 127
states, 209
Steinmuller, 13
Sterling, 74
Stewart, 13, 124
structural dynamics, 176
Struhl, 125
studies, 71
Sugisawa, 92
Sullivan, 176
sun photometer, 14
SUNCODE, 288
Sungain, 272
sunspace, 271
surface
  selective, 271
  selective cooling, 250
Sussott, 100

T

Tang, 71
Temps, 13
test boxes, 276

test modules, 265, 269
Texxor, 272
thermal capacitance, 210
thermal network, 269
thermo-tolerance, 118
Thermoanaerobium brockii, 117
thermochemical, 62
Thompson, 92
Thresher, 176, 179
time constant, 212, 287
time shift, 8
tower, 181
transduction, 120
transfer function, 212
transformation, 119
transglycosylation formation, 78
transposons, 120
trap levels, 141
Trombe wall, 266
turbidity, 14

U

United States Department of Energy, 4, 16
ultra-fast quenching, 163
uncertainty, 222
Union Carbide process, 153-156
University of California-San Diego, 269
University Research and Training Sites, 4, 14

V

Vail, 76
validation, 288
Valko, 13
Vittoe, 11
Voeikov Main Geophysical Observatory, 14

W

Wallasey School, 267
Waller, 91
waterwall, 270
Watt, 11
weather data, 15
Weathershield Quad Pane, 272

web-dendrites, 161, 166
Weber, 278
Weinstein, 74
Wells, 31
Wendler, 5
Wenzl, 64
White, 255
Wiegerink, 91
Williams, 62
wind
    characteristics, 178
    structure, 194
    systems, 175
wind turbine
    control, 176
    design criteria, 179
    development costs, 183
    large systems, 175, 188
    maintenance, 182
    materials, 194
    national programs, 186
    problems, 189
    reliability, 182
    safety, 177
Winn, 226
Wise, 64
Witwer, 62
Wolfrom, 92
World Meteorological Organization, 14, 25
world radiation reference scale, 23

X

xylose, 64

Y

Yanigamachi, 242
year, typical meteorological, 8, 9
Yellott, 267

Z

Zonen, 32
Zymomonas mobilis, 114

AUTHOR INDEX

Antal Jr., Michael Jerry	61
Bahm, Raymond J.	1
Balcomb, J. Douglas	265
Eveleigh, D. E.	113
Lerner, James J.	175
Picataggio, S.	113
Rodot, M.	133
Selzer, H.	188
Stokes, H. W.	113
Winn, C. Byron	209
Yellott, John I.	241
Zerlaut, G. A.	19